Worksheets For Anatomy and Physiology Text and Laboratory Workbook
(Volume 2)

Homeostasis, the maintenance of our internal physiological processes. Kayaking the inner passage, Alaska.

Art Department
  Director - Terry H. Davenport, R.T.R.M.
    Medical illustrations and art layout review by Terry H. Davenport, R.T.R.M

Second Edition : Two Volume Edition
ISBN: 0-9637670-1-1
  Volume 1: *Anatomy and Physiology Text and Laboratory Workbook*
  Volume 2: *Worksheets for Anatomy and Physiology Text and Laboratory Workbook*

Copyright 2007 by Stephen G. Davenport, B.A., M.A.

All rights reserved. No part of this publication may be reproduced, stored in a retrieval system, or transmitted in any form or by any means, electronic or mechanical, including photocopy, recording, or otherwise, without permission in writing from the publisher.

Published by Link Publishing, P. O. Box 15562, San Antonio, TX. 78212
Web site: www.linkpublishing.com

Table of Contents

CHAPTER 1 - INTRODUCTION TO A & P
Workbook (Volume 1) - Pages 1 - 16
Worksheets - Pages 1-12

CHAPTER 2 - MICROSCOPE
Workbook (Volume 1) - Pages 17 - 32
Worksheets - Pages 13 -20

CHAPTER 3 - CHEMISTRY INTRODUCTION
Workbook (Volume 1) - Pages 33 - 92
Worksheets for Introduction - Pages 21 - 31
Worksheets for Carbohydrate - Pages 33 - 41
Worksheets for Amino Acids - Pages 43 - 45
Worksheets for Chemistry of Lipids - Pages 47 - 52
Worksheets for Chemistry of Nucleotides - Pages 53 - 57
Worksheets for Chemistry of AMP, ADP, ATP - Pages 59 -62

CHAPTER 4 - THE CELL
Worksbook (Volume 1) - Pages 93 - 110
Worksheets - Pages 63 - 74

CHAPTER 5 - TRANSPORT PROCESSES
Workbook (Volume 1) - Pages 111 - 146
Worksheets - Pages 75 - 90

CHAPTER 6 - CELL DIVISION
Workbook (Volume 1) - Pages 147 - 162
Worksheets - Pages 91 - 102

CHAPTER 7 - PROTEIN SYNTHESIS
Workbook (Volume 1) - Pages 163 - 170
Worksheets - Pages 103 -106

CHAPTER 8 - TISSUES
Workbook (Volume 1) - Pages 171 - 196
Worksheets - Pages 107 - 122

CHAPTER 9 - INTEGUMENTARY SYSTEM
Workbook (Volume 1) - Pages 197 - 206
Worksheets - Pages 123 - 130

CHAPTER 10 - BONE TISSUE AND BONES
Workbook (Volume 1) - Pages 207 - 224
Worksheets - Pages 131 - 146

CHAPTER 11 - THE SKELETON
Workbook (Volume 1) - Pages 225 - 232
Worksheets - Pages 147 - 154

CHAPTER 12 - AXIAL SKELETON
Workbook (Volume 1) - Pages 233 - 258
Worksheets - Pages 155 - 174

CHAPTER 13 - APPENDICULAR SKELETON
Workbook (Volume 1) - Pages 259 - 284
Worksheets - Pages 175 - 200

CHAPTER 14 - ARTICULATIONS
Workbook (Volume 1) - Pages 285 - 306
Worksheets - Pages 201 - 218

CHAPTER 15 - MUSCULAR SYSTEM
Workbook (Volume 1) - Pages 307 - 316
Worksheets - Pages 219 - 228

CHAPTER 16 - MUSCLES AND CONTRACTION
Workbook (Volume 1) - Pages 317 - 338
Worksheets - Pages229 - 246

CHAPTER 17 - NERVOUS SYSTEM
Workbook (Volume 1) - Pages 339 - 368
Worksheets - Pages 247 - 268

CHAPTER 18 - THE EYE
Workbook (Volume 1) - Pages 369 - 390
Worksheets - Pages 269 - 286

CHAPTER 19 - THE EAR
Workbook (Volume 1) - Pages 391 - 400
Worksheets - Pages 287 - 294

CHAPTER 20 - THE ENDOCRINE SYSTEM
Workbook (Volume 1) - Pages 401 - 416
Worksheets - Pages 295 - 306

CHAPTER 21 - BLOOD
Workbook (Volume 1) - Pages 417 - 428
Worksheets - Pages 307 - 312

CHAPTER 22 - LYMPHATIC SYSTEM
Workbook (Volume 1) - Pages 429 - 436
Worksheets - Pages 313 - 318

CHAPTER 23 - THE HEART
Workbook (Volume 1) - Pages 437 - 456
Worksheets - Pages 319 - 332

CHAPTER 24 - BLOOD VESSELS AND CIRCULATION
Workbook (Volume 1) - Pages 457 - 472
Worksheets - Pages 333- 342

CHAPTER 25 - THE RESPIRATORY SYSTEM
Workbook (Volume 1) - Pages 473 - 488
Worksheets - Pages 343 - 352

CHAPTER 26 - THE DIGESTIVE SYSTEM
Workbook (Volume 1) - Pages 489 - 510
Worksheets - Pages 353 - 370

CHAPTER 27 - THE URINARY SYSTEM
Workbook (Volume 1) - Pages 511 - 524
Worksheets - Pages 371 - 382

CHAPTER 28 - THE REPRODUCTIVE SYSTEM
Workbook (Volume 1) - Pages 525 - 552
Worksheets - Pages 383 - 407

CHAPTER 29 - DISSECTION OF THE FETAL PIG
Workbook (Volume 1) - Pages 553 - 574
Worksheets - Pages 409 - 427

www.linkpublising.com
"STUDY TOOLS"
are available at our web site and currently include:
GRADED INTERACTIVE EXAMS with VIDEOS and LINKS.
The interactive exams are designed from the worksheet questions.
The following interactive exams with videos and links were available at time of publication.
Check the site for updates.

INTERACTIVE EXAMS
for Anatomy and Physiology Text and Laboratory Workbook, Stephen G Davenport, Link Publishing, Copyright 2006

Chapter 1: Introduction to Anatomy and Physiology
1. Anatomy and Physiology
2. Organizational Levels
3. Characteristics of Human Life
4. Homeostasis
5. Organization and Terminology
6. Directional Terminology
7. Planes and Sections
8. Body Cavities and Membranes
9. Lungs: Membranes and Cavities
10. Heart: Membranes and Cavities
11. Abdomino-pelvic: Membranes and Cavities
12. Abdomino-pelvic Regions and Quadrants
13. Imaging

Chapter 2: Microscope
1. Introduction to the Microscope
2. Compound Biological Microscope
3. Working Distance and Focal Point of Objectives, Depth of Field, Field of View, and Magnification
4. Stage, Condenser, and Diaphragm
5. Care and Use of the Microscope and Specimens
6. Focusing the Microscope
7. Lab Activities - Millimeter Ruler, Letter "e" (or printed letter), Silk Fibers (or colored fibers), and Cheek Cells

Chapter 3: Chemistry and Metabolism - Introduction
1. Periodic Law and Table, Atoms and Atomic Structure
2. Energy and Matter
3. Chemical Reactions
4. Combining Matter
5. Combining Matter - Chemical Equations and Reactions
6. Energy of Chemical Reactions and Enzymes
7. Chemical Bonds - Ionic Bonds
8. Chemical Bonds - Covalent Bonds
9. Chemical Bonds - Hydrogen Bonds
10. Amphiphilic Molecule
11. Water Surface Tension - Demonstrations
12. Mixtures
13. Biochemistry - Introduction, and Water
14. Acids and Bases
15. pH, Buffers, and Salts
16. Organic Chemistry - Introduction

Chapter 3A: Chemistry and Metabolism - Carbohydrates
1. Introduction to Carbohydrates and Monosaccharides
2. Disaccharides, Oligosaccharides, and Polysaccharides
3. Dehydration Synthesis and Hydrolysis
4. Glycoaminoglycans (GACs) and Glycoproteins
5. Carbohydrate Metabolism - Introduction
6. Glucose Catabolism - Introduction
7. Glucose Catabolism - Overview: Glycolysis, NADH, Acetyl-CoA, Carbon Dioxide, and Coenzyme A (CoA)
8. Krebs Cycle
9. Electron Transport System and Summary of Energy Capture from Glucose Catabolism
10. Glucose Anabolism

Chapter 3B: Amino Acids and Proteins
1. Amino Acids - Introduction
2. Protein Structure and Denaturation

Chapter 3C: Lipids
1. Lipids - Introduction, and Fatty Acids
2. Glycerides
3. Phospholipids
4. Steroids
5. Lipid Metabolism - Introduction, and Fatty Acid Anabolism (triglycerides)
6. Mobilization of Triglycerides
7. Beta-oxidation of Fatty Acids
8. Ketone Bodies
9. Synthesis of Fatty Acids

Chapter 3D: Nucleic Acids
1. Nucleic Acids - Introduction
2. Deoxyribonucleic Acid (DNA) - Introduction
3. Deoxyribonucleic Acid (DNA) - Structure
4. Ribonucleic Acid (RNA) - Introduction
5. Ribonucleic Acid (RNA) - Structure
6. Messenger, Ribosomal, and Transfer RNAs - Introduction

Chapter 3E: Adenosine Phosphate - AMP, ADP, & ATP
1. Adenosine Triphosphate (ATP) - Introduction
2. Adenosine Monophosphate (AMP)
3. Adenosine Diphosphate (ADP)
4. Adenosine Triphosphate (ATP)

Chapter 4: Cytology
1. Introduction
2. Exfoliated Cheek (Squamous) Cells with Lab Activity
3. Stratified Squamous Epithelium with Lab Activity
4. Adipose Tissue with Lab Activity
5. Blood Cells with Lab Activity
6. Sperm with Lab Activity
7. Columnar Ciliated Cells and Goblet Cells with Lab Activity
8. Smooth Muscle Cells with Lab Activity
9. Cell Overview - Observations with Lab Activities
10. Plasma Membrane - Introduction
11. Plasma Membrane - Lipids
12. Plasma Membrane - Proteins
13. Plasma Membrane Junctions
14. Microvilli
15. Cytoplasm
16. Nucleus
17. Mitochondria
18. Ribosomes
19. Endoplasmic reticulum
20. Golgi Apparatus
21. Lysosomes
22. Centrosome and Centrioles
23. Cilia and Flagella
24. Inclusions

Chapter 5: Transport
1. Body Fluids and Compartments
2. Mixtures
3. Molecular and Particle Movements with Lab Activity
4. Passive Movements - Diffusion
5. Molecular Motion and Weight with Lab Activity
6. Simple Diffusion Across the Plasma Membrane
7. Movement of Water by Hydrostatic Pressure
8. Filtration with Lab Activity
9. Movement of Water by Osmosis
10. Effects of Osmotic Solutions and Tonicity
11. Osmotic Pressure
12. Osmotic Pressure and Tonicities of Solutions
13. Osmosis- Osmometer with Lab Activity
14. Osmosis- Red Blood Cells with Lab Activity
15. Osmosis- Potato Cells - Lab Activity
16. Osmosis- Elodea with Lab Activity
17. Osmosis- Paramecium with Lab Activity
18. Dialysis
19. Dialysis and Osmosis with Lab Activity using Tubing Membrane
20. Dialysis and Osmosis with Lab Activity using Unshelled Egg
21. Fluid Movement Across the Capillary
22. Fluid Movement at the Arterial End of Capillary
23. Fluid Movement at the Venous End of Capillary
24. Active Transport Processes
25. Active Transport and Membrane Potentials
26. Yeast Active Transport with Lab Activity
27. Vesicular Transport
28. Phagocytosis, Kupffer Cells with Lab Activity
29. Phagocytosis, Amoeba with Lab Activity
30. Phagocytosis, Paramecium with Lab Activity
31. Pinocytosis and Receptor-Mediated Endocytosis

Chapter 6: Cell Division
1. Cell Division - Introduction
2. Cell Life Cycle
3. Cell Life Cycle - DNA Replication
4. Interphase
5. Mitosis - Prophase
6. Mitosis - Metaphase
7. Mitosis - Anaphase
8. Mitosis - Telophase
9. Daughter Cells - Cytokinesis
10. Mitosis with Lab Activity (Whitefish Blastula and Allium Root Tip)
11. Meiosis and Gamete Production - Overview
12. Meiosis - Descriptive
13. Meiosis - Spermatogenesis with Lab Activity
14. Meiosis - Oogenesis with Lab Activity

Chapter 7: Protein Synthesis
1. Protein Synthesis - Introduction
2. Messenger RNA (mRNA)
3. Ribosomal RNA (rRNA)
4. Transfer RNA (tRNA)
5. DNA
6. Transcription
7. Translation - Initiation
8. Translation - Elongation
9. Translation - Termination
10. Polyribosome

Chapter 8: Tissues
1. Tissues - Introduction
2. Tissues - Classification
3. Simple Squamous Epithelium with Lab Activity
4. Simple Cuboidal Epithelium with Lab Activity
5. Simple Columnar Epithelium with Lab Activity
6. Pseudostratified Ciliated Columnar Epithelium with Lab Activity
7. Review of Simple Epithelia
8. Stratified Squamous Epithelium with Lab Activity
9. Transitional Epithelium with Lab Activity
10. Review of Stratified Epithelia
11. Classification of Epithelium According to Location
12. Glandular Epithelium and Secretion (Glands) with Lab Activity
13. Connective Tissue - Introduction
14. Areolar Tissue with Lab Activity
15. Adipose Tissue with Lab Activity
16. Reticular Tissue with Lab Activity
17. Loose Connective Tissues - Review
18. Dense Regular Connective Tissue with Lab Activity
19. Dense Irregular Connective Tissue with Lab Activity
20. Elastic Connective Tissue with Lab Activity
21. Dense Connective Tissue - Review
22. Cartilage - Introduction

23. Hyaline Cartilage with Lab Activity
24. Fibrocartilage with Lab Activity
25. Elastic Cartilage with Lab Activity
26. Cartilage - Review
27. Bone
28. Blood
29. Muscle - Introduction
30. Skeletal Muscle with Lab Activity
31. Cardiac Muscle with Lab Activity
32. Smooth Muscle with Lab Activity
33. Muscle Tissue - Review
34. Nerve Tissue

Chapter 9: Integumentary System
1. Integumentary System - Introduction
2. Cells of the Epidermis
3. Layers of the Epidermis
4. Dermis
5. Skin Hairless with Lab Activity
6. Skin Hairy with Lab Activity
7. Skin Color with Lab Activity
8. Accessory Structure - Hair with Lab Activity
9. Accessory Structures - Glands with Lab Activity
10. Accessory Structures - Nails
11. Skin Innervation

Chapter 10: Bone Tissue and Bones
1. Bone Structure and Classification
2. Gross Anatomy of a Long Bone
3. Gross Anatomy of a Flat Bone
4. Membranes of Bone
5. Bone Tissue and Structure with Lab Activity
6. Compact Bone
7. Cells of Bone Tissue with Lab Activity
8. Bone Development and Growth - Introduction
9. Intramembranous Ossification with Lab Activity
10. Intramembranous Ossification - Fetal Skull Lab Activity
11. Endochondral Ossification - Primary Ossification with Lab Activities
12. Endochondral Ossification - Secondary Ossification with Lab Activities
13. Endochondral Ossification - Longitudinal Growth with Lab Activities
14. Endochondral Ossification - Appositional Growth with Lab Activities
15. Bone Dynamics - Introduction
16. Bone Dynamics - Calcitonin
17. Bone Dynamics - Parathyroid hormone
18. Bone Dynamics - Bone Remodeling with Lab Activities

Chapter 11: The Skeleton
1. The Skeleton - Introduction, Terminology, Divisions
2. Skull and Associated Bones with Lab Activity
3. Vertebral Column with Lab Activity
4. Thorax (Thoracic Cage) with Lab Activity
5. Pectoral (Shoulder) Girdle with Lab Activity
6. Upper Limb with Lab Activity
7. Pelvic Hip Girdle with Lab Activity
8. Lower Limb with Lab Activity

Chapter 12: The Axial Skeleton
1. Skull with Lab Activity
2. Frontal Bone with Lab Activity
3. Parietal Bones with Lab Activity
4. Occipital Bones with Lab Activity
5. Temporal Bones with Lab Activity
6. Sphenoid Bone with Lab Activity
7. Ethmoid Bone with Lab Activity
8. Mandible with Lab Activity
9. Maxillary Bones (Maxillae) with Lab Activity
10. Nasal Bones, Lacrimal Bones, and Inferior Nasal Conchae with Lab Activity
11. Zygomatic Bones with Lab Activity
12. Palatine and Vomer Bones with Lab Activity
13. Foramina, Canals, and Fissures
14. Paranasal Sinuses with Lab Activity
15. Skull`s Associated Bones - Hyoid and Ossicles with Lab Activity
16. Vertebral Column - Introduction
17. Typical Vertebral Structure
18. Cervical Division with Lab Activity
19. Atlas and Axis with Lab Activity
20. Cervical Articulations with Lab Activity
21. Thoracic Division with Lab Activity
22. Lumbar Division with Lab Activity
23. Sacrum with Lab Activity
24. Thorax (Thoracic Cage) with Lab Activity
25. Ribs with Lab Activity
26. Sternum with Lab Activity

Chapter 13: The Appendicular Skeleton
1. Clavicles with Lab Activity
2. Scapulas with Lab Activity
3. Humerus - Proximal with Lab Activity
4. Humerus - Distal with Lab Activity
5. Forearm - Introduction
6. Ulna - Proximal with Lab Activity
7. Ulna - Distal with Lab Activity
8. Radius - Proximal with Lab Activity
9. Radius - Distal with Lab Activity
10. Articulations of the Radius and Ulna with Lab Activity
11. Hand with Lab Activity
12. Articulations of the Carpus with Lab Activity
13. Pelvic Girdle - Coxal Bones with Lab Activity
14. Pelvis with Lab Activity
15. Male and Female Pelvic Structure
16. Femur - Proximal with Lab Activity
17. Femur - Distal with Lab Activity
18. Patella with Lab Activity
19. Bones of the Leg
20. Tibia - Proximal with Lab Activity
21. Tibia - Distal with Lab Activity
22. Fibula with Lab Activity
23. Fibula - Proximal with Lab Activity
24. Fibula - Distal with Lab Activity
25. Foot with Lab Activity

Chapter 14: Articulations
1. Articulations - Introduction
2. Fibrous Joints - Sutures with Lab Activity
3. Fibrous Joints - Gomphoses with Lab Activity
4. Fibrous Joints - Syndesmoses with Lab Activity
5. Cartilaginous Joints - Synchondroses with Lab Activity
6. Cartilaginous Joints - Symphyses with Lab Activity
7. Synovial Joints - Introduction to Structure and Function
8. Angular Movements: Flexion, Extension, Adduction, Abduction, Circumduction
9. Circular Movements: Rotation, Supination, and Pronation
10. Special Movements: Elevation, Depression, Retraction, and Protraction
11. Synovial Joints - Types - Introduction
12. Synovial Joints - Nonaxial Gliding Joints with Lab Activity
13. Synovial Joints - Uniaxial Hinge Joints with Lab Activity
14. Synovial Joints - Uniaxial Pivot Joints with Lab Activity
15. Synovial Joints - Biaxial Condyloid Joints with Lab Activity
16. Synovial Joints - Biaxial Saddle Joints with Lab Activity
17. Synovial Joints - Multiaxial Ball-and-Socket Joints with Lab Activity
18. Shoulder Joint and Skeletal Articulations
19. Elbow Joint and Skeletal Articulations
20. Hip (Coxal) Joint and Skeletal Articulations
21. Knee Joint and Skeletal Articulations

Chapter 15: The Muscular System
1. Functional Groups and Naming Muscles
2. Types of Movements
3. Principal Superficial Muscles

Chapter 16: Muscles and Contraction
1. Types and Function of Muscle Tissue
2. Skeletal Muscle
3. Study of Sleletal Muscle, l.s. with Lab Activity 1
4. Study of Sleletal Muscle, x.s. with Lab Activity 2 -
5. Attachment of Skeletal Muscles
6. Microanatomy and Neuromuscular Junction
7. Sarcolemma, Sarcoplasm, and T Tubules
8. Sarcoplasmic Reticulum
9. Myofibrils and A Band
10. I Bands
11. Sarcomere
12. Study of Neuromuscular Junctions and Muscle Spindles with Lab Activity 3 -
13. Study of the Synapse (Neuromuscular Junction) with Lab Activity 4 -
14. Excitation - Contraction of Skeletal Muscle Fiber
15. Sliding Filaments - Contraction of Skeletal Muscle Fiber
16. Twitch
17. Varying Contraction by Motor Unit Recruitment
18. Motor Unit Recruitment with Lab Activity 5
19. Treppe - Staircase Effect
20. Isotonic Contraction
21. Isometric Contraction
22. Varying Contraction by Frequency of Stimulation
23. Frog Gastrocnemius Muscle with Lab Activity 6
24. Muscle Energetics
25. Muscle Energetics and Types of Fibers
26. Smooth muscle
27. Study of Smooth muscle with Lab Activity 7
28. Contraction and Features of Smooth Muscle
29. Cardiac Muscle
30. Study of Cardiac Muscle with Lab Activity 8

Chapter 17: The Nervous System
1. The Nervous System, Introduction
2. Neurons
3. Neuroglia
4. Myelinated and Unmyelinated Axons of PNSc
5. Myelinated and Unmyelinated Axons of CNS
6. Nerve
7. Specialized Neuron Endings
8. Human Brain
9. Cerebrum (silver stain - section) and Cerebral cortex
10. Diencephalon
11. Brain Stem
12. Cerebellum
13. Cerebellum (silver stain - section)
14. Ventricles
15. Meninges
16. Cerebrospinal Fluid
17. Cranial Nerves
18. Dissection of Sheep Brain
19. Spinal Cord

Chapter 18: The Eye
1. Accessory Structures
2. General Structure of the Eyeball
3. Fibrous Tunic
4. Vascular Tunic
5. Sensory Tunic
6. Chambers, Cavities, Fluids, and Lens
7. Dissection of Sheep Eye - External Anatomy
8. Dissection of Sheep Eye - Anterior Portion
9. Dissection of Sheep Eye - Posterior Portion
10. Light
11. Light Refraction
12. Light Refraction onto the Retina
13. Visual Acuity
14. Visual Tests
15. Retina
16. Visual Pigments
17. Transduction of Light
18. Optic Nerve Pathway and Visual Fields

Chapter 19: The Ear
1. General Structure of the Ear
2. Outer Ear, Middle Ear, and Ossicles
3. Inner Ear
4. Cochlea
5. Vestibule
6. Semicircular Canals

7. Cochlea
8. Mechanism of Hearing
9. Pathways of Sound Conduction
10. Hearing Tests
11. Equilibrium
12. Pathways to the Brain

Chapter 20: The Endocrine System
1. Introduction to the Endocrine System
2. Introduction to the Endocrine System - Web Study
3. Pituitary Gland
4. Anterior Pituitary Gland
5. Thyroid Stimulating Hormone (TSH)
6. Adrenocorticotropic hormone (ACTH)
7. Follicle stimulating hormone (FSH)
8. Luteinizing hormone (LH)
9. Growth hormone (GH)
10. Prolactin (PRL)
11. Pars intermedia
12. Posterior Pituitary
13. Hypothalamus
14. Pituitary Gland Disorders - Web Study
15. Thyroid Gland
16. Thyroid Gland Disorders - Web Study
17. Parathyroid
18. Parathyroid Gland Disorders - Web Study
19. Adrenal Gland
20. Adrenal Gland Disorders - Web Study
21. Pancreas
22. Pancreas Disorders - Web Study
23. Testes
24. Ovaries
25. Thymus
26. Kidneys

Chapter 21: The Blood
1. Blood
2. Plasma
3. Formed Elements
4. Erythrocytes and Hemoglobin
5. Leukocytes and Platelets
6. Differential Count
7. ABO Blood Groups - Typing
8. Rh Blood Typing and Hemolytic Disease of the Newborn
9. Blood Typing with Videos and Lab Activities

Chapter 22: The Lymphatic System
1. Introduction to the Lymphatic System
2. Pathway of Lymphatic vessels
3. Introduction to the Lymphatic System - Web Study
4. The Lymphatic System Disorders - Web Study
5. Introduction to Lymphatic Tissues and Organs
6. Tonsils
7. Lymph Node
8. Spleen
9. Thymus
10. Peyer's Patches

Chapter 23: The Heart
1. Introduction to the Heart and Heart Coverings
2. Gross Anatomy of the Heart and Valves
3. Gross Anatomy of the Heart, Valves, and Vessels
4. Route of Blood Flow Through the Heart
5. Coronary Circulation
6. Sheep Heart Dissection - Introduction and Coronal Section
7. Video and Exam on External Anatomy
8. Sheep Heart Dissection - Right Side of the Heart
9. Video and Exam on the Dissection of Right Side of Heart
10. Sheep Heart Dissection - Left Side of the Heart
11. Video and Exam on the Dissection of the Left Side of Heart
12. Sheep Heart Dissection - Valves and Ventricular Walls
13. Electrical Events of the Heart
14. Cardiac Cycle
15. Auscultation of the Heart
16. Cardiac Muscle - Introduction
17. Cardiac Muscle - Action Potential
18. Cardiac Conduction fibers
19. Cardiac Output - ESV, EDV, and Rate

Chapter 24: Blood Vessels and Circulation
1. Introduction to Blood Vessels - Arteries and Arterioles
2. Introduction to Blood Vessels - Capillaries
3. Introduction to Blood Vessels - Venules and Veins
4. Artery and Vein Structure
5. Arteries and Veins Structure with Lab Activity 1
6. Valves and Valve Structure with Lab Activity 2
7. Circulation Pathways
8. Systemic Blood Pressure and Measurement with Lab Activity 3
9. Regulation of Blood Pressure and Flow - Introduction
10. Regulation of Blood Pressure and Flow - Peripheral Resistance
11. Regulation of Blood Pressure and Flow - Additional Factors
12. Fluid Movement at the Capillary

Chapter 25: Respiratory System
1. Introduction to the Respiratory System with Lab Activity 1
2. Upper Respiratory Tract - Nose and Pharynx
3. Upper Respiratory Tract - Pharynx
4. Upper Respiratory Tract - Larynx and Trachea with Lab Activity 2
5. Lungs
6. Lungs: Trachea, Bronchial Tree, Lung with Lab Activities - 3 & 4
7. Pulmonary Ventilation
8. Respiratory Volumes and Capacities with Lab Activity 5
9. Gas Movement - Respiratory Membrane
10. Transport of Respiratory Gases
11. Mechanisms Controlling Respiration

Chapter 26: Digestive System
1. Introduction to the Digestive System with Lab Activity 1
2. Mucosa
3. Submucosa and Muscularis Externa
4. Serosa and the Peritoneum
5. Mouth
6. Tongue with Lab Activities 2 & 3
7. Salivary Glands with Lab Activity 4
8. Pharynx and Esophagus with Lab Activity 5
9. Stomach with Lab Activity 6
10. Control of Gastric Activity
11. Small Intestine
12. Small Intestine - Mucosa, Submucosa, Muscularis, and Serosa
13. Regions of the Small Intestine with Lab Activities 7 & 8
14. Pancreas with Lab Activity 9
15. Liver with Lab Activities 10 - 13
16. Large Intestine with Lab Activity 14

Chapter 27: Urinary System
1. Introduction to the Urinary System with Lab Activity 1
2. Video - Dissection of the Kidney
3. Anatomy of the Kidney - Dissection Lab Activity 2
4. Nephrons
5. Nephrons with Lab Activity 3
6. Histology of the Kidney and Nephrons with Lab Activity 4
7. Processes of Urine Formation
8. Filtration at the Glomerulus: Filtration Membrane and Pressures
9. Glomerular Filtration - Control Mechanisms
10. Reabsorption
11. Secretion
12. Production of Medullary Osmotic Gradient
13. Final Regulation of Urine Water Volume
14. Ureters, Urinary Bladder, and Urethra
15. Micturation

Chapter 28: The Reproductive System
1. Male Reproductive System - Introduction
2. Testes with Lab Activity 1
3. Testes - Meiosis
4. Meiosis - Spermatogenesis With Lab Activity 2
5. Sustentacular Cells
6. Male Hormonal Regulation - FSH
7. Male Hormonal Regulation - LH
8. Spermatozoa (sperm) with Lab Activity 3
9. Epididymis with Lab Activity 4
10. Ductus (Vas) Deferens with Lab Activity 5
11. Seminal Vesicles with Lab Activity 6
12. Prostate Gland with Lab Activity 7
13. Penis with Lab Activity 8
14. Female Reproductive system - Introduction
15. Female`s External Genitalia
16. Female`s Internal Reproductive Organs
17. Ovary with Lab Activity 9
18. Corpus Luteum and Corpus Albicans with Lab Activities 10 & 11
19. Meiososis and Oogenesis with Lab Activity 12
20. Ovarian Cycle
21. Uterine Tube (Fallopian) with Lab Activity 13
22. Uterus with Lab Activities 14 - 16
23. Uterine Cycle
24. Vagina with Lab Activity 17
25. Mammary Glands with Lab Activities 18 & 19

Chapter 29: Dissection of the Fetal Pig
1. Video and Test on General External Structure
2. Video and Test on Intro-Incisions
3. Video and Test on Overview of Abdominal Organs
4. Video and Test on Overview Superficial Structures of Neck
5. Video and Test on Overview of Deep Structures of Neck
6. Video and Test on Overview of External Anatomy of Heart and Veins of the Neck
7. Video and Test on Overview of Vessels of Heart and Arteries of Neck
8. Video and Test on Overview of Fetal Heart and Circulation
9. Video and Test on Overview of Abdominal Dissection
10. Video and Test on Fetal Pig Male Reproductive System - Introduction
11. Video and Test on Fetal Pig Male Reproductive System - Dissection
12. Video and Test on Fetal Pig Female Reproductive System - Dissection

INTRODUCTION TO ANATOMY AND PHYSIOLOGY WORKSHEETS

Anatomy

1. Define anatomy. _____
_____

2. Define cellular anatomy. _____
_____

3. Define cytology. _____
_____

4. Define developmental anatomy (embryology). _____
_____

5. Define gross anatomy. _____
_____

6. Define histological anatomy. _____
_____

7. Define histology. _____
_____

8. Define microscopic anatomy. _____
_____

9. Define regional anatomy. _____
_____

10. Define systemic anatomy. _____
_____

11. Define surface anatomy. _____
_____

Physiology

12. Define physiology. _____

13. Define cell physiology. _____
_____

14. Define pathology. _____
_____

15. Define systemic physiology. _____
_____

16. Define special (organ) physiology. _____
_____

Complementarity

17. What does complementarity of anatomy and physiology refer to? _____
_____

Organizational Levels

18. List in sequence (lowest first) the six hierarchical levels of anatomy and physiology.
    (1) _____
    (2) _____
    (3) _____
    (4) _____
    (5) _____
    (6) _____

19. How does the chemical level (atoms, molecules, and their interactions) relate to cells? _____
_____

20. Cells are built on the _____ level and are organized into the _____ level.

21. What are the three components of the cell theory?
    (1) _____
    (2) _____
    (3) _____

22. Tissues are built on the _____ level and are organized into the _____ level.

23. What are the four fundamental groups of tissues?
    (1) _____
    (2) _____
    (3) _____
    (4) _____

24. Organs are built on the _____ level and are organized into the _____ level.

25. Organ systems are built on the _____ level and are organized into the _____ level.

26. Match the following systems with their components:

Cardiovascular system Muscular system
Digestive system Nervous system
Endocrine system Respiratory system
Female reproductive system Skeletal system
Integumentary system Urinary system
Lymphatic system
Male reproductive system

_____ Brain, spinal cord, nerves, and receptors
_____ Heart, blood vessels, and blood
_____ Kidneys, ureters, urinary bladder, and urethra
_____ Lymph nodes, lymphatic vessels and their fluid called lymph, tonsils, spleen, and thymus
_____ Mouth, esophagus, stomach, small intestine, large intestine, anus, and accessory
_____ Nasal cavity, voice box (larynx), windpipe (trachea), and lungs
_____ Organs such as salivary gland, pancreas, liver and gallbladder
_____ Organs which produce hormones (chemical messengers) which include pituitary, testes, ovaries, thymus, thyroid
_____ Ovaries, fallopian tubes, uterus, and vagina
_____ Skeletal muscles
_____ Skeleton
_____ Skin, hair, nails, sweat glands and oil glands
_____ Testes, ductus (vas) deferens, prostate, seminal vesicles, and penis

Chapter 1 Introduction to Anatomy and Physiology

27 Match the following systems with their functions:
   Cardiovascular system
   Digestive system
   Endocrine system
   Female reproductive system
   Integumentary system
   Lymphatic system
   Male reproductive system
   Muscular system
   Nervous system
   Respiratory system
   Skeletal system
   Urinary system

   _____ Delivery of air to lungs for oxygen and carbon dioxide exchange between air and blood
   _____ Immediate control of systems, personality, emotions, etc.
   _____ Includes the production, storage, and elimination of urine, which involves regulation of water, electrolytes, and blood pH.
   _____ Includes the skeleton which supports, protects, provides for storage of calcium, and serves as a site of blood cell production
   _____ Long-term regulation of systems by production and release of hormones
   _____ Movement of the body and involved in body temperature regulation
   _____ Processing and absorption of nutrients
   _____ Production of egg, implantation and development
   _____ Production of lymphocytes for immunity, and collects, filters, and transports fluid (lymph)
   _____ Production of sperm
   _____ Protection (by skin, hair, etc.), site of sensory receptors, involved in body temperature control, etc.
   _____ Transport of blood; including cells, nutrients, wastes, gases, hormones, etc.

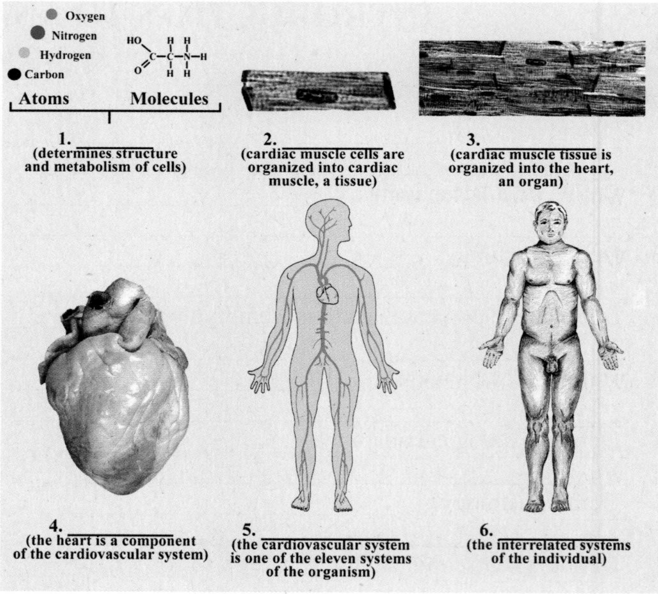

Figure 1.1

28 In reference to **Figure 1.1**, identify levels #1 - #6.
   1_____ 4_____
   2_____ 5_____
   3_____ 6_____

Characteristics of Human Life

29 Define metabolism. _____

30 What are the two major divisions of metabolism?
   (1)_____
   (2)_____

31 Define catabolism. _____

32 Define anabolism. _____

33 What are three ways growth may occur?
   (1)_____
   (2)_____
   (3)_____

34 Define differentiation. _____

35 Define responsiveness. _____

36 Motion begins with controlled molecular actions within the _____.

37 List five areas where movements are seen.
   (1)_____
   (2)_____
   (3)_____
   (4)_____
   (5)_____

38 What are two processes of cell reproduction?
   (1)_____
   (2)_____

39 What does meiosis give rise to? _____

40 What are three things cell division provides? _____

Chapter 1 Introduction to Anatomy and Physiology   **3**

Homeostasis

41 Define homeostasis. _____
_____
42 What are three components of a control mechanism? _____
_____
43 What is the function of receptors? _____
_____
44 What is the function of the control center? _____
_____
_____
_____
45 What is the function of effectors? _____
_____
_____
46 What is a negative feedback mechanism? _____
_____
47 What is a positive feedback mechanism? _____
_____

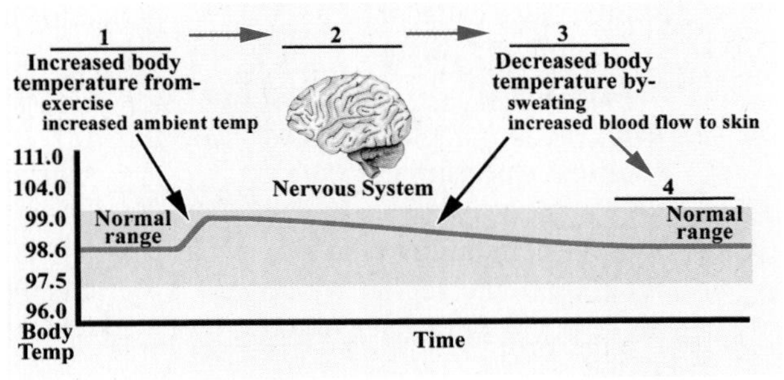

Figure 1.2

48 In reference to **Figure 1.2**, identify #1 - #3.
1 _____ 3 _____
2 _____

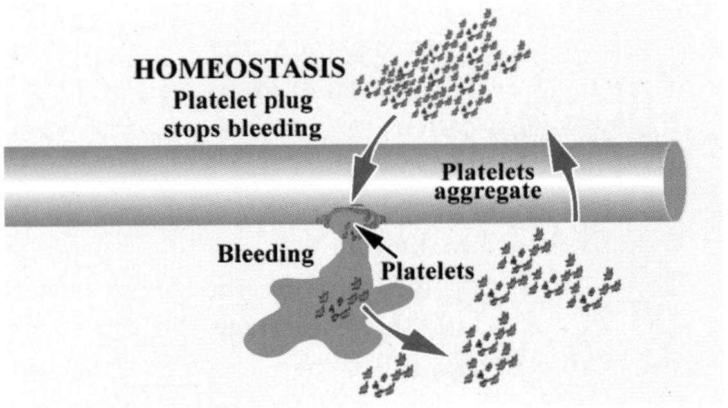

Figure 1.3

49 In reference to **Figure 1.3**, is a negative or positive feedback mechanism shown? _____

Organization and Terminology

50 Describe anatomical position. _____
_____
_____
_____

51 Define supine position. _____
_____

52 Define prone position. _____
_____

Regional Terminology

53 What are the two major regions of the body?
(1) _____
(2) _____
54 What does the axial region include? _____
_____
55 What does the appendicular region include? _____
_____

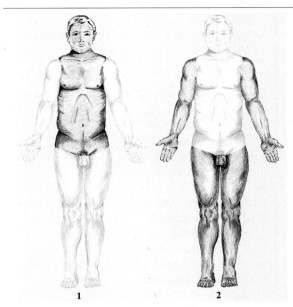

Figure 1.4

56 In reference to **Figure 1.4**, what region is highlighted in "1"? _____ What region is highlighted in "2"? _____

57 Define the following terms:
- **Abdomen** - pertaining to the part of the body between the _____ and the _____
- **Antibrachial**, also called the _____, pertaining to the part of the upper extremity between the _____ and the _____
- **Anticubital** - pertaining to the area at the _____ of the _____.
- **Arm**, also called the _____ region, the region from the _____ to the _____
- **Axilla** - the small _____ area under the _____, where the _____ joins to the _____
- **Brachial**, also called the _____, pertaining to the region from the _____ to the _____
- **Buccal**, also called the _____, pertaining to fleshy _____ on each side of the _____
- **Buttock**, also called the _____ region, the fleshy portion of the _____ on which a person _____
- **Calcaneus**, also called the _____, pertaining to the _____
- **Calf**; also called the _____ region, the _____ compartment of the _____
- **Carpus**, also called the _____, the region of _____ from the _____ to the metacarpals (bones of the palm)

Chapter 1 Introduction to Anatomy and Physiology

- **Cervical** - pertaining to the _____
- **Cheek**, also called the _____ region, the fleshy _____ on each side of the _____
- **Chin**, also called the _____ region, the portion of the _____ below the lower _____ and including the anterior projection of the _____
- **Cranium**, also called the _____ region, the bones of the _____ that enclose the _____
- **Crural**, also called the _____, pertaining to the region from the _____ to the _____
- **Digit** - the _____ or _____
- **Elbow** - the region of the _____ between the _____ and the _____
- **Face**, also called the _____ region, the region of the _____ from _____ to the _____
- **Femoral** - pertaining to the _____
- **Flank** - the fleshy part of the _____ between the _____ and _____
- **Forearm**, also called the _____ region, the part of the upper _____ between the _____ and the _____
- **Forehead**, also called the _____, or _____ region, the part of the _____ above the _____
- **Frons** - pertaining to the _____, or _____ region
- **Genitalia** - the _____ organs; especially the _____ genital (sexual) organs
- **Gluteus** - the _____
- **Groin**, also called the _____ region, the _____ region between the _____ and the _____
- **Hallux** - the _____ _____, the first _____ of the foot
- **Hand**, also called the _____ region, the region from the _____ to the _____
- **Heel** - the _____ of the _____, also called _____ region.
- **Inguinal**, also called the _____, pertaining to the _____ region between the _____ and the _____
- **Knee** - the region of the _____ between the _____ and _____
- **Leg**, also called the _____ region, the region from the _____ to the _____
- **Loin**, also called the _____ region, the part of the _____ on each side of the spinal column between the _____ and the _____
- **Lumbar** - pertaining to the _____
- **Manus**, also called the _____, the region from the _____ to _____
- **Mental**, also called the _____, the anterior projecting portion of the _____ beneath the lower _____

- **Mouth**, also called the _____ region, the opening at the _____ of the _____ system
- **Neck**, also called the _____ region, the region connecting the _____ and _____
- **Nose**, also called the _____ region, the part of the _____ that houses the _____ and covers the anterior nasal _____
- **Occiput** - the _____ of the _____
- **Oral** - pertaining to the _____
- **Orbit**, also called the _____ region, the bony _____ which contains the _____
- **Palm**, also called the _____ region, the _____ of the _____
- **Pelvis**, also called the _____ region, pertaining to the part of the body under the _____, bounded by the _____, the _____, and the _____.
- **Pinna** - the projecting portion of the _____
- **Plantar**, also called the _____, pertaining to the _____ of the _____
- **Pollex** - the _____, the first _____ of the hand
- **Popliteal** - pertaining to the _____ surface of the _____
- **Sacral** - pertaining to the area at the _____
- **Shoulder** - the region where the _____ attaches to the _____ or the area of the _____ which covers the _____ _____
- **Sural**, also called the _____, pertaining to the muscular _____ portion of the _____
- **Sole**, also called the _____ region, the _____ of the _____
- **Tarsus**, also called the _____, the part of the _____ between the _____ and the _____ (bones of arch)
- **Thigh**, also called the _____ region, the portion of the lower _____ between the _____ and the _____
- **Thorax**, also called the _____ region, the _____, the part of the body between the _____ and the respiratory _____
- **Umbilicus**, also called the _____, the abdominal _____ that marks the point of former attachment of the umbilical _____

Name _____
Class _____

Chapter 1 Introduction to Anatomy and Physiology   **5**

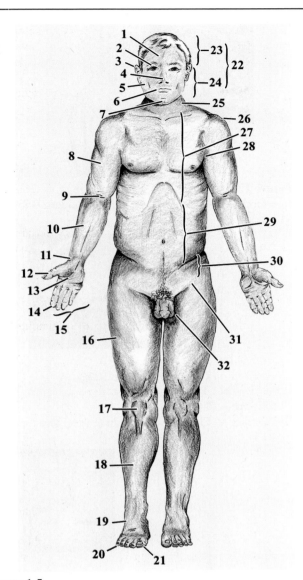

Figure 1.5
58   In reference to **Figure 1.5**, identify #1 - #32.

1 _____	17 _____
2 _____	18 _____
3 _____	19 _____
4 _____	20 _____
5 _____	21 _____
6 _____	22 _____
7 _____	23 _____
8 _____	24 _____
9 _____	25 _____
10 _____	26 _____
11 _____	27 _____
12 _____	28 _____
13 _____	29 _____
14 _____	30 _____
15 _____	31 _____
16 _____	32 _____

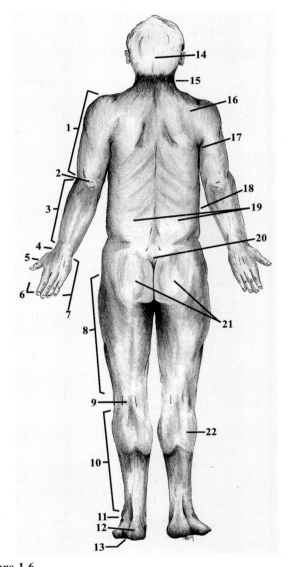

Figure 1.6
59   In reference to **Figure 1.6**, identify #1 - #20.

1 _____	11 _____
2 _____	12 _____
3 _____	13 _____
4 _____	14 _____
5 _____	15 _____
6 _____	16 _____
7 _____	17 _____
8 _____	18 _____
9 _____	19 _____
10 _____	20 _____

Chapter 1 Introduction to Anatomy and Physiology

Name _____
Class _____

DIRECTIONAL TERMINOLOGY

60 Define the following terms:
- Superior _____
_____
_____

- Inferior _____
_____
_____

- Anterior (ventral) _____
_____

- Posterior (dorsal) _____
_____

- Medial _____

- Lateral _____

- Intermediate _____

- Proximal _____

- Distal _____

- Superficial _____
_____

- Deep _____
_____

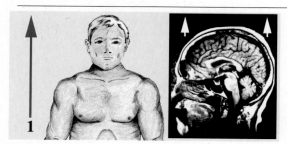

Figure 1.7

61 In reference to **Figure 1.7**, the directional term shown for #1 is _____

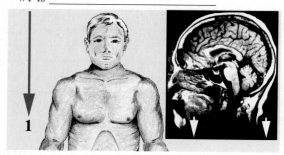

Figure 1.8

62 In reference to **Figure 1.8**, the directional term shown for #1 is _____

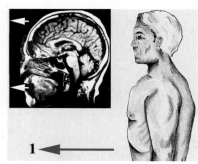

Figure 1.9

63 In reference to **Figure 1.9**, the directional term shown for #1 is _____

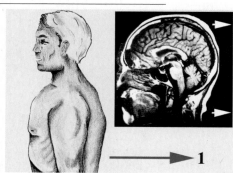

Figure 1.10

64 In reference to **Figure 1.10**, the directional term shown for #1 is _____

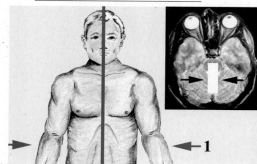

Figure 1.11

65 In reference to **Figure 1.11**, the directional term shown for #1 is _____

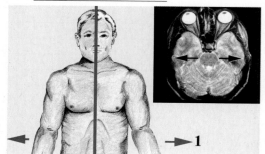

Figure 1.12

66 In reference to **Figure 1.12**, the directional term shown for #1 is _____

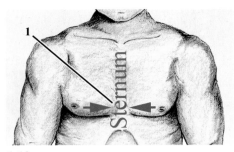

Figure 1.13

67 In reference to **Figure 1.13**, the directional term shown for #1 is _____

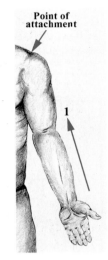

Figure 1.14

68 In reference to **Figure 1.14**, the directional term shown for #1 is _____>

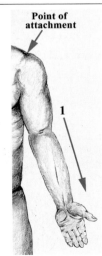

Figure 1.15

69 In reference to **Figure 1.15**, the directional term shown for #1 is _____ .

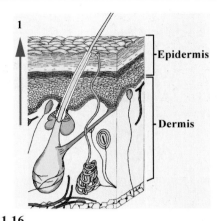

Figure 1.16

70 In reference to **Figure 1.16**, the directional term shown for #1 is _____ .

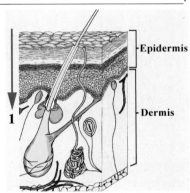

Figure 1.17

71 In reference to **Figure 1.17**, the directional term shown for #1 is _____

Body Planes and Sections

72 Define the following terms:
- Plane _____
- Sagittal plane _____
- Midsagittal plane _____
- Parasagittal plane _____
- Frontal (coronal) plane _____
- Transverse (horizontal) plane _____

8 Chapter 1 Introduction to Anatomy and Physiology

Name _____
Class _____

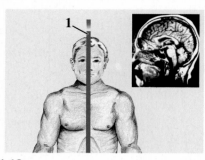

Figure 1.18
73 In reference to **Figure 1.18**, the plane shown for #1 is
_____

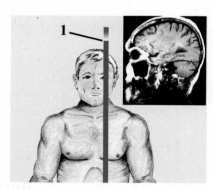

Figure 1.19
74 In reference to **Figure 1.19**, the plane shown for #1 is
_____

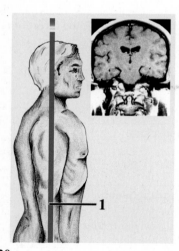

Figure 1.20
75 In reference to **Figure 1.20**, the plane shown for #1 is
_____

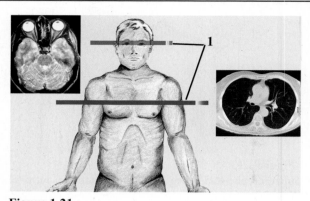

Figure 1.21
76 In reference to **Figure 1.21**, the plane shown for #1 is
_____

Body Cavities and Membranes
77 What are the two largest cavities of the body?
(1)_____
(2)_____

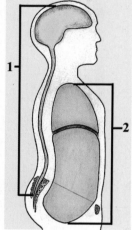

Figure 1.22
78 In reference to **Figure 1.22**, identify #1 & #2.
1 _____ 2 _____

79 What are the two divisions of the dorsal body cavity?
(1)_____
(2)_____
80 What does the cranial cavity contain? _____
81 What does the vertebral cavity contain? _____

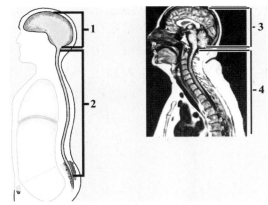

Figure 1.23

82. In reference to **Figure 1.23**, identify #1- #4.
    1 _____ 3 _____
    2 _____ 4 _____

83. What are the two divisions of the ventral body cavity?
    (1) _____
    (2) _____

84. What are the three divisions of the thoracic cavity?
    (1) _____
    (2) _____
    (3) _____

85. What separates the thoracic cavity from the abdomino-pelvic cavity? _____

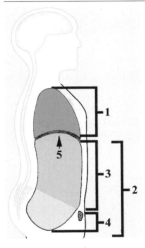

Figure 1.24

86. In reference to **Figure 1.24**, identify #1 - #5.
    1 _____ 4 _____
    2 _____ 5 _____
    3 _____

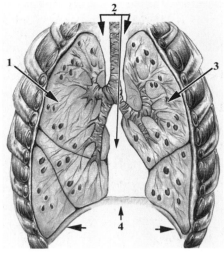

Figure 1.25

In reference to **Figure 1.xx**, identify #1 - #4.
    1 _____ 3 _____
    2 _____ 4 _____

87. What is the function of serous fluid? _____
    _____

Lungs; Membranes and Cavities

88. What name is given to the portion of the serous membrane that forms the lining of the outer wall of the pleural cavity? _____

89. What name is given to the portion of the serous membrane that forms the lining of the inner wall of the pleural cavity? _____

90. What is found within the pleural cavities? _____
    _____

91. Where is the mediastinum located? _____
    _____

92. What organs are located in the mediastinum? _____
    _____

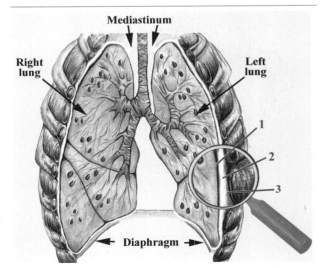

Figure 1.26

93. In reference to **Figure 1.26**, identify #1 - #3.
    1 _____ 3 _____
    2 _____

Chapter 1 Introduction to Anatomy and Physiology

Laboratory Animal Demonstration- Thorax

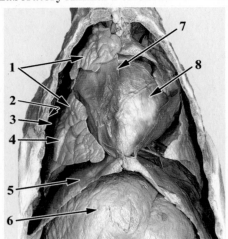

94 In reference to **Figure 1.27**, identify #1 - #8.
1 _____ 5 _____
2 _____ 6 _____
3 _____ 7 _____
4 _____ 8 _____

Heart: Membranes and Cavity

95 What name is given to the portion of the serous membrane that forms the lining of the outer wall of the pericardial cavity? _____
96 What name is given to the portion of the serous membrane that forms the lining of the inner wall of the pericardial cavity? _____
97 What is found within the pericardial cavity? _____

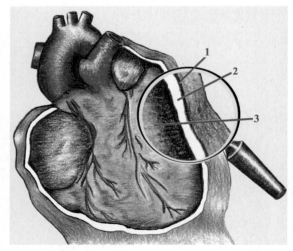

Figure 1.28
98 In reference to **Figure 1.28**, identify #1 - #3.
1 _____ 3 _____
2 _____

Laboratory Animal Demonstration- Heart

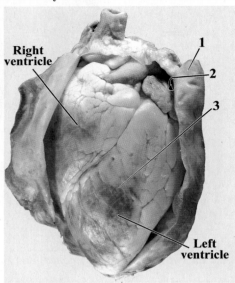

Figure 1.29
99 In reference to **Figure 1.29**, identify #1 - #3.
1 _____ 3 _____
2 _____

Abdomino-pelvic Cavity

100 List some of the organs the abdominal cavity contains.
_____
101 List some of the organs the pelvic cavity contains. _____
102 What name is given to the portion of the serous membrane which forms the lining of the outer wall of the peritoneal cavity? _____
103 What name is given to the portion of the serous membrane that forms the lining of the inner wall of the peritoneal cavity? _____

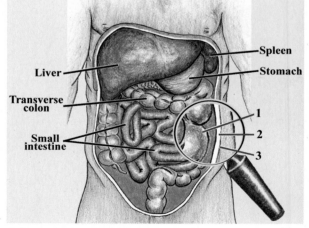

Figure 1.30
104 In reference to **Figure 1.30**, identify #1 - #3.
1 _____ 3 _____
2 _____

Chapter 1 Introduction to Anatomy and Physiology

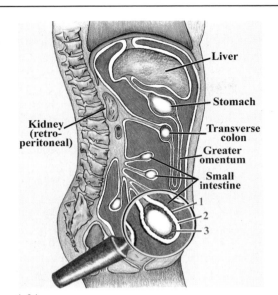

Figure 1.31
105 In reference to **Figure 1.31**, identify #1 - #3.
1 _____ 3 _____
2 _____

Laboratory Animal Demonstration- Abdomino-Pelvic

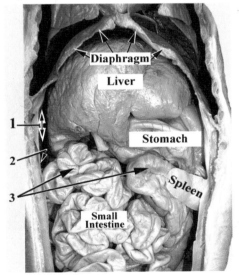

Figure 1.32
106 In reference to **Figure 1.32**, identify #1 - #3.
1 _____ 3 _____
2 _____

Abdomino-Pelvic Regions
107 Fill in the blanks to match the following regions with their locations:

    Epigastric   Iliac
    Hypochondriac   Lumbar
    Hypogastric   Umbilical

_____ Superior to the umbilical region and between the hypochondriac regions

_____ Middle abdominal region; inferior to the epigastric, superior to the hypogastric and between the lumbar regions.

_____ Inferior to the umbilical region and between the iliac regions

_____ Beneath the cartilages of the false ribs, superior to the lumbar regions, lateral to the epigastric region and

_____ Between the hypochondriac and the iliac and lateral to the umbilical

_____ Inferior to the lumbar and lateral to the hypogastric region

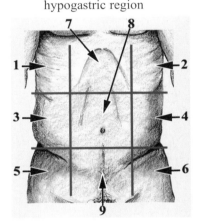

Figure 1.33
108 In reference to **Figure 1.33**, identify #1 - #9.
1_____ 6_____
2_____ 7_____
3_____ 8_____
4_____ 9_____
5_____

Chapter 1 Introduction to Anatomy and Physiology

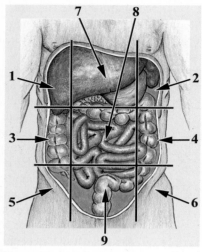

Figure 1.34

109 In reference to **Figure 1.34**, identify #1 - #9
1 _____ 6 _____
2 _____ 7 _____
3 _____ 8 _____
4 _____ 9 _____
5 _____

ABDOMINO-PELVIC QUADRANTS

110 Where are the lines drawn that divide the abdomino-pelvic cavity into quadrants? _____

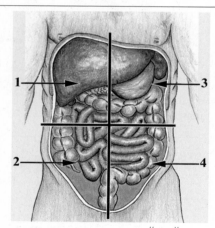

111 In reference to **Figure 1.35**, identify #1 - #4.
1 _____ 3 _____
2 _____ 4 _____

IMAGING

112 What are two types of imaging that use electromagnet energy (x-rays)?
(1) _____
(2) _____

113 The type of imaging that uses electromagnetic energy (x-rays) projected in a 360-degree motion around the body is _____.

114 The type of imaging that uses high frequency sound is called _____.

115 The type of imaging that uses radio frequency radiation and a strong magnetic field is _____.

116 The type of imaging that uses a radiopharmaceutical is _____.

117 The type of imaging commonly used in orthopedics (skeletal system) and orthodontics (irregularities in teeth) is _____.

118 The type of imaging commonly used in studying the fetus, gallbladder, heart, spleen, pancreas, and urinary bladder is the _____.

119 Match the following procedures with their images:

Procedure	Figure
Computerized tomography (CT)	_____
Magnetic resonance imaging (MRI)	_____
Nuclear medicine scan	_____
Radiogrpah (X-ray)	_____
Sonography (Ultrasound, US)	_____

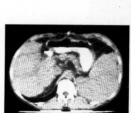 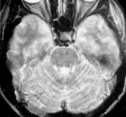 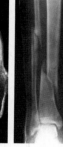

Figure 1.36 **Figure 1.37** **Fig. 1.38**

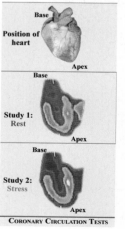 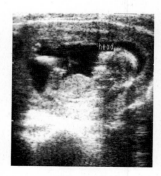

Figure 1.39 **Figure 1.40**

Microscopes and Microscopy

1. What are the most common types of microscopes used in the biological sciences? _____

2. Essentially, what does a compound microscope consist of? _____

3. What is the magnification range of compound microscopes? _____

4. What is a dissecting microscope? _____

5. What is an electron microscope? _____

6. What magnification is commonly obtained by transmission electron microscopes? _____

COMPOUND MICROSCOPE

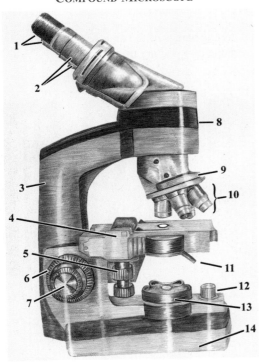

Figure 2.1

7. In reference to **Figure 2.1**, identify #1 - #14.
   1 _____ 8 _____
   2 _____ 9 _____
   3 _____ 10 _____
   4 _____ 11 _____
   5 _____ 12 _____
   6 _____ 13 _____
   7 _____ 14 _____

BASE, LIGHT SOURCE, AND ARM

8. What are three functions of the base of the microscope? _____

9. Check the features of the light source that applies to the microscope you are using.

Feature	Yes	No
Mirror		
Light bulb (plugs in)		
Variable control switch (dimmer switch)		

10. What is the arm of the microscope? _____

FOCUSING KNOBS

11. What are the names of the two focusing knobs? _____

12. What focusing knob is always the larger of the two? _____

13. Does the microscope you are using have both focusing knobs as a single mechanism, or are the two focusing knobs separate? _____

    If the two knobs are separate, which focusing knob is the uppermost in position? _____

Figure 2.2

14. In reference to **Figure 2.2**, identify #1 & #2.
    1 _____ 2 _____

Figure 2.3

15. In reference to **Figure 2.3**, identify #1 & #2.
    1 _____ 2 _____

16. Where is the body of the microscope located? _____

17. What is located on the lower surface of the microscope's body? _____

14 Chapter 2 Microscope

Name _____
Class _____

18. What is located on the upper surface of the microscope's body? _____

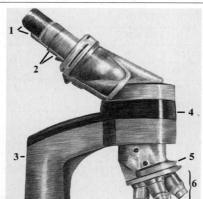

Figure 2.4

19. In reference to **Figure 2.4**, identify #1-#6.
    1 _____ 4 _____
    2 _____ 5 _____
    3 _____ 6 _____

OCULAR

20. What is an ocular? _____

21. A microscope with one ocular is called a _____, with two oculars a _____, and with three oculars a _____.

22. The most common ocular magnification is _____.

23. Considering the microscope you are using, what is the magnification of the ocular? _____

24. Oculars with pointers are designed to _____ within the body tube.

25. To clean the ocular, it (should or should not be) removed from the body tube. The ocular should only be cleaned with _____.

Figure 2.5

26. In reference to **Figure 2.5**, what are the two structures shown? _____

27. In reference to **Figure 2.5**, what do the numbers at #1 represent? _____

NOSEPIECE AND OBJECTIVES

28. Where are the objectives located? _____

29. When the nosepiece has two or more objectives, the nosepiece will _____, allowing selection of the objectives.

30. Each objective has an alignment stop and must "click" into the stop for perfect _____.

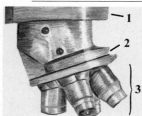

Figure 2.6

31. In reference to **Figure 2.6**, identify #1 - #3.
    1 _____ 3 _____
    2 _____

OBJECTIVES

32. What is the function of the microscope's objective? _____

Name of Objectives

33. What are the names of the four objectives that are usually supplied on a compound microscope? _____

34. What are the names of the objectives that are found on the microscope that you are using? _____

Figure 2.7

35. In reference to **Figure 2.7**, name each objective.
    1 _____ 3 _____
    2 _____ 4 _____

Magnifications

36. Give the magnification for each of the objectives of the microscope you are using. _____

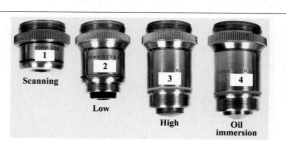

Figure 2.8

37. In reference to **Figure 2.8**, give the magnification of each objective.
    1 _____ 3 _____
    2 _____ 4 _____

38. As the magnification of the objective increases, the length of the objective _____.

Working Distance and Focal Point of Objectives

39 Define working distance. _____

40 What happens to the working distance as magnification increases? _____

41 Most modern microscopes are designed to be parfocal, each objective has its focal point in the same _____.

42 When using parfocal microscopes, the objective that is in use (should or should not) be moved away from the slide before the next objective is positioned.

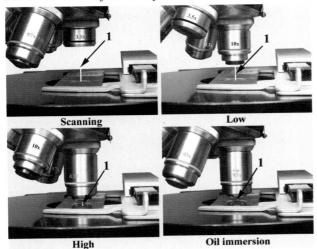

Figure 2.9

43 In reference to **Figure 2.9**, #1 is the _____.

44 In reference to **Figure 2.9**, if the specimen is in focus with the scanning power objective, the specimen (will or will not) be in focus when the low power objective is moved into position.

Depth of Field

45 Define depth of field. _____

46 What happens to the depth of field as magnification increases? _____

47 Considering all four objectives, which objective has the deepest depth of field? _____

48 Considering the microscope that you are using, which objective has the deepest depth of field? _____

49 Considering all four objectives, which objective has the shallowest depth of field? _____

50 Considering the microscope that you are using, which objective has the shallowest depth of field? _____

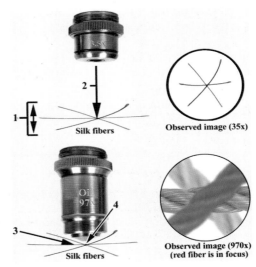

Figure 2.10

51 In reference to **Figure 2.10**, identify #1 - #4.
   1 _____ 3 _____
   2 _____ 4 _____

Field of View (Field)

52 Define field of view (field). _____

53 As the magnification increases, the field of view _____.

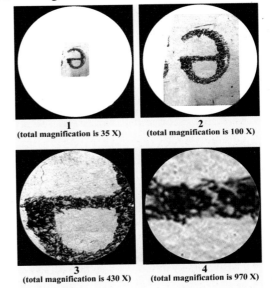

Figure 2.11

54 In reference to **Figure 2.11**, identify the objectives used for each field, #1 - #4.
   1 _____ 3 _____
   2 _____ 4 _____

Chapter 2 Microscope

Total Magnification Power

55 How is the total magnification determined? _____

56 Complete the following table for a microscope that has the listed lens systems.

Lens system	Ocular Magnification	Objective Magnification	Total Magnification
Ocular	10 X	----------	----------
Scanning power	10 X	3.5 X	
Low power	10 X	10 X	
High power	10 X	43 X	
Oil immersion	10 X	97 X	

57 Complete the following table for the microscope you are using (use NA if not applicable).

Lens system	Ocular Magnification	Objective Magnification	Total Magnification
Ocular	____ X	----------	----------
Scanning power	____ X	____ X	
Low power	____ X	____ X	
High power	____ X	____ X	
Oil immersion	____ X	____ X	

Stage

58 What is the function of the stage? _____

59 What are two functions of a mechanical stage? _____

60 What movements are allowed by the stage control knobs? _____

61 When using a mechanical stage, what are two ways to avoid damaging microscope slides? _____

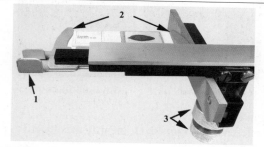

Figure 2.12

62 In reference to **Figure 2.12**, identify #1 - #3.
1 _____ 3 _____
2 _____

Condenser Lens

63 What is the function of the condenser lens? _____

64 Does the microscope you are using have a fixed or adjustable condenser lens? _____

Figure 2.13

65 In reference to **Figure 2.13**, identify #1 - #2.
1 _____ 2 _____

Diaphragm - Iris or Disk

66 What is the function of the diaphragm? _____

67 How is the iris diaphragm adjusted? _____

68 Why is it generally better to keep the amount of light slightly decreased? _____

69 What are two reasons that light adjustments are made? _____

70 Describe the disk diaphragm. _____

71 Where is usual location for the access point for moving the disk diaphragm? _____

72 Considering the microscope you are using, does it have an iris or a disk diaphragm? _____

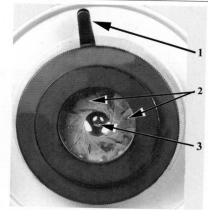

Figure 2.14

73 In reference to **Figure 2.14**, identify #1 - #3.
1 _____ 3 _____
2 _____

Figure 2.15

74 In reference to **Figure 2.15**, identify #1 - #2.
    1 _____ 2 _____

CARE AND USE OF THE MICROSCOPE
Setting-up the Microscope

75 When you carry the microscope one hand should hold the microscope's _____, and the other hand should hold the _____.
76 Always sit the microscope down _____ and away from the _____ of the table.
77 Clean all optical surfaces with _____.
78 Before using the microscope inspect it and make sure that _____.

Putting away the microscope

79 Make sure that the microscope slide is not on the _____ and either the _____ or _____ objective is in position.
80 If immersion oil was used it must be removed from the objective with _____ and oil must be cleaned _____.
81 If the microscope has an electrical cord, the cord should be _____.

MICROSCOPIC SPECIMENS

82 Prepared microscope slides are _____ and must be handled with _____ by the microscopist.
83 The preparation that utilizes water (or appropriate liquid) for suspension of the specimen is called a _____.
84 The thin glass that is placed on top of a microscopic specimen to protect the specimen and the high magnification objectives is called a _____.
85 When observing a wet mount the objective lenses should never contact the _____ or the _____.

FOCUSING THE MICROSCOPE

86 Either the _____ or _____ objective should be positioned before placing the microscope slide on the stage.
87 When placing the microscope slide, it must be moved (by the mechanical stage) so that the specimen is _____.

88 When starting to focus with scanning power (or low power), the objective should be moved to its _____ position and then raised until an image is observed.
89 The total magnification of the specimen is determined by _____.
90 The diaphragm should be adjusted to obtain the _____.
91 Before moving to the low power objective, the specimen should be centered in the _____.
92 With parfocal microscopes, before moving to the low power objective, the scanning power objective should (be moved away / should not be move away) from the specimen.
93 Before moving to the high power objective, the specimen should be centered in the _____.
94 With parfocal microscopes, before moving to the high power objective, the low power objective should (be moved away / should not be move away) from the specimen.
95 Focusing when using the high power objective must be done with the _____ adjustment knob, and care must be taken to avoid contact with the _____ of the microscope slide.
96 When using the oil immersion objective, a drop of _____ must be placed onto the microscope slide before moving the objective into position.
97 When using the oil immersion objective, focusing must be done with the _____ adjustment knob, and care must be taken to avoid contact with the _____ of the microscope slide.
98 After observation with the oil immersion objective the _____ objective should be moved into position. The _____ objective should never be moved back into position while immersion oil is on the microscope slide.
99 When observations are finished, all immersion oil must be removed from the oil immersion objective and the microscope slide with _____.

OBSERVATION OF MILLIMETER RULER
DIAMETER OF FIELD AND WORKING DISTANCE

Scanning Objective

100 For the microscope you are using, what is the diameter of the field for the scanning power objective? _____
101 As magnification increases the diameter of the field _____.
102 Before moving to higher magnification objectives, the specimen should always be _____ in the field of view.

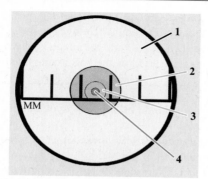

Figure 2.16

103 In reference to **Figure 2.16**, identify the objectives that correspond to the fields #1 - #4.
1 _____ 3 _____
2 _____ 4 _____

104 For the microscope you are using, what is the approximate working distance of the scanning power objective? _____

Low Power Objective

105 For the microscope you are using, what is the diameter of the field for the low power objective? _____

106 For the microscope you are using, what is the working distance of the low power objective? _____

High Power Objective

107 For the microscope you are using, what is the diameter of the field for the low power objective? _____

108 For the microscope you are using, what is the working distance of the low power objective? _____

LAB ACTIVITY

LETTER "E" (OR PRINTED LETTERS)

Scanning Power Objective

109 The lens systems of the compound microscope produce an image that appears _____ and _____.

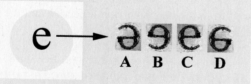

Figure 2.17

110 In reference to **Figure 2.17**, which illustration (A, B, C, or D) is the correct microscopic appearance for the normally positioned letter "e"? _____

111 If the microscope slide of the letter "e" is moved upward, what direction does the letter "e" appear to move as visualized through the microscope? _____
_____
_____

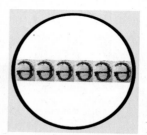

Figure 2.18

112 Calculate the width of the letter "e" if the diameter the field in **Figure 2.18** is 9 mm. _____ Convert your answer to micrometers (µm). _____ µm

113 For the microscope you are using and with the diameter of the scanning field that you measured with the mm ruler, calculate the width of the letter e. _____ mm. Convert your answer to micrometers (µm). _____ µm

114 What is the working distance of the scanning power objective? _____

Low Power Objective

Figure 2.19

115 Calculate the width of the letter e if the diameter the field in **Figure 2.19** is 3 mm. _____ mm Convert your answer to micrometers (µm). _____ µm

116 For the microscope you are using and with the diameter of the low power field that you measured with the mm ruler, calculate the width of the letter e. _____ mm. Convert your answer to micrometers (µm). _____ µm

117 What is the working distance of the low power objective? _____

High Power Objective

118 When using the high power objective, the _____ adjustment knob is NOT used.

Figure 2.20

119 Calculate the width of the letter e if the diameter the field in **Figure 2.20** is 1.5 mm. _____ mm Convert your answer to micrometers (µm). _____ µm

120 For the microscope you are using and with the diameter of the high power field that you measured with the mm ruler, calculate the width of the letter e. _____ mm. Convert your answer to micrometers (µm). _____ µm

121 What is the working distance of the high power objective? _____

Oil Immersion Objective
122 When using the oil immersion objective, the _____ adjustment knob is NOT used.
123 What is the working distance of the oil immersion objective? _____

LAB ACTIVITY

SILK FIBERS (CROSS COLORED THREADS)
124 What is the purpose of observation of the silk fibers (colored threads)? _____
_____
_____
125 The objective that has the _____ depth of field is best for determining the sequence of the fibers.
126 Define depth of field. _____
_____
127 For the microscope that you are using, which objective has the deepest depth of field? _____
128 For the microscope that you are using, which objective has the shallowest depth of field? _____
129 For the microscope slide you are using, what is the sequence from top to bottom of the silk thread (colored threads). _____
_____

LAB ACTIVITY

EXFOLIATED SQUAMOUS CELLS

Low Power Objective
130 For the microscope you are using and using the diameter of the scanning power field that you measured with the mm ruler, calculate the diameter of a cheek cell. _____ mm. Convert your answer to micrometers (µm). _____ µm

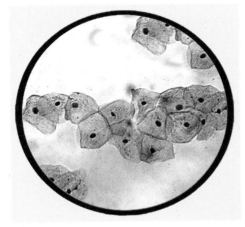

Figure 2.21

131 Calculate the diameter of a cheek cell if the diameter the field in **Figure 2.21** is 1.5 mm. _____ mm Convert your answer to micrometers (µm). _____

High Power Objective
130 For the microscope you are using and with the diameter of the low power field that you measured with the mm ruler, calculate the diameter of a cheek cell. _____ mm. Convert your answer to micrometers (µm). _____ µm

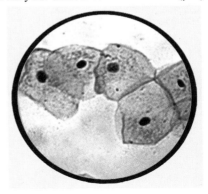

Figure 2.22

131 Calculate the diameter of a cheek cell if the diameter the field in **Figure 2.21** is 0.5mm. _____ mm Convert your answer to micrometers (µm). _____

Oil Immersion Objective
132 For the microscope you are using and with the diameter of the high power field that you measured with the mm ruler, calculate the diameter of a cheek cell. _____ mm. Convert your answer to micrometers (µm). _____ µm
133 Which objective, the low power or the high power is more accurate in determining the size of the cheek cell? _____

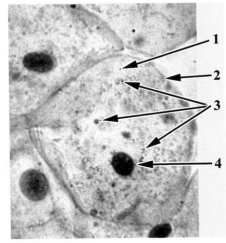

Figure 2.23

134 In reference to **Figure 2.22**, label #1 - #4.
  1 _____ 3 _____
  2 _____ 4 _____
135 Calculate the diameter of a cheek cell if the diameter the field in **Figure 2.23** is 0.25 mm. _____
_____
Convert your answer to micrometers (µm). _____

Name ______________________________
Class ______________________________

LAB ACTIVITY

Pond Water - Preparation of a Wet Mount

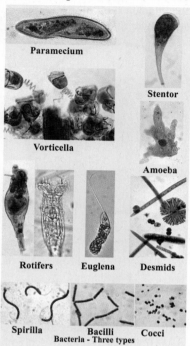

Figure 2.24

136 In reference to **Figure 2.24**, circle the organisms that you observed. In the Figure 2.25 , sketch any other organisms that you observed.

Sketches of Organisms from Pond Water

Figure 2.25

Sketch any other organisms that you observed.

Name _____
Class _____

Chapter 3 Chemistry and Metabolism   21

CHEMISTRY AND METABOLISM

1 Define matter. _____
_____
2 Define chemical elements. _____
_____

PERIODIC LAW AND TABLE

3 Define periodic law. _____
_____
_____
4 What is the periodic table of the chemical elements? ___
_____
_____
5 Define atomic number. _____
_____
_____

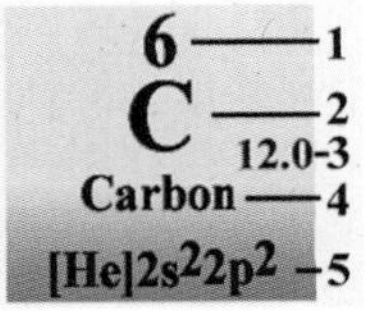

Figure 3.1

6 In reference to **Figure 3.1**, what is the atomic number of carbon? _____
7 In reference to **Figure 3.1**, what is the chemical symbol of carbon? _____
8 What are the names and symbols of the four major elements of the human body?
   1 _____ 3 _____
   2 _____ 4 _____
9 What are the names and symbols of the additional nine elements that comprise about 3.9% of the body?
   1 _____ 6 _____
   2 _____ 7 _____
   3 _____ 8 _____
   4 _____ 9 _____
   5 _____
10 What does electron configuration describe? _____
_____

ATOMS and ATOMIC STRUCTURE

11 What is an atom? _____
_____
12 What are the two basic organizational areas of an atom?
_____
_____

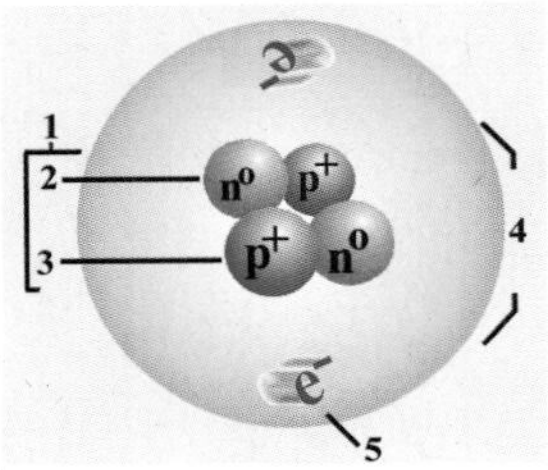

Figure 3.2
13 In reference to **Figure 3.2**, identify #1 - #5
   1 _____ 4 _____
   2 _____ 5 _____
   3 _____

14 Where is the nucleus of an atom located? _____
_____
15 What do the energy levels of an atom describe? _____
_____
16 What are the two major components contained in the nucleus? _____
_____
17 What is the difference between an orbit and an orbital?
_____
_____
18 What is the charge of the nucleus? _____
_____
19 Why does the nucleus have this charge? _____
_____
20 Where is most of the mass of an atom found? _____
_____
21 What is the symbol and charge of a proton? _____
_____
22 What does atomic number describe? _____
_____
23 Referring to the periodic table of the elements, what is the name of the element with atomic number 8? _____
_____
24 Where are neutrons located? _____
_____
What is the symbol and charge for a neutron? _____
_____
25 Define isotope? _____
_____

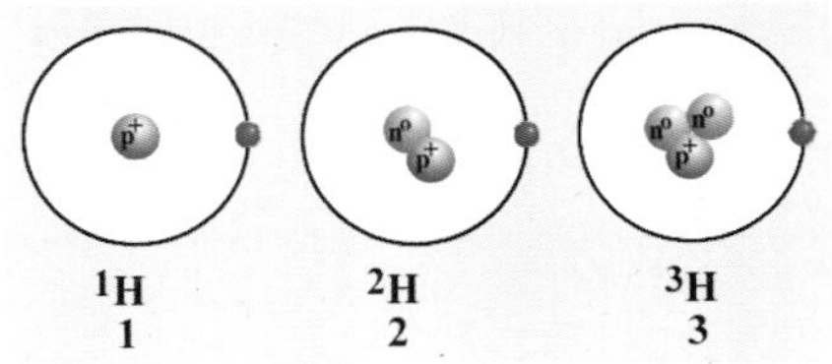

Figure 3.3
26 Figure 3.3 shows three isotopes of what element? _____
_____
27 Define mass number. _____
_____
_____
28 Define atomic weight _____
_____
_____
29 Use the periodic table and give the atomic number and weight of carbon and of lead. Carbon _____
Lead _____
30 What are energy levels? _____
_____ How are energy levels identified? _____
31 Where are electrons located? _____
_____
32 How many electrons does an atom of a naturally occurring element have? _____
_____

Chapter 3 Chemistry and Metabolism

Name _____
Class _____

33. What is the symbol and charge of an electron? _____

34. Using the periodic table, what are the atomic number, atomic weight, number of protons, and number of electrons of an atom of oxygen? _____

35. What is the atomic charge of an atom of oxygen? _____

36. If an atom of oxygen gained two electrons, how would its symbol and associated charge be written? _____

ENERGY and MATTER

37. Define energy. _____

38. Define potential energy _____

39. Define kinetic energy _____

40. What are four forms of energy? _____

41. Define chemical energy. _____

42. Give an example of potential chemical energy. _____

43. Give an example of kinetic chemical energy. _____

44. Define electrical energy. _____

45. Give an example of potential electrical energy. _____

46. Give an example of kinetic electrical energy. _____

47. Define mechanical energy. _____

48. Define radiant energy. _____

49. Give an example of radiant energy _____

Figure 3.4

50. In reference to **Figure 3.4**, identify the forms of energy:
    1 _____ 6 _____
    2 _____ 7 _____
    3 _____ 8 _____
    4 _____ 9 _____
    5 _____ 10 _____

CHEMICAL REACTIONS
ROLE OF ELECTRONS

51. What is the symbol for the four major electron orbitals? _____

52. What is the maximum number of electrons contained in each of the first three major electron orbitals? _____

53. Which orbitals are found in each of the first three energy levels? #1 _____ #2 _____
    #3 _____

54. How many electrons can each energy level contain?
    #1 _____ #2 _____ #3 _____

55. Refer to the periodic table and give the electron configurations for:
    phosphorus _____
    sulfur _____

56. As members of the main-group elements that conform to the "rule of eight", how many electrons does phosphorus and sulfur each need for stability and in what level would they be placed?
    phosphorus _____
    sulfur _____

57. Even though orbital *d* of energy level 3, can hold a maximum of 10 electrons, sometimes it only holds _____.

58. All atoms (except the noble, or inert, gases) gain, lose, or share their outer electrons in a way that is _____

59. What does a straight line drawn between two atoms represent? _____

MOLECULAR MODEL KIT

60. Fill in the following table by identifying the **elements in the molecular model kit**. Use the number of bonding sites, **NOT** the color, to determine the elements. If the element is not represented in the kit, write "NA" (not applicable).

Name of element	Symbol	# of bonds	Color
Carbon			
Oxygen			
Hydrogen			
Nitrogen			
Phosphorus			
Chlorine			

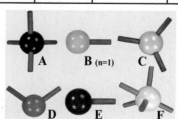

Figure 3.5

61. In reference to **Figure 3.5**, identify the atomic models that match the following elements:
    Carbon _____   Hydrogen _____
    Chlorine _____   Oxygen _____

Name _____
Class _____

Chapter 3 Chemistry and Metabolism   23

COMBINING MATTER

62 Why do atoms react? _____

63 What is a chemical bond? _____

64 Define molecule. _____

65 Define compound? _____

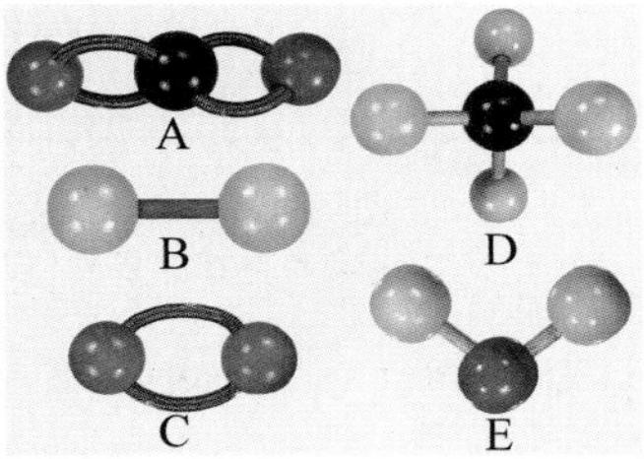

Figure 3.6

66 In reference to **Figure 3.6**, which are molecules? _____

67 In reference to **Figure 3.6**, which are compounds? _____

68 Define empirical formula. _____

69 Define chemical formula. _____

70 Define structural formula. _____

Chemical Equations

71 What is a chemical equation? _____

72 In molecules, what do numbers written as subscript to the element's symbol show? _____

73 In molecules, what do numbers written as a prefix to the molecule show? _____

74 What does it mean to "balance an equation?" _____

75 Considering this equation,
$6CO_2 + 12H_2O \longrightarrow C_6H_{12}O_6 + 6O_2 + 6H_2O$
How many molecules of water are produced? _____
What is the total number of hydrogen atoms in the products? _____ What is the total number of oxygen atoms in the reactants? _____ Is this equation balanced _____

Chemical Reactions

76 Define metabolism. _____

77 Define anabolism. _____

78 Define catabolism _____

79 List five general categories of chemical reactions. _____

80 What is a synthesis reaction? _____

81 In a dehydration synthesis reaction, where does the molecule of water originate? _____

82 What is a decomposition reaction? _____

83 What is the function of the molecule of water used in a hydrolysis reaction? _____

84 What is an exchange reaction? _____

85 What is an oxidation-reduction (redox) reaction? _____

86 Define oxidation. _____

87 Define reduction. _____

88 What is an oxidizing agent? _____

89 What is a reducing agent? _____

90 What is a reversible reaction? _____

$$A + B \longrightarrow AB$$

Figure 3.7

91 Name the type of reaction shown in **Figure 3.7**. _____

$$A + B \longrightarrow AB + H_2O$$

Figure 3.8

92 Name the type of reaction shown in **Figure 3.8**. _____

$$AB \longrightarrow A + B$$

Figure 3.9

93 Name the type of reaction shown in **Figure 3.9**. _____

$$AB + H_2O \longrightarrow A + B$$

Figure 3.10

94 Name the type of reaction shown in **Figure 3.10**. _____

$$AB + CD \longrightarrow AD + BC$$

Figure 3.11

95 Name the type of reaction shown in **Figure 3.11**. _____

Chapter 3 Chemistry and Metabolism

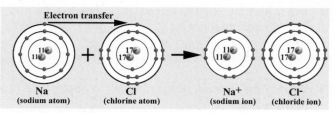

Figure 3.12

96 Name the type of reaction shown in **Figure 3.12**. _____

Which atom is oxidized? _____ Which atom is reduced? _____ Which atom is the oxidizing agent? _____ Which atom is the reducing agent? _____

$$A + B \rightleftharpoons AB$$

Figure 3.13

97 Name the type of reaction shown in **Figure 3.13**. _____

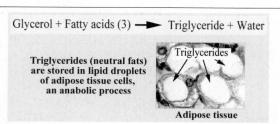

Figure 3.14

98 Name the type of reaction shown in **Figure 3.14**. _____

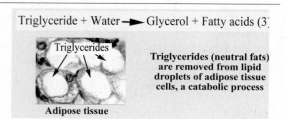

Figure 3.15

99 Name the type of reaction shown in **Figure 3.15**. _____

ENERGY OF CHEMICAL REACTIONS

100 Breaking chemical bonds of reactants can either occur spontaneously or _____.
101 Forming new bonds of the products _____ energy.
102 How is the overall energy of a reaction determined? _____

103 What characterizes exergonic reactions? _____

104 What are three possible utilizations for the energy that is released by exergonic reactions? _____

105 What characterizes endogonic reactions? _____

106 How is the amount of absorbed energy determined? _____

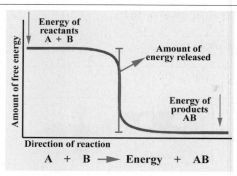

Figure 3.16

107 Does **Figure 3.16** represent an exergonic or an endogonic reaction? _____

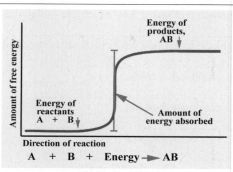

Figure 3.17

108 Does **Figure 3.17** represent an exergonic or an endogonic reaction? _____

ENZYMES

109 What is the function of enzymes? _____

110 How do enzymes affect the activation energy of a reaction? _____

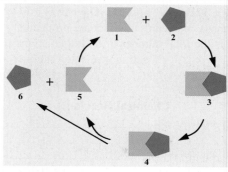

Figure 3.18

111 In **Figure 3.18** identify numbers 1 - 6.
1 _____
2 _____
3 _____
4 _____
5 _____
6 _____

CHEMICAL BONDS
IONIC BONDS

112 What is the charge of the atoms of the naturally occurring elements (from the periodic table)? _____

113 What is an ion? _____

114 Define cation. _____

115 Define anion _____

116 Define ionic bond _____

117 Considering the ionic bonding of a sodium ion and a chlorine (chloride) ions, which ion is the cation and the which is the anion? _____
Which atom is oxidized? _____
Which atom is reduced? _____

118 How are crystals of salts formed? _____

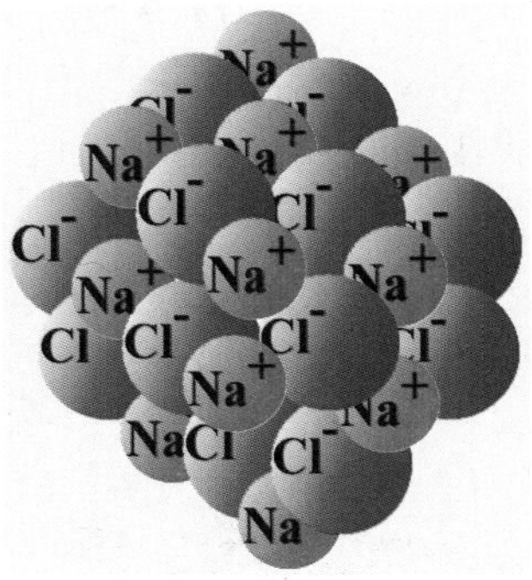

Figure 3.19

119 In reference to **Figure 3.19**, what is illustrated? _____
What type of bonds organize this structure? _____

COVALENT BONDS

120 What is a covalent bond? _____

121 How many pairs of electrons are shared in a single covalent bond? _____

122 What are three ways molecules can be written? _____

123 What is electron dot structure? _____

124 In a structural formula what does a straight line between atoms represent? _____

125 How many pairs of electrons are shared in a double covalent bond? _____

Figure 3.20

126 In reference to **Figure 3.20**, what is the name of this molecule? _____

127 In reference to **Figure 3.20**, what type of bond is represented? _____ How many pairs of electrons are found in this type of bond? _____

Figure 3.21

128 In reference to **Figure 3.21**, what is the name of this molecule? _____

129 In reference to **Figure 3.21**, what types of bonds are represented? _____ How many pairs of electrons are found between the two atoms? _____

COVALENT BONDING: CARBON AS CENTRAL ATOM

130 How many single covalent bonds can be formed by an atom of carbon? _____

131 What type of organic molecules (carbon based) are formed, if carbon atoms are central and their bonding is repeated? _____

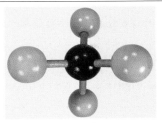

Figure 3.22

132 In reference to **Figure 3.22**, what is the name of this molecule? _____

133 Does **Figure 3.21** show the maximum number of single covalent bonds formed by an atom of carbon? _____

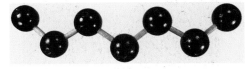

Figure 3.23

134 In reference to **Figure 3.23**, what type of chain is formed? _____

135 What is a hydrocarbon chain? _____

Chapter 3 Chemistry and Metabolism

Covalent Bonding: Polar and Non-polar Molecules

136 What are two features that are used in determining if a molecule is polar or nonpolar? _____

137 Describe non-polar molecules. _____

138 Describe polar molecules. _____

139 Define electronegative. _____

140 Define electropositive. _____

141 In a molecule of water, which atom is electronegative? _____ How does this characteristic influence the molecule of water? _____

142 How is shape important in determining whether a molecule is polar or non-polar? _____

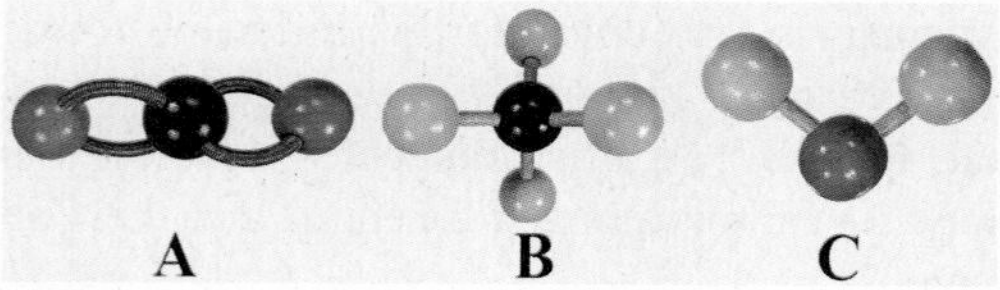

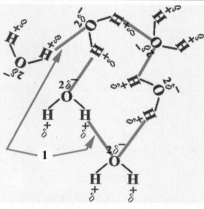

Figure 3.24

143 Name the three molecules shown in **Figure 3.24**.
A _____
B _____
C _____

144 In reference to **Figure 3.24**, which molecule/s is:
nonpolar? _____
polar? _____

Polar Characteristics of Water - Electrical Demonstration

Figure 3.25

145 In reference to **Figure 3.25**, what is the charge of the balloon? _____ Why is the stream of water bending toward the balloon? _____

HYDROGEN BONDS

146 Define hydrogen bond. _____

147 What is an important function of hydrogen bonds? _____

Figure 3.26

148 In reference to **Figure 3.26**, identify the bonds at #1. _____

149 In a molecule of water, what are the atoms (and their charges) that produce the hydrogen bonds? _____

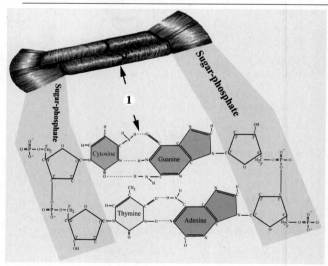

Figure 3.27

150 In reference to **Figure 3.27**, what is the name of the illustrated molecule? _____

In reference to **Figure 3.27**, what is the name of the bonds at #1? _____

151 Define denature. _____

AMPHIPHILIC MOLECULES

152 Define amphiphilic molecule. _____

153 Define hydrophobic. _____

154 Define hydrophilic. _____

155 What portion of the phospholipid molecule is hydrophobic? _____

156 What portion of the phospholipid molecule is hydrophilic? _____

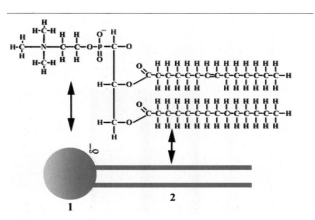

Figure 3.28

157 In reference to **Figure 3.28**, which region (#1 or #2) is hydrophilic? _____ Which region is hydrophobic? _____

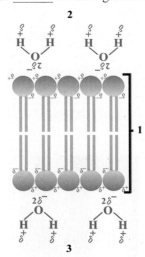

Figure 3.29

158 In reference to **Figure 3.29**, identify #1. _____

159 What portions of the phospholipids forms the central region of the plasma membrane? _____

160 What portions of the phospholipids faces the intracellular and extracellular environments. _____

161 What is a surfactant? _____

162 How do surfactants affect the molecular cohesion of water? _____

163 Why is surfactant found in the air sacs of the lung? _____

164 What are micelles? _____

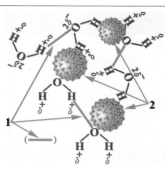

Figure 3.30

165 In reference to **Figure 3.30**, what happened to the hydrogen bonds at #2? _____

166 In reference to **Figure 3.30**, what is the name of the structures at #2? _____

167 In addition to breaking the hydrogen bonds of water (making water more interactive), what else do micelles dissolve? _____

168 What is the function of bile salts? _____

WATER SURFACE TENSION DEMONSTRATIONS

Figure 3.31

169 In reference to **Figure 3.31**, what property of water allows the paper clips to float on the surface? _____

Why didn't the large (2 inch) paper clip rest on the surface of the water? _____

170 Why did the paper clips sink when detergent (surfactant) was added to the water? _____

28 Chapter 3 Chemistry and Metabolism

Name _____
Class _____

Figure 3.32

171 In reference to **Figure 3.32**, what is the substance being sprinkled onto the surface of water? _____

172 What doesn't the material sink? _____

Figure 3.33

173 In reference to **Figure 3.33**, what was added to the water to cause the material at the surface to start sinking? _____

174 How did the added substance cause this effect? _____

MIXTURES

175 Define mixture. _____

176 List three common mixtures. _____

177 Define solution. _____

178 Define solvent. _____

179 Define solute. _____

Figure 3.34

180 In reference to **Figure 3.34**, what type of mixture is represented? _____

181 What is the solvent? _____

182 What is the solute? _____

Figure 3.35

183. In reference to **Figure 3.35,** what type of mixture is represented in tube B? _____

184 Define colloid. _____

185 What are blood colloidal proteins? _____

Figure 3.36

186 In reference to **Figure 3.36**, what type of a solution is represented in tube A? _____

187 Define suspension. _____

Chapter 3 Chemistry and Metabolism

BIOCHEMISTRY

188 Define biochemistry. _____
_____

189 Define inorganic chemistry. _____
_____

190 Define organic chemistry. _____
_____
_____

WATER

191 What are four features of water that its functions revolve around? _____
_____
_____

Solvent

192 What is water an excellent solvent of? _____
_____

193 Why is a molecule of water polar? _____
_____
_____

194 What produces the charges associated with a molecule of water? _____
_____

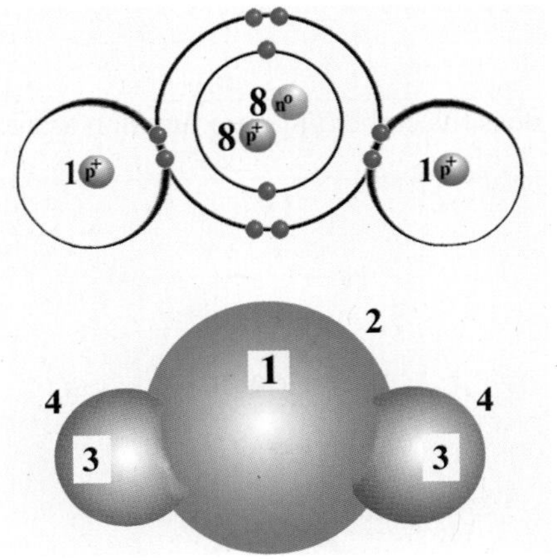

Figure 3.37

195 In reference to **Figure 3.37**, identify #1 - #4.
1 _____ 3 _____
2 _____ 4 _____

196 What property of water gives it the ability to dissociate certain substances? _____
_____

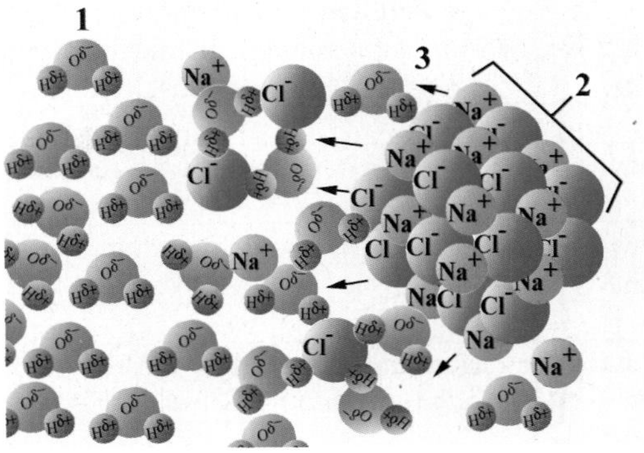

Figure 3.38

197 In reference to **Figure 3.38**, identify #1 - #3.
1 _____ 3 _____
2 _____

198 What is a hydration sphere? _____
_____

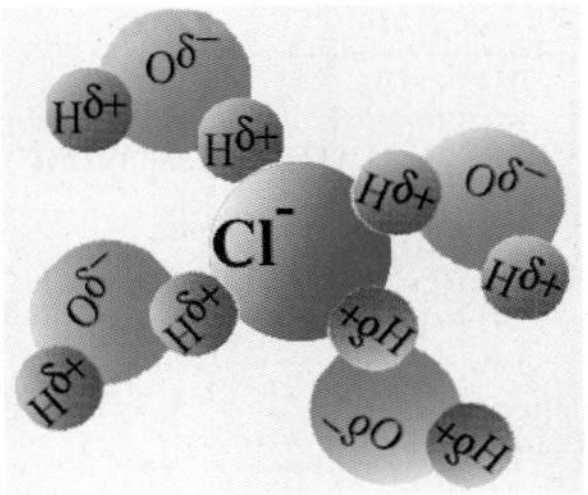

Figure 3.39

199 In reference to **Figure 3.39**, what is this unit of organization called? _____ What produces its organization? _____

200 What are electrolytes? _____
_____

201 What are electrolytes of our body mostly concerned with? _____
_____

202 Name and give the chemical symbol of five common electrolytes in the body. _____
_____

Specific Heat

203 What does it mean to say that water has a high specific heat? _____
_____

204 Why is water an excellent stabilizer of body heat? _____
_____

205 What are two ways the body can use water to transfer heat in becoming cooler? _____
_____
_____
_____

206 What are two ways that the body can conserve heat? _____
_____
_____

Chapter 3 Chemistry and Metabolism

Chemical Reactivity

207 What are two common biological reactions that involve water? _____

208. How is water involved in the hydrolysis reaction? _____

209 How is water involved in the dehydration synthesis reaction? _____

210 Identify the reaction represented by the equation:
AB + Water —> A + B. _____

211 Identify the reaction represented by the equation:
A + B —> AB + Water. _____

Lubricant

212 What allows water to function as an effective lubricant? _____

213 The interaction of water and what substance (found in synovial fluid) is an example of water functioning as a lubricant due to its polar nature? _____

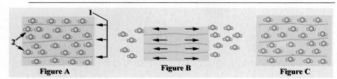

Figure 3.40

214 In reference to **Figure 3.40**, identify #1 - #2.
1 _____ 2 _____
Is **Figure B** compressed or non-compressed? _____

ACIDS, BASES, and BUFFERS

ACIDS

215 What is a general definition of an acid? _____

216 In aqueous solutions, what do acids dissociate into? _____

Inorganic acid

217 What two ions are formed when hydrochloric acid (HCl) dissociates in an aqueous solution? _____

218 How is the proton formed? _____

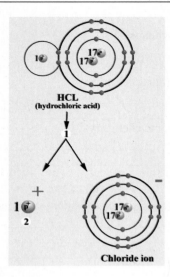

Figure 3.41

219 In reference to **Figure 3.41**, identify #1 - #2.
1 _____ 2 _____

220 What is formed when an acid (such as HCl) is combined with a hydroxide-containing base (such as NaOH)? _____

221 Describe neutralization. _____

Organic acid

222 Give an example of an organic group which functions as an _____ acid.

223 How does the carboxyl group function as an acid? _____

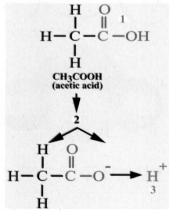

Figure 3.42

224 In reference to **Figure 3.42**, identify #1 - #3.
1 _____ 3 _____
2 _____

225 What is a strong acid? _____

226 What is a weak acid? _____

227 Identify the equations as to a strong or weak acid.
$HCl \longrightarrow H^+ + Cl^-$ _____
$CH_3COOH + H_2O \longleftrightarrow H^+ + CH_3COO^-$ _____

Name _____
Class _____

Chapter 3 Chemistry and Metabolism   31

BASES

228 What is a general definition of a base? _____
_____

229 In aqueous solutions what do bases dissociate into? _____
_____

230 What group characterizes many common bases? _____

231 In aqueous solutions, what do hydroxide-containing bases dissociate into? _____

232 What happens when a hydroxide-containing base (such as Na**OH**) combines with an acid (such as HCl)? _____
_____

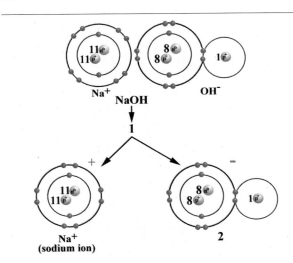

Figure 3.43

233 In reference to **Figure 3.43** identify #1 - #2.
  1 _____ 2 _____

234 What is a strong base? _____
_____

235 What is a weak base? _____
_____

236 Identify the equations as to a strong or weak base.
  NaOH —> Na$^+$ + OH$^-$ _____

  NH$_3$ + H$_2$O <—> NH$_4^+$ + OH$^-$ _____

pH OF A SOLUTION

237 What does the pH of a solution describe? _____
_____

238 What does the abbreviation pH stand for? _____
_____

239 What is the range of the pH scale? _____
_____

240 What number represents neutral pH? _____
_____

241 What do numbers lower than pH 7 indicate? _____
_____

242 What do numbers higher than pH 7 indicate? _____
_____

243 A substance with a pH of 7.35 is alkaline or acidic? _____
_____

244 A substance with a pH of 7.35 changed to a pH of 7.3, did the substance became more acidic or more alkaline? _____

245 Mathematically, how are the numbers of the pH scale expressed? _____
_____

246 A substance with a pH of 4 is _____ times more _____ (acidic or alkaline) than a substance of pH 6.

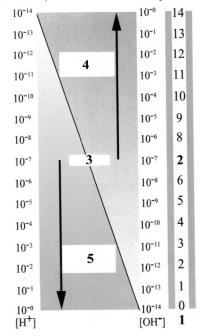

Figure 3.44

247 In reference to **Figure 3.44**, identify #1 - #5.
  1 _____ 4 _____
  2 _____ 5 _____
  3 _____

32 Chapter 3 Chemistry and Metabolism

BUFFERS

248 What is the function of buffers? _____

249 How do weak acids and weak bases function as buffers? _____

250 Carbonic acid is a weak acid because it undergoes _____ dissociation and exists in _____ with its dissociation products.

251 When an acid (increased hydrogen ions, H$^+$) is added to a solution of carbonic acid, which of the following equations occurs?

$H_2CO_3 \rightarrow H^+ + HCO_3^-$ _____

$H^+ + HCO_3^- \rightarrow H_2CO_3$ _____

Does increased amounts of the un-dissociated carbonic acid change the pH? _____

252 When a base (increased hydrogen ion acceptors such as OH$^-$) is added to a solution of carbonic acid, which of the following equations occurs?

$H_2CO_3 \rightarrow H^+ + HCO_3^-$
$H^+ + HCO_3^- \rightarrow H_2CO_3$

Why doesn't the pH become more alkaline with the addition of the base? _____

SALTS

253 Define a salt. _____

254 What do salts dissociate into? _____

255 What are two functions of ions? _____

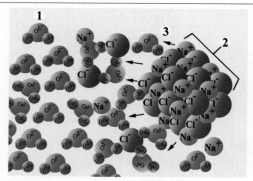

Figure 3.45

256 In reference to **Figure 3.45**, identify #1 - #3.
1 _____ 3 _____
2 _____

257 Ions used in the conduction of electrical activity (such as in nervous and muscle tissues) are called _____.

258 Name five common electrolytes and give their chemical symbol. _____

ORGANIC CHEMISTRY

259 What is organic chemistry? _____

260 What are two common reactions of organic chemistry that involve water? _____

261 Define dehydration synthesis. _____

262 Define hydrolysis. _____

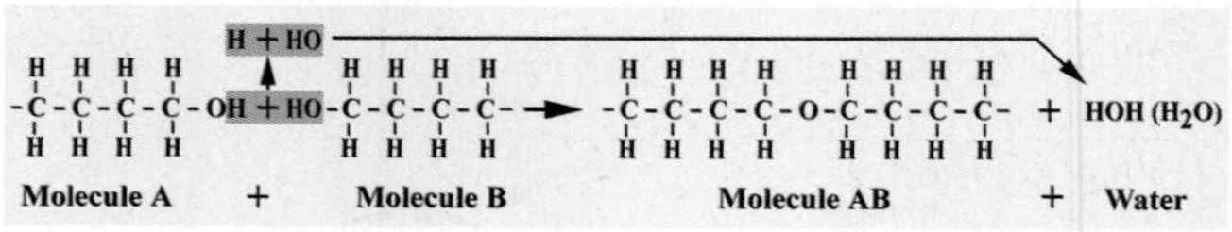

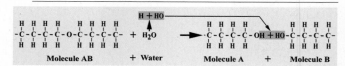

Figure 3.46

263 In reference to **Figure 3.46**, what is the reaction? _____

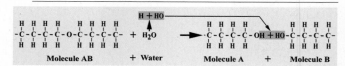

Figure 3.47

264 In reference to **Figure 3.47**, what is the reaction? _____

Name _____
Class _____   Chapter 3-A Chemistry and Metabolism - Carbohydrates   33

CARBOHYDRATES - WORKSHEETS

1 What are carbohydrates commonly called? _____

2 What are three common uses of carbohydrates? _____

3 What are the three elements of carbohydrates and what is their usual composition ratio? _____

4 What are four groups of carbohydrates? _____

MONOSACCHARIDES

5 How are monosaccharides described? _____

6 How can monosaccharides be named? _____

7 What are two examples of pentose carbon sugars? _____

8 What are three examples of hexose carbon sugars? _____

Chemical formulas

9 Define empirical formula. _____

10 Define chemical formula. _____

11 Define structural formula. _____

12 Define isomer and name four examples of isomers. _____

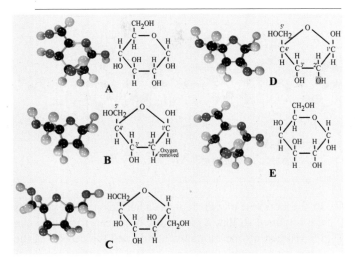

Figure 3-A.1

13 In reference to **Figure 3-A.1**, list the letters of the pentose carbon sugars: _____, the hexose sugars: _____.

14 In reference to **Figure 3-A.1**, match the letters with the names: glucose _____, fructose _____, ribose _____, galactose _____, deoxyribose _____.

15 In reference to **Figure 3-A.1**, list the letters of the isomers: _____. What is their chemical formula _____.

Lab Activities

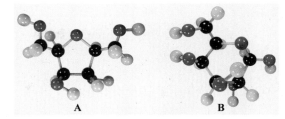

Figure 3-A.2.

16 In reference to **Figure 3-A.2**, which letters represent the molecules of glucose and of fructose? _____

DISACCHARIDE

16 How are disaccharides described? _____

17 Name three common disaccharides. _____

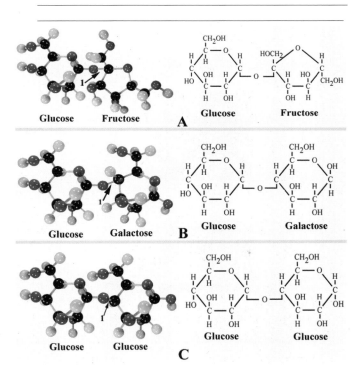

Figure 3-A.3.

18 In reference to **Figure 3-A.3**, match the letters with the names: maltose _____, sucrose, _____, lactose _____.

OLIGOSACCHARIDES

19 How are oligosaccharides described? _____

20 Where are oligosaccharides commonly found? _____

21 What are two functions are associated with membrane bound oligosaccharide-glycoproteins? _____

Chapter 3-A Chemistry and Metabolism - Carbohydrates

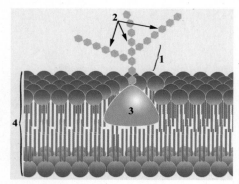

Figure 3-A.4

22 In reference to **Figure 3-A.4**, identify #1 -#4.
   1 _____ 3 _____
   2 _____ 4 _____

POLYSACCHARIDES

23 How are polysaccharides described? _____

24 Name two common polysaccharides. _____

25 In plants such as corn and potato, how is glucose typically stored? _____

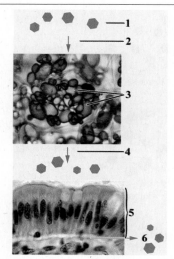

Figure 3-A.5

26 In reference to **Figure 3-A.5**, identify #1 - #6.
   1 _____ 4 _____
   2 _____ 5 _____
   3 _____ 6 _____

27 What is a common name for hydrolysis? _____

28 The carbohydrate that functions as a storage form of glucose in plants is _____.

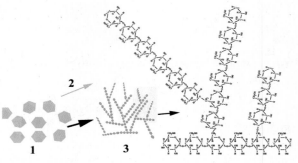

Figure 3-A.6

29 In reference to **Figure 3-A.6**, identify #1 -#3.
   1 _____ 3 _____
   2 _____

30 In animals, what is the carbohydrate storage form of glucose? _____

31 In what two tissues is glycogen commonly stored? _____

32 What is the function of glycogen in muscle? _____

33 During short periods of fasting or prolonged exercise, how is blood sugar maintained? _____

CARBOHYDRATE DEHYDRATION SYNTHESIS

34 Define dehydration synthesis. _____

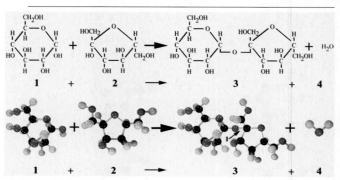

Figure 3-A.7

35 In reference to **Figure 3-A.7**, identify #1 -#4.
   1 _____ 3 _____
   2 _____ 4 _____

36 In the bonding of two monosaccharides, what two parts are removed so that a glycosidic bond can form between the oxygen of one monosaccharide and the carbon of the other monosaccharide? _____

Lab Activity

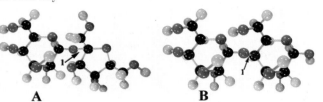

Figure 3-A.8

37 In reference to **Figure 3-A.8**, which letter represents the molecule of sucrose? _____

38 What two monosaccharides were used to construct the molecular model of sucrose? _____

39 Beside the molecule of sucrose, what other molecule was produced by the dehydration synthesis reaction? _____

Carbohydrate Hydrolysis

40 Define hydrolysis. _____

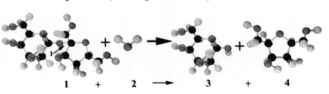

Figure 3-A.9

41 In reference to **Figure 3-A.9**, identify #1 -#4.
  1 _____ 3 _____
  2 _____ 4 _____

42 In hydrolysis, what two products are formed from the splitting of the molecule of water? _____

43 In hydrolysis, what happens to the two products formed from the water molecule? _____

Lab Activity

44 What monosaccharides were produced by the hydrolysis of sucrose? _____

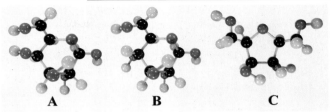

Figure 3-A.10

45 In reference to **Figure 3-A.10**, identify the letters that represent the molecules: glucose _____ and fructose _____.

46 What molecule was split so that the monosaccharides could be correctly and completely reassembled? _____

Glycoaminoglycans (GAGs)

47 What are two properties of mucus do the glycoaminoglycans produce? _____

48 What property do they give to the ground substance of connective tissues? _____

49 Describe the structure of the glycoaminoglycans. _____

50 Proteoglycan monomers typically have glycoaminoglycans bonded to a _____.

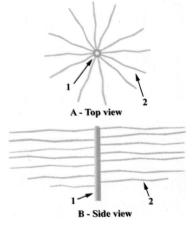

Figure 3-A.11

51 In reference to **Figure 3-A.11**, what type of molecule is shown? _____

52 In reference to **Figure 3-A.11**, identify #1 -#2.
  1 _____ 2 _____

53 Proteoglycan aggregates typically have proteoglycan monomers bonded to _____.

36 Chapter 3-A Chemistry and Metabolism - Carbohydrates

Name _____
Class _____

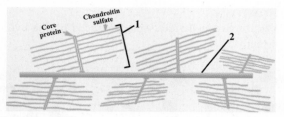

Figure 3-A.12

54 In reference to **Figure 3-A.12**, identify #1 -#2.
1 _____ 2 _____

55 Where are two locations of proteoglycan aggregates? ___
_____

56 What are two functions of proteoglycan aggregates? ___
_____
_____
_____

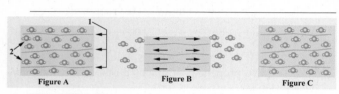

Figure 3-A.13

57 In reference to Figure 3-A.13, identify #1 and #2 in A.
1 _____ 2 _____

58 In reference to **Figure 3-A.13**, why does water associate with the glycoaminoglycans? _____
_____
_____

59 In reference to **Figure 3-A.13**, identify B and C as to either compressed or non-compressed states. _____
_____
_____

60 What characteristics do GAGs give to the ground substance of connective tissues? _____
_____
_____

GLYCOPROTEINS

61 Describe glycoproteins. _____
_____

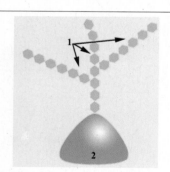

Figure 3-A.14

62 In reference to **Figure 3-A.14**, identify #1 -#2.
1 _____ 2 _____

63 Where are glycoproteins commonly found? _____
_____
_____

64 What are two functions of membrane-associated glycoproteins? _____
_____
_____

65 What are two functions of extracellular glycoproteins? __
_____
_____

CARBOHYDRATE METABOLISM

66 What are three ways the body utilizes glucose? _____
_____
_____

Lab Activity

67 What is the major source of dietary carbohydrates? ____
_____

68 What are two major dietary sources of starch? _____
_____

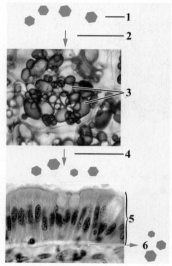

Figure 3-A.15

69 In reference to **Figure 3-A.15**, identify #1 - #6.
1 _____ 4 _____
2 _____ 5 _____
3 _____ 6 _____

Chapter 3-A Chemistry and Metabolism - Carbohydrates

DIETARY CARBOHYDRATES

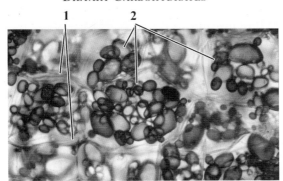

Figure 3-A.16
70 In reference to **Figure 3-A.16**, identify #1 -#2.
   1 _____ 2 _____

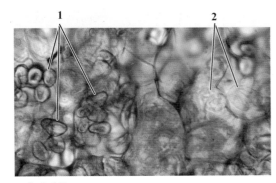

Figure 3-A.17
71 In reference to **Figure 3-A.17**, identify #1 -#2.
   1 _____ 2 _____

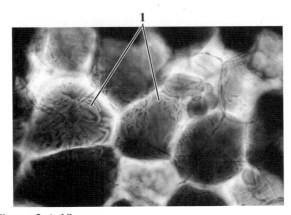

Figure 3-A.18
72 In reference to **Figure 3-A.18**, identify #1.
   1 _____

LIVER GLYCOGEN

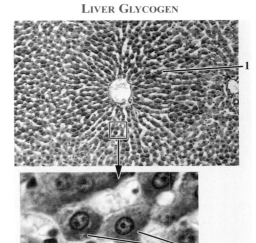

Figure 3-A.19
73 In reference to **Figure 3-A.19**, identify #1.
   1 _____

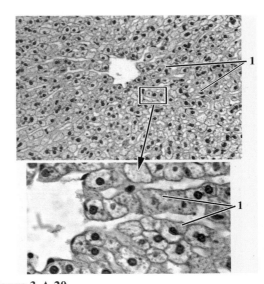

Figure 3-A.20
74 In reference to **Figure 3-A.20**, identify #1.
   1 _____

Chapter 3-A Chemistry and Metabolism - Carbohydrates

GLUCOSE CATABOLISM

75 In the well-fed state, what is the major source of glucose? _____

76 In the fasting state, what are two sources of glucose? _____

77 What is the catabolism of glucose mostly for? _____

78 Besides being used as a source of energy, what are two substances the intermediaries of glucose may produce? _____

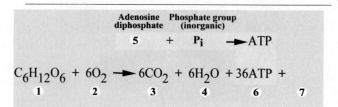

Figure 3-A.21

79 In reference to **Figure 3-A.21**, identify #1 - #7.
   1 _____ 5 _____
   2 _____ 6 _____
   3 _____ 7 _____
   4 _____

ACETYL-CoA

80 What is the pivotal molecule in the metabolism of fuels? _____

81 What are three generalized steps in the catabolism of fuel molecules? _____

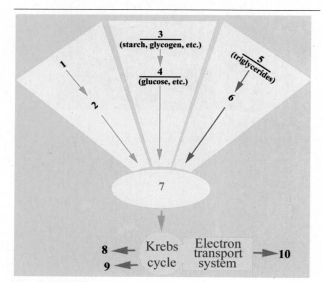

Figure 3-A.22

82 In reference to **Figure 3-A.22**, identify #1 - #10.
   1 _____ 6 _____
   2 _____ 7 _____
   3 _____ 8 _____
   4 _____ 9 _____
   5 _____ 10 _____

GLUCOSE CATABOLISM OVERVIEW
GLYCOLYSIS

83 Define glycolysis. _____

Aerobic glycolysis

84 What is the coenzyme that must be rapidly resupplied in the glycolytic pathway? _____

85 In aerobic glycolysis, once the coenzyme NAD^+ is reduced to NADH, where is it oxidized back to NAD^+? _____

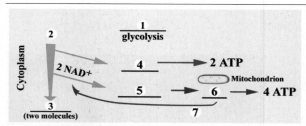

Figure 3-A.23

86 In reference to **Figure 3-A.23**, identify #1 - #7.
   1 _____ 5 _____
   2 _____ 6 _____
   3 _____ 7 _____
   4 _____

Anaerobic glycolysis

87 When does anaerobic glycolysis occur? _____

88 In anaerobic glycolysis, once the coenzyme NAD^+ is reduced to NADH, where is it oxidized back to NAD^+? - _____

89 What are the advantages of anaerobic glycolysis? _____

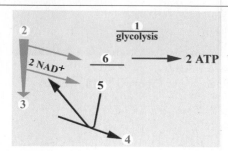

Figure 3-A.24

90 In reference to **Figure 3-A.24**, identify #1 - #6.
   1 _____ 4 _____
   2 _____ 5 _____
   3 _____ 6 _____

Name _____
Class _____

Chapter 3-A Chemistry and Metabolism - Carbohydrates   39

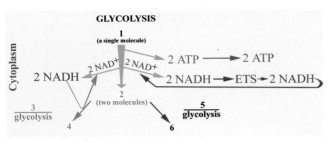

Figure 3-A.25

91 In reference to **Figure 3-A.25**, identify #1 -#6.
    1 _____   4 _____
    2 _____   5 _____
    3 _____   6 _____

92 In cells with mitochondria, is anaerobic or aerobic glycolysis continuous? _____ When does the other "type" of glycolysis begin to operate? _____
    _____

Energy associated with NADH

93 How many molecules of NADH are formed in glycolysis? _____

94 In aerobic glycolysis, where is the NADH oxidized to NAD^+? _____

95 For most cells of the body, how many ATP are produced by the electron transport system (ETS) from the 2 NADH from aerobic glycolysis? _____ Certain cells such as the heart, kidney, and liver produce _____ ATP from the 2 NADH from aerobic glycolysis.

Formation of ATP

96 How many ATP are used in the glycolytic pathway? ____ How many ATP are produced by the glycolytic pathway? _____. What is the total net gain of ATP from glycolysis? _____

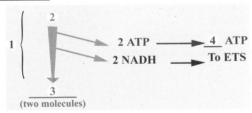

Figure 3-A.26

97 In reference to **Figure 3-A.26**, identify #1 -#4.
    1 _____   3 _____
    2 _____   4 _____

98 Fill in the following **summary of glycolysis** (for one molecule of glucose) as total number and kinds of molecules directly produced. If the molecule is not produced, answer with "NA" (not applicable).
    • _____ pyruvic acid molecules
    • _____ ATP (net gain)
    • _____ NADH
    • _____ CO_2

Formation of Acetyl Groups

99 Describe the formation of acetyl groups. _____
    _____

Energy associated with NADH

100 How many NADH are produced in the production of the two acetyl groups from the two pyruvic acids? _____
    _____

101 Where are the NADH oxidized to NAD^+? _____

102 For the two pyruvic acids (converted to two acetyl groups), what is the total number of ATP produced by the ETS? _____

Production of Carbon Dioxide

103 What is decarboxylation? _____

104 How many molecules of carbon dioxide (CO_2) are produced in the conversion of the two pyruvic acids to two acetyl groups? _____
    _____

105 What happens to the carbon dioxide? _____
    _____
    _____

Coenzyme A (CoA)

106 What is the function of CoA? _____
    _____

107 What are two other names for Krebs cycle? _____
    _____
    _____

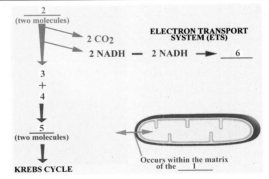

Figure 3-A.27

108 In reference to **Figure 3-A.27**, identify #1 -#6.
    1 _____   4 _____
    2 _____   5 _____
    3 _____   6 _____

40 Chapter 3-A Chemistry and Metabolism - Carbohydrates

Name _____
Class _____

109 Fill in the following **summary for the conversion of two pyruvic acids to two acetyl-CoA** (total number and kinds of molecules directly produced). If the molecule is not produced, answer with "NA" (not applicable).
- _____ Acetyl-CoA
- _____ ATP
- _____ NADH
- _____ CO_2

KREBS CYCLE

110 Describe Krebs cycle. _____
_____
_____
_____

Acetyl-CoA converted to acetyl group and CoA

111 What happens to CoA produced from the separation from acetyl-CoA? _____
_____

112 Name the intermediary of Krebs cycle that the acetyl group joins? _____
_____

113 What molecule is the first in the series of oxidation-reduction reactions of Krebs cycle? _____
_____

114 The series of reactions of Krebs cycles ends with a molecule of _____, to accept another acetyl group.

Formation of NADH, $FADH_2$, and ATP

115 How many NADH are produced for each acetyl group that enters Krebs cycle? _____
_____

116 The NADH from Krebs cycle enters the _____, and each produces _____ ATP. Thus, the total number of _____ NADH from Krebs cycle produces a total of _____ ATP in ETS.

117 How many $FADH_2$ are produced for each acetyl group that enters Krebs cycle? _____

118 Each $FADH_2$ enters the _____ where it produces _____ ATP. Thus, for the total of _____ $FADH_2$ produced by Krebs cycle, the ETS produces _____ ATP.

119 How many ATP are produced from the two acetyl groups that enter Krebs cycle? _____
_____

120 Krebs cycle results in the total production of _____ ATP.

Formation of CO_2

121 Krebs cycle results in the total production of _____ CO_2.

Summary

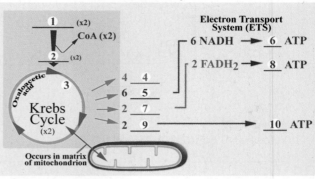

Figure 3-A.28

122 In reference to **Figure 3-A.28**, identify #1 -#6.
1 _____ 6 _____
2 _____ 7 _____
3 _____ 8 _____
4 _____ 9 _____
5 _____ 10 _____

123 Fill in the following **summary for Krebs cycle** (for one molecule of glucose). Give the total number of molecules directly produced. If the molecule is not produced, answer with "NA" (not applicable).
- _____ Citric acid   _____ $FADH_2$
- _____ ATP   _____ CO_2
- _____ NADH   _____ H_2O

ELECTRON TRANSPORT SYSTEM (ETS)

124 Describe the electron transport system. _____
_____

125 What molecules enter the ETS? _____
126 Where does the ETS occur? _____
127 What becomes concentrated in the inter-membrane space of the mitochondrion? _____
128 Where is ATP synthase? _____ What directly drives the production of ATP? _____
129 Why is molecular oxygen necessary? _____
_____

Chapter 3-A Chemistry and Metabolism - Carbohydrates

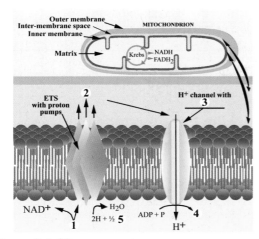

Figure 3-A.29

130 In reference to **Figure 3-A.29**, identify #1 -#5.
1 _____ 4 _____
2 _____ 5 _____
3 _____

ENERGY TRANSFER IN ETS

131 Fill in the following **summary for total energy capture by ETS** (for one molecule of glucose)
   Glycolysis:
   contributes ___ NADH, ETS produces ___ ATP
   Conversion of pyruvic acid to acetyl-CoA
   contributes ___ NADH, ETS produces ___ ATP
   Krebs cycle
   contributes ___ NADH, ETS produces ___ ATP
   contributes ___ FADH2, ETS produces ___ ATP
   Total number of ATP by ETS: _____ from NADH + ____ from FADH2 = _____ ATP

132 Write a "U" next to each molecule **used,** write a "P" next to each molecule **produced,** and write "NA" for each molecule **not applicable** in ETS: ____ NADH, ____ FADH2, ____ CO_2, ____ ATP, ____ H_2O, ____ O_2.

SUMMARY OF ENERGY CAPTURE

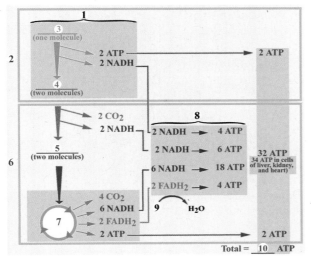

Figure 3-A.30

133 In reference to **Figure 3-A.30**, identify #1 -#10.
1 _____ 6 _____
2 _____ 7 _____
3 _____ 8 _____
4 _____ 9 _____
5 _____ 10 _____

GLUCOSE ANABOLISM

134 What does glucose anabolism refer to? _____

135 What are three anabolic process of glucose? _____

GLUCONEOGENESIS

136 Describe gluconeogenesis. _____

137 Why is the maintenance of blood glucose level extremely important to some tissues? _____

138 When would gluconeogenesis become an important way of maintaining blood glucose? _____

139 What is the main organ for gluconeogenesis? _____

140 What are some common precursors for gluconeogenesis? _____

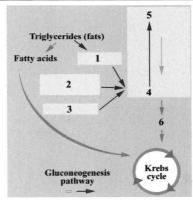

Figure 3-A.31

141 In reference to **Figure 3-A.31**, identify #1 -#6.
1 _____ 4 _____
2 _____ 5 _____
3 _____ 6 _____

GLYCOGENESIS

142 Describe glycogenesis. _____

143 What are the two organs of the body where glycogenesis is dominate? _____

144 When is liver glycogenesis promoted? _____

145 When is muscle glycogenesis promoted? _____

Chapter 3-A Chemistry and Metabolism - Carbohydrates

146 What hormone promotes overall glycogenesis? _____

147 When is the hormone that promotes glycogenesis produced? _____

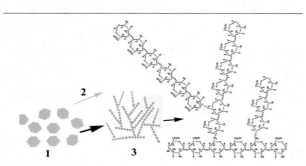

Figure 3-A.32

148 In reference to **Figure 3-A.32**, identify #1 -#3.
 1 _____ 3 _____
 2 _____

Lab activity

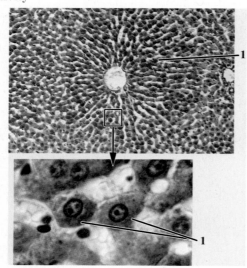

Figure 3-A.33

149 In reference to **Figure 3-A.33**, identify #1.
 1 _____

GLYCOGENOLYSIS

150 Describe glycogenolysis. _____

151 When does glycogenolysis from the liver dominate? _____

152 What is the primary hormone that promotes liver glycogenolysis? _____

153 What hormone promotes glycogenolysis during the fight-or-flight response? _____

154 How is the glycogen found in skeletal muscle typically used? _____

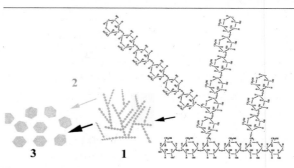

Figure 3-A.34

155 In reference to **Figure 3-A.34**, identify #1 -#3.
 1 _____ 3 _____
 2 _____

Lab activity

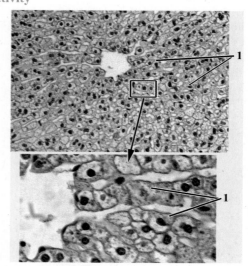

Figure 3-A.35

156 In reference to **Figure 3-A.35**, identify #1.
 1 _____

157 Assuming the animal was living, how could glycogen be removed from the liver? _____

AMINO ACIDS AND PROTEINS - WORKSHEETS

1. What are the building blocks of peptides and proteins? _____
_____

2. What is the name of the covalent bond formed between amino acids? _____

3. What are peptides? _____
_____
_____
_____

4. What are proteins? _____
_____

5. What part of the structure of the amino acids accounts for the differences in the amino acids? _____
_____

6. What are the four site that are bonded to the central carbon of an amino acid?
   1. _____
   2. _____
   3. _____
   4. _____

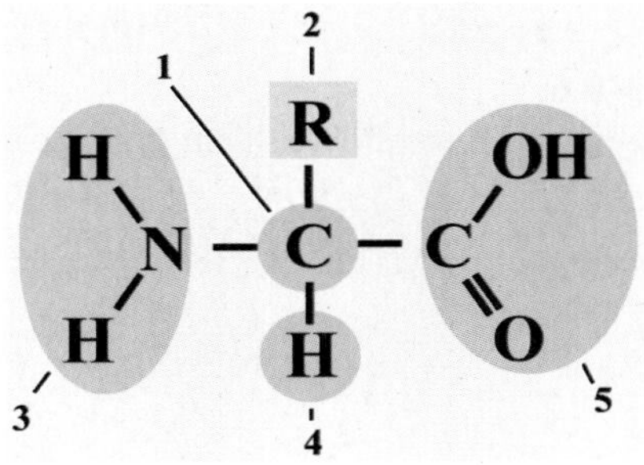

Figure 3-B.1

7. In reference to **Figure 3-B.1**, identify #1 - #5.
   1. _____ 4. _____
   2. _____ 5. _____
   3. _____

MOLECULAR MODEL KIT

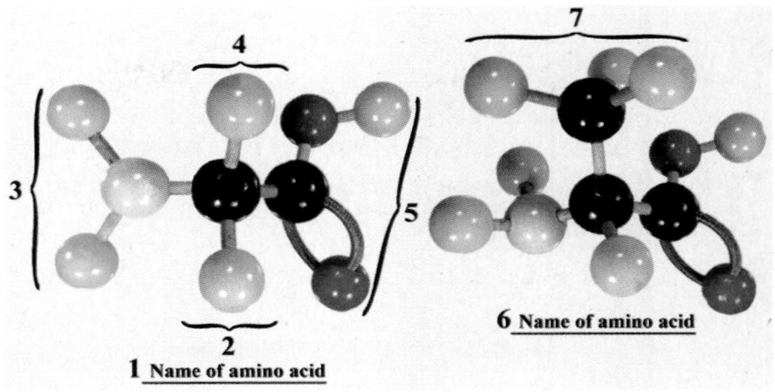

Figure 3-B.2

8. In reference to **Figure 3-B.2**, identify #1 - #7.
   1. _____ 5. _____
   2. _____ 6. _____
   3. _____ 7. _____
   4. _____

UTILIZATION OF AMINO ACIDS

9. What are three ways amino acids are utilized by the body?
   1. _____
   2. _____
   3. _____

Protein Synthesis

10. What are the two process of protein synthesis? _____
_____

11. What are examples of the structural function of proteins?
_____
_____
_____

12. What are examples of the dynamic function of proteins?
_____
_____
_____

44 Chapter 3-B Chemistry and Metabolism - Amino Acids

Name _____
Class _____

Synthesis of Non-protein Compounds

13 What are several examples of compounds made from amino acids? _____
_____
_____

Amino acids as a Source of Energy

14 About what percentage of the body's energy is derived from the catabolism of amino acids? _____
_____

Dehydration Synthesis

15 What does protein dehydration synthesis produce?
_____
_____
_____

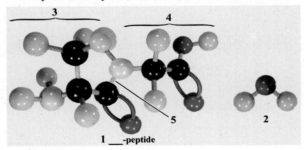

Figure 3-B.3

16 In reference to **Figure 3-B.3**, identify #1 - #4.
1 _____ 3 _____
2 _____ 4 _____

MOLECULAR MODEL KIT
Dehydration Synthesis and Molecular Models

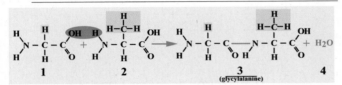

Figure 3-B.4

17 In reference to **Figure 3-B.4**, identify #1 - #2.
1 _____ 2 _____

18 In reference to **Figure 3-B.4**, the amino acid sequence is glycine-alanine or alanine-glycine?
1 _____

19 In reference to **Figure 3-B.4**, identify #3 - #5.
3 _____ 5 _____
4 _____

Hydrolysis

20 What is produced by the hydrolysis of proteins? _____
_____

21 During hydrolysis, why is water split? _____
_____
_____

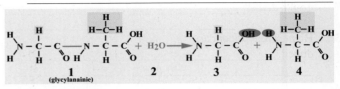

Figure 3-B.5

22 In reference to **Figure 3-B.5**, identify #1 - #4.
1 _____ 3 _____
2 _____ 4 _____

MOLECULAR MODEL KIT
Hydrolysis and Molecular Models

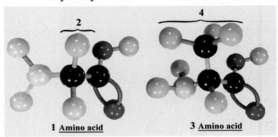

Figure 3-B.6

23 In reference to **Figure 3-B.6**, identify #1 - #4.
1 _____ 3 _____
2 _____ 4 _____

PROTEIN STRUCTURE

24 From simple to complex, list the four levels of protein organization.
    1 _____ 3 _____
    2 _____ 4 _____

25 What is the basis of the primary level of protein organization? _____

26 How is the secondary level of protein organization produced? _____

27 How is the tertiary level of protein organization produced? _____

28 How is the quaternary level of protein organization produced? _____

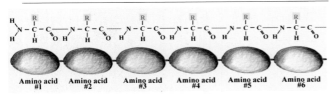

Figure 3-B.7

29 In reference to **Figure 3-B.7**, what level of organization is shown? _____

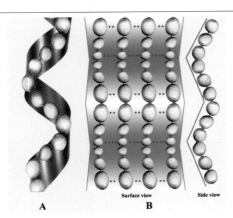

Figure 3-B.8

30 In reference to **Figure 3-B.8**, what level of organization is shown? _____

31 In reference to **Figure 3-B.8**, what structural organization is shown in A _____ B _____

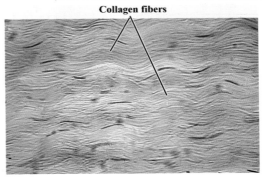

Figure 3-B.9

32 In reference to **Figure 3-B.9**, what level of organization is shown? _____

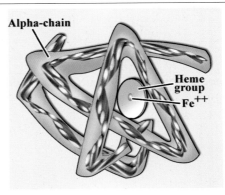

Figure 3-B.10

33 In reference to **Figure 3-B.10**, what level of organization is shown? _____

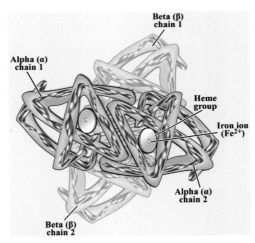

Figure 3-B.11

34 In reference to **Figure 3-B.11**, what level of organization is shown? _____

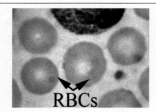

Figure 3-B.12

35 In reference to **Figure 3-B.12**, what level of organization is shown in hemoglobin of RBCs? _____

PROTEIN DENATURATION

36 What characterizes the denaturation of a peptide or a protein? _____

37 What are two ways that proteins are easily denatured? _____

LIPIDS - WORKSHEETS

1. What are four functions of lipids? _____
_____
_____

2. What are four common lipid molecules? _____
_____
_____

FATTY ACIDS

3. Describe a fatty acid. _____
_____

4. What is the general formula for fatty acids? _____

5. What in the formula $CH_3(CH_2)_nCOOH$, what does the subscript "n" represent? _____

6. Identify the carboxyl group in the formula $CH_3(CH_2)_nCOOH$. _____ What can it function as? _____

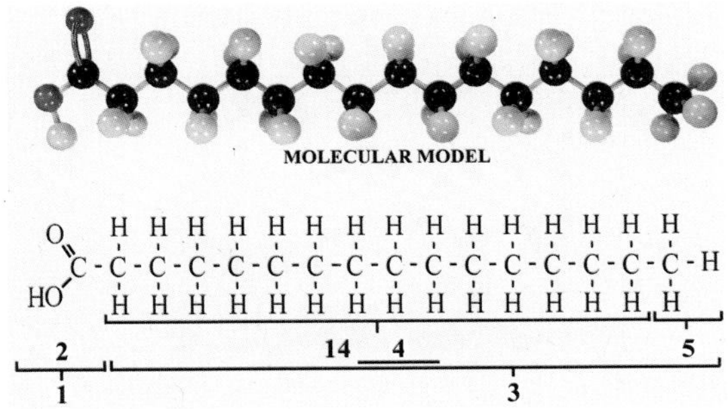

Figure 3-C.1

7. In reference to **Figure 3-C.1**, identify #1 - #5.
   1 _____ 4 _____
   2 _____ 5 _____
   3 _____

8. In the blood, how do most short chain fatty acids exist? __
_____

   How are they transported? _____
_____

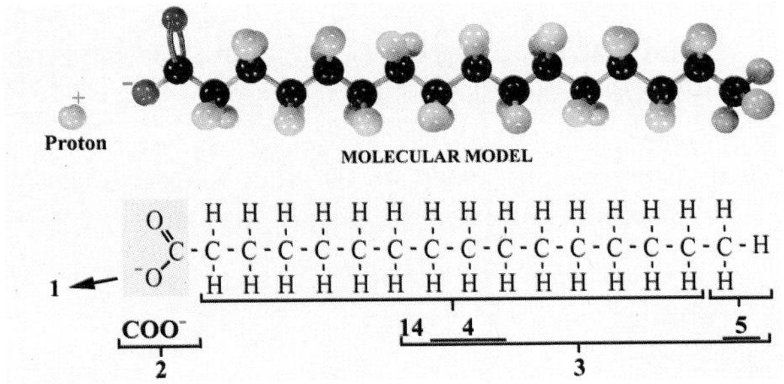

Figure 3-C.2

9. In reference to **Figure 3-C.2**, identify #1 - #5.
   1 _____ 4 _____
   2 _____ 5 _____
   3 _____

10. Are long chain fatty acids amphiphilic or hydrophobic? _____

11. How are long chain fatty acids transported in the blood? _____

12. Which are most commonly found in the blood, short chain or long chain fatty acids? _____

13. Are the long chain fatty acids transported as free units, or are they usually constituents of other lipid molecules such as triglycerides? _____
_____

Saturated fatty acid

14. Describe a saturated fatty acid. _____
_____

15. In a saturated fatty acid, what types of bonds are found in the CH_2 chain? _____
_____

Unsaturated Fatty Acid

16. Describe an unsaturated fatty acid. _____
_____
_____

17. In an unsaturated fatty acid, what type of bond must be present in the CH_2 chain? _____
_____

LAB ACTIVITY

18. Distinguish which one of the following formulas represents a saturated fatty acid and which represents an unsaturated fatty acid:
    $CH_3(CH_2)_2CH=CHCH_2COOH$ _____
    $CH_3(CH_2)_5COOH$ _____

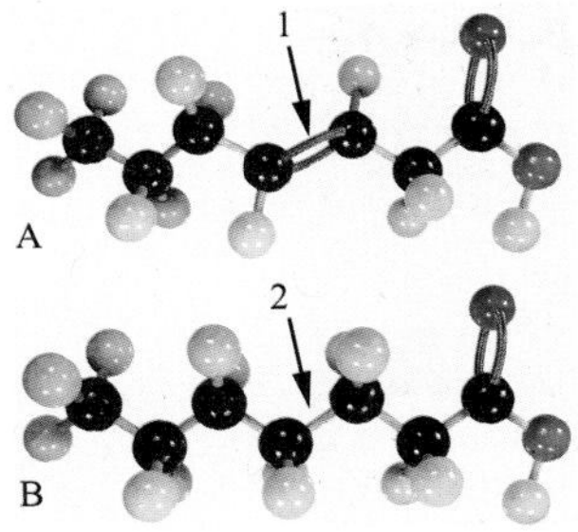

Figure 3-C.3

19. 7. In reference to **Figure 3-C.3**, identify "A" or "B" as
    saturated fatty acid _____
    unsaturated fatty acid _____

20. In reference to **Figure 3-C.3**, identify #1 - #2.
    1 _____ 2 _____

21. In reference to **Figure 3-C.3**, which figure is represented by each of the following formulas:
    $CH_3(CH_2)_2CH=CHCH_2COOH$ _____
    $CH_3(CH_2)_5COOH$ _____

48 Chapter 3-C Chemistry and Metabolism - Lipids

GLYCERIDES

22 What are glycerides commonly called? _____

23 What are the two components of glycerides? _____

24 What is glycerol? _____

LAB ACTIVITY

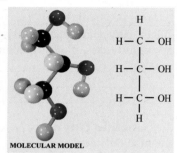

Figure 3-C.4

25 Name of the molecule shown in **Figure 3-C.4**. _____
26 Describe a monoglyceride. _____
27 Describe a diglyceride. _____
28 Describe a triglyceride. _____
29 How are different varieties of glycerides produced? _____

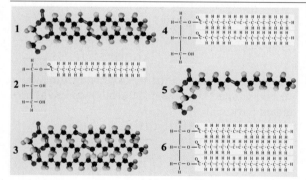

Figure 3-C.5

30 In reference to **Figure 3-C.5**, identify #1 - #6.
    1 _____   4 _____
    2 _____   5 _____
    3 _____   6 _____

GLYCERIDE DEHYDRATION SYNTHESIS

31 Describe a dehydration synthesis reaction. _____

32 What molecules are combined to produce glycerides? _____ What else is produced by the reaction? _____

33 What are the products of a dehydration synthesis reaction involving glycerol and three fatty acids? _____

Figure 3-C.6

34 In reference to **Figure 3-C.6**, which illustration, **A** or **B**, represents a dehydration synthesis reaction? _____
35 Where are triglycerides commonly stored? _____

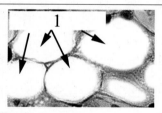

Figure 3-C.7

36 What is the tissue shown in **Figure 3-C.7**? _____
    Identify #1. _____

LAB ACTIVITY

Figure 3-C.8

37 In reference to **Figure 3-C.8**, which illustrated product, **1** or **2**, was assembled in the laboratory activity? _____

38 In reference to **Figure 3-C.8**, what is the primary structural difference between the fatty acids shown in **1** and **2**? _____

GLYCERIDE HYDROLYSIS

39 Describe a hydrolysis reaction. _____

40 Name the molecules required for the hydrolysis of a triglyceride molecule? _____

41 Name the molecules produced by the hydrolysis of a triglyceride molecule? _____
_____
_____

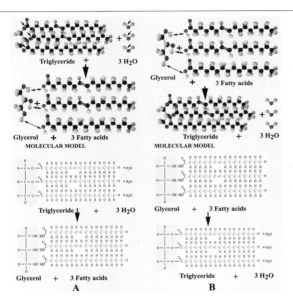

Figure 3-C.9

42 In reference to **Figure 3-C.9**, which illustration, **A** or **B**, represents a hydrolysis reaction? _____
_____

LAB ACTIVITY

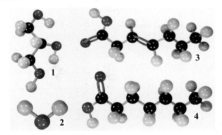

Figure 3-C.10

43 In reference to **Figure 3-C.10**, which of the above was produced by the hydrolysis of the monoglyceride? _____
_____
_____

PHOSPHOLIPIDS

44 Describe a phospholipid. _____
_____

45 Define amphiphilic molecule. _____
_____

46 Which portion of a phospholipid is hydrophilic? _____
_____

47 Which portion of a phospholipid is hydrophobic? _____
_____

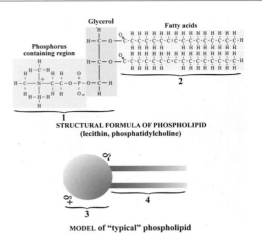

Figure 3-C.11

48 In reference to **Figure 3-C.11**, identify #1 - #4.
1 _____ 3 _____
2 _____ 4 _____

49 As a constituent of the plasma membrane, how abundant are phospholipids? _____
_____

How many phospholipid layers are formed in the plasma membrane? _____

50 How are the hydrophobic and hydrophilic regions of the phospholipids organized in the cell membrane? _____
_____

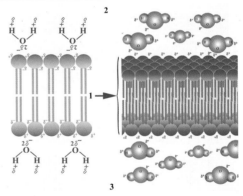

Figure 3-C.12

51 In reference to **Figure 3-C.12**, identify #1 - #3.
1 _____ 3 _____
2 _____

STEROIDS

52 Describe the structure of steroids. _____
_____

53 What is a steroid of major importance to the body? _____
_____

54 What are several functions of cholesterol? _____
_____

55 What is an important group of steroids? _____

56 Name three of the steroid sex hormones. _____
_____

Chapter 3-C Chemistry and Metabolism - Lipids

57 What is the general function of the sex hormones? _____

58 Name two glucocorticoids. _____

59 What is the general function of the glucocorticoids? _____

60 Name a mineralocorticoid. _____

61 What is the general function of a mineralocorticoid? _____

62 What is the general function of bile salts? _____

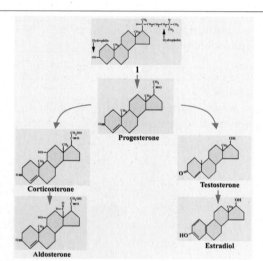

Figure 3-C.13

63 In reference to **Figure 3-C.13**, identify the steroid at #1. _____

LIPID METABOLISM

64 Besides carbohydrates, what is the other major source of fuel for the body? _____

65 What are two major sources for fatty acids? _____

FATTY ACID ANABOLISM (IN FORMATION OF TRIGLYCERIDES)

66 What molecules are digested fats first synthesized into by the absorptive cells of the intestine? _____

67 What are chylomicrons? _____

68 What is the primary target of chylomicrons? _____

69 What are lipid droplets of adipose tissue? _____

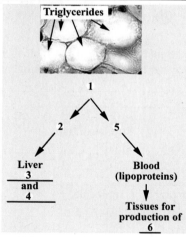

Figure 3-C.14

70 In reference to **Figure 3-C.14**, identify #1 - #4.
  1 _____ 3 _____
  2 _____ 4 _____

71 About what percent of the body fuel reserves is stored as triglycerides? _____

72 What are two important features of triglycerides that make them ideal as an energy reserve? _____

MOBILIZATION OF TRIGLYCERIDES

73 When are triglycerides mostly released from adipose tissue? _____

74 What are triglycerides catabolized into? _____

75 What are the primary hormones that promote the catabolism of triglycerides? _____

76 What are the fatty acids mostly used to produce? _____

77 What happens to glycerol? _____

Figure 3-C.15

78 In reference to **Figure 3-C.15**, identify #1 - #6.
  1 _____ 4 _____
  2 _____ 5 _____
  3 _____ 6 _____

ß-OXIDATION OF FATTY ACIDS

79 What is ß-oxidation? _____

80 Where does ß-oxidation occur? _____

81 What is the two-carbon product formed from the ß-oxidation of fatty acids? _____

82 What happens to the acetyl-CoA produced by ß-oxidation? _____

83 For each acetyl-CoA that enters Krebs cycle, how many ATP are produced? _____

84 What are the reduced coenzymes produced by ß-oxidation? _____

85 What happens to NADH and $FADH_2$? _____

86 How many ATP are produced for each NADH and $FADH_2$? _____

KETONE BODIES

88 What two products are produced by mobilization of triglycerides? _____

89 What happens to glycerol? _____

90 What happens to the fatty acids? _____

91 When does the liver convert acetyl-CoA into ketone bodies? _____

92 Essentially, what are ketone bodies? _____

93 What happens to ketone bodies once they leave the liver? _____

94 What happens when ketone production exceeds ketone use? _____

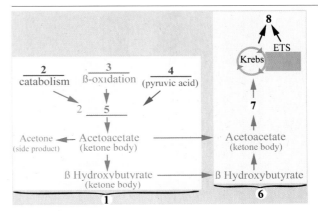

Figure 3-C.17

95 In reference to **Figure 3-C.17**, identify #1 - #8.
1 _____ 5 _____
2 _____ 6 _____
3 _____ 7 _____
4 _____ 8 _____

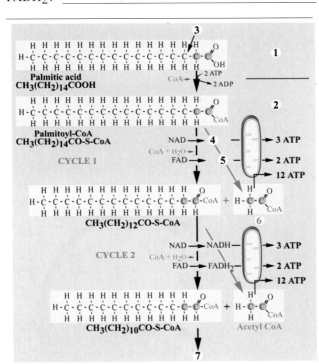

Figure 3-C.16

87 In reference to **Figure 3-C.16**, identify #1 - #7.
1 _____ 5 _____
2 _____ 6 _____
3 _____ 7 _____
4 _____

52 Chapter 3-C Chemistry and Metabolism - Lipids

SYNTHESIS OF FATTY ACIDS

96 What is the origin of most of the fatty acids used by the body? _____

97 If not supplied in the diet, where in the body are the non-essential fatty acids synthesized? _____

98 In the well-fed state, why does fatty acid synthesis occur in the liver? _____

99 Where in the cell does fatty acid synthesis occur? _____

100 What molecules are used in the synthesis of fatty acids? _____

101 What is used for the two carbon groups in the assembly of the fatty acids? _____

102 For the newly produced fatty acids, what is one common modification? _____

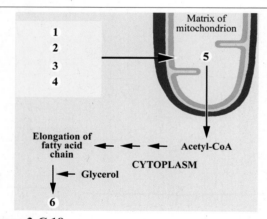

Figure 3-C.18

103 In reference to **Figure 3-C.18**, identify #1 - #6.

1 _____   4 _____
2 _____   5 _____
3 _____   6 _____

NUCLEIC ACIDS - WORKSHEETS

NUCLEOTIDES

1. What are the two nucleic acids? _____

2. What is the name of the building blocks of the nucleic acids? _____

3. What are the three basic components of a nucleotide?
   1. _____
   2. _____
   3. _____

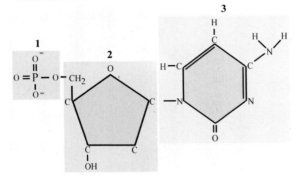

Figure 3-D.1

4. In reference to **Figure 3-D.1**, identify # 1 - #3.
   1. _____ 3. _____
   2. _____

PHOSPHATE GROUPS

5. What portions of adjacent nucleotides bond to produce a polynucleotide strand? _____

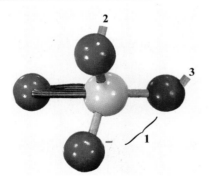

Figure 3-D.2

6. In reference to **Figure 3-D.2**, what is the name of the molecular group? _____

7. In reference to **Figure 3-D.2**, what bonds to #2? _____

8. In reference to **Figure 3-D.2**, what bonds to #3? _____

PENTOSE SUGAR

9. What is the name of the pentose sugar of RNA? _____

10. What is the name of the pentose sugar of DNA? _____

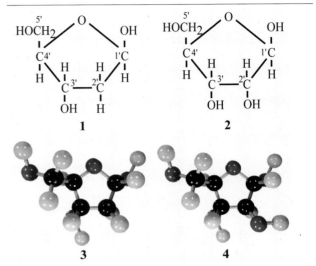

Figure 3-D.3

11. In reference to **Figure 3-D.1**, identify # 1 - #4.
    1. _____ 3. _____
    2. _____ 4. _____

NITROGEN BASES

12. What are nitrogen bases? _____

13. What are the two groups of nitrogen bases? _____

14. What nitrogen bases are found in the purines? _____

15. What nitrogen bases are found in the pyrimidines? _____

16. In the structure of a nucleotide, where are the nitrogen bases covalently bonded? _____

Chapter 3-D Chemistry and Metabolism - Nucleic Acids

NITROGEN BASES OF DNA
17 What are the four nitrogen bases of DNA? _____
_____
18 What are the complementary nitrogen base pairs of DNA?
_____

NITROGEN BASES OF RNA
19 What are the four nitrogen bases of RNA? _____
_____

20 What are the complementary nitrogen base pairs of RNA?
_____

21 What type of bonds associate a nitrogen base with its complement? _____

DEOXYRIBONUCLEIC ACID - DNA
CHROMOSOME
22 What is the most organized form of DNA? _____
_____ How is the DNA of chromosomes organized? ___
_____

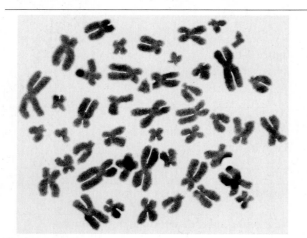

Figure 3-D.4
23 What is Figure 3-D.4? _____
_____
24 What is a karyotype? _____
_____
25 How many chromosomes are found in a human body (somatic) cell? _____ How many chromosomes are shown in Figure 3-D.4? _____

CHROMATIN
26 How is chromatin organized? _____
_____
27 What does chromatin mostly consist of? _____
_____

MOLECULAR DNA
28 Organizationally, how is molecular DNA ranked? _____
_____
29 What are two uses of the genetic code of molecular DNA?
1 _____
2 _____

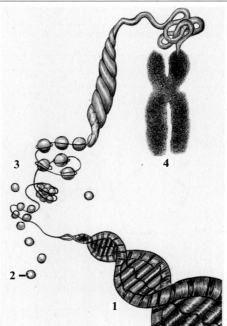

Figure 3-D.5
30 In reference to Figure 3-D.5, identify # 1 - #4.
1 _____ 3 _____
2 _____ 4 _____
31 How is a molecule of DNA organized? _____
_____
_____

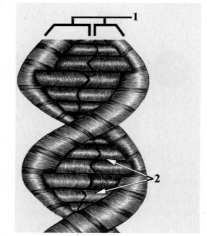

Figure 3-D.6
32 In reference to Figure 3-D.6, identify # 1 - #2.
1 _____ 2 _____

Chapter 3-D Chemistry and Metabolism - Nucleic Acids

STRUCTURE OF DNA
NUCLEOTIDES

33 What are the building blocks of molecular DNA? _____

34 What are the three components of a DNA nucleotide?
   1 _____
   2 _____
   3 _____

35 What are the four nitrogen bases of DNA?
   1 _____ 3 _____
   2 _____ 4 _____

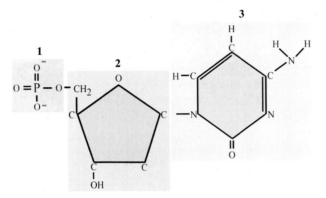

Figure 3-D.7

36 In reference to **Figure 3-D.7**, identify # 1 - #3.
   1 _____ 3 _____
   2 _____

PHOSPHATE GROUP

37 At what sites do the sequential nucleotides bond to produce a polynucleotide strand? _____

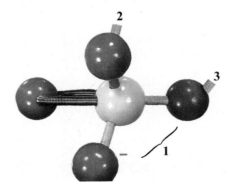

Figure 3-D.8

38 In reference to **Figure 3-D.8**, identify # 1 - #3.
   1 _____
   2 _____
   3 _____

DEOXYRIBOSE

39 What two molecular groups are bonded to the deoxyribose of the nucleotide? _____

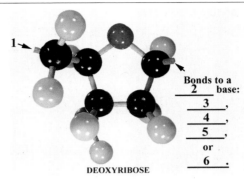

Figure 3-D.9

40 In reference to **Figure 3-D.9**, identify # 1 - #6.
   1 _____ 4 _____
   2 _____ 5 _____
   3 _____ 6 _____

NITROGEN BASES

41 What are the complementary base pairs of DNA? _____
   _____ What type of bonding occurs between the complementary pairs? _____

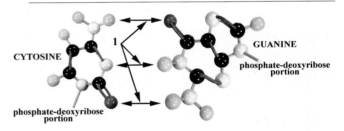

Figure 3-D.10

42 In reference to **Figure 3-D.10**, what types of bonds are found at number 1? _____

DNA STRUCTURE

43 Bonding of the nitrogen bases produces a molecule that consists of _____ strands.

44 What does "the strands are antiparallel" mean? _____

45 What is the final structural form of DNA? _____

56 Chapter 3-D Chemistry and Metabolism - Nucleic Acids

RIBONUCLEIC ACID - RNA
NUCLEOTIDES

46 What are the structural building blocks of RNA? _____

47 What are the three components of a nucleotide?
1 _____
2 _____
3 _____

48 What are the four nitrogen bases of RNA?
1 _____   4 _____
2 _____   5 _____

RIBOSE SUGAR

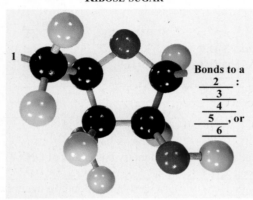

Figure 3-D.11

49 In reference to Figure 3-D.11, what is shown? _____

50 In reference to Figure 3-D.11, identify # 1 - #6.
1 _____   4 _____
2 _____   5 _____
3 _____   6 _____

NITROGEN BASES

51 What is the name of the nitrogen base of RNA that is not found in DNA? _____

RNA MOLECULAR STRUCTURE

52 Describe the molecular structure of RNA. _____

53 What are the three types of RNA?
1 _____
2 _____
3 _____

MOLECULAR MODEL KIT

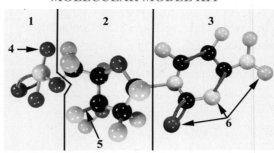

Figure 3-D.12

54 What is the shown in Figure 3-D.12? _____

55 In reference to Figure 3-D.12, what are the major groups shown at numbers:
1 _____
2 _____
3 _____

56 In reference to Figure 3-D.12, what structural units bond at numbers:
4 _____
5 _____
6 _____

57 In reference to Figure 3-D.12, what types of bonds are found at #6? _____

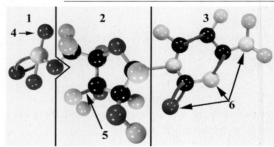

Figure 3-D.13

58 What is the shown in Figure 3-D.13? _____

36. In reference to Figure 3-D.13, what are the major groups shown at numbers:
1 _____
2 _____
3 _____

59 In reference to Figure 3-D.13, what structural units bond at numbers:
4 _____
5 _____
6 _____

60 In reference to Figure 3-D.13, what types of bonds are found at #6? _____

Name _____
Class _____

Chapter 3-D Chemistry and Metabolism - Nucleic Acids   **57**

Messenger RNA (mRNA)

61 How is mRNA structured? _____
_____

62 Where does mRNA originate? _____
_____

63 What is a codon? _____
_____

64 What is the function of mRNA? _____
_____
_____

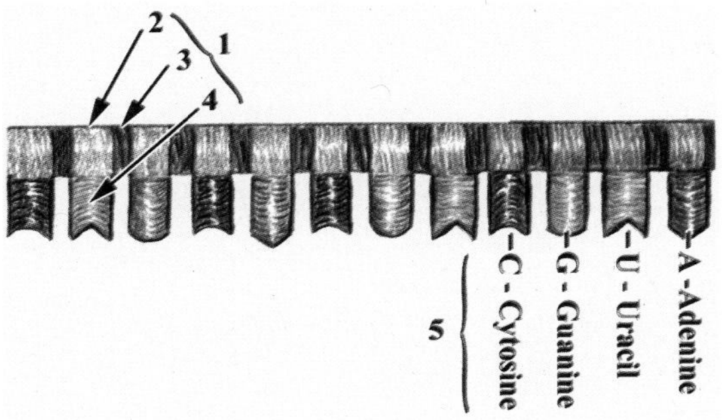

Figure 3-D.14

65 In reference to **Figure 3-D.14**, identify # 1 - #5.
1 _____ 4 _____
2 _____ 5 _____
3 _____

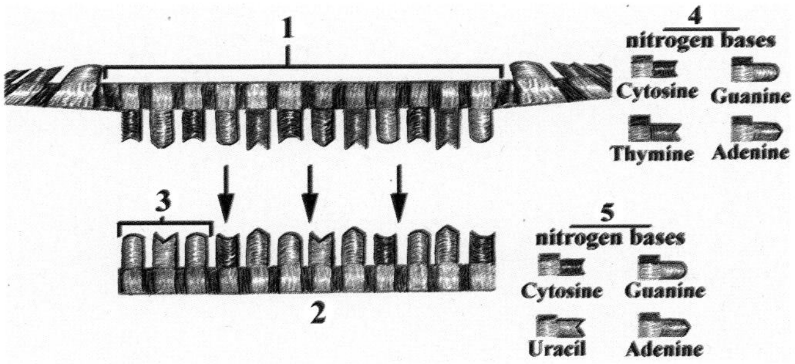

Figure 3-D.15

66 In reference to **Figure 3-D.15**, identify # 1 - #5.
1 _____ 4 _____
2 _____ 5 _____
3 _____

Ribosomal RNA (rRNA)

67 Where does rRNA originate? _____
_____

68 What is the function of ribosomes? _____
_____
_____

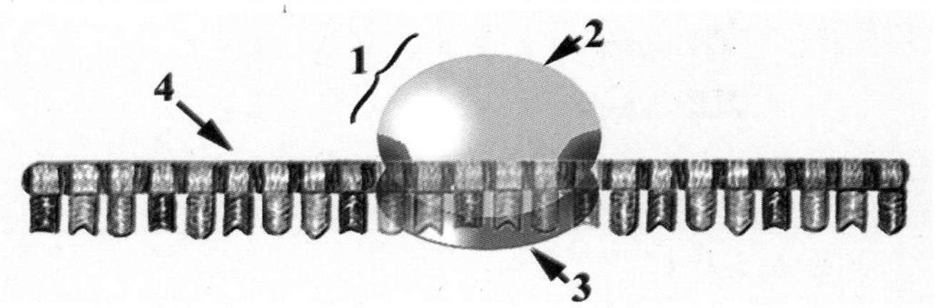

Figure 3-D.16

69 In reference to **Figure 3-D.16**, identify # 1 - #4.
1 _____ 3 _____
2 _____ 4 _____

Transfer RNA (tRNA)

70 How is tRNA structured? _____
_____

71 What are the two functional sites of tRNA?
1 _____
2 _____

72 What does an anticodon bond to? _____
_____
_____

73 What is the function of tRNA? _____
_____
_____

Chapter 3-D Chemistry and Metabolism - Nucleic Acids

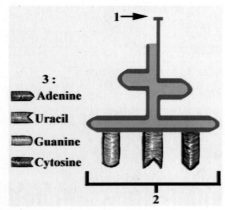

Figure 3-D.17

74 Identify the structure shown in **Figure 3-D.17**. _____
_____
_____

75 In reference to **Figure 3-D.17**, what are the regions at numbers:
1 _____
2 _____

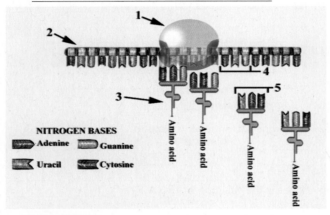

Figure 3-D.18

76 In reference to **Figure 3-D.18**, identify # 1 - #5.
1 _____ 4 _____
2 _____ 5 _____
3 _____

ADENOSINE PHOSPHATE - AMP, ADP, & ATP - WORKSHEETS

ADENOSINE TRIPHOSPHATE (ATP)

1. What molecule is the most abundant of the immediately usable energy rich molecules in the cell? _____

2. When energy rich foods are catabolized for fuel, what are two final destinations for the transfer of energy? _____

3. What is the equation for the oxidation of one molecule of glucose? _____

4. What molecule is energy transferred to for the production of ATP? _____

Figure 3-E.1

5. In reference to **Figure 3-E.1**, identify #1 - #7.
   1 _____   5 _____
   2 _____   6 _____
   3 _____   7 _____
   4 _____

ADENOSINE PHOSPHATES: AMP, ADP, & ATP
ADENOSINE MONOPHOSPHATE, AMP

6. What is the simplest building block for the adenosine phosphates? _____

7. What group is added to reduce both AMP and ADP? _____

8. What molecule is formed by the reduction of adenosine monophosphate? _____

9. What molecule is formed by the reduction of adenosine diphosphate? _____

10. What are the products of the oxidation of ATP? _____

11. What are the products of the oxidation of ADP? _____

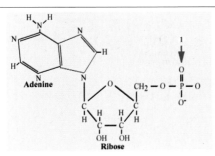

Figure 3-E.2

12. In reference to **Figure 3-E.2**, what molecule is shown? _____

13. In reference to **Figure 3-E.2**, identify #1.
    1 _____

14. What molecule is formed by the **reduction** of adenosine monophosphate (AMP)? _____

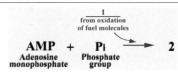

Figure 3-E.3

15. In reference to **Figure 3-E.3**, identify #1 - #2.
    1 _____   2 _____

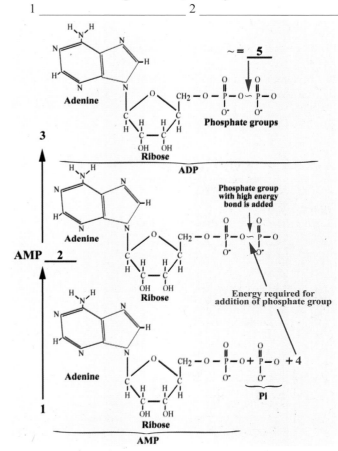

Figure 3-E.4

16. In reference to **Figure 3-E.4**, identify #1 - #5.
    1 _____   4 _____
    2 _____   5 _____
    3 _____

ADENOSINE DIPHOSPHATE, ADP

17. In addition to a phosphate group, what is required for the reduction of AMP to ADP? _____

Chapter 3-E Chemistry and Metabolism - ATP

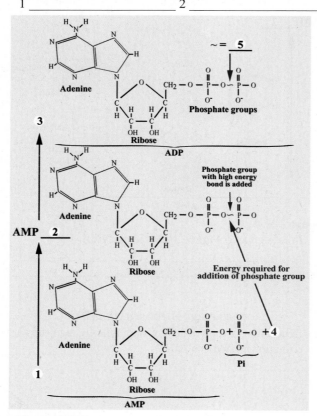

Figure 3-E.5

18 In reference to **Figure 3-E.5**, what molecule is shown?
_____

19 In reference to **Figure 3-E.5**, identify #1-#2.
1 _____ 2 _____

20 In addition to a phosphate group, what are the products of the **oxidation** of ADP? _____
_____

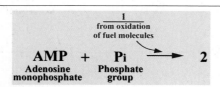

Figure 3-E.6

21 In reference to **Figure 3-E.6**, identify #1 - #2.
1 _____ 2 _____

Figure 3-E.7

22 In reference to **Figure 3-E.7**, identify #1 - #5.
1 _____ 4 _____
2 _____ 5 _____
3 _____

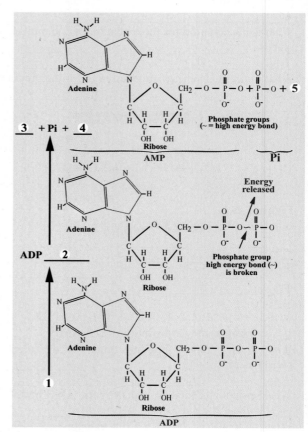

Figure 3-E.8

23 In reference to **Figure 3-E.8**, identify #1 - #2.
1 _____ 2 _____

Figure 3-E.9

24 In reference to **Figure 3-E.9**, identify #1 - #5.
1 _____ 4 _____
2 _____ 5 _____
3 _____

25 What molecule is formed by the **reduction** of adenosine diphosphate (ADP)? _____
_____

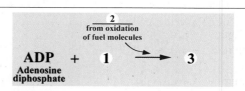

Figure 3-E.10

26 In reference to **Figure 3-E.10**, identify #1 - #3.
1 _____ 3 _____
2 _____

Name _____
Class _____

Chapter 3-E Chemistry and Metabolism - ATP   61

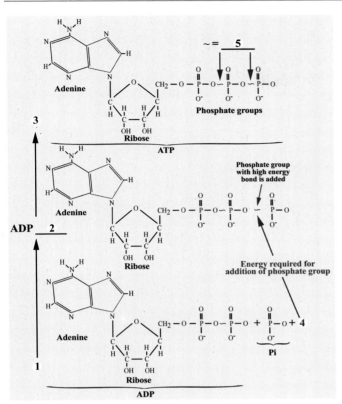

Figure 3-E.11

27 In reference to **Figure 3-E.11**, identify #1 - #5.
    1 _____ 4 _____
    2 _____ 5 _____
    3 _____

ADENOSINE TRIPHOSPHATE

28 In addition to a phosphate group, what is required for the reduction of ADP to ATP? _____
_____

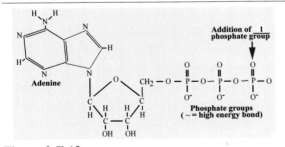

Figure 3-E.12

29 In reference to **Figure 3-E.12**, what molecule is shown? _____

30 In reference to **Figure 3-E.12**, identify #1. _____

31 In addition to a phosphate group, what are the products of the oxidation of ATP? _____
_____

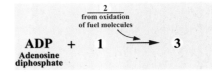

Figure 3-E.13

32 In reference to **Figure 3-E.13**, identify #1 - #3.
    1 _____ 3 _____
    2 _____

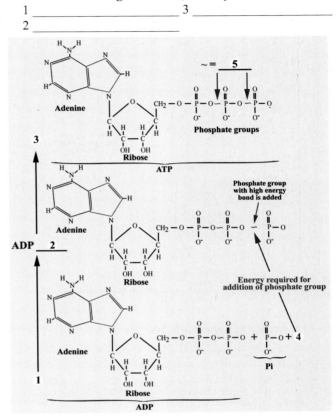

Figure 3-E.14

33 In reference to **Figure 3-E.14**, identify #1 - #5.
    1 _____ 4 _____
    2 _____ 5 _____
    3 _____

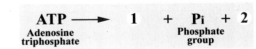

Figure 3-E.15

34 In reference to **Figure 3-E.15**, identify #1 - #2.
    1 _____ 2 _____

Name ______________________
Class ______________________

3

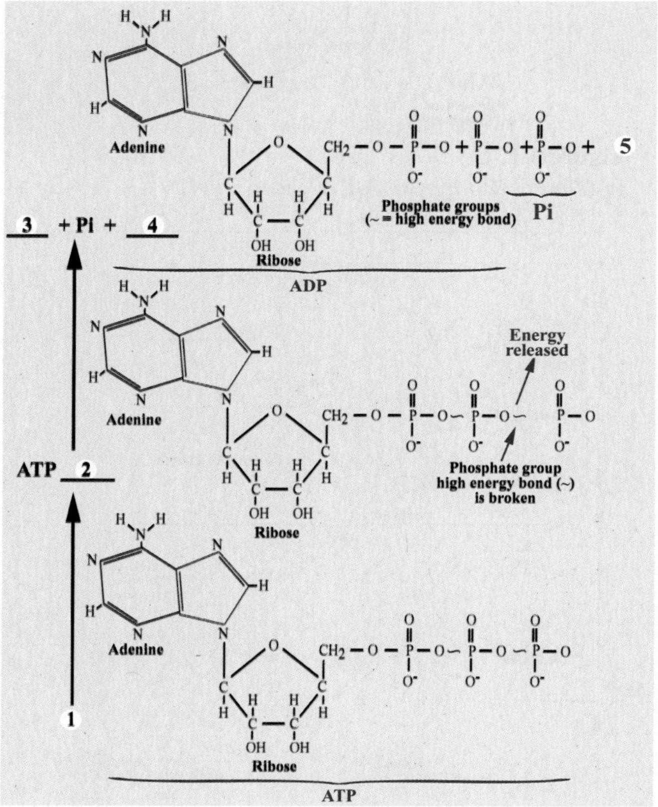

Figure 3-E.16

35 In reference to **Figure 3-E.16**, identify #1 - #5.

1 ______________________ 4 ______________________

2 ______________________ 5 ______________________

3 ______________________

36 What is the most abundant immediately usable source of cellular energy? ______________________

____

37 What are some of the work processes driven by energy released from ATP? ______________________

____

____

____

____

Name _____
Class _____

Chapter 4 - Cytology   63

Cytology- Worksheets

1 Define cytology. _____
   _____
   _____

   What is the name of photographs taken with a light microscope? _____
   _____

   What is the name of photographs taken with an electron microscope? _____
   _____

2 What are the three major parts of a "generalized" cell?
   (1) _____
   (2) _____
   (3) _____

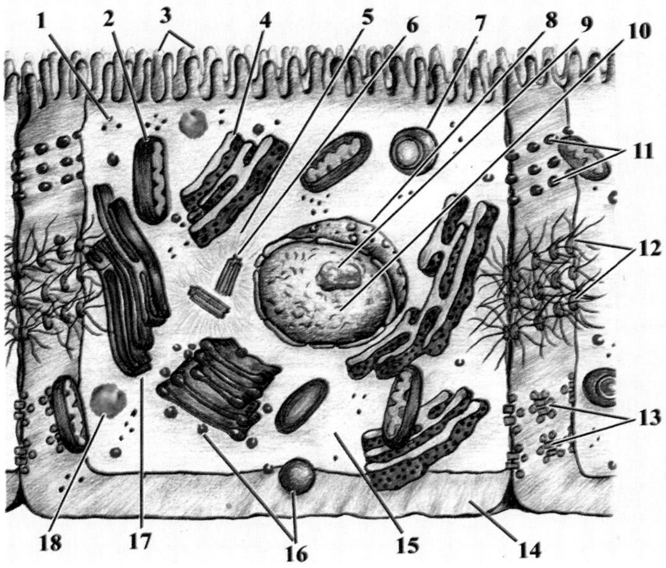

Figure 4.1

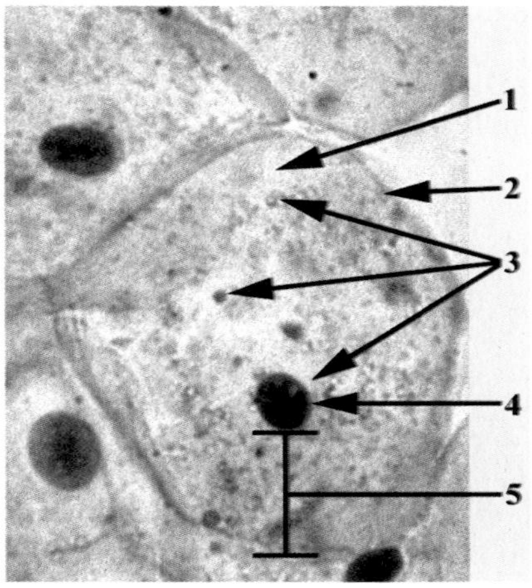

Figure 4.2

6 In reference to **Figure 4.2**, identify #1 - #5.
   1. _____   4. _____
   2. _____   5. _____
   3. _____

Lab activity

7 Was chromatin observed in the nucleus? _____
8 Were any specific organelles or inclusions observed? If so, which? _____
9 Would you describe the plasma membrane as rigid or pliable? _____
10 Describe the plasma membrane. _____
   _____

11 What is cytoplasm? _____
   _____

12 What is cytosol? _____
   _____

3 In reference to **Figure 4.1**, identify #1 - #18.
   1. _____   10. _____
   2. _____   11. _____
   3. _____   12. _____
   4. _____   13. _____
   5. _____   14. _____
   6. _____   15. _____
   7. _____   16. _____
   8. _____   17. _____
   9. _____   18. _____

13 What are organelles? _____
   _____

14 Give an examples of an organelle. _____
   _____

15 What are inclusions? _____
   _____

16 Give an example of an inclusion. _____
   _____

EXFOLIATED EPITHELIAL CELLS

4 What type of tissue lines the mouth? _____
   _____

5 What is a wet mount? _____
   _____
   _____

STRATIFIED SQUAMOUS EPITHELIUM

17 What is a tissue? _____
   _____

18 Describe stratified squamous epithelium. _____
   _____

19 What is the function of stratified squamous epithelium? _____
   _____
   _____

64 Chapter 4 - Cytology

20 Name three locations of stratified squamous epithelium.
_____
_____
_____

21 What are two important cellular regions of stratified squamous epithelium? _____
_____

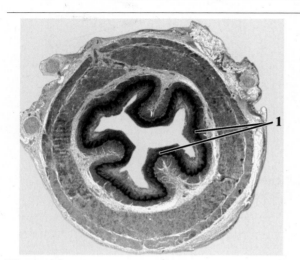

Figure 4.3
22 In reference to **Figure 4.3**, identify #1. _____
_____

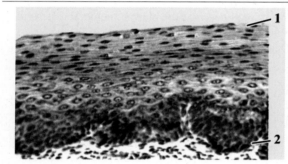

Figure 4.4
23 In reference to **Figure 4.4**, identify #1 & #2.
1 _____ 2 _____
24 What is the function of the squamous surface cells? _____
_____
25 What is the function of the mitotic basal cells? _____
_____

ADIPOSE TISSUE

26 What is the name of the cells that are organized to form adipose tissue? _____ In what form are fats (triglycerides) stored in adipocytes? _____
_____
_____

27 What are three functions of adipose tissue?
1. _____
2. _____
3. _____

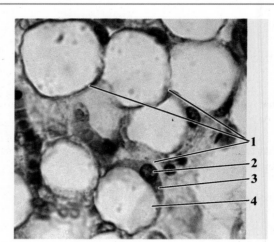

Figure 4.5
28 In reference to **Figure 4.5**, identify #1 - #4.
1. _____ 3. _____
2. _____ 4. _____

BLOOD CELLS

29 What is another name for red blood cells? _____
_____
30 Are circulating erythrocytes nucleated? _____
_____
31 What is the shape of an erythrocyte? _____
_____
32 What is another name for white blood cells? _____
_____
33 Are leukocytes nucleated? _____
_____
34 How does the shape and location of nuclei vary among the different leukocytes? _____
_____
_____

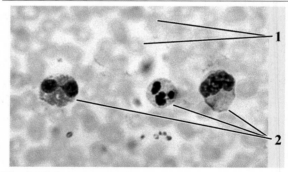

Figure 4.6
35 In reference to **Figure 4.6**, identify #1 & #2.
1. _____ 2. _____

SPERM

36 What are the three distinctive regions of a sperm cell?
1. _____ 3. _____
2. _____
37 What is the function of each region of sperm?
1. _____
2. _____
3. _____

Name _____
Class _____

Chapter 4 - Cytology

38 What is the function of sperm? _____
_____

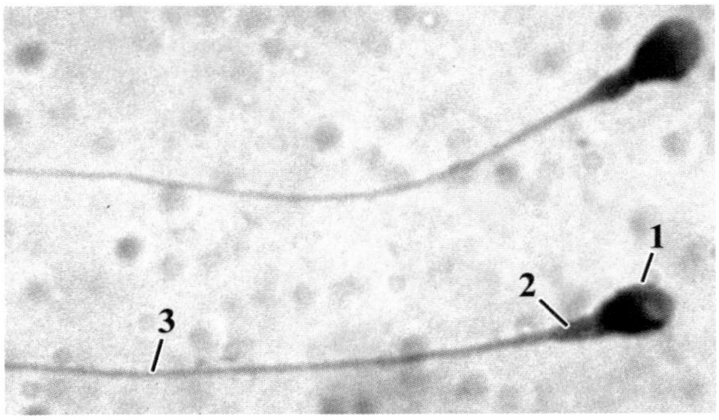

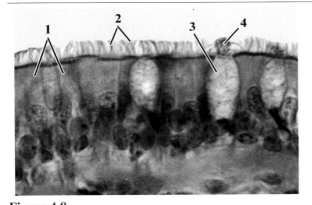

Figure 4.7

39 In reference to **Figure 4.7**, identify #1 - #3.
1. _____   3. _____
2. _____

COLUMNAR CILIATED CELLS AND GOBLET (MUCOUS) CELLS

40 Describe columnar ciliated cells. _____
_____

41 What is the function of the epithelial lining that is formed by columnar ciliated cells? _____
_____

42 Where is an ideal location for the study of ciliated columnar cells? _____
_____

43 What are goblet cells? _____
_____

Figure 4.8

44 In reference to **Figure 4.8**, identify #1 - #4.
1. _____   3. _____
2. _____   4. _____

SMOOTH MUSCLE CELLS

45 Why are smooth muscle cells called "smooth?" _____
_____

46 Describe the shape of smooth muscle cells. _____

47 Where is smooth muscle commonly found? _____

48 What is the function of smooth muscle? _____
_____

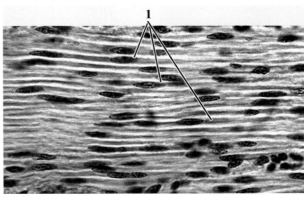

Figure 4.9

49 In reference to **Figure 4.9**, identify the tissue shown at #1.
_____

Figure 4.10

50 In reference to **Figure 4.10**, identify #1 & #2.
1. _____   2. _____

OVERVIEW: CELL OBSERVATIONS

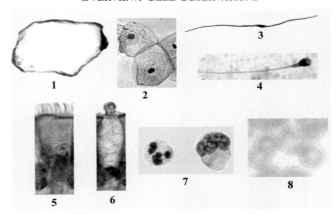

Figure 4.11

51 In reference to **Figure 4.11**, identify the cells #1 - #8.
1. _____   5. _____
2. _____   6. _____
3. _____   7. _____
4. _____   8. _____

52 In reference to **Figure 4.11** (cellular commonality), what features do all cells share in common? _____
_____
_____

53 In reference to **Figure 4.11** (cellular diversity) match the functions with the cell's number:
____ defense against disease
____ production of mucus
____ movement of mucus over surface of cells
____ transport of respiratory gases (especially oxygen)
____ regulation of material movement in viscera
____ protection of surface from abrasion
____ storage of lipid (triglycerides)
____ motile gamete for fertilization of oocyte

Chapter 4 - Cytology

STRUCTURE AND FUNCTION OF A GENERALIZED CELL

PLASMA MEMBRANE

54 What does the plasma membrane form? _____
_____

55 What two fluids environments does the plasma membrane separate? _____
_____

56 Is the plasma membrane usually a homogenous structure? _____ Explain. _____
_____
_____

57 What are five functions of the plasma membrane?
(1) _____
(2) _____
(3) _____
(4) _____
(5) _____

58 What molecule forms the framework of the plasma membrane? _____
_____

59 Name two other lipid molecules associated with the plasma membrane.
(1) _____
(2) _____

60 Proteins associated with the plasma membrane are located both _____ and _____.

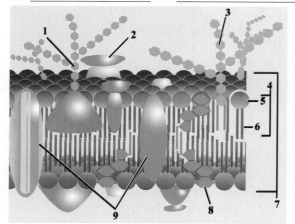

Figure 4.12

61 In reference to **Figure 4.12**, identify #1 - #9.
1. _____   6. _____
2. _____   7. _____
3. _____   8. _____
4. _____   9. _____
5. _____

LIPIDS OF THE PLASMA MEMBRANE
LIPID MOLECULES - STRUCTURE

62 Name three lipid molecules of the plasma membrane.
(1) _____
(2) _____
(3) _____

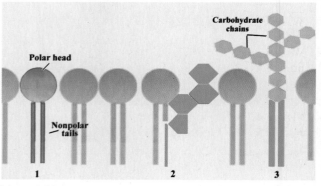

Figure 4.13

63 In reference to **Figure 4.13**, identify the lipid molecules shown at #1 - #3.
(1) _____
(2) _____
(3) _____

64 Which part of the phospholipid molecule is hydrophilic?
_____

65 Which part of the phospholipid molecule is hydrophobic?
_____

66 How are the phospholipids arranged to form the framework of the plasma membrane? _____
_____

67 Where do the carbohydrate chains of the glycolipids project? _____
_____

68 How is cholesterol distributed in the plasma membrane?
_____

69 Which portion of cholesterol is hydrophobic? _____

70 Which portion of cholesterol is hydrophilic? _____

LIPID MOLECULES - FUNCTIONS

70 What are general characteristics of substances that can diffuse through the phospholipid bilayer? _____
_____

72 What are general characteristics of substances that don't diffuse through the phospholipid bilayer? _____
_____

73 What are functions of the glycocalyx? _____
_____
_____

74 What is a function of cholesterol that is found in the plasma membrane? _____
_____
_____

Name _____
Class _____

Chapter 4 - Cytology   67

PROTEINS OF THE PLASMA MEMBRANE
PROTEIN STRUCTURE

75 What two names may be applied to protein molecules according to their location?
    1 _____ 2 _____

76 Where are integral proteins located? _____

77 Where are peripheral proteins located? _____

78 Where do the carbohydrate chains of the glycoproteins project? _____

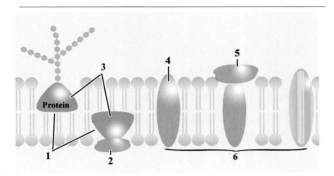

Figure 4.14

79 In reference to **Figure 4.14**, identify the cells #1 - #6.
    1. _____ 4. _____
    2. _____ 5. _____
    3. _____ 6. _____

PROTEIN FUNCTIONS
Transmembrane protein channels

80 What is the function of transmembrane protein channels? _____

81 What is the difference between passive and active transport? _____

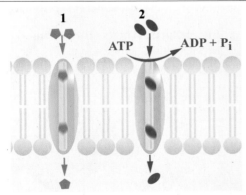

Figure 4.15

82 In reference to **Figure 4.15**, what type of channel proteins is shown? _____

83 In reference to **Figure 4.15** identify #1 & #2.
    1. _____ 2. _____

Transmembrane protein carrier molecules

84 What is the function of transmembrane protein carrier molecules? _____

85 Is facilitated diffusion active or passive? _____

86 What direction are substances transported in relation to concentration gradient? _____

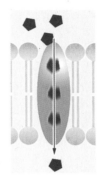

Figure 4.16

87 In reference to **Figure 4.16**, what type of protein is shown? _____

Glycoproteins

88 What are three functions of glycoproteins?
    1 _____
    2 _____
    3 _____

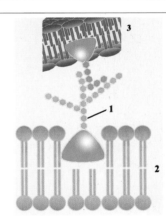

Figure 4.17

89 In reference to **Figure 4.17**, what type of protein is shown at #1? _____

90 In reference to **Figure 4.17**, and in reference to cell recognition, which cell is being recognized, #2 or #3? _____

91 In reference to **Figure 4.17**, if the glycoprotein is functioning as a receptor, which cell is being bound, #2 or #3? _____

68 Chapter 4 - Cytology

Name _____
Class _____

MEMBRANE ENZYMES

92 What is the function of an enzyme? _____
_____
_____

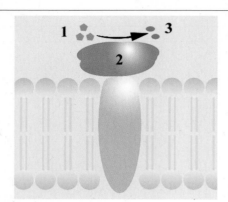

Figure 4.18

93 In reference to **Figure 4.18**, identify #1 - #3.
1 _____ 3 _____
2 _____

STRUCTURAL PROTEINS

94 What is the function of structural membrane proteins?
_____

Figure 4.19

95 In reference to **Figure 4.19**, identify #1 & #2.
1 _____ 2 _____

MEMBRANE JUNCTIONS

96 What are three membrane junctions?
1 _____ 3 _____
2 _____

GAP JUNCTION

97 How is a gap junction formed? _____
_____

98 What is the function of gap junctions? _____
_____

99 Where are gap junctions found? _____
_____

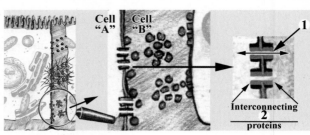

Figure 4.20

100 In reference to **Figure 4.20**, what type of membrane junctions are shown? _____

101 In reference to **Figure 4.20**, identify #1 & #2.
1 _____ 2 _____

DESMOSOME

102 How is a desmosome formed? _____
_____

103 What is the function of a desmosome? _____
_____

104 Where are desmosomes commonly found? _____
_____

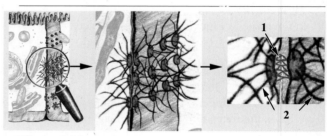

Figure 4.21

105 In reference to **Figure 4.21**, what type of membrane junction is shown? _____
In reference to **Figure 4.21**, identify #1 & #2.
1 _____ 2 _____

Intercellular Bridges (Desmosomes)

106 Where are intercellular bridges found? _____
_____
_____ What is the function of intercellular bridges? _____
_____

107 What are membrane junctions? _____
_____

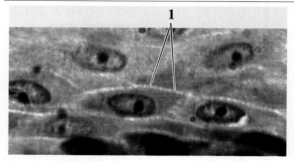

Figure 4.22

107 In reference to **Figure 4.22**, identify #1. _____
_____

Chapter 4 - Cytology

TIGHT JUNCTIONS
108 How is a tight junction formed? _____

109 What is the function of tight junctions? _____

110 Where are tight junctions located? _____

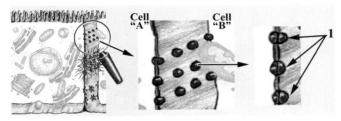

Figure 4.23
111 In reference to **Figure 4.23**, identify #1. _____

MICROVILLI
112 How are microvilli formed? _____

113 What is the function of microvilli? _____

114 Where are microvilli found? _____

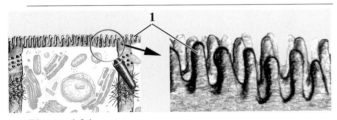

Figure 4.24
115 In reference to **Figure 4.24**, identify #1. _____

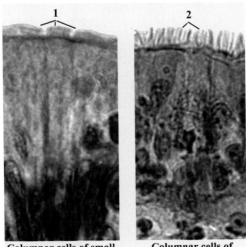

Figure 4.25
116 In reference to **Figure 4.25**, identify #1 & #2.
1. _____ 2. _____

CYTOPLASM
117 What region of the cell does cytoplasm describe? _____

118 What is cytosol? _____

119 What are the major constituents of cytosol? _____

120 What are organelles? _____

121 Give three examples of organelles. _____

122 What are inclusions? _____

123 Give three examples of inclusions. _____

Organelles
124 Since most organelles are too small to be easily visualized with the light microscope, what instrument is used for their visualization? _____

Nucleus
125 What is the name of the membrane that surrounds the nucleus? _____

126 What are three major components of the nucleus?
1 _____   3 _____
2 _____

127 What is the function of the nucleus? _____

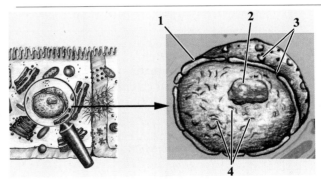

Figure 4.26
128 In reference to **Figure 4.26**, identify #1 - #4.
1. _____   3. _____
2. _____   4. _____

70   Chapter 4 - Cytology

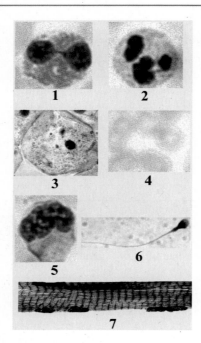

Figure 4.27

129 Match #1 - #7 of **Figure 4.27** with the following descriptions.
   _____ bilobed nucleus of WBC
   _____ mulilobed nucleus of WBC
   _____ indented nucleus of WBC
   _____ head houses nucleus
   _____ multinucleate skeletal muscle
   _____ spherical nucleus of squamous cell
   _____ anucleate RBC

Nuclear Envelope

130 How many layers comprise the nuclear envelope? _____

131 What is the function of nuclear pores? _____

132 What molecule does not move through nuclear pores? _____

133 List several substances that freely move through nuclear pores. _____

Chromatin

134 What is the composition of chromatin? _____

135 Describe chromatin. _____

136 When is chromatin organized into chromosomes? _____

137 What are chromosomes? _____

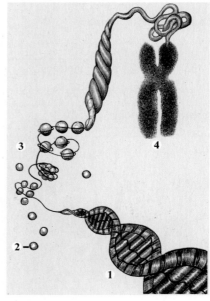

Figure 4.28

138 In reference to **Figure 4.28**, identify #1 - #4.
   1._____   3._____
   2._____   4._____

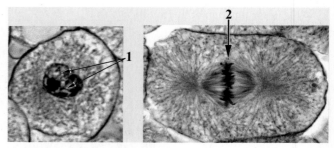

Figure 4.29

139 In reference to **Figure 4.29**, identify #1 & #2.
   1._____   2._____

Nucleoli

140 Describe the appearance of nucleoli. _____

141 What do nucleoli contain? _____

142 What is the function of nucleoli? _____

Chapter 4 - Cytology

MITOCHONDRIA

143 Describe the appearance of mitochondria. _____

144 How many membranes surround a mitochondrion? _____

145 What are cristae? _____

146 What is matrix? _____

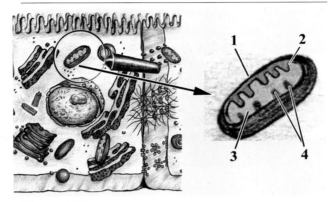

Figure 4.30

147 In reference to **Figure 4.30**, identify #1 - #4.
   1. _____ 3. _____
   2. _____ 4. _____

148 What is the function of mitochondria? _____

149 What two catabolic processes occur within the mitochondria? _____

150 In the catabolism of a molecule of glucose, how many ATPs are produced by the Krebs cycle and electron transport system? _____

RIBOSOMES

151 Where does ribosomal RNA originate? _____

152 What are the two possible locations for ribosomes?
   (1) _____
   (2) _____

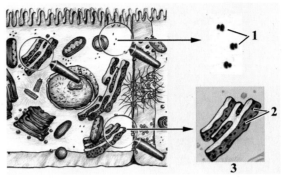

Figure 4.32

153 In reference to **Figure 4.32**, identify #1 - #3.
   1. _____ 3. _____
   2. _____

154 What is the function of ribosomes? _____

155 Where are polypeptides and proteins produced by the free ribosomes released? _____

156 Where are polypeptides and proteins produced by attached ribosomes released? _____

157 What happens to polypeptides and proteins that enter the endoplasmic reticulum? _____

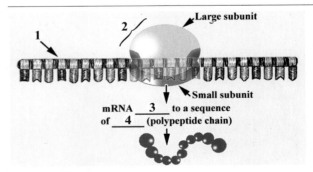

Figure 4.33

158 In reference to **Figure 4.33**, identify #1 - #5.
   1. _____ 3. _____
   2. _____ 4. _____

Chapter 4 - Cytology

ENDOPLASMIC RETICULUM (ER)

159 How is the endoplasmic reticulum structured? _____

160 Where is the endoplasmic reticulum located? _____

161 What two types of endoplasmic reticulum can be described depending upon the association with ribosomes?
1. _____ 2. _____

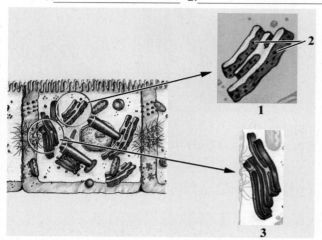

Figure 4.34

162 In reference to **Figure 4.34**, identify #1 - #3.
1. _____ 3. _____
2. _____

163 What type of molecules are released by ribosomes into the cavity of the rough endoplasmic reticulum? _____

164 What is the function of transport vesicles? _____

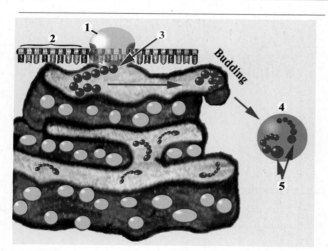

Figure 4.35

165 In reference to **Figure 4.35**, identify #1 - #5.
1. _____ 4. _____
2. _____ 5. _____
3. _____

166 What is the function of the smooth endoplasmic reticulum? _____

166 What is the function of smooth endoplasmic reticulum (sarcoplasmic reticulum) of skeletal and cardiac muscle? _____

GOLGI APPARATUS

168 What is the structure of the Golgi apparatus? _____

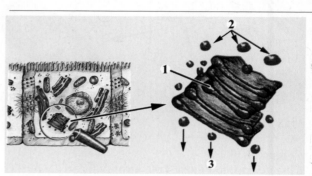

Figure 4.36

169 In reference to **Figure 4.36**, identify #1 - #3.
1. _____ 3. _____
2. _____

170 What is the function of the Golgi apparatus? _____

171 What is delivered to the Golgi apparatus in the transport vesicles? _____

172 What are the three possible destinations for processed and packaged proteins?
1 _____
2 _____
3 _____

Chapter 4 - Cytology

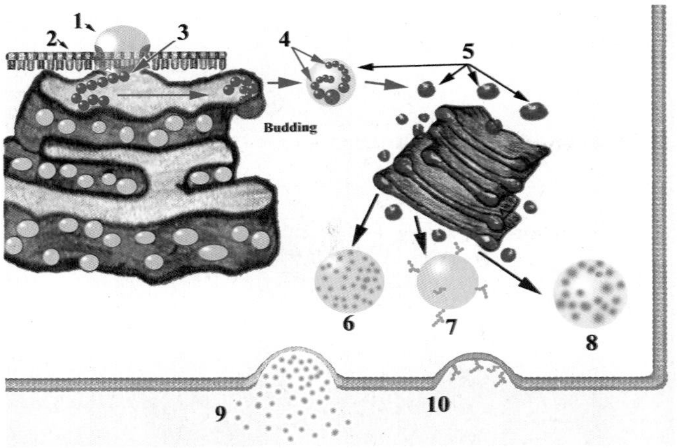

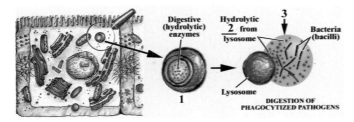

Figure 4.37
173 In reference to **Figure 4.37**, identify #1 - #10.
1._____ 6._____
2._____ 7._____
3._____ 8._____
4._____ 9._____
5._____ 10._____

LYSOSOMES
174 How are lysosomes formed?_____

175 What do lysosomes contain?_____

176 What is a phagosome?_____

177 What is the function of lysosomes?_____

Figure 4.38
178 In reference to **Figure 4.38**, identify #1 - #3.
1._____ 3._____
2._____

CENTROSOME AND CENTRIOLES
179 What forms the centrosome?_____

180 What is the function of the centrosome in the nondividing cell?_____

181 What is the function of the centrosome in the dividing cell?_____

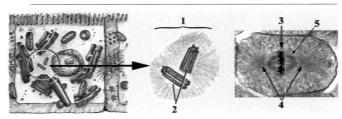

Figure 4.39
182 In reference to **Figure 4.39**, identify #1 - #5.
1._____ 4._____
2._____ 5._____
3._____

CILIA AND FLAGELLA
183 What are cilia and flagella composed of?_____

184 What is a basal body?_____

185 Where are basal bodies located?_____

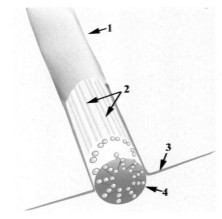

Figure 4.40
186 In reference to **Figure 4.40**, identify #1 - #4.
1._____ 3._____
2._____ 4._____

74 Chapter 4 - Cytology

187 What is the function of cilia? _____
_____
_____

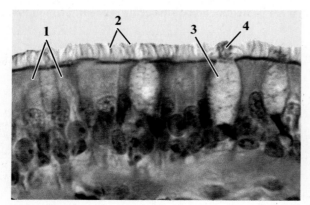

Figure 4.41
188 In reference to **Figure 4.41**, identify #1 - #4.
    1. _____ 3. _____
    2. _____ 4. _____
189 What is the function of flagella? _____
_____
_____

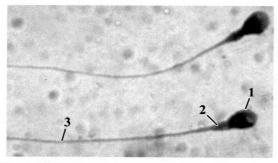

Figure 4.42
190 In reference to **Figure 4.42**, identify #1 - #3.
    1. _____ 3. _____
    2. _____

INCLUSIONS

191 What are inclusions? _____
_____
_____

192 List several common inclusions. _____
_____
_____

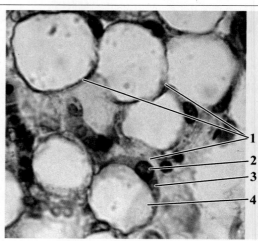

Figure 4.43
193 In reference to **Figure 4.43**, identify #1 - #4.
    1. _____ 3. _____
    2. _____ 4. _____

Transport Processes - Worksheets

1. Into what two anatomic compartments do cells organize the body?
   1. _____
   2. _____
2. What structure must fluids (water and solutes) pass through in order to move from compartment to compartment? _____

INTRACELLULAR FLUID COMPARTMENT

3. What forms the intracellular compartment? _____
4. What does the intracellular compartment contain? _____
5. What percentage of the body's total fluid is intracellular fluid? _____

EXTRACELLULAR FLUID COMPARTMENT

6. What forms the extracellular compartment? _____
7. What does the extracellular compartment contain? _____
8. What are the two major subdivisions of the extracellular compartment? (1) _____ (2) _____
9. What is the function of the extracellular fluid compartments? _____
10. What forms the interstitial compartment? _____
11. What is located within the interstitial compartment? _____
12. What percentage of the body's total fluid is interstitial fluid? _____
13. What forms the intravascular compartment? _____
14. What percentage of the body's total fluid is intravascular fluid? _____
15. What are plasma and lymph? _____

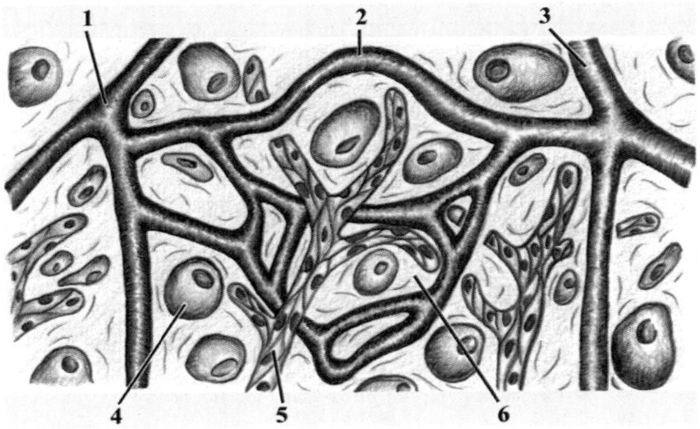

Figure 5.1

16. In reference to **Figure 5.1**, what compartment is identified by #s 1 - 3. _____
17. In reference to **Figure 5.1**, identify #1 - #4.
    1. _____   4. _____
    2. _____   5. _____
    3. _____   6. _____
18. In reference to **Figure 5.1**, what compartment is shown at #6? _____
19. In reference to **Figure 5.1**, what compartment is shown at #4? _____
20. What two processes are involved in transport across the plasma membrane? _____
21. What do all active processes require? _____
22. List five passive processes?
    1. _____   4. _____
    2. _____   5. _____
    3. _____
23. What are two active transport processes? _____

MIXTURES

24. What is a mixture? _____
25. What are three mixtures?
    1. _____   3. _____
    2. _____
26. A mixture of water and other small molecules and particles are in a constant state of _____ motion.
27. In a mixture of water and other small molecules and particles, the solutes (dissolved substances) _____ (will or will not) settle out and their movement will _____.
28. In a suspension, the large particles will _____.

Solutions

29. What is a solution? _____
30. What is a solute? _____
31. What is a solvent? _____
32. Do both homogenous and single-phase systems apply to solutions? _____

Colloid

33. What is a colloid? _____
34. Do colloid sized molecules and particles normally pass through the plasma membrane? _____
35. What are blood colloidal proteins? _____
36. Do both homogenous and single-phase systems apply to colloids? _____

Suspension

37. What is a suspension? _____
38. What is a two-phase system? _____

76   Chapter 5 - Transport Processes

39 Over time, what happens to the solid phase? _____
_____

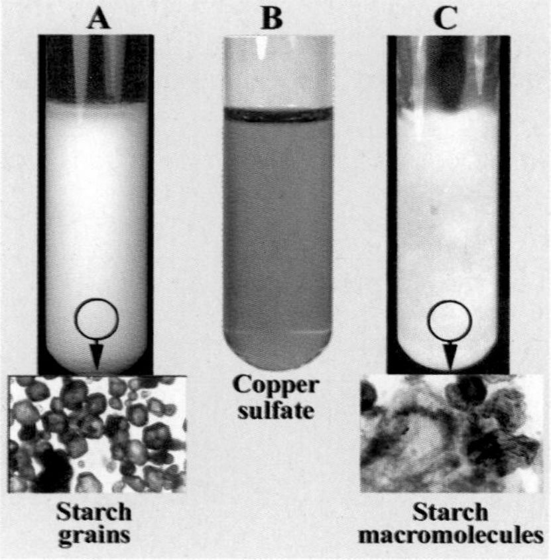

Figure 5.2

40 In reference to **Figure 5.2**, match the letter with the type of water-based mixture:
Solution _____
Suspension _____
Colloid _____

41 In reference to **Figure 5.2**, match the letter (may have more than one) to the description of the mixture
Contains a solute _____
Contains smallest solutes or particles _____
Solutes or particles larger than those of solution _____
Solutes or particles larger than those of colloid _____
Single-phase system _____
Dual-phase system _____
Homogenous system _____
Settles out _____
Does not settle out _____

LAB ACTIVITY

Molecular and Particle Movement

42 Why is it usually necessary to let the microscope slide preparation of milk rest several minutes before observation? _____

43 Describe the movement of the individual milk particles.

44 Did all of the particles in the milk move at the same rate? _____ Explain you answer. _____
_____

PASSIVE MOVEMENT ACROSS THE PLASMA MEMBRANE

45 Define diffusion. _____
_____

46 The process of equalization always proceeds from an area of _____ concentration to an area of _____ concentration. What is net diffusion? _____
_____

47 What are four factors that influence the movement of molecules (and particles) through their environment?
1 _____   3 _____
2 _____   4 _____

48 Increasing temperature _____ the rate of diffusion. In the same system, larger molecules move _____ than smaller molecules.

49 In environments of increasing permeability, molecules diffuse _____ .

50 Molecules of like charge _____, and molecules of unlike charges _____ .

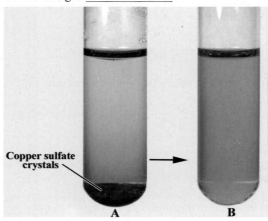

Figure 5.3

51 In reference to test tube A in **Figure 5.3**, what process is occurring? _____

52 Which tube, A or B, has reached concentration equilibrium? _____

LAB ACTIVITY

Molecular Motion and weight

53 In a mixture, (of equal temperature) which molecules or particles move the fastest? _____
_____

54 What is the molecular weight of methylene blue? _____
_____ What is the molecular weight of potassium permanganate? _____

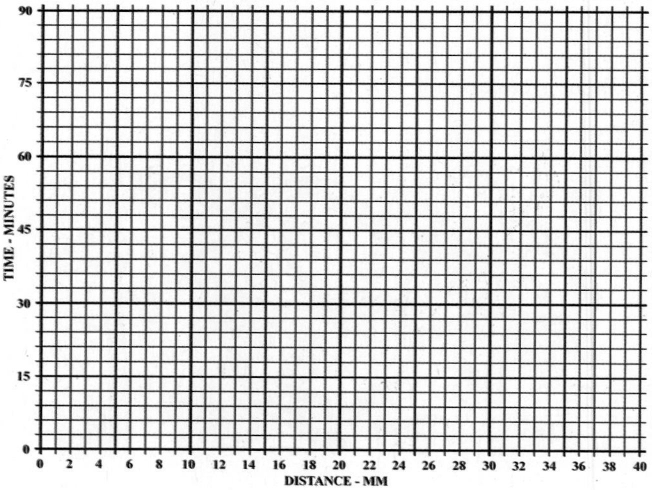

Figure 5.4

Plot of diffusion rates of methylene blue and potassium permanganate.

55 Which dye, the methylene blue or the potassium permanganate, diffused at the fastest rate? _____
_____

56 Why did one dye diffuse faster than the other? _____
_____

Simple Diffusion Across the Plasma Membrane

57 What characteristic must a substance have in order to diffuse across the plasma membrane? _____

58 What are three common characteristics that give substances permeability to the plasma membrane?
1 _____
2 _____
3 _____

59 What characteristics do substances that diffuse through the phospholipid bilayer exhibit? _____
_____

60 What is the general function of membrane protein channels? _____
_____

61 What is facilitated diffusion? _____
_____

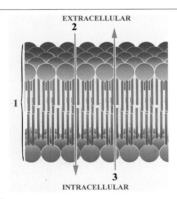

Figure 5.5

62 In reference to **Figure 5.5**, identify #1 - #3.
1 _____   3 _____
2 _____

63 In reference to **Figure 5.5**, why do #2 and #3 diffuse in the direction indicated by the arrows? _____
_____

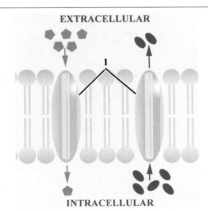

Figure 5.6

64 In reference to **Figure 5.6**, identify #1.
1 _____

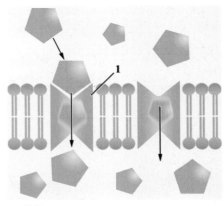

Figure 5.7

65 In reference to **Figure 5.7**, identify #1.
1 _____

MOVEMENT OF WATER BY HYDROSTATIC PRESSURE

66 Define hydrostatic pressure. _____
_____

67 What are three sources of the body's hydrostatic pressure?
1 _____
2 _____
3 _____

68 A change in the net water movement into or out of a cell changes the _____ and results in a change of the cell's _____.

69 Net water movement into or out of blood vessels causes a change in the blood _____. Increased fluid volume _____ pressure, and decreased fluid volume _____ pressure.

70 Increased fluid in the interstitial spaces causes _____ _____. The loss of tissue turgor is caused by a _____ of interstitial fluid volume.

Filtration

71 Define- filtration. _____
_____
_____

72 Define- filter. _____
_____

73 What characteristic of a filter determines whether or not a substance will pass? _____

LAB ACTIVITY

Filtration

74 What determines if a substance will pass through the filter paper? _____
_____

75 What produces the filtration pressure in the filtration apparatus? _____

76 Did the filtration rate change as the funnel emptied? ____ Explain your answer. _____

77 How was the presence (or absence) of copper sulfate tested? _____

78 Chapter 5 - Transport Processes

78 How was the presence (or absence) of unboiled corn starch tested? _____

Chemicals:	Presence of copper sulfate	Presence of starch
Filtrate		

Figure 5.8
79 Test results for filtration of solution of copper sulfate and starch.

80 In reference to **Figure 5.9**, what does test result indicate about "A," the **filtrate** of copper sulfate and unboiled corn starch? _____

81 In reference to **Figure 5.9**, what does test result indicate about "B," the **solution** of copper sulfate and unboiled corn starch? _____

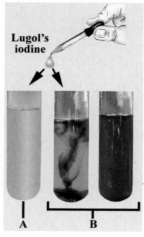

Figure 5.9

82 What does selective permeability mean? _____

83 What is the driving force for capillary filtration? _____

84 In which direction are substances filtered across a capillary? _____

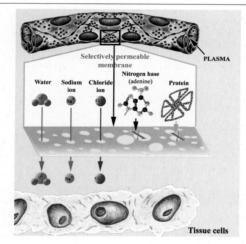

Figure 5.10

85 In reference to **Figure 5.10**, what is the driving force for filtration? _____

86 In reference to **Figure 5.10**, into what space does the filtrate move? _____ What is the filtrate called? _____

87 What is the function of interstitial fluid? _____

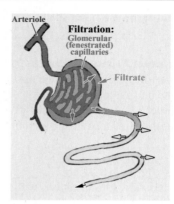

Figure 5.11
88 In reference to **Figure 5.11**, what happens to the filtrate that is produced at the glomerular capillaries? _____

MOVEMENT OF WATER BY OSMOSIS
Osmosis

89 Define - osmosis. _____

90 What happens when the solute concentrations of impermeable solutes differ between the intracellular and extracellular fluid?? _____

91 Water osmotically moves from a region of _____ concentration to a region of _____ concentration.

92 If a solution outside the cell has a higher concentration of impermeable solutes than within the cell, then the water concentration outside the cell is _____ than in the cell, and water will diffuse _____ the cell.

93 If there is a net osmotic water movement out of the cell, the cell's hydrostatic pressure _____.

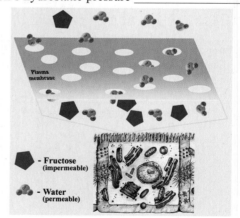

Figure 5.12
94 In reference to **Figure 5.12**, the membrane is described as _____.

95 In reference to **Figure 5.12**, fructose
  A net diffusion is into the cell
  B net diffusion is out of the cell
  C does not diffuse through the membrane
  D More than one of the above choices is correct. The correct choices are _____.

96 In reference to **Figure 5.12**, water
   A. net diffusion is into the cell
   B. net diffusion is out of the cell
   C. does not diffuse through the membrane
   D. More than one of the above choices is correct. The correct choices are _____.
97 In reference to **Figure 5.12,** the cell _____ water, its hydrostatic pressure _____, which causes the cell to _____.

Solute Concentration

98 When does net diffusion of water occur? _____

99 What determines the concentration of water? _____

100 What are two requirements of the solute in order for osmosis to occur? _____

101 What is percent concentration as applied to a solution? __

102 What is molarity? _____

103 What is a mole? _____

104 How would you calculate a mole of NaCl? _____

105 How would you make a one molar solution of NaCl? _____

106 How many particles does a mole of any substance contain? _____

107 What is the difference between osmolality and osmolarity? _____

Effects of Osmotic Solutions
Osmolality

108 What does osmolality of a solution a measure? _____

109 What is required to a solution to be osmotically effective? _____

110 What happens when the solutes (and water) are permeable to the membrane? _____

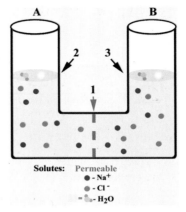

Figure 5.13

111 In reference to **Figure 5.13**, identify #1. _____

112 In reference to **Figure 5.13**, does either water level #2 or #3 move? ____ Explain your answer. _____

Tonicity

113 What is tonicity? _____

114 What particles are considered in the tonicity of a solution? _____

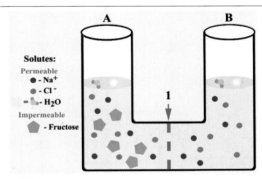

Figure 5.14

115 In reference to **Figure 5.14**, identify #1. _____ What substance is impermeable to the membrane? _____

116 In reference to **Figure 5.14**, which side "A" or "B" has the highest tonicity? _____ Explain your answer. _____

117 In reference to **Figure 5.14**, which side "A" or "B" has the greatest amount of water? _____ Explain your answer. _____

118 In reference to **Figure 5.14**, water will
   A. show net diffusion from side "A" to side "B"
   B. show net diffusion from side "B" to side "A"
   C. not show any net diffusion
119 In reference to **Figure 5.14**, the hydrostatic pressure in side "A" will
   A. decrease
   B. increase
   C. remain the same

Chapter 5 - Transport Processes

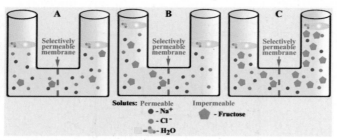

Figure 5.15

120 In reference to **Figure 5.15**, which apparatus (or apparatuses) show(s) the **correct results** for the osmotic effects of its solution:
A. only apparatus "A"
B. only apparatus "B"
C. only apparatus "C"
D. apparatuses "B" and "C"
E. apparatuses "A," "B," and "C"

OSMOTIC PRESSURE

121 Define osmotic pressure. _____
_____

122 What is a solutions osmotic pressure proportional to? ____
_____

123 Which solution, a solution with 50% NaCl or a solution with 10% NaCl, would have the greatest osmotic pressure when compared to a red blood cell (1% NaCl)? _____

124 Osmotic pressure results because of the osmotic movement of water and is the pressure required to _____ water's movement.

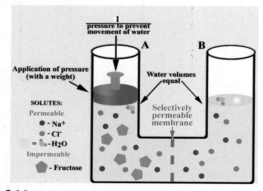

Figure 5.16

125 In reference to **Figure 5.16**, what does #1 represent? ____
_____

126 In reference to **Figure 5.16**, if additional impermeable solutes were added to side "A", would a larger or smaller "weight" be required to stop the osmotic movement of water? _____ Explain your answer. _____
_____
_____

TONICITIES OF SOLUTIONS

127 What are the three possible tonicities of solutions?
1 _____ 3 _____
2 _____

128 Define isotonic solution. _____
_____

129 Do isotonic solutions result in a net osmotic movement of water? _____ What effect, if any, would an isotonic solution have on the shape of a plant or animal cell? ____
_____

130 Define hypotonic solution. _____
_____

131 Do hypotonic solutions result in a net osmotic movement of water? ____ Explain your answer. _____
_____

132 What happens to the hydrostatic pressure of a cell placed in a hypotonic solution? _____
_____

133 What effect, if any, would a hypotonic solution have on the shape of an animal cell? _____
_____

134 What would be the effect of a hypotonic solution on a plant cell? _____
_____

135 Define hypertonic solution. _____

136 Do hypertonic solutions result in a net osmotic movement of water? ____ Explain your answer. _____
_____

137 What happens to the hydrostatic pressure of a cell placed in a hypertonic solution? _____

138 What effect, if any, would a hypertonic solution have on the shape of an animal cell? _____
_____

139 What would be the effect of a hypertonic solution on a plant cell? _____
_____

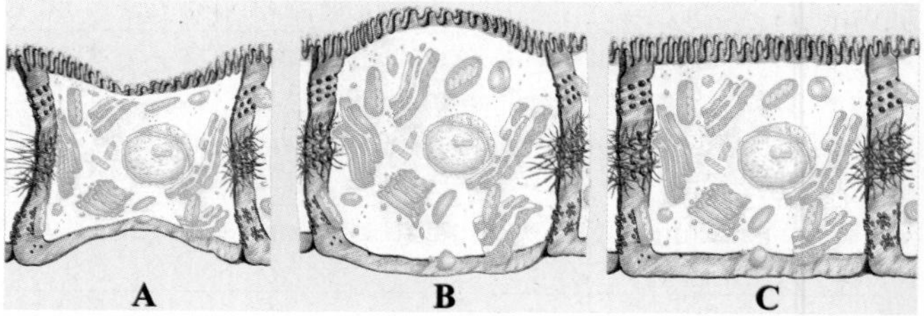

Figure 5.17

140 In reference to **Figure 5.17**, identify the tonicity of the solutions the cells were placed into:
A _____
B _____
C _____

141 In reference to **Figure 5.17** which cell has the greatest hydrostatic pressure? _____ Which cell has the least hydrostatic pressure? _____ Which cell has its normal hydrostatic pressure? _____

Chapter 5 - Transport Processes

LAB ACTIVITY

Osmometer

142 What is an osmometer? _____

143 The typical thistle tube osmometer is filled with _____.

144 The selectively permeable membrane typically used with a thistle tube osmometer is permeable to _____ but not to larger molecules such as _____.

145 The thistle tube osmometer is inverted into a beaker filled with _____.

146 The osmotic environment, the distilled water, is a _____ tonic environment.

147 If the distilled water environment was replaced with a solution of 70% impermeable solute (fructose), the environment would be _____ tonic.

148 If the distilled water environment was replaced with a solution of 50% impermeable solute (fructose), the environment would be _____ tonic.

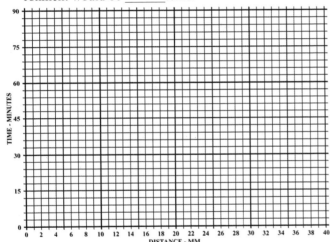

Figure 5.18

149 Graph of column movement (mm) every 15 minutes.

150 In reference to Figure 5.18, your results showed that over time, the rate of the net diffusion of water:
   A. increased
   B. increased, then decreased, then increased
   C. decreased
   D. decreased, then increased, then decreased
   E. remained the same
   F. was sporadic
   G. did not occur as the column did not move

151 Is your answer to the preceding question what was expected from an analysis of the experiment's materials and setup?   Explain your answer. _____

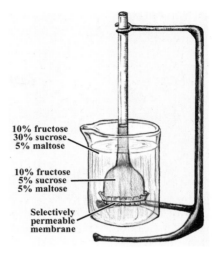

Figure 5.19

152 In reference to Figure 5.19, the thistle tube is immersed in a _____ tonic solution.

153 In reference to Figure 5.19, the column of the solution in the stem of the tube should _____, as hydrostatic pressure _____.

LAB ACTIVITY

Red Blood Cells

154 Describe the appearance of red blood cells in an isotonic solution. _____

155 Describe the appearance of red blood cells in a hypertonic solution. _____

156 Describe the appearance of red blood cells in a hypotonic solution. _____

157 With respect to red blood cells, match the letter of the following osmotic solutions with the following tonicities:
   A isotonic   B hypertonic   C hypotonic
   ____ 5% NaCl solution   ____ 0.9% NaCl solution
   ____ 0.1% NaCl solution   ____ 10% NaCl solution
   ____ 1.5 % NaCl solution   ____ 0.8% NaCl solution

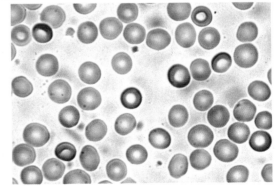

Figure 5.20

158 In reference to Figure 5.20, what is the tonicity of the solution used for the dilution of the red blood cells? _____ Explain your answer. _____

82 Chapter 5 - Transcript Processes

Name _____
Class _____

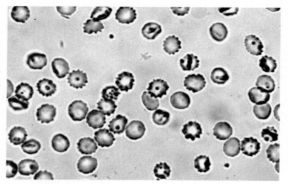

Figure 5.21

159 In reference to **Figure 5.21**, what is the tonicity of the solution used for the dilution of the red blood cells? _____ Explain your answer. _____

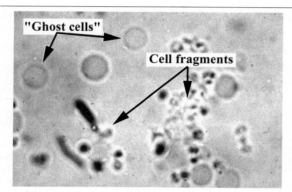

Figure 5.22

160 In reference to **Figure 5.22**, what is the tonicity of the solution used for the dilution of the red blood cells? _____ Explain your answer. _____

LAB ACTIVITY

Osmosis and Potato Cells

161 Define turgor. _____

162 Osmotic solutions of different tonicities than the cell alter the cell's water content. The resulting changes in turgidity changes the _____ of the tissue. Plant cells that lose water become _____. Plant cells that gain water become _____.

163 Define plasmolysis. _____

164 What tonicity of osmotic solution causes plasmolysis? _____

165 What tonicity of osmotic solution causes increased turgor in flaccid plant cells? _____

166 How did the rigidity of the slices of potato placed in the distilled water change? _____

167 What is the tonicity of the distilled water? _____

168 How did the rigidity of the slice of potato placed in the 10% NaCl change? _____

169 What is the tonicity of the 10% NaCl solution? _____

LAB ACTIVITY

Osmosis and Elodea

170 Elodea is commonly found growing in fresh water streams and ponds. What is the tonicity of these environments? _____

171 What tonicity of solution would cause plasmolysis of Elodea? _____

172 What physical changes result to the cells of Elodea when plasmolysis occurs? _____

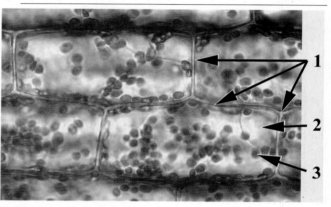

Figure 5.23

173 In reference to **Figure 5.23**, what is the tonicity of the solution the Elodea cells are subjected to? _____
174 In reference to **Figure 5.23**, identify #1 - #3.
1 _____
2 _____
3 _____

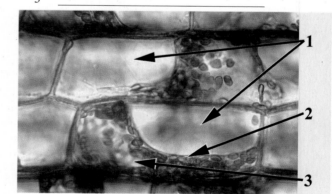

Figure 5.24

175 In reference to **Figure 5.24**, what is the tonicity of the solution the Elodea cells are subjected to? _____
176 In reference to **Figure 5.24**, identify #1 - #3.
1 _____
2 _____
3 _____

Name _____
Class _____

Chapter 5 - Transport Processes **83**

LAB ACTIVITY

Osmosis and Paramecium

177 Fresh water Protozoa such as Paramecia and Amoeba live in what type of osmotic environment? _____

178 In which direction does fluid osmotically move, into or out of, the Paramecium? _____

179 What organelles maintain cytoplasmic osmolarity? _____

180 When are contractile vacuoles easiest to observe? _____

181 Is the rate of contractile vacuole fluid expulsion related to the tonicity of the environment? _____ Explain your answer. _____

182 What would happen to the shape of Paramecium if contractile activity ceased? _____

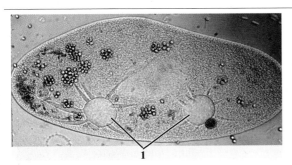

Figure 5.25
183 In reference to **Figure 5.25**, identify #1. _____

DIALYSIS

184 Define dialysis. _____

185 What characteristic of the membrane determines if a solute will pass or be restricted? _____

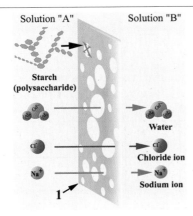

Figure 5.26
186 In reference to **Figure 5.26**, identify #1. _____

187 In reference to **Figure 5.26**, why doesn't starch diffuse through the membrane? _____

188 In reference to **Figure 5.26**, what **solutes** are being separated from solution "A?" _____

LAB ACTIVITY

Osmosis and Dialysis

Osmosis (using dialysis tubing - membrane)

189 What is dialysis tubing? _____

190 The dialysis bag containing 10% starch and 10% NaCl is placed in an environment of distilled water. What is the tonicity of the distilled water? _____

191 Explain why the weight of the bag changed. _____

Starting weight	15 minutes	30 minutes	45 minutes	60 minutes	75 minutes	90 minutes

Figure 5.27
192 Table of weight changes
193 Record and graph the weight changes for the egg placed into distilled water in the following time-weight graph.

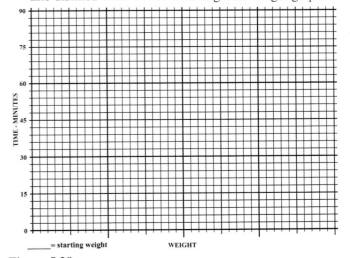

_____ = starting weight   WEIGHT

Figure 5.28
Graph of results of time-weight change.

194 In reference to **Figure 5.28**, your results showed that over time, the rate of the net diffusion of water:
A. increased
B. increased, then decreased, then increased
C. decreased
D. decreased, then increased, then decreased
E. remained the same
F. was sporadic

84 Chapter 5 - Transport Processes

195 Is your answer to the preceding question what was expected from an analysis of the experiment's materials and setup?   Explain your answer. _____
_____
_____

Dialysis (using dialysis tubing - membrane)

196 What is a control test? _____
_____

197 Which **solute** passed from the dialysis bag into the distilled water? _____

198 Which **solute** was restricted from passing out of the dialysis bag into the distilled water? _____

199 Complete **Figure 5.29**, reaction results - dialysis of NaCl.

NaCl Dialysis	-Control- distilled water	-Control- 10% NaCl	Solution from beaker
Addition of silver nitrate ($AgNO_3$)	White precipitate Yes / No	White precipitate Yes / No	White precipitate Yes / No
Presence of NaCl	Yes / No	Yes / No	Yes / No

Figure 5.29
Dialysis - table of reaction results for dialysis of NaCl.

200 Complete **Figure 5.30**, reaction results - dialysis of starch

Starch Dialysis	-Control- distilled water	-Control- 10% Starch	Solution from beaker
Addition of Lugol's iodine	Dark blue product Yes / No	Dark blue product Yes / No	Dark blue product Yes / No
Presence of starch	Yes / No	Yes / No	Yes / No

Figure 5.30
Table of reaction results for dialysis of starch.

LAB ACTIVITY

OSMOSIS AND DIALYSIS

Osmosis (using unshelled egg)

201 What tonicity is the storage solution to the egg? _____
202 Compared an egg's normal hydrostatic pressure, the hydrostatic pressure of an egg in the storage solution is _____ (less, the same, more).
203 What tonicity is the distilled water to the egg? _____
204 What happens to the hydrostatic pressure of an egg placed in distilled water? _____
205 Record the weight changes for an egg placed into distilled water in the following table.

Starting weight	15 minutes	30 minutes	45 minutes	60 minutes	75 minutes	90 minutes

Figure 5.31
Table of results of time-weight changes

206 Record and graph the weight changes for the egg placed into distilled water in the following time-weight graph.

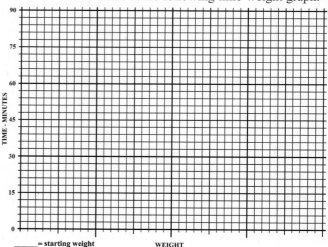

Figure 5.32
Graph of results of time-weight changes.

207 In reference to **Figure 5.32**, your results showed that over time, the rate of the net diffusion of water:
   A. increased
   B. increased, then decreased, then increased
   C. decreased
   D. decreased, then increased, then decreased
   E. remained the same
   F. was sporadic

208 Is your answer to the preceding question what is expected from an analysis of the experiment's materials and setup?_____ Explain your answer. _____
_____
_____

Dialysis (using unshelled egg)

209 During the time the egg was in the storage solution, various _____ from the solution _____ through the egg's shell membrane into the egg.

210 During the time the egg was placed in distilled water, various _____ diffused from the egg into the distilled water.

211 Define dialysis. _____
_____
_____

212 What is a control test? _____
_____

Test for sugar (monosaccharide, fructose)

213 Record in the following table the results for the test for sugar.

Test Experiment	Storage solution	Eggwhite	Egg's test solution after one hour
Sugar	Present Absent	Present Absent	Present Absent

Figure 5.33
Test results for sugar

214 The test results indicate that sugar was
 A. present in the storage solution
 B. present in eggwhite
 C. permeable to the egg's shell membrane
 D. impermeable to the egg's shell membrane
 E. Two answers are correct: _____ and _____

Test for starch (boiled corn starch)

215 Record in the following table the results for the test for starch.

Test Experiment	Storage Solution	Eggwhite	Egg's solution after one hour
Starch	Present Absent	Present Absent	Present Absent

Figure 5.34
Test results for starch

216 The test results indicate that starch was
 A. present in the storage solution
 B. present in eggwhite
 C. permeable to the egg's shell membrane
 D. impermeable to the egg's shell membrane
 E. Two answers are correct: _____ and _____

Test for NaCl

217 Record in the following table the results for the test for NaCl.

Test Experiment	Storage Solution	Eggwhite	Egg's solution after one hour
NaCl	Present Absent	Present Absent	Present Absent

Figure 5.35
Test results for NaCl

218 The test results indicate that NaCl was
 A. present in the storage solution
 B. present in eggwhite
 C. permeable to the egg's shell membrane
 D. impermeable to the egg's shell membrane
 E. Two answers are correct: _____ and _____

Test for albumin

219 Record in the following table the results for the test for albumin.

Test Experiment	Storage Solution	Eggwhite	Egg's solution after one hour
Albumin	Present Absent	Present Absent	Present Absent

Figure 5.36
Test results for albumin

220 The test results indicate that albumin was
 A. present in the storage solution
 B. present in eggwhite
 C. permeable to the egg's shell membrane
 D. impermeable to the egg's shell membrane
 E. Two answers are correct: _____ and _____

FLUID MOVEMENT ACROSS THE CAPILLARY

221 What are the smallest blood vessels? _____

222 What is the function of blood capillaries? _____

223 What are two major forces for the exchange of water between the plasma (at the capillary) and the interstitial fluid _____

224 Define hydrostatic pressure. _____

225 What is blood pressure? _____

226 What is the vascular route from the heart? _____

227 What happens to blood pressure as it flows along the vascular route? (Increases, decreases, remains the same) Explain your answer. _____

228 Which end of the capillary has a slightly higher blood pressure? _____

229 What does hydrostatic pressure of the capillary (capillary blood pressure) promote? _____

230 What does hydrostatic pressure of the interstitial fluid promote? _____

231 What impermeable solute is a major contributor to osmotic pressure? _____

232 What does blood colloid osmotic pressure promote? _____

233 What does interstitial fluid osmotic pressure promote? _____

234 How is net filtration pressure determined? _____

235 What is the major force that produces net filtration pressure at the arterial end of the capillary? _____

236 Filtration pressure at the arterial end of the capillary results in the movement of fluid _____ the capillary.

237 What is the major force that produces net filtration pressure at the venous end of the capillary? _____

238 Filtration pressure at the venous end of the capillary results in the movement of fluid _____ the capillary.

86 Chapter 5 - Transport Processes

Fluid Movement at the Arterial End of the Capillary

239 At the arterial end of the capillary, which of the two hydrostatic pressures is the greatest? _____

240 In what direction does blood hydrostatic pressure (capillary blood pressure) promote fluid movement? _____

241 At the arterial end of the capillary, which of the two osmotic pressures is the greatest? _____

242 In what direction does blood colloid osmotic pressure promote fluid movement? _____

243 How is the net filtration pressure at the arterial end of the capillary determined? _____

244 What is the largest contributor to net filtration pressure? _____

245 In what direction does net filtration pressure promote fluid movement? _____

Fluid Movement at the Venous End of the Capillary

246 At the venous end of the capillary, which of the two hydrostatic pressures is the greatest? _____

247 In what direction does blood hydrostatic pressure (capillary blood pressure) promote fluid movement? _____

248 At the venous end of the capillary, which of the two osmotic pressures is the greatest? _____

249 In what direction does blood colloid osmotic pressure promote fluid movement? _____

250 How is the net filtration pressure at the venous end of the capillary determined? _____

251 At the venous end of the capillary, what is the largest contributor to net filtration pressure? _____

252 In what direction does net filtration pressure promote fluid movement? _____

253 What happens to fluids that are not returned into the venous end of the capillary? _____

254 Where do fluids of the lymphatic system drain? _____

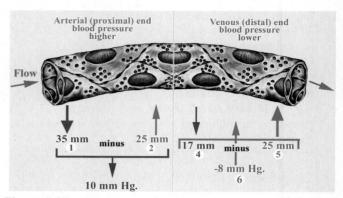

Figure 5.37

255 In reference to **Figure 5.37**, identify the "pressure" and "value" for #1 - 6.

	Pressure	Value
1	_____	_____
2	_____	_____
3	_____	_____
4	_____	_____
5	_____	_____
6	_____	_____

ACTIVE TRANSPORT PROCESSES

256 When the cell expends _____ for transport, then the process is called _____.

257 Active transport requires _____ to provide the mechanism of solute movement across the plasma membrane.

258 Vesicular transport requires that the substances be moved across the plasma membrane in _____ (sacs) called _____.

259 What are the two types of vesicular transport? _____

260 What is a function of solute pumps? _____

261 With respect to a diffusion gradient, solute pumps typically transport their solutes in which direction? _____

262 What produces obligatory water movements? _____

Membrane Potentials

263 What effect do passive processes have upon a concentration gradient? _____

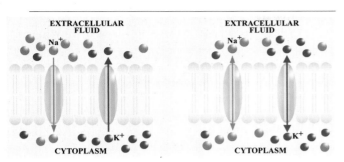

Figure 5.38

264 In reference to **Figure 5.38**, identify the process shown at #1. _____ Is this process active or passive?

265 In reference to **Figure 5.38**, what happens to the concentration gradient shown in Fig. A? _____

266 Excitable cells such as neurons and muscles, utilize energy (ATP) to actively maintain _____ _____ by membrane solute pumps.

267 What is the function of the sodium-potassium pump? _____

268 What concentration gradients are maintained by the sodium-potassium pump?

269 What do nerve cells maintain to produce the electrical signals of the nervous system, nerve impulses? _____

270 A EKG, or ECG, shows the _____ activity of the heart.

271 The maintenance of electrolyte gradients needed to produce and maintain the heart's electrical gradients are due to _____ _____ pumps.

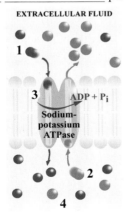

Figure 5.39

272 In reference to **Figure 5.39**, identify #1 - #6.
1 _____
2 _____
3 _____
4 _____
5 _____
6 _____

LAB ACTIVITY

Active Transport in Yeast

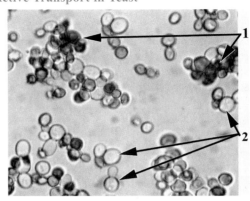

Figure 5.40

273 In reference to **Figure 5.40**, identify #1 & #2.
1 _____
2 _____

274 In reference to **Figure 5.40**, why didn't all the yeast stain with congo red? _____

Active Processes
Vesicular Transport

275 _____ requires that substances be moved either into or out of the cell in membranous pouches (sacs) called vesicles.

276 Vesicular transport is a (an) _____ (active or passive) process.

Exocytosis

277 What is exocytosis? _____

278 What are common substances that are exocytosed? _____

279 What is secretion? _____

280 What are two ways a cell can secrete substances?
1 _____
2 _____

Chapter 5 - Transport Processes

281 List four substances that are secreted by exocytosis?
   1 _____
   2 _____
   3 _____
   4 _____

282 Name a location in the body where cellular ion secretion, often hormonally controlled, maintains blood pH and electrolyte concentration. _____

283 What is excretion? _____

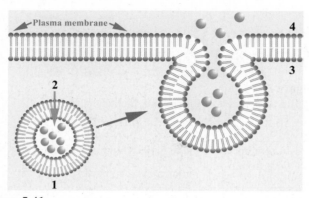

Figure 5.41

284 In reference to **Figure 5.41**, identify #1 - #4.
   1 _____
   2 _____
   3 _____
   4 _____

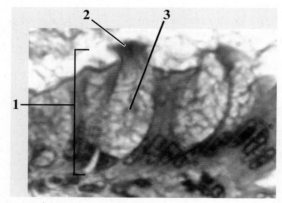

Figure 5.42

285 In reference to **Figure 5.42**, identify #1 - #3.
   1 _____
   2 _____
   3 _____

286 In reference to **Figure 5.42**, where did the substances found in #3 most likely originate? _____

ENDOCYTOSIS

287 What is endocytosis? _____

288 What are three types of endocytosis?
   1 _____
   2 _____
   3 _____

PHAGOCYTOSIS

289 What is phagocytosis? _____

290 What are pseudopods? _____

291 What are phagosomes? _____

292 What happens to the contents of phagosomes? _____

293 Name a phagocyte found in the body. _____

294 How do phagocytes move? _____

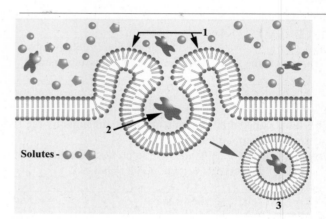

Figure 5.43

295 In reference to **Figure 5.43**, identify #1 - #3.
   1 _____
   2 _____
   3 _____

LAB ACTIVITY

Microscopic observation of macrophages of the liver, Kupffer's cells

296 What are Kupffer's cells? _____

297 What is the function of Kupffer's cells? _____

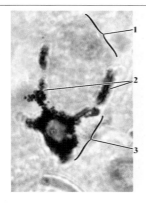

Figure 5.44
298 In reference to **Figure 5.44**, identify #1 - #3.
  1 _____
  2 _____
  3 _____

LAB ACTIVITY

Microscope observation of living amoeba for amoeboid movement and phagocytosis.

299 What are amoebas? _____

300 How do amoebas move? _____

301 How do amoebas obtain their food? _____

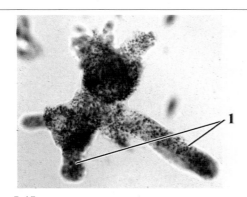

Figure 5.45
302 In reference to **Figure 5.45**, identify #1. _____

LAB ACTIVITY

Microscope observation of Paramecium for phagocytosis.

303 What are Paramecia? _____

304 How do Paramecia obtain their food? _____

305 After a short time, the yeasts contained within the food vacuoles (phagosomes) turn to what color? _____
What does the change of color indicate? _____

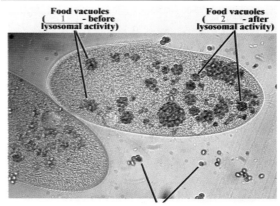

Figure 5.46
306 In reference to **Figure 5.46**, identify #1 - #3.
  1 _____
  2 _____
  3 _____

90 Chapter 5 - Transport Processes

PINOCYTOSIS

307 What is pinocytosis? _____

308 Is pinocytosis specific or nonspecific? _____
309 What are endosomes? _____

310 What is transcytosis? _____

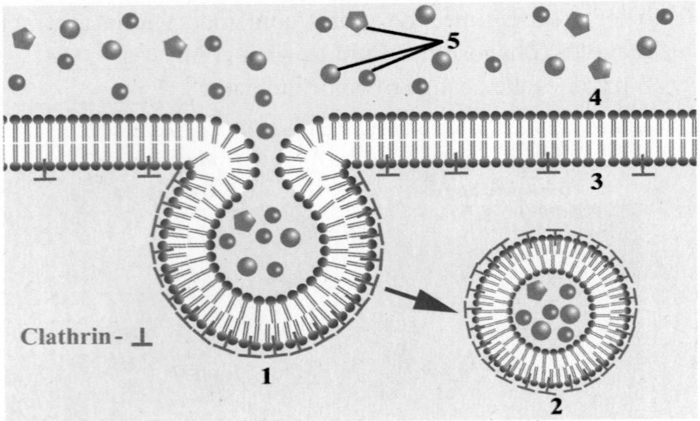

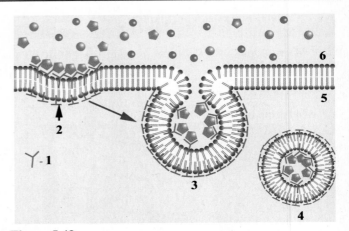

Figure 5.47
311 In reference to **Figure 5.47**, identify #1 - #5.
    1 _____   4 _____
    2 _____   5 _____
    3 _____

RECEPTOR-MEDIATED ENDOCYTOSIS

312 What determines which substances a cell will engulf? ___
_____

313 What are two common cell membrane receptors? _____
_____

314 After the cells receptors are bound to their specific molecules, the coated pit _____ to form a membranous _____.

315 What is an endosome? _____
_____

316 What are two possible destinations for endosomes?
    1 _____
    _____
    2 _____

Figure 5.48
317 In reference to **Figure 5.48**, identify #1- #6.
    1 _____   4 _____
    2 _____   5 _____
    3 _____   6 _____

Name _____
Class _____

Chapter 6 - Cell Division   91

Cell Division - Worksheets

1. What two cell components does cell division involve? _____

2. What are two types of nuclear division? _____

3. How many homologous chromosome pairs do human somatic cells contain? _____

4. What was the origin of each chromosome of a homologous pair? _____

5. What are homologous chromosomes? _____

6. What is a diploid (2n) cell? _____

7. What is a haploid (n) cell? _____

8. In what part of the cell does mitosis occur? _____

9. What is the result of mitosis? _____

10. What is cytokinesis? _____

11. What is meiosis? _____

12. What happens to the total number of chromosomes as a result of meiosis? _____

13. Where does meiosis occur, and what does it produce? _____

14. As a result of meiosis, gametes contain only chromosomes that are _____ (same or different).

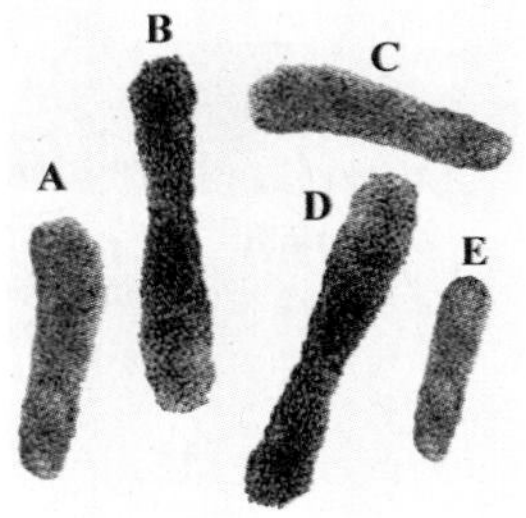

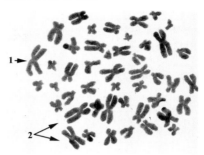

Figure 6.2

17. In reference to **Figure 6.2**, human chromosomes from a cell arrested in metaphase, how many chromosomes are shown? _____ Is it possible to determine if this cell was from a male or female? _____ Explain. _____

18. In reference to **Figure 6.2**, identify #1 - #2.
    1 _____
    2 _____

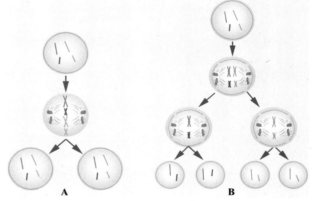

Figure 6.3

19. In reference to **Figure 6.3**, which illustration, **A** or **B** shows meiosis? _____

20. In reference to **Figure 6.3**, which illustration, A or B shows mitosis? _____

21. In reference to **Figure 6.3**, are the parent cells haploid or diploid? _____ Are the daughter cells of illustration **A** haploid or diploid? _____ Are the daughter cells of illustration **B** haploid or diploid? _____

Figure 6.1

15. In reference to **Figure 6.1**, identify the pairs of homologous chromosomes by matching their letters (such as F-G). _____

16. One member of each homologous chromosome was inherited from each _____.

CELL LIFE CYCLE

22. What are the two major parts of the life cycle of a cell? _____

23. What is interphase? _____

24. If a cell divides, it nuclear division will be by either _____.

25. What are the four life cycle phases of a mitotic cell? _____ What type of cells does G_0 describe? _____

26. What happens during the G_1 phase? _____

92 Chapter 6 - Cell Division

27 What happens during the S phase? _____
_____
28 What happens during the G$_2$ phase? _____
_____
29 What happens during the M phase? _____
_____
30 Name two types of cells that are in the G$_0$ phase. _____

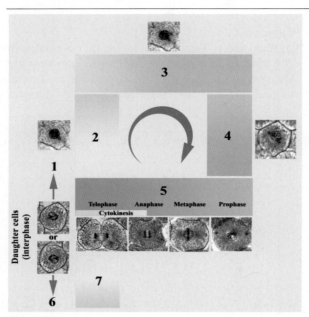

Figure 6.4

31 In reference to **Figure 6.4**, what is the destiny of cells at #1? _____
32 In reference to **Figure 6.4**, what phase is #2? _____
This phase can be simply described as_____.
33 In reference to **Figure 6.4**, what phase is #3? _____
This phase can be simply described as_____.
34 In reference to **Figure 6.4**, what phase is #4? _____
This phase can be simply described as_____.
35 In reference to **Figure 6.4**, what phase is #5? _____
This phase can be simply described as_____.
36 In reference to **Figure 6.4**, what is the destiny of cells at #6? _____
37 In reference to **Figure 6.4**, what phase is #7? _____
This phase can be simply described as_____.

DNA Replication

38 In what phase of interphase is DNA replicated? _____
39 Why is DNA replicated? _____
40 What are the three building blocks of a DNA nucleotide?
1 _____
2 _____
3 _____
41 What are the four nitrogen bases of DNA?
1 _____ 3 _____
2 _____ 4 _____
42 What are the complementary base pairs? _____
_____
43 A molecule of DNA consists of _____ polynucleotide strands that are _____ bonded at their complementary bases.
44 How does DNA replication begin? _____
_____
45 After separation of the two DNA polynucleotide strands, each strand is then used as a _____
_____
46 Enzymes called DNA _____ bond the nucleotides to form each of the two _____
_____
47 What is a chromatid? _____
_____
48 Each chromatid consists of an original strand of the DNA molecule and a _____.
49 What is the function of the centromere? _____

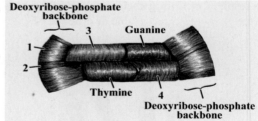

Figure 6.5

50 In reference to **Figure 6.5**, identify #1 - 4.
1 _____ 3 _____
2 _____ 4 _____
51 A replicated chromosome consists of two _____ attached at the _____.
52 When are the chromatids referred to as chromosomes? ___
_____
_____

Name _____
Class _____

Chapter 6 - Cell Division 93

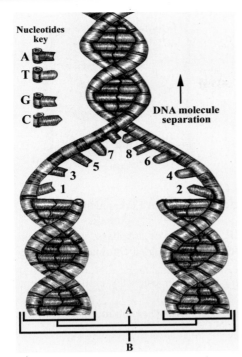

Figure 6.6

53 In reference to **Figure 6.6**, use the "nucleotides key," in the upper left of the illustration and identify the nucleotides for # 1 - 8.
   1 _____ 2 _____
   3 _____ 4 _____
   5 _____ 6 _____
   7 _____ 8 _____
54 In reference to **Figure 6.6**, identify "**A**." _____
55 In reference to Figure **6.6**, identify "**B**." _____
56 When are the chromatids referred to as chromosomes? _____
_____

CELL INTERPHASE AND MITOSIS
INTERPHASE

57 In preparation for mitosis, during interphase chromosomes and centrioles are _____.

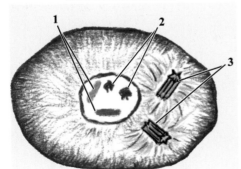

Figure 6.7

58 In reference to **Figure 6.7**, is the cell in early or late interphase? _____
59 In reference to **Figure 6.7**, identify #1 - #3.
   1 _____ 3 _____
   2 _____

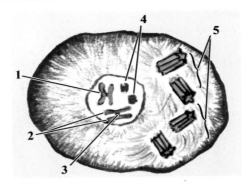

Figure 6.8

60 In reference to **Figure 6.8**, is the cell in early or late interphase? _____
61 In reference to **Figure 6.8**, identify #1 - #5.
   1 _____ 4 _____
   2 _____ 5 _____
   3 _____

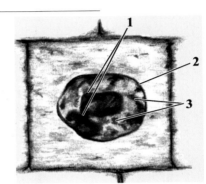

Figure 6.9

62 In reference to **Figure 6.9**, is the cell from a plant or animal? _____
63 In reference to **Figure 6.9**, identify #1 - #3.
   1 _____ 3 _____
   2 _____

MITOSIS
PROPHASE

64 During prophase what happens to the nuclear membrane? _____
_____

65 During prophase, how are the chromosomes arranged? ___
_____

66 During prophase, what happens to the nucleoli? _____

67 During prophase, the centrioles (centrosomes) move to form the _____ of the mitotic spindle.

68 What are the two types of fibers that appear from the centrosomes? _____
_____

69 What are two types of spindle fibers? _____

70 Where are kinetochore microtubules located? _____

71 What is the function of kinetochore microtubules? _____

72 Where are nonkinetochore microtubules located? _____
_____

94 Chapter 6 - Cell Division

73 Where are aster fibers located? _____
_____

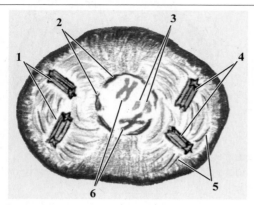

Figure 6.10

74 In reference to **Figure 6.10**, identify #1 - #6.
  1 _____   4 _____
  2 _____   5 _____
  3 _____   6 _____

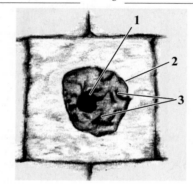

Figure 6.11

75 In reference to **Figure 6.11**, identify #1 - #3.
  1 _____
  2 _____
  3 _____

METAPHASE

76 During metaphase, how are the chromosomes arranged?
_____
_____

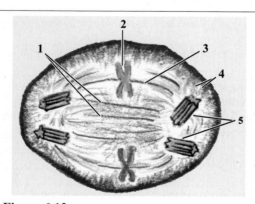

Figure 6.12

77 In reference to **Figure 6.12**, identify #1 - #5.
  1 _____   4 _____
  2 _____   5 _____
  3 _____

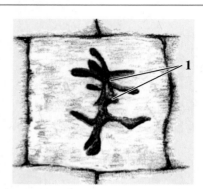

Figure 6.13

78 In reference to **Figure 6.13**, identify #1. _____

ANAPHASE

79 At the beginning of anaphase, what happens to the centromeres and associated chromatids? _____
_____

80 During anaphase, when the chromatids separate they are called _____.

81 During anaphase, where do the sister chromosomes move?
_____
_____

82 When does anaphase end? _____
_____

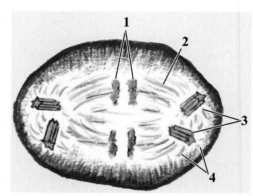

Figure 6.14

83 In reference to **Figure 6.14**, identify #1 - #4.
  1 _____   3 _____
  2 _____   4 _____

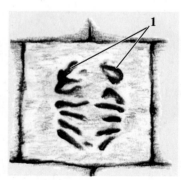

Figure 6.15

84 In reference to **Figure 6.15**, identify #1. _____

TELOPHASE

85 When does telophase begin? _____
_____
86 During telophase, the nuclear membrane _____.
87 During telophase, the chromosomes become dispersed into an organized form called _____.
88 During telophase nucleoli are _____.
89 During telophase, the mitotic spindle and aster fibers _____.

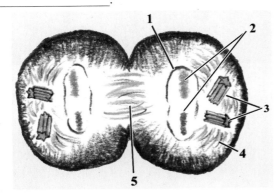

Figure 6.16

90 In reference to **Figure 6.16**, identify #1 - #5.
    1 _____   4 _____
    2 _____   5 _____
    3 _____

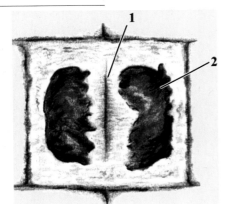

Figure 6.17

91 In reference to **Figure 6.17**, identify #1 - #2.
    1 _____   2 _____

DAUGHTER CELLS

92 What are daughter cells? _____
_____
93 What is cytokinesis? _____
_____
94 When does cytokinesis begin? _____
_____
95 A plant cell produces two daughter cells by the formation of a _____.

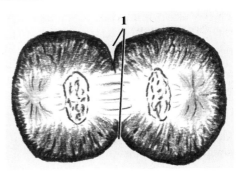

Figure 6.18

96 In reference to **Figure 6.18**, the cell shown is from a/an _____ (plant or animal).
97 In reference to **Figure 6.18**, identify #1. _____

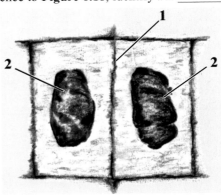

Figure 6.19

98 In reference to **Figure 6.19**, the cell shown is from a/an _____ (plant or animal).
99 In reference to **Figure 6.19**, identify #1 - #2.
    1 _____
    2 _____

100 In reference to **Figure 6.20**, identify each phase of the cell's life cycle and explain the **observed** characteristics that the phase exhibits.
    A Identify phase 1 _____
       Characteristics shown: _____
       _____

    B Identify phase 2 _____
       Characteristics shown: _____
       _____

    C Identify phase 3 _____
       Characteristics shown: _____
       _____

    D Identify phase 4 _____
       Characteristics shown: _____
       _____

    E Identify phase 5 _____
       Characteristics shown: _____
       _____

96   Chapter 6 - Cell Division

F   Identify phase 6 _____
    Characteristics shown: _____
    _____

G   Identify phase 7 _____
    Characteristics shown: _____
    _____

H   Identify phase 8 _____
    Characteristics shown: _____
    _____

I   Identify phase 9 _____
    Characteristics shown: _____
    _____

J   Identify phase 10 _____
    Characteristics shown: _____
    _____

K   Identify phase 11 _____
    Characteristics shown: _____
    _____

L   Identify phase 12 _____
    Characteristics shown: _____
    _____

M   Identify phase 13 _____
    Characteristics shown: _____
    _____

N   Identify phase 14 _____
    Characteristics shown: _____
    _____

O   Identify phase 15 _____
    Characteristics shown: _____
    _____

P   Identify phase 16 _____
    Characteristics shown: _____
    _____

Q   Identify phase 17 _____
    Characteristics shown: _____
    _____

R   Identify phase 18 _____
    Characteristics shown: _____
    _____

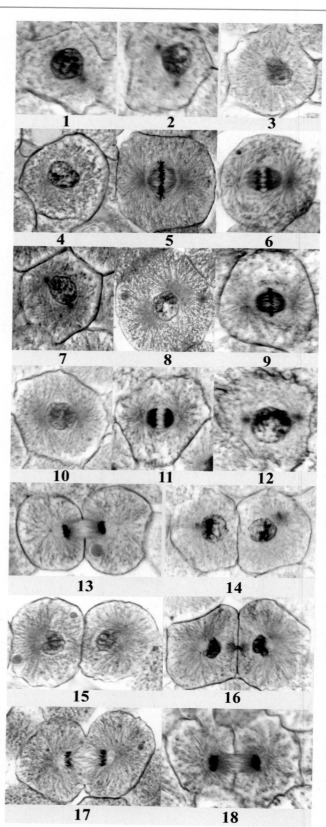

Figure 6.20

Name _____
Class _____

Chapter 6 - Cell Division   97

PLANT CELLS

101 In reference to **Figure 6.21**, identify each phase of the cell's life cycle and explain the **observed** characteristics that the phase exhibits.

A   Identify phase 1 _____
    Characteristics shown: _____
    _____
    _____

B   Identify phase 2 _____
    Characteristics shown: _____
    _____
    _____

C   Identify phase 3 _____
    Characteristics shown: _____
    _____
    _____

D   Identify phase 4 _____
    Characteristics shown: _____
    _____
    _____

E   Identify phase 5 _____
    Characteristics shown: _____
    _____
    _____

F   Identify phase 6 _____
    Characteristics shown: _____
    _____
    _____

G   Identify phase 7 _____
    Characteristics shown: _____
    _____
    _____

H   Identify phase 8 _____
    Characteristics shown: _____
    _____
    _____

I   Identify phase 9 _____
    Characteristics shown: _____
    _____
    _____

J   Identify phase 10 _____
    Characteristics shown: _____
    _____
    _____

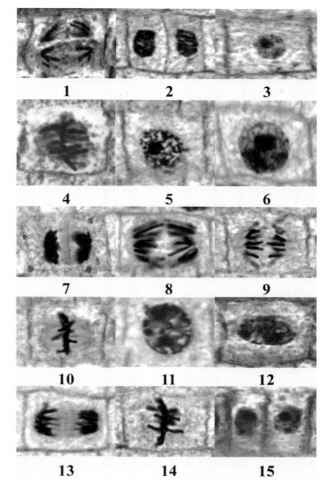

Figure 6.21

K   Identify phase 11 _____
    Characteristics shown: _____
    _____
    _____

L   Identify phase 12 _____
    Characteristics shown: _____
    _____
    _____

M   Identify phase 13 _____
    Characteristics shown: _____
    _____
    _____

N   Identify phase 14 _____
    Characteristics shown: _____
    _____
    _____

O   Identify phase 15 _____
    Characteristics shown: _____
    _____
    _____

Chapter 6 - Cell Division

MEIOSIS AND GAMETE PRODUCTION

102 What is meiosis? _____

103 In meiosis, which chromosomes are distributed to the daughter cells? _____

104 What is the function of the haploid (n) cells formed by meiosis? _____

105 What does the term 'diploid' mean? _____

106 What is the diploid number of chromosomes for humans? _____

107 What is a homologous chromosome? _____

108 How many pairs of homologous chromosomes are found in humans? _____

109 How is the haploid number of chromosomes described? _____

110 What is the haploid chromosome number in humans? _____

111 Before meiosis begins what happens to the chromosomes? _____

112 What are the two stages of meiosis? _____

113 In meiosis I, the replicated chromosomes are first paired, in an event called _____.

114 During synapsis _____ of portions of the chromosomes occurs.

115 After crossing-over, one replicated chromosome (a homologous chromosome) is distributed to _____.

116 The daughter cells that enter meiosis II have one of _____ of chromosome in the _____ form.

117 In meiosis II the replicated chromosomes are _____ and one of each sister chromosome is distributed to a _____, which function as _____.

118 Fusion of gametes, a sperm and an egg, produces a _____ number of chromosomes in a cell called a _____.

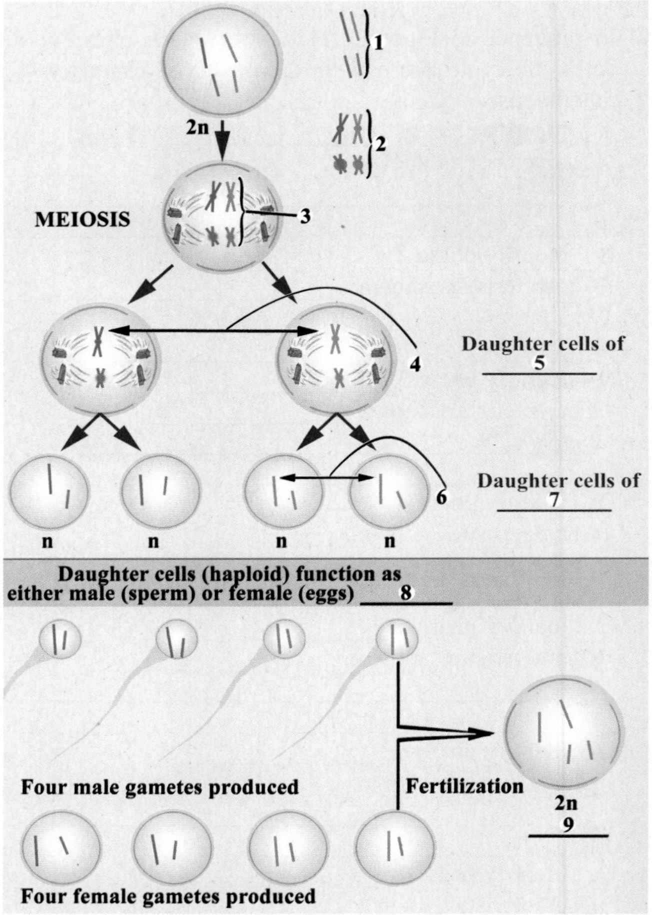

Figure 6.22

119 In reference to **Figure 6.22**, identify #1 - #9.
1 _____ 6 _____
2 _____ 7 _____
3 _____ 8 _____
4 _____ 9 _____
5 _____

MEIOSIS - DESCRIPTIVE

120 Meiosis I produces ____ daughter cells. The daughter cells are _____ (haploid or diploid) and contain one each of the _____.

121 What are two events that lead to genetic variations?
1 _____

2 _____

Name _____
Class _____

Chapter 6 - Cell Division   99

122 Meiosis I is divided into the stages called:
1 _____ 2 _____
3 _____ 4 _____
123 What are six events of prophase I?
1 _____
2 _____
3 _____
4 _____
5 _____
6 _____
124 What is a synapse? _____
_____
125 What is a tetrad? _____
_____
126 What is crossing-over? _____
_____
_____
127 What characterizes metaphase I? _____
_____
128 What characterizes anaphase I? _____
_____
129 What characterizes telophase I? _____
_____
_____
130 What characterizes the daughter cells of meiosis I? _____
_____
_____
131 Meiosis II can produce a total of how many daughter cells?
_____ How many functional daughter cells are produced by the female? _____
132 Each daughter cell is _____ (haploid or diploid) and contains _____ of each homologous chromosome.
133 Meisosis II is divided into the stages called:
1 _____ 2 _____
3 _____ 4 _____
134 What are five events of prophase II?
1 _____
2 _____
3 _____
4 _____
5 _____
135 What characterizes metaphase II? _____
_____
136 What characterizes anaphase II? _____
_____
137 What characterizes telophase II? _____
_____
138 What characterizes the daughter cells of meiosis II? _____
_____
_____

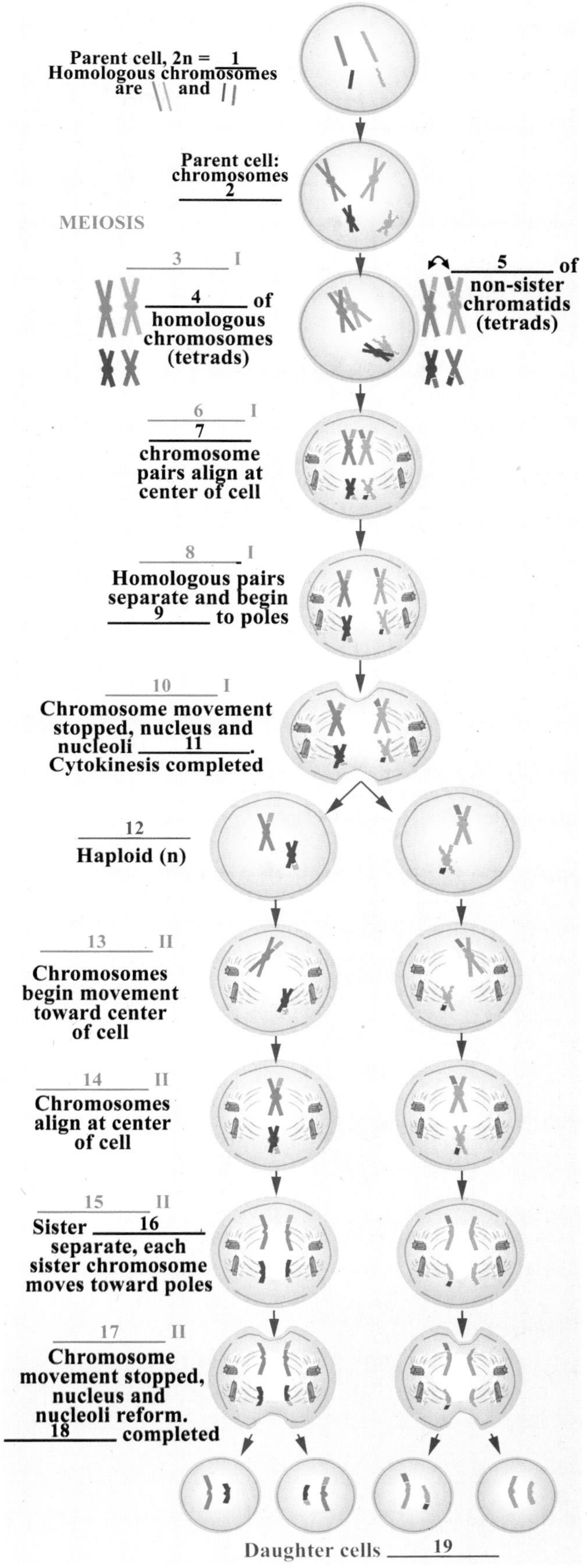

Parent cell, 2n = __1__
Homologous chromosomes are \\ and //

Parent cell: chromosomes __2__

MEIOSIS

__3__ I
__4__ of homologous chromosomes (tetrads)

__5__ of non-sister chromatids (tetrads)

__6__ I
__7__ chromosome pairs align at center of cell

__8__ I
Homologous pairs separate and begin __9__ to poles

__10__ I
Chromosome movement stopped, nucleus and nucleoli __11__. Cytokinesis completed

__12__
Haploid (n)

__13__ II
Chromosomes begin movement toward center of cell

__14__ II
Chromosomes align at center of cell

__15__ II
Sister __16__ separate, each sister chromosome moves toward poles

__17__ II
Chromosome movement stopped, nucleus and nucleoli reform. __18__ completed

Daughter cells __19__

Figure 6.23

100 Chapter 6 - Cell Division

139 In reference to **Figure 6.23**, identify #1 - #19.

1 _____ 11 _____
2 _____ 12 _____
3 _____ 13 _____
4 _____ 14 _____
5 _____ 15 _____
6 _____ 16 _____
7 _____ 17 _____
8 _____ 18 _____
9 _____ 19 _____
10 _____

Meiosis - Spermatogenesis

140 Precursor (stem) cells are located at the periphery of the _____ and undergo _____ divisions.

141 Some of the daughter spermatogonia differentiate into primary _____.

142 Primary spermatocytes enter the stage of division called _____ and produce two (haploid / diploid) daughter cells called _____.

143 The two secondary spermatocytes enter the stage of division called _____ and each produces two (haploid / diploid) cells called _____.

144 During _____ spermatids are transformed into _____.

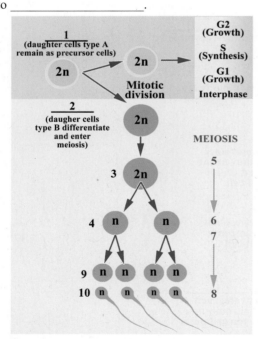

Figure 6.24

145 In reference to **Figure 6.24**, identify #1 - #10.

1 _____ 6 _____
2 _____ 7 _____
3 _____ 8 _____
4 _____ 9 _____
5 _____ 10 _____

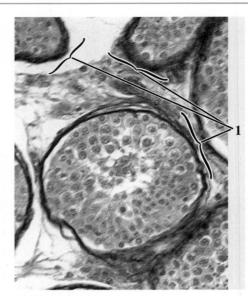

Figure 6.25

146 In reference to **Figure 6.25**, identify #1.

1 _____

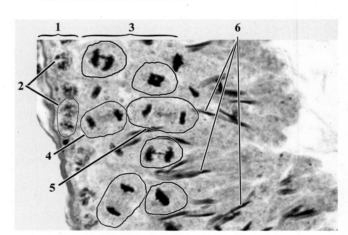

Figure 6.26

147 In reference to **Figure 6.26**, identify #1 - #6.

1 _____ 4 _____
2 _____ 5 _____
3 _____ 6 _____

Chapter 6 - Cell Division   101

148 From the spermatogonium, list in sequence the stages of spermatogenesis. _____
_____

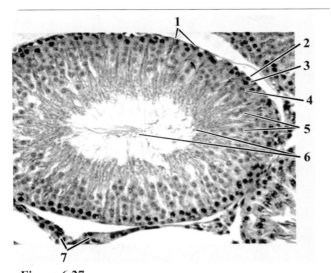

Figure 6.27
149 In reference to **Figure 6.27**, identify #1 - #7.
1 _____ 5 _____
2 _____ 6 _____
3 _____ 7 _____
4 _____

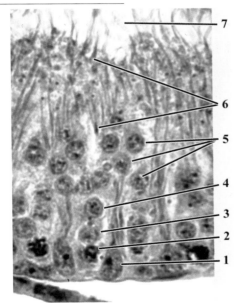

Figure 6.28
150 In reference to **Figure 6.28**, identify #1 - #7.
1 _____ 5 _____
2 _____ 6 _____
3 _____ 7 _____
4 _____

MEIOSIS - OOGENESIS

151 What is oogenesis? _____
_____

152 When does oogenesis begin? _____
_____

153 At birth and until puberty, the ovary contains several million _____.

154 What hormones target receptive primordial follicles and stimulate their development? _____
_____

155 Primordial follicles develop into _____ follicles, characterized by several layers of _____ cells surrounding a _____ oocyte.

156 A primary oocyte completes meiosis I to form two cells, the first polar body and the _____ oocyte.

157 The secondary oocyte arrested in meiosis II is ovulated, and completes meiosis II if _____ occurs.

158 Nuclear union between the ovum and the sperm produces a _____.

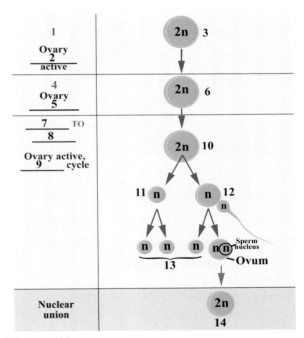

Figure 6.29
159 In reference to **Figure 6.29**, identify #1 - #14.
1 _____ 8 _____
2 _____ 9 _____
3 _____ 10 _____
4 _____ 11 _____
5 _____ 12 _____
6 _____ 13 _____
7 _____ 14 _____

102   Chapter 6 - Cell Division

Name _____
Class _____

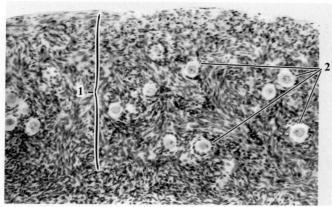

Figure 6.30

160 Is the ovary shown in **Figure 6.30** mature or immature?
_____

161 In reference to **Figure 6.30**, identify #1 - #2.
   1 _____ 2 _____

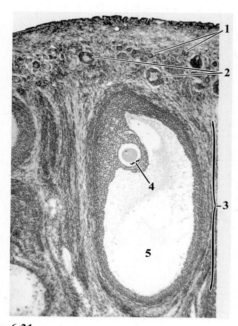

Figure 6.31

162 Is the ovary shown in **Figure 6.31** mature or immature?
_____

163 In reference to **Figure 6.31**, identify #1 - #5.
   1 _____ 4 _____
   2 _____ 5 _____
   3 _____

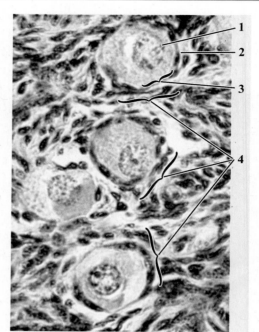

Figure 6.32

164 In reference to **Figure 6.32**, identify #1 - #4.
   1 _____ 3 _____
   2 _____ 4 _____

Name _____
Class _____

Protein Synthesis - Worksheets

1. How does DNA determine the structure and activity of the cell? _____

2. What are three functions of proteins?
   1 _____
   2 _____
   3 _____

3. List and briefly describe the two stages of protein synthesis.
   1 _____
   _____
   2 _____
   _____

RIBONUCLEIC ACIDS (RNAs)

4. What are the three RNAs that participate in protein synthesis? _____
   _____

MESSENGER RNA (mRNA)

5. Where is mRNA assembled? _____
   _____
   What is transcription? _____
   _____

6. Messenger RNA consists of a _____ strand of _____.

7. Which nitrogen base is not found in RNA? _____.

8. What is an exon? _____
   _____

9. What is an intron? _____

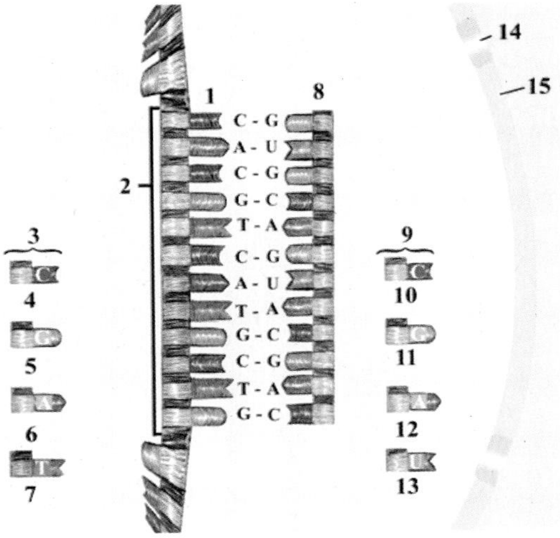

Figure 7.1

10. In reference to **Figure 7.1**, identify #1 - #15.
    1 _____   9 _____
    2 _____   10 _____
    3 _____   11 _____
    4 _____   12 _____
    5 _____   13 _____
    6 _____   14 _____
    7 _____   15 _____
    8 _____

11. Functional mRNA leaves the nucleus through _____ and attaches to a _____.

12. At the ribosome, how is the mRNA used? _____
    _____

13. What are groups of three nitrogen bases of mRNA called?
    _____

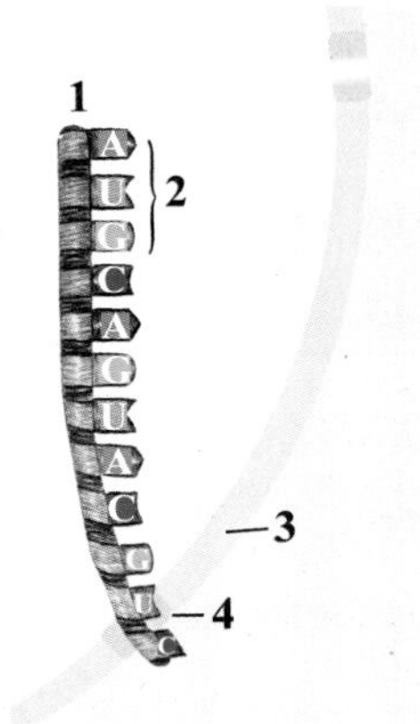

Figure 7.2

14. In reference to **Figure 7.2**, identify # 1 - 4.
    1 _____   3 _____
    2 _____   4 _____

RIBOSOMAL RNA (rRNA)

15. Where does translation occur? _____

16. How many subunits are found in a ribosome? _____

17. What are the subunits called? _____

18. Where are ribosomal subunits made? _____

19. What is a function of the small ribosomal subunit? _____

20. What are three functions of the large ribosomal subunit?
    1 _____
    2 _____
    3 _____

21. What are the names of the two sites of the large ribosomal subunit? _____

22. How was the name P-site derived? _____

23. How was the name of the A-site derived? _____

24. Where is mRNA positioned for translation? _____
    _____

104 Chapter 7 - Protein Synthesis

Name _____
Class _____

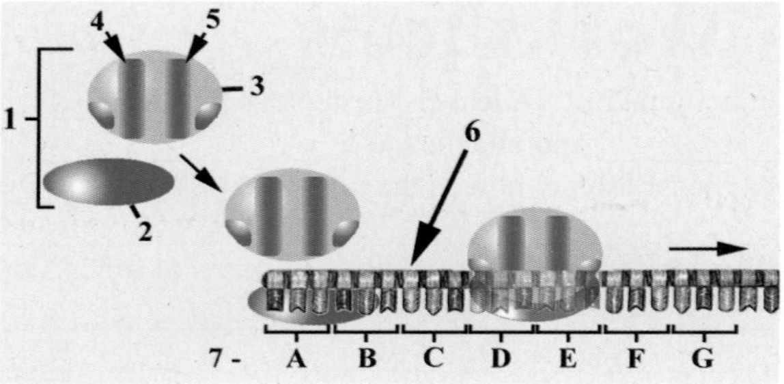

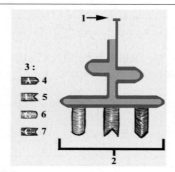

Figure 7.3

25 In reference to **Figure 7.3**, identify #1 - #7.

1 _____   5 _____
2 _____   6 _____
3 _____   7 _____
4 _____

TRANSFER RNA (tRNA)

26 What is the function of tRNA? _____

27 How many tRNAs are found in the cell? _____

28 How many amino acids are used in protein synthesis? ___

29 How many functional sites does tRNA have? _____

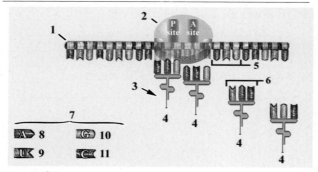

Figure 7.4

30 In reference to **Figure 7.4**, identify #1 - #7.

1 _____   5 _____
2 _____   6 _____
3 _____   7 _____
4 _____

31 What is the function of the anticodon site? _____

32 What is the function of the amino acid site? _____

Figure 7.5

33 In reference to **Figure 7.5**, identify #1 - #11.

1 _____   7 _____
2 _____   8 _____
3 _____   9 _____
4 _____   10 _____
5 _____   11 _____
6 _____

DNA

34 How do the nucleotides of DNA differ? _____

35 What are the four nitrogen bases of DNA?

1 _____   3 _____
2 _____   4 _____

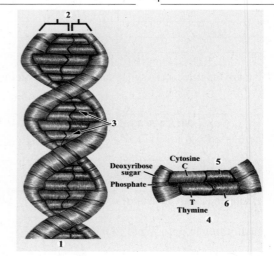

Figure 7.6

36 In reference to **Figure 7.6**, identify #1 - #6.

1 _____   4 _____
2 _____   5 _____
3 _____   6 _____

37 What are the functional units of the chromosome? _____

38 How is the "language" of DNA organized and read? _____

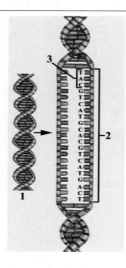

Figure 7.7

39 In reference to **Figure 7.7**, identify #1 - #3.

1 _____   3 _____
2 _____

Name _____
Class _____

Chapter 7 - Protein Synthesis 105

TRANSCRIPTION

40 What is transcription? _____

41 Where does transcription occur? _____

42 Which of the four nitrogen bases of DNA, is not found in RNA? _____ Instead of this base, RNA has a substituted base called _____.

43 When does transcription begin? _____

44 Which DNA strand, the sense or the antisense, is used for transcription? _____

45 What happens when RNA polymerase reaches the terminator site? _____

46 What are the groups of three nitrogen bases of mRNA called? _____

47 What is a codon table? _____

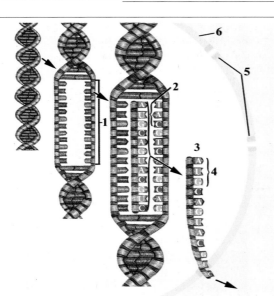

Figure 7.8

48 In reference to **Figure 7.8**, identify #1 - #6.
    1 _____   4 _____
    2 _____   5 _____
    3 _____   6 _____

49 In reference to **Figure 7.8**, complete the nitrogen base (**NB**) sequence for the mRNA strand.
    AUG, _____, _____, _____.

50 In reference to **Figure 7.8**, use the codon table and identify the amino acids that the mRNA strand codes for.
    codon 1 codes for _____
    codon 2 codes for _____
    codon 3 codes for _____
    codon 4 codes for _____

TRANSLATION

51 What is translation? _____

52 What are the three events of translation?
    1 _____   3 _____
    2 _____

INITIATION

53 What happens first in initiation? _____

54 What happens after a small ribosomal subunit binds to mRNA? _____

55 What happens after the tRNA carrying methionine binds to its codon? _____

56 After binding of the large ribosomal subunit, where is the mRNA positioned? _____

57 What are the two sites of the large ribosomal subunit called? _____

58 Which site of the large ribosomal subunit is occupied by the first tRNA carrying methionine? _____

59 When does the A-site become occupied? _____

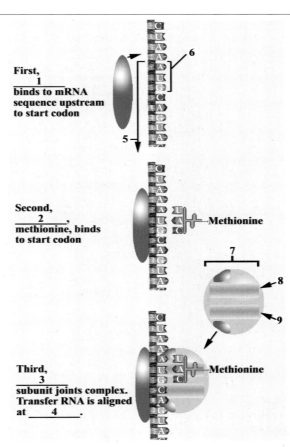

Figure 7.9

60 In reference to **Figure 7.9**, complete #1 - #4.
    1 _____
    2 _____
    3 _____
    4 _____

61 In reference to **Figure 7.9**, identify #5 - #9.
    5 _____   8 _____
    6 _____   9 _____
    7 _____

Chapter 7 - Protein Synthesis

ELONGATION

62. When does elongation begin? _____

63. What happens after a tRNA arrives and occupies the A-site of the large ribosomal subunit? _____

64. After the formation of a peptide bond, what happens to the amino acid free tRNA occupying the P-site? _____

65. After the amino acid free tRNA is ejected from the P-site, what does the ribosome do? _____

66. When the ribosome moves down one codon, what site on the large ribosomal subunit is made available? _____

67. Once the large ribosomal subunit's A-site is made available, what happens? _____

68. After the A-site is occupied what happens between the two associated amino acids? _____

69. The process of elongation ends when the ribosome reaches a _____ _____.

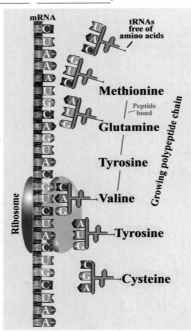

Figure 7.10

70. In reference to **Figure 7.10**, what is the next event in elongation? _____

71. In reference to **Figure 7.10**, after formation of the peptide bond between valine and tyrosine, what happens to the amino acid free tRNA? _____

72. In reference to **Figure 7.10**, what happens after the tRNA is ejected from the P-site? _____

73. In reference to **Figure 7.10**, what is the next amino acid that will be bonded to tyrosine? _____

74. In reference to **Figure 7.10**, what does the codon UAA translate to? _____

TERMINATION

75. When is the process of translation terminated? _____

76. What are three events that result from release factors occupying the A-site?
    1. _____
    2. _____
    3. _____

77. What happens to the ribosomal subunits after they separate? _____

POLYRIBOSOME

78. What is a polyribosome? _____

79. What is the advantage of having a cluster of ribosomes attached to a single mRNA strand? _____

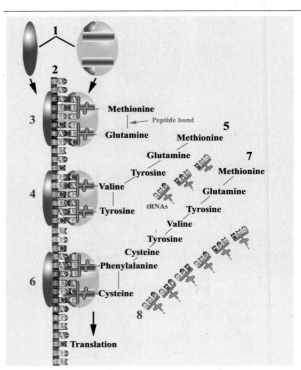

Figure 7.11

80. In reference to **Figure 7.11**, identify #1 - #8.
    1. _____ 5. _____
    2. _____ 6. _____
    3. _____ 7. _____
    4. _____ 8. _____

Tissues - Worksheets

TISSUES

1. What are tissues? _____
   _____

2. What is histology? _____
   _____

3. What are the four basic tissues of the body?
   1 _____ 3 _____
   2 _____ 4 _____

4. What are epithelia? _____
   _____
   _____

5. Give three general locations of epithelia.
   1 _____
   2 _____
   3 _____

6. Give six general functions of epithelia.
   1 _____ 4 _____
   2 _____ 5 _____
   3 _____ 6 _____

7. Give two examples of structural modifications of epithelia that determine epithelial function? _____
   _____
   _____

8. Give two examples of surface modifications of epithelial cells that determine epithelial function. _____
   _____
   _____

9. What are seven characteristics of epithelia?
   1 _____ 5 _____
   2 _____ 6 _____
   3 _____ 7 _____
   4 _____

10. What does cellularity refer to? _____

11. How are epithelia described as to cellularity? _____

12. What two cell junctions are commonly found in epithelia?
    1 _____ 2 _____

13. Epithelial cells are organized into _____.

14. What does polarity refer to? _____
    _____

15. Describe how epithelia exhibit polarity. _____
    _____
    _____

16. Where is the basement membrane located? _____

17. What are the structural components of the basement membrane? _____
    _____
    _____

18. Where is the vascular supply for epithelia? _____
    _____

19. Most epithelial cells are rapidly _____ when they are abrated or die.

CLASSIFICATION OF EPITHELIA

20. What are ways used to classify epithelia?
    1 _____
    2 _____
    3 _____

21. What two criteria are used to classify tissue according to their structure?
    1 _____
    2 _____

22. Epithelia that form membrane in specific areas are classified according to _____.

23. Epithelia that form glands and produce a secretion are called _____.

24. What are three shapes and descriptions of epithelial cells?
    1 _____
    2 _____
    3 _____

25. What are the two possible arrangements for the number of cellular layers?
    1 _____
    2 _____

26. Describe simple epithelia. _____
    _____

27. Describe pseudostratified epithelia. _____
    _____

28. Describe stratified epithelia. _____
    _____

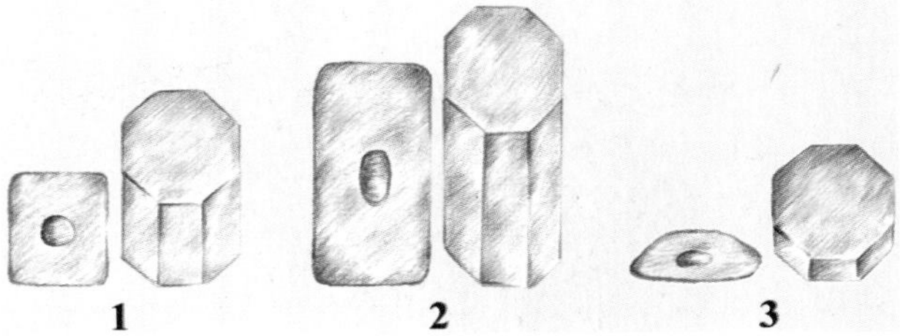

Figure 8.1

29. In reference to **Figure 8.1**, identify the shapes #1 - #3.
    1 _____ 3 _____
    2 _____

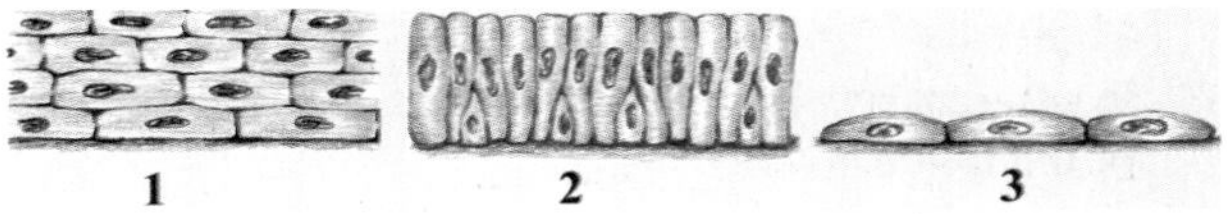

Figure 8.2

30. In reference to **Figure 8.2**, identify the layers #1 - #3.
    1 _____ 3 _____
    2 _____

108 **Chapter 8 - Tissues**

Name _____
Class _____

SIMPLE EPITHELIA
SIMPLE SQUAMOUS EPITHELIUM

31 Describe the structure of simple squamous epithelium?
_____
_____

32 List five locations of simple squamous epithelium.
1 _____ 4 _____
2 _____ 5 _____
3 _____

33 What are three general functions of simple squamous epithelium?

34 1 _____ 3 _____
    2 _____

35 Simple squamous epithelium that is located in the cardiovascular and lymphatic systems is called _____.

36 What is the function of simple squamous epithelium that lines blood vessels? _____
_____
_____

37 In the capillaries, the thin endothelium of the walls promotes the exchange of permeable materials by the processes of _____ and _____.

38 In the lung's air sacs (alveoli), simple squamous epithelium facilitates the _____ of respiratory gases. Simple squamous epithelium that forms the lining of the ventral body cavities is called _____.

39 Mesothelium functions in _____
_____
_____

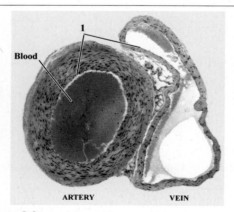

Figure 8.3

40 In reference to **Figure 8.3**, what is the name and location of the tissue at #1? _____
_____

41 In reference to **Figure 8.3**, what is the function of the tissue at #1. _____
_____
_____

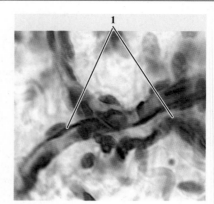

Figure 8.4

42 In reference to **Figure 8.4**, what is the name and location of the tissue? _____
_____

43 In reference to **Figure 8.4**, what does the thin structure of the epithelium at #1 facilitate? _____
_____

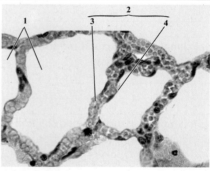

Figure 8.5

44 In reference to **Figure 8.5**, what is the name and location of the tissue? _____
_____

45 In reference to **Figure 8.5**, identify #1 - #4.
1 _____ 3 _____
2 _____ 4 _____

46 In reference to **Figure 8.5**, what does the thin structure of the epithelium at #3 & #4 facilitate? _____
_____

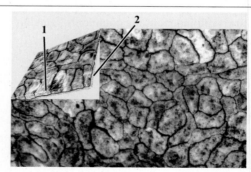

Figure 8.6

47 In reference to **Figure 8.6**, what is the name and location of the tissue? _____
_____

48 In reference to **Figure 8.6**, what is the function of the tissue? _____
_____

Name _____
Class _____

Chapter 8 - Tissues   109

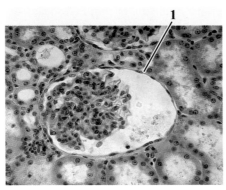

Figure 8.7

49 In reference to **Figure 8.7**, what is the name and location of the tissue? _____
    _____

50 In reference to **Figure 8.7**, what is the function of #1? _____

SIMPLE CUBOIDAL EPITHELIUM

51 Describe the structure of simple cuboidal epithelium. ____
    _____

52 Give three locations of simple cuboidal epithelium.
    1 _____ 3 _____
    2 _____

53 Give two functions of simple cuboidal epithelium.
    1 _____ 2 _____

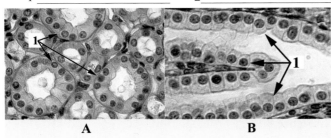

Figure 8.8

54 In reference to **Figure 8.8**, what is the name and location of the tissue? _____
    _____

55 In reference to **Figure 8.8**, give the function of the tissue at A -#1 and B-#1. _____
    _____

SIMPLE COLUMNAR EPITHELIUM

56 Describe the structure of simple columnar epithelium. ____
    _____

57 Give three locations of simple columnar epithelium.
    1 _____
    2 _____
    3 _____

58 Give two functions of simple columnar epithelium.
    1 _____ 2 _____

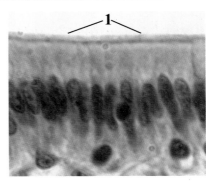

Figure 8.9

59 In reference to **Figure 8.9**, what is the name and location of the tissue? _____
    _____

60 In reference to **Figure 8.9**, give the function of the tissue.
    _____
    _____

61 In reference to **Figure 8.9**, the structures seen at #1 are _____ and function in _____
    _____

PSEUDOSTRATIFIED CILIATED COLUMNAR EPITHELIUM

62 Describe the structure of pseudostratified ciliated columnar epithelium. _____

63 Give four locations of pseudostratified ciliated columnar epithelium.
    1 _____
    2 _____
    3 _____
    1 _____

64 Give three functions of pseudostratified ciliated columnar epithelium.
    1 _____ 3 _____
    2 _____

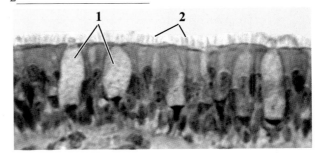

Figure 8.10

65 In reference to **Figure 8.10**, what is the name and location of the tissue? _____
    _____

66 In reference to **Figure 8.10**, give the function of the tissue.
    _____
    _____

67 In reference to **Figure 8.10**, identify and give the function of #1 & #2.
    1 _____
    2 _____

110 Chapter 8 - Tissues

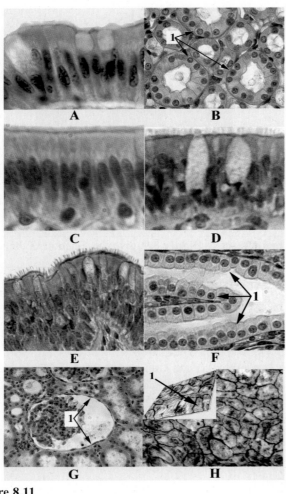

Figure 8.11

68 In reference to **Figure 8.11**, identify the tissues:
    A _____
    B _____
    C _____
    D _____
    E _____
    F _____
    G _____
    H _____

STRATIFIED EPITHELIA
STRATIFIED SQUAMOUS EPITHELIUM

69 Describe the structure of stratified squamous epithelium.

70 What are two variations that may be shown by the squamous surface cells of stratified squamous epithelium?
    1 _____
    2 _____

71 What are the two varieties of stratified squamous epithelium? 1 _____ 2 _____

72 Give a location of keratinized stratified squamous epithelium. _____

73 Give four locations of nonkeratinized stratified squamous epithelium.
    1 _____ 3 _____
    2 _____ 4 _____

74 The primary function of stratified squamous epithelium is _____. Additionally, the keratinized variety helps protect the body from _____
    _____

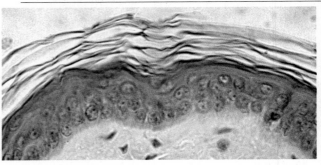

Figure 8.12

75 In reference to **Figure 8.12**, what is the name and location of the tissue? _____
    _____

76 In reference to **Figure 8.12**, give the function of the tissue. _____
    _____

77 In reference to **Figure 8.12**, the surface cells are _____ (dead or alive) and are _____ (keratinized or nonkeratinized.)

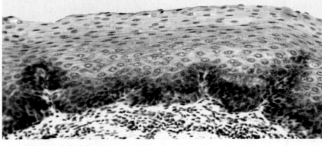

Figure 8.13

78 In reference to **Figure 8.13**, what is the name and location of the tissue? _____
    _____

79 In reference to **Figure 8.13**, give the function of the tissue.
    _____

80 In reference to **Figure 8.13**, the surface cells are _____ (dead or alive) and are _____ (keratinized or nonkeratinized.)

TRANSITIONAL EPITHELIUM

81 Describe the structure of transitional epithelium. _____
    _____

82 Give three locations of transitional epithelium.
    1 _____ 3 _____
    2 _____

83 Give a function of transitional epithelium.
    _____

Name _____
Class _____

Chapter 8 - Tissues   111

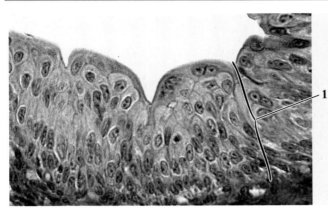

Figure 8.14

84 In reference to **Figure 8.14**, what is the name and location of the tissue? _____
_____

85 In reference to **Figure 8.14**, give the function of the tissue.
_____
_____

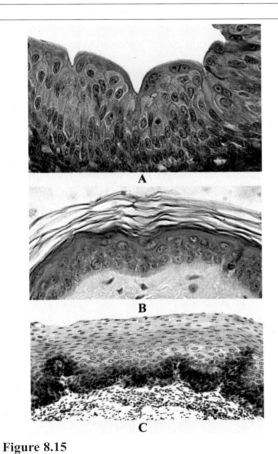

Figure 8.15

86 In reference to **Figure 8.15**, identify A - C.
A _____
B _____
C _____

CLASSIFICATION OF EPITHELIA
ACCORDING TO LOCATION

87 According to location epithelia are classified as
   1 _____ 2 _____

ENDOTHELIUM

88 Describe the structure of endothelium. _____
_____

89 Where is endothelium located? _____
_____

90 What is the general function of endothelium? _____
_____
_____ What additional function of endothelium applies to the cardiovascular system? _____
_____

EPITHELIAL MEMBRANES

91 Describe the structure of epithelial membranes. _____
_____

92 Name three epithelial membranes and give locations.
   1 _____ Location _____
   _____
   2 _____ Location _____
   _____
   3 _____ Location _____
   _____

MUCOUS MEMBRANES

93 What are mucous membranes? _____
_____

94 Describe the structure of a mucous membrane. _____

95 Give three locations of mucous membranes.
   1 _____ 3 _____
   2 _____

96 Give three functions of mucous membranes.
   1 _____ 3 _____
   2 _____

SEROUS MEMBRANES

97 What is a serous membrane? _____
_____

98 Describe the structure of serous membranes. _____
_____
_____

99 Give three serous membranes.
   1 _____ 3 _____
   2 _____

100 Give a function of the serous membranes.
_____
_____

CUTANEOUS MEMBRANES

101 What is the cutaneous membrane? _____
_____

102 Describe the structure of the cutaneous membrane. _____
_____
_____

103 Give the location of the cutaneous membrane.
_____

112 Chapter 8 - Tissues

104 Give three functions of the cutaneous membrane.
  1 _____
  2 _____
  3 _____

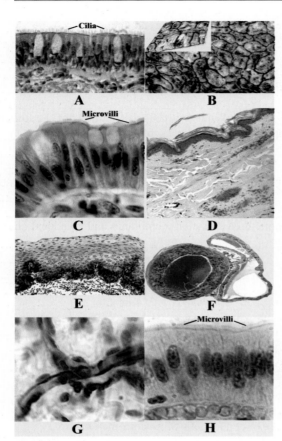

Figure 8.16

105 In reference to **Figure 8.16**, match the letters of the illustrated tissues to the following classification:
  Endothelium _____
  Mucous membrane _____
  Serous membrane _____
  Cutaneous membrane _____

GLANDULAR EPITHELIA

106 What is a gland? _____
_____

107 Depending upon where their secretion is released, glands are either classified as
  1 _____ 2 _____

EXOCRINE GLANDS

108 Where is the secretion of exocrine glands released? _____
_____

109 What are three examples of exocrine glands?
  1 _____ 3 _____
  2 _____

ENDOCRINE GLANDS

110 Where is the secretion of endocrine glands released? _____
_____
_____

111 What are the secretions of endocrine glands called?
_____

112 What is a hormone? _____
_____
_____

113 What are four examples of endocrine glands?
  1 _____ 3 _____
  2 _____ 4 _____

GLANDULAR SECRETION

114 Based upon the method of secretion, what are the three classifications of glands?
  1 _____ 3 _____
  2 _____

MEROCRINE GLANDS

115 What is the method of secretion of merocrine glands? _____
_____

116 What are three locations of merocrine glands?
  1 _____ 3 _____
  2 _____

APOCRINE GLANDS

117 What is the method of secretion of apocrine glands? _____
_____

118 What are two locations of apocrine glands?
  1 _____ 2 _____

HOLOCRINE GLANDS

119 What is the method of secretion of holocrine glands? _____
_____

120 What is an example of a holocrine gland?
  1 _____

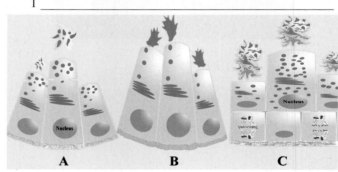

Figure 8.17

121 In reference to **Figure 8.17**, identify the types of glands:
  A _____ B _____
  C _____

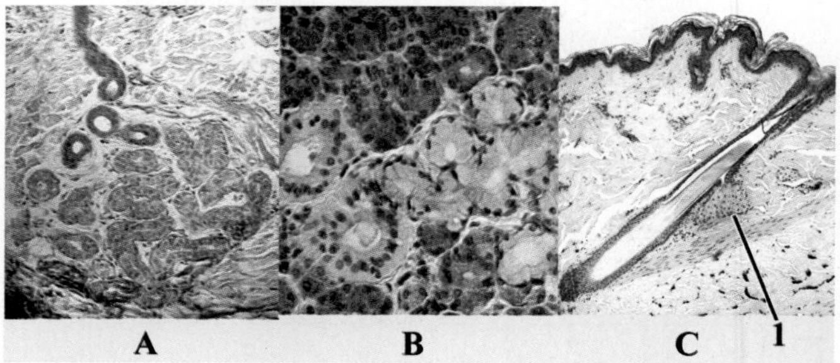

Figure 8.18

122 In reference to **Figure 8.18**, identify the glands and give their secretion type.
  A Name _____ Type _____
  B Name _____ Type _____
  C Name _____ Type _____

CONNECTIVE TISSUE

123 What are six general functions of connective tissues?
  1 _____ 4 _____
  2 _____ 5 _____
  3 _____ 6 _____

124 What are three characteristics of connective tissues?
  1 _____ 3 _____
  2 _____

125 What is the matrix? _____

126 What are two components of the matrix? _____

127 What is the ground substance? _____

128 Where is the ground substance located? _____

129 What are fibers? _____

130 What are three common fibers found in connective tissue?
  1 _____ 3 _____
  2 _____

131 What is the function of collagen fibers? _____

132 What is the function of elastic fibers? _____

133 What is the function of reticular fibers? _____

134 What type of cell is always found in a specific connective tissue? _____

135 What is the difference between the suffixes "cyte" and "blast"? _____

136 What three characteristics of the matrix are used in the classification of connective tissues?
  1 _____
  2 _____
  3 _____

137 Using structural characteristics, what are four types of connective tissues?
  1 _____ 3 _____
  2 _____ 4 _____

CONNECTIVE TISSUE PROPER

138 What are three characteristics of the matrix of connective tissue proper?
  1 _____
  2 _____
  3 _____

139 What are the structural cells of connective tissue proper? _____

140 What are the two subclasses of connective tissue proper?
  1 _____
  2 _____

141 How are the fibers arranged in loose connective tissue? _____

142 What are three examples of loose connective tissue?
  1 _____ 3 _____
  2 _____

143 How are the fibers arranged in dense connective tissue? _____

144 What are three examples of dense connective tissue?
  1 _____ 3 _____
  2 _____

CARTILAGE

145 What are three characteristics of the matrix of cartilage?
  1 _____ 3 _____
  2 _____

146 What are the structural cells of cartilage? _____

147 What are the three types of cartilage?
  1 _____ 3 _____
  2 _____

BONE

148 What are four characteristics of the matrix of bone?
  1 _____ 3 _____
  2 _____ 4 _____

149 What are the structural cells of bone? _____

150 What are two types of bone tissue?
  1 _____ 2 _____

BLOOD

151 What are two characteristics of the matrix of blood?
  1 _____ 2 _____

152 What are hemocytoblasts? _____

153 Where are leukocytes and erythrocytes found? _____

LOOSE CONNECTIVE TISSUES
AREOLAR CONNECTIVE TISSUE

154 What are the three fibers found in areolar tissue?
  1 _____ 3 _____
  2 _____

155 What does "viscous ground substance" mean? _____

156 What are the structural cells of areolar tissue? _____

157 Among the types of cells found associated with areolar tissue are mast cells and macrophages. Give their functions:
  mast cells _____

  macrophages _____

158 Where is areolar tissue located? _____

159 What are two general functions of areolar tissue?
  1 _____ 2 _____

114   Chapter 8 - Tissues

Name _____
Class _____

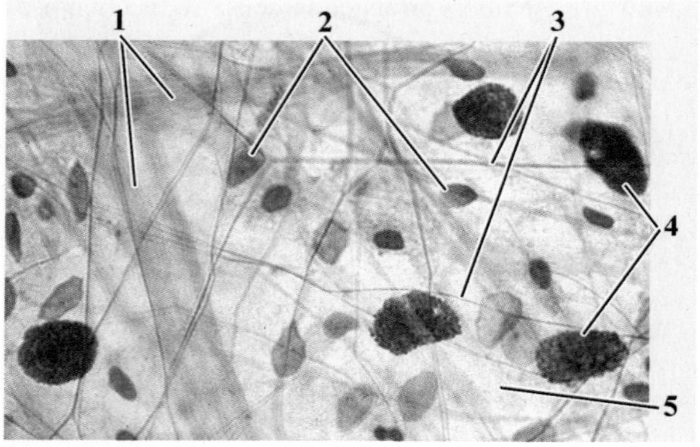

Figure 8.19

160 In reference to **Figure 8.19**, what is the name and general location of the tissue? _____

161 In reference to **Figure 8.19**, give the function of the tissue. _____

162 In reference to **Figure 8.19**, identify and give the function of #1 - #5.
1 _____
2 _____
3 _____
4 _____
5 _____

ADIPOSE TISSUE

163 The matrix in mature adipose tissue is _____ (small or large) in quantity? _____ What types of fibers are present in adipose tissue? _____

164 In mature adipose tissue, where are the fibers located? _____

165 What cells dominate adipose tissue? _____

166 Where is adipose tissue located? _____

167 What are four functions of adipose tissue?
1 _____
2 _____
3 _____
4 _____

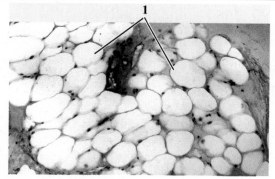

Figure 8.20

168 In reference to **Figure 8.20**, what is the name and general location of the tissue? _____

169 In reference to **Figure 8.20**, identify and give the function of #1. _____

170 The cleared areas of the adipocytes were storage sites for _____.

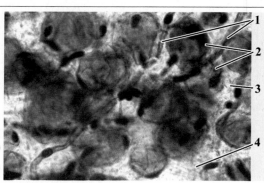

Figure 8.21

171 In reference to **Figure 8.21**, what is the name and location of the tissue? _____

172 In reference to **Figure 8.21**, identify and give the function of #1 - #3.
1 _____
2 _____
3 _____

RETICULAR CONNECTIVE TISSUE

173 Describe the structure of the matrix of reticular connective tissue. _____

174 What is the name of the structural cells of reticular tissue? _____

175 Give four locations of reticular tissue.
1 _____ 2 _____
3 _____ 4 _____

176 What is the function of reticular tissue? _____

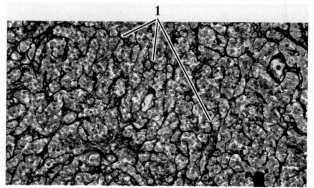

Figure 8.22

177 In reference to **Figure 8.22**, what is the name and location of the tissue? _____

178 In reference to **Figure 8.22**, identify and give the function of #1. _____

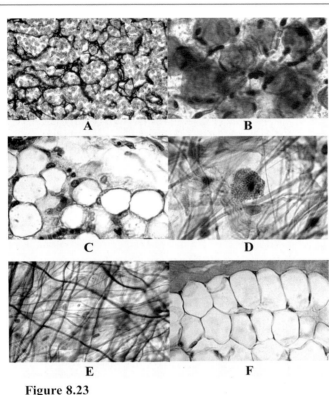

Figure 8.23

179 In reference to **Figure 8.23**, identify tissues A - F.
A _____ B _____
C _____ D _____
E _____ F _____

DENSE CONNECTIVE TISSUES
DENSE REGULAR CONNECTIVE TISSUE

180 Describe the matrix of dense regular connective tissue. ___
_____
181 What are the structure cells of dense regular connective tissue? _____
182 Where is dense regular connective tissue commonly found? _____
183 What are two functions of dense regular connective tissue?
   1 _____
   2 _____
184 What do tendons attach? _____
_____
185 What do ligaments attach? _____
_____

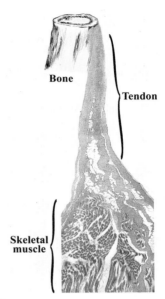

Figure 8.24

186 In reference to **Figure 8.24**, what is the function of the tendon? _____
_____
_____

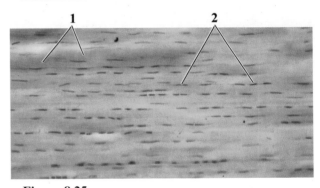

Figure 8.25

187 In reference to **Figure 8.25**, what is the name and location of the tissue? _____
_____
188 In reference to **Figure 8.25**, identify and give the function of #1 & #2.
   1 _____
   2 _____

DENSE IRREGULAR CONNECTIVE TISSUE

189 Describe the matrix of dense irregular connective tissue.
_____
_____
190 Give three locations of dense irregular connective tissue.
   1 _____ 2 _____
   3 _____
191 Give three functions of dense irregular connective tissue.
   1 _____
   2 _____
   3 _____
192 What are the structural cells of dense irregular connective?
_____

116   Chapter 8 - Tissues

Name _____
Class _____

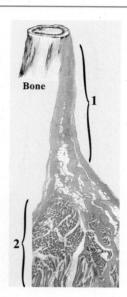

Figure 8.26

193 In reference to **Figure 8.26**, what type of connective tissue surrounds #2? _____
_____

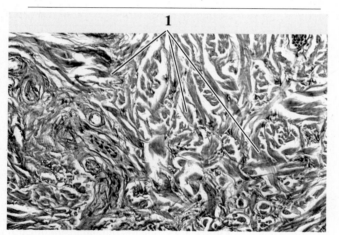

Figure 8.27

194 In reference to **Figure 8.27**, what is the name and location of the tissue? _____
_____

195 In reference to **Figure 8.27**, identify and give the function of #1. _____
_____

ELASTIC CONNECTIVE TISSUE

196 Describe the matrix of elastic connective tissue. _____
_____
_____

197 Give four locations of elastic connective tissue.
   1 _____   3 _____
   2 _____   4 _____

198 Give three functions of elastic connective tissue.
   1 _____   3 _____
   2 _____

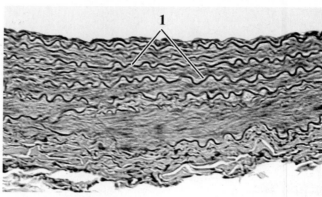

Figure 8.28

199 In reference to **Figure 8.28**, what is the name and location of the tissue? _____
_____

200 In reference to **Figure 8.28**, identify and give the function of #1. _____
_____
_____

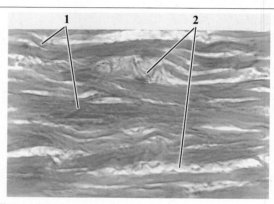

Figure 8.29

201 In reference to **Figure 8.29**, what is the name and location of the tissue? _____
_____

202 In reference to **Figure 8.29**, identify and give the functions of #1 & #2.
   1 _____
   2 _____

Name _____
Class _____

Chapter 8 - Tissues   **117**

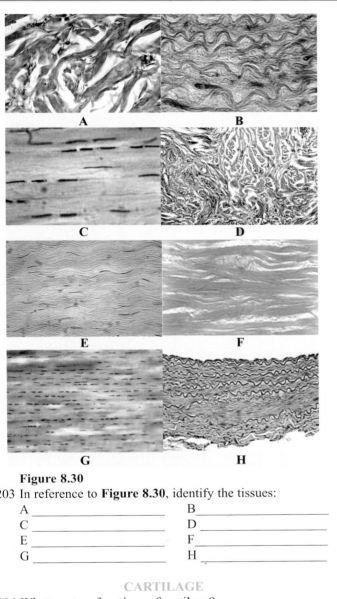

Figure 8.30

203 In reference to **Figure 8.30**, identify the tissues:
A _____   B _____
C _____   D _____
E _____   F _____
G _____   H _____

CARTILAGE

204 What are two functions of cartilage? _____
_____

205 Is cartilage vascular or avascular? _____
   Describe the matrix of cartilage. _____
_____

206 What mostly forms the matrix of cartilage? _____
_____

207 What are the structural cells of cartilage? _____
_____

HYALINE CARTILAGE

208 What forms the matrix of hyaline cartilage? _____

209 What does "amorphous matrix" mean? _____
_____

210 Give four locations of hyaline cartilage.
   1 _____
   2 _____
   3 _____
   4 _____

211 What are three functions of hyaline cartilage?
   1 _____   3 _____
   2 _____

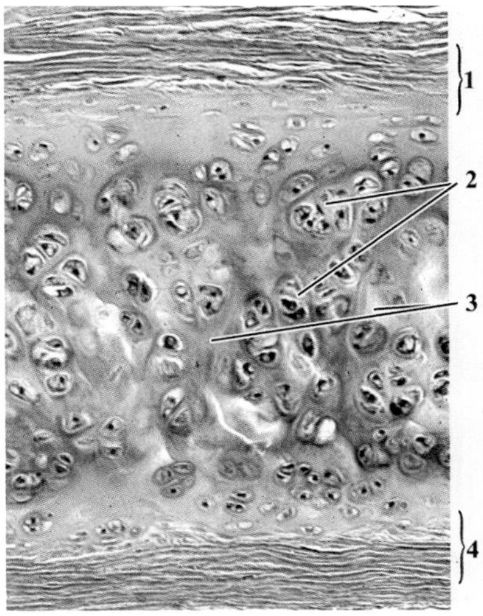

Figure 8.31

212 In reference to **Figure 8.31**, what is the name and location of the tissue? _____
_____

213 In reference to **Figure 8.31**, identify and give the function of # 1 - #3.
   1 _____
   2 _____
   3 _____

FIBROCARTILAGE

214 What forms the matrix of fibrocartilage? _____
_____

215 Give three locations of fibrocartilage.
   1 _____
   2 _____
   3 _____

216 What are two functions of fibrocartilage?
   1 _____   2 _____

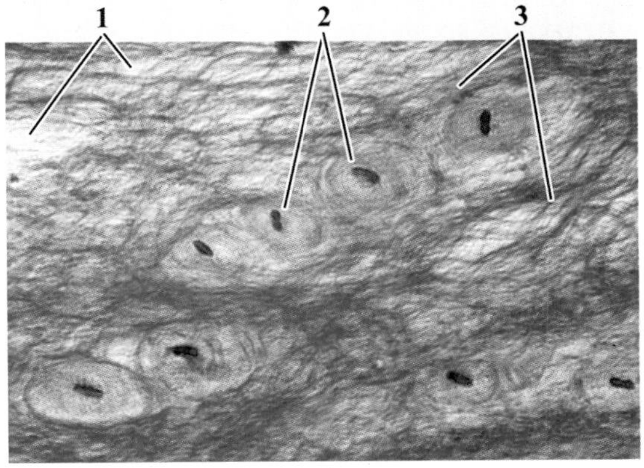

Figure 8.32

217 In reference to **Figure 8.32**, what is the name and location of the tissue? _____
_____

218 In reference to **Figure 8.32**, identify and give the function of # 1 - #3.
   1 _____
   2 _____
   3 _____

118   Chapter 8 - Tissues

ELASTIC CARTILAGE

219 What forms the matrix of elastic cartilage? _____
_____

220 Give two locations of elastic cartilage.
   1 _____
   2 _____

221 What are three functions of elastic cartilage.
   1 _____   3 _____
   2 _____

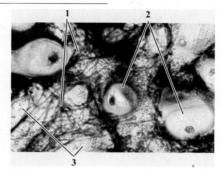

Figure 8.33

222 In reference to **Figure 8.33**, what is the name and location of the tissue? _____
_____

223 In reference to **Figure 8.33**, identify and give the function of # 1 - #3.
   1 _____
   2 _____
   3 _____

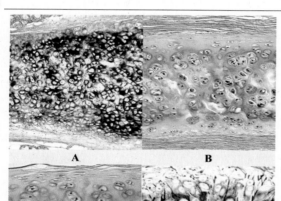
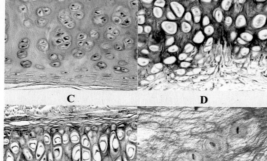
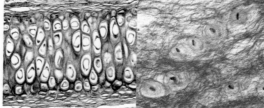

Figure 8.34

224 In reference to **Figure 8.34**, identify the tissues:
   A _____   B _____
   C _____   D _____
   E _____   F _____
   G _____   H _____

BONE

225 Describe the matrix of bone. _____
_____

226 In bone tissue, what is the function of calcium salts?
_____
_____

227 In bone tissue, what is the function of collagenous fibers?
_____
_____

228 What are osteocytes? _____
_____

229 What are osteoblasts? _____
_____

230 What are osteoclasts? _____
_____

231 What are two structural types of bone tissue?
   1 _____   2 _____

232 What are the organizational units of compact bone tissue?
_____

233 What is found in the central (Haversian) canal? _____
_____

234 What are lamellae? _____
_____

235 What is the function of the canaliculi? _____
_____

236 What are trabeculae? _____
_____

237 Where is bone tissue located? _____
_____ Give five functions of bones.
   1 _____   4 _____
   2 _____   5 _____
   3 _____

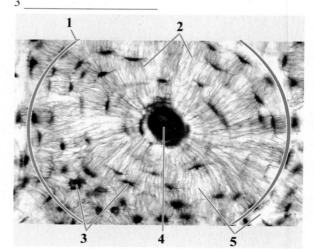

Figure 8.35

238 In reference to **Figure 8.35**, what is the name and location of the tissue? _____
_____

Name _____
Class _____

Chapter 8 - Tissues   119

239 In reference to **Figure 8.35**, identify and give the function of # 1 - #5.
   1 _____
   2 _____
   3 _____
   4 _____
   5 _____

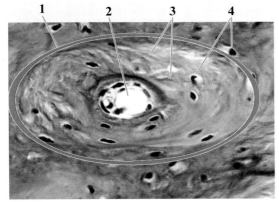

Figure 8.36

240 In reference to **Figure 8.36**, what is the name of the tissue? _____

241 In reference to **Figure 8.36**, identify and give the function of # 1 - #4.
   1 _____
   2 _____
   3 _____
   4 _____

BLOOD

242 Describe the matrix of blood. _____

243 What is the function of plasma? _____

244 What are the cells of blood? _____

245 Where is blood located? _____

246 What are arteries? _____ What are veins? _____

247 What is the function of capillaries? _____

248 Give three functions of blood.
   1 _____
   2 _____
   3 _____

249 What is the function of erythrocytes? _____

250 What is the function of leukocytes? _____

251 What is the function of platelets? _____

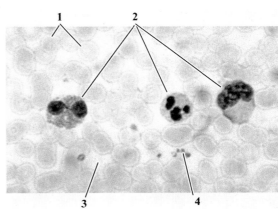

Figure 8.37

252 In reference to **Figure 837**, what is the name and function of the tissue? _____

253 In reference to **Figure 8.37**, identify and give the function of # 1 - #4.
   1 _____
   2 _____
   3 _____
   4 _____

MUSCLE TISSUE

254 What are the cells of muscle tissue called? _____

255 List each type of muscle tissue and give a location.
   1 _____
   2 _____
   3 _____

256 What is the function of the body's muscle tissues? _____

257 What is the primary control system of muscle contraction? _____ Besides this system, what other system regulates the contraction of cardiac and smooth muscle? _____

Skeletal Muscle Tissue

258 What are four characteristics of skeletal muscle fibers?
   1 _____   3 _____
   2 _____   4 _____

259 What are myofibrils? _____

260 What produces the striations? _____

261 What are the light cross-bands called? _____

262 What are the dark cross-bands called? _____

263 What is the name of the structure centered in the I band? _____

264 What is the name of the functional unit of contraction and where is it located? _____

120 Chapter 8 - Tissues

Name _____
Class _____

265 Where is skeletal muscle tissue located? _____

266 What are the functions of skeletal muscle tissue? _____
_____

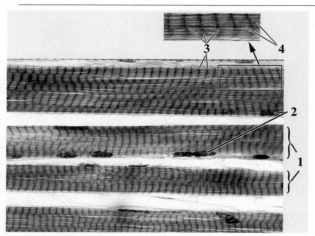

Figure 8.38

267 In reference to **Figure 8.38**, what is the name and location of the tissue? _____
_____

268 In reference to **Figure 8.38**, give the function of the tissue.
_____

269 In reference to **Figure 8.38**, identify #1 - #4.
1 _____ 3 _____
2 _____ 4 _____

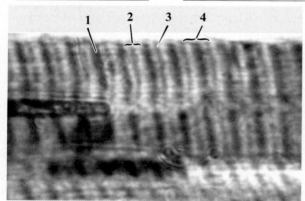

Figure 8.39

270 In reference to **Figure 8.39**, identify #1 - #4.
1 _____ 3 _____
2 _____ 4 _____

CARDIAC MUSCLE TISSUE

271 What are four characteristics of cardiac muscle fibers?
1 _____ 3 _____
2 _____ 4 _____

272 What are intercalated discs? _____
_____

273 Where is cardiac muscle tissue located? _____
_____

274 What is the function of cardiac muscle tissue? _____
_____

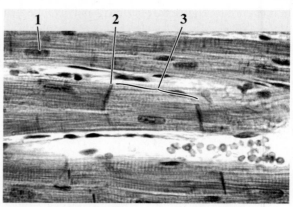

Figure 8.40

275 In reference to **Figure 8.40**, what is the name and location of the tissue? _____
_____

276 In reference to **Figure 8.40**, give the function of the tissue.
_____

277 In reference to **Figure 8.40**, identify #1 - #3.
1 _____ 3 _____
2 _____

SMOOTH MUSCLE TISSUE

278 How is smooth muscle organized? _____

279 What are four characteristics of smooth muscle fibers?
1 _____ 3 _____
2 _____ 4 _____

280 Where is smooth muscle tissue located? _____
_____

281 What is the function of smooth muscle tissue? _____
_____

282 What is plasticity? _____
_____

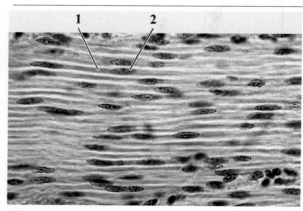

Figure 8.41

283 In reference to **Figure 8.41**, what is the name and location of the tissue? _____
_____

284 In reference to **Figure 8.41**, give the function of the tissue.
_____

285 In reference to **Figure 8.41**, identify #1 - #2.
1 _____ 2 _____

Name _____
Class _____

Chapter 8 - Tissues **121**

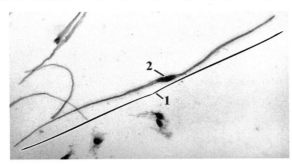

Figure 8.42

286 In reference to **Figure 8.42**, what is the name of this tissue preparation? _____

287 In reference to **Figure 8.42**, identify #1 - #2.
1 _____ 2 _____

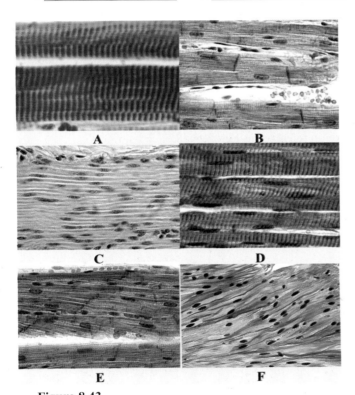

Figure 8.43

288 In reference to **Figure 8.43**, identify the tissues:
A _____ B _____
C _____ D _____
E _____ F _____

NERVE TISSUE

289 What cells are the principal cells of nerve tissue? _____
_____

290 What are three functions of neurons? _____
_____

291 What is the general function of neuroglia? _____
_____

292 Where is nerve tissue located? _____
_____
_____

293 What forms the central nervous system? _____
_____

294 What forms the peripheral nervous system? _____
_____
_____

295 What is the function of nerve tissue? _____
_____
_____

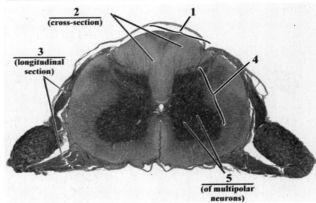

Figure 8.44

296 In reference to **Figure 8.44**, identify #1 - #5.
1 _____ 4 _____
2 _____ 5 _____
3 _____

297 Where is the gray matter of the spinal cord located? _____
_____

298 The gray matter mostly consist of _____
_____
_____

122 Chapter 8 - Tissues

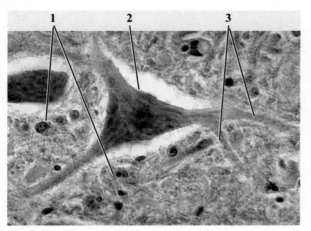

Figure 8.45
299 In reference to **Figure 8.45**, identify #1 - #3.
   1 _____   3 _____
   2 _____
300 The white matter is located to the outside of the spinal cord's _____.
301 What does the white matter mostly contain? _____
_____
_____
302 In a cross section of the spinal cord, most axons of the white matter are seen in _____ section.

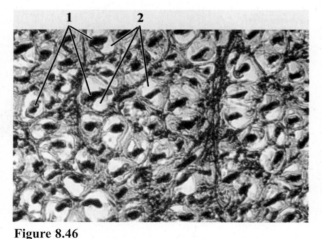

Figure 8.46
303 In reference to **Figure 8.46**, identify #1 - #2.
   1 _____   2 _____
304 In a cross section of the spinal cord, the roots of the spinal cord show fibers (axons) in _____ section.

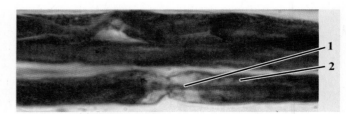

Figure 8.47
305 In reference to **Figure 8.47**, identify #1 - #2.
   1 _____   2 _____

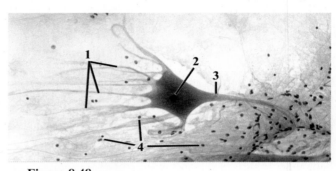

Figure 8.48
306 In reference to **Figure 8.48**, identify #1 - #5.
   1 _____   3 _____
   2 _____   4 _____

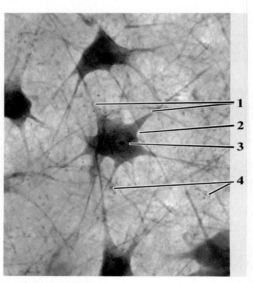

Figure 8.49
307 In reference to **Figure 8.49**, identify #1 - #4.
   1 _____   3 _____
   2 _____   4 _____

The Integumentary System - Worksheets

1. What are five components of the integumentary system?
   1. _____   4. _____
   2. _____   5. _____
   3. _____

   FUNCTIONS OF THE INTEGUMENTARY SYSTEM

2. What are the three barriers provided by the integumentary system and how does each function to protect the body?
   1. _____
   _____
   2. _____
   _____
   3. _____

3. How does the integumentary system provide temperature regulation for the body? _____
   _____

4. How does the integumentary system provide the body with sensory information? _____
   _____

5. How does the integumentary system function in excretion? _____
   _____

6. How does the integumentary system function in the production of vitamin D? _____
   _____
   _____ What is the function of vitamin D? _____
   _____

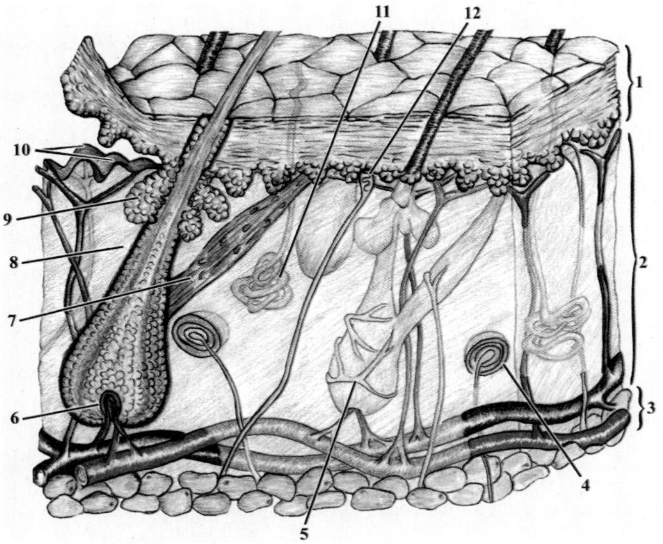

Figure 9.1

7. In reference to **Figure 9.1**, identify #1 - #12.
   1. _____   7. _____
   2. _____   8. _____
   3. _____   9. _____
   4. _____   10. _____
   5. _____   11. _____
   6. _____   12. _____

Structure of the Skin

8. What are the two regions of the skin?
   1. _____   2. _____

9. What is the name of the layer underlying the skin and what are three of its functions? Layer _____
   1. _____
   2. _____
   3. _____

   EPIDERMIS
   CELLS OF THE EPIDERMIS

10. What are the structural cells of the epidermis? _____
    _____

11. Besides the structural cells, name three other cells found in the epidermis. 1 _____
    2. _____ 3. _____

12. What is the name and location of the protein that makes the structural cells tough? _____
    _____

13. What substance makes the skin almost waterproof? _____
    _____ Where is this substance located? _____
    _____

14. What is the function of desmosomes? _____
    _____

15. Where are melanocytes located? _____
    _____

16. Where do keratinocytes store melanin? _____
    _____ Why is melanin stored in this location? _____
    _____

17. What is the function of Langerhan's cells? _____
    _____

    LAYERS OF THE EPIDERMIS

18. From inner to outer list the five layers of the epidermis.
    1. _____   4. _____
    2. _____   5. _____
    3. _____

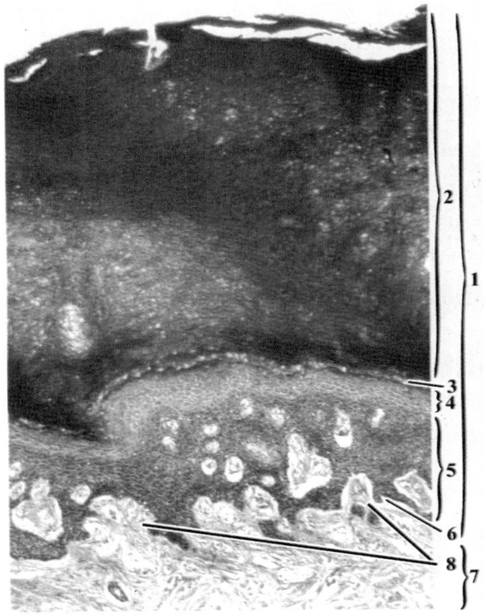

Figure 9.2

Chapter 9 - Integumentary System

19 In reference to **Figure 9.2**, identify #1 - #8.
    1 _____ 5 _____
    2 _____ 6 _____
    3 _____ 7 _____
    4 _____ 8 _____

Stratum basale

20 What characterizes the stratum basale? _____

21 Besides keratinocytes, what other type of cell is found in the stratum basale? _____

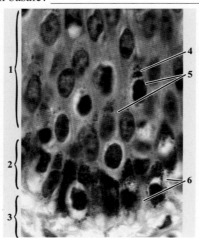

Figure 9.3

22 In reference to **Figure 9.3**, identify #1 - #6.
    1 _____ 4 _____
    2 _____ 5 _____
    3 _____ 6 _____

Stratum spinosum

23 What are the most numerous cells of the stratum spinosum? _____

24 What is formed by the twisting of the helical keratin molecules? _____

25 What cell junctions are associated with the intermediate filaments? _____

26 What function is provided by the association between intermediate filaments and desmosomes? _____

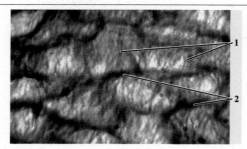

Figure 9.4

27 In reference to **Figure 9.4**, identify #1 & #2.
    1 _____
    2 _____

28 In reference to **Figure 9.4**, what are the filaments at #1 make of and what is their function? _____

29 In reference to **Figure 9.4**, what is the function of #2? _____

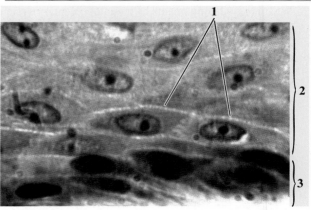

Figure 9.5

30 In reference to **Figure 9.5**, identify #1 - #3.
    1 _____ 3 _____
    2 _____

31 In reference to **Figure 9.5**, why do the cell appear "spiny?" _____

Stratum granulosum

32 What characterizes the stratum granulosum? _____

33 What happens to the cells as they are pushed toward the more superficial layer of the stratum spinosum? _____

34 What are two events that result due to cell death?
    1 _____
    2 _____

35 What is the function of keratohyaline? _____

36 What is the function of the glycolipids? _____

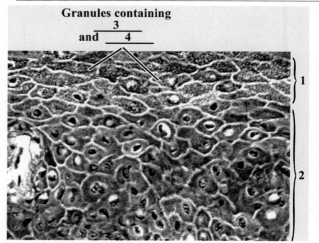

Figure 9.6

37 In reference to **Figure 9.6**, identify #1 - #4.
    1 _____ 3 _____
    2 _____ 4 _____

38 What two substances are released by the disintegration of the granules? 1 _____ 2 _____

Name _____
Class _____

Chapter 9 - Integumentary System

Stratum lucidum
39 What epidermal layer is only found in thick skin? _____
    _____

Stratum corneum
40 Describe the cells of the stratum corneum. _____
    _____

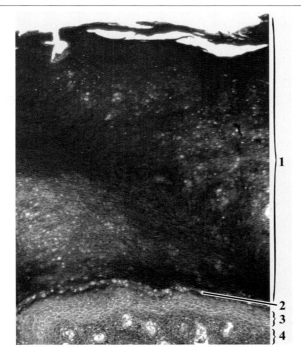

Figure 9.7

41 In reference to **Figure 9.7**, identify #1 - #4.
    1 _____ 3 _____
    2 _____ 3 _____
42 What happens to the surface cells of the stratum corneum?
    _____
    _____

DERMIS
43 What type of tissue is the dermis? _____
    _____
44 What are the structural cells of the dermis? _____
    _____
45 What is the most abundant connective tissue fiber? _____
    _____ What is the function of this type of fiber? _____
    _____
46 What is the function of elastic fibers? _____
    _____
47 From superficial to deep, what are the two layers of the dermis? 1 _____ 2 _____
48 Where is the papillary layer located? _____
    _____
49 What type of tissue is found in the papillary layer? _____
    _____
50 What is the function of dermal papillae? _____
    _____
    _____

51 What are four benefits of increasing the surface area between the dermis and epidermis?
    1 _____
    2 _____
    3 _____
    4 _____
52 Where is the reticular layer located? _____
    _____
53 What type of tissue is found in the reticular layer? _____
    _____
54 What is the function of the dense irregular connective tissue of the reticular layer? _____
    _____
55 How are epidermal ridges produced? _____
    _____
56 List five structures that can be found in the reticular layer of the dermis.
    1 _____ 4 _____
    2 _____ 5 _____
    3 _____
57 What are the two types of skin and where is each located?
    _____
    _____
    _____

LAB ACTIVITY 1

SKIN, Hairless

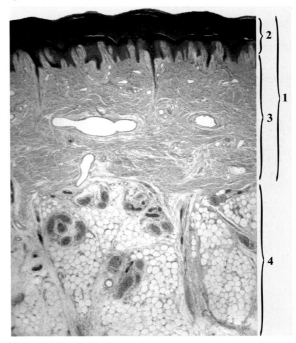

Figure 9.8

58 **Figure 9.8** shows a skin preparation from the _____. In reference to **Figure 9.8**, identify #1 - #4
    1 _____ 3 _____
    2 _____ 4 _____

126   Chapter 9 - Integumentary System

Name _____
Class _____

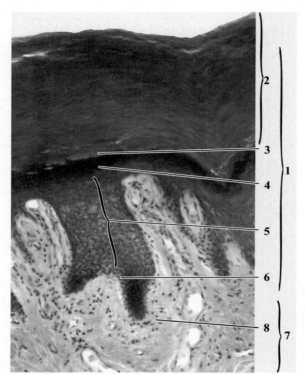

Figure 9.9

59 In reference to **Figure 9.9**, identify #1 - #8.
    1 _____   5 _____
    2 _____   6 _____
    3 _____   7 _____
    4 _____   8 _____

60 In reference to **Figure 9.9**, what type of tissue is #1? ___
    _____

61 In reference to **Figure 9.9**, what is the function of #1? ___

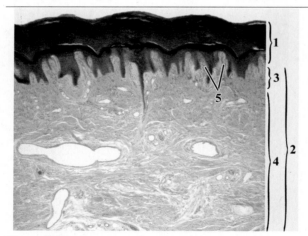

Figure 9.10

62 In reference to **Figure 9.10**, identify #1 - #5.
    1 _____   4 _____
    2 _____   5 _____
    3 _____

63 In reference to **Figure 9.10**, what is the function of #2? ___
    _____

64 In reference to **Figure 9.10**, what is the function of the capillaries found in #3? _____
    _____

65 In reference to **Figure 9.10**, what is the function of #5? ___
    _____

66 In reference to **Figure 9.10**, what type of tissue is found at #3? _____

67 In reference to **Figure 9.10**, what is the function of #4? ___
    _____

 LAB ACTIVITY 2

SKIN, Hairy

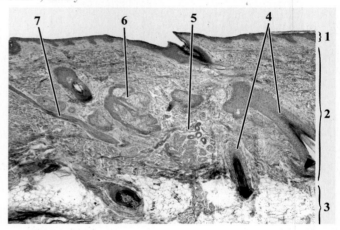

Figure 9.11

68 In reference to **Figure 9.11**, identify #1 - #7.
    1 _____   5 _____
    2 _____   6 _____
    3 _____   7 _____
    4 _____

69 What is a pilosebaceous unit? _____
    _____

70 What is the origin of a pilosebaceous unit and when does it begin to form? _____
    _____
    _____

Chapter 9 - Integumentary System 127

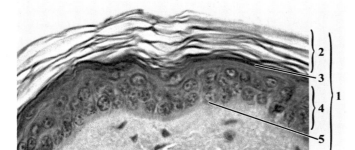

Figure 9.12

71 In reference to **Figure 9.12**, identify #1 - #5.
1 _____ 4 _____
2 _____ 5 _____
3 _____

72 In reference to **Figure 9.12**, is this preparation of pigmented or non-pigmented skin? _____

73 Which epidermal layer is not observed in thin skin? _____
_____

SKIN COLOR

74 What are the two primary pigments that contribute to skin color? 1 _____ 2 _____

75 In addition to pigmentation, what are two additional features that contribute to skin color?
1 _____
2 _____

76 What is the function of melanocytes? _____
_____

77 Where do keratinocytes store melanin? _____
_____

LAB ACTIVITY 3

SKIN, Pigmented

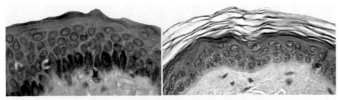

Figure A   Figure B

Figure 9.13

78 In reference to **Figure 9.13**, which illustration "A" or "B" shows pigmented skin? _____

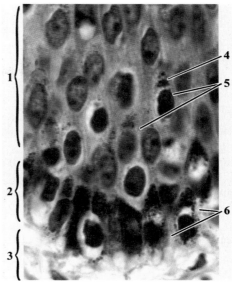

Figure 9.14

79 In reference to **Figure 9.14**, identify #1 - #6.
1 _____ 4 _____
2 _____ 5 _____
3 _____ 6 _____

80 In reference to **Figure 9.14**, at what cellular location do the keratinocytes store melanin? _____
_____

81 Where does carotene originate? _____
_____

82 Which layer of the epidermis accumulates carotene? _____
_____ In addition to this epidermal layer, carotene is found in the _____.

83 Why does skin that has a small amount of melanin normally appear pinkish? _____
_____

84 What does cyanosis mean? _____
_____

85 What produces pallor, or paleness, of the skin? _____
_____

Chapter 9 - Integumentary System

ACCESSORY STRUCTURES OF THE SKIN

86 What are the accessory structures of the skin?
1 _____ 3 _____
2 _____ 4 _____

HAIR

87 What is the primary function of hair? _____
_____

Hair structure

88 What is a pilosebaceous unit? _____
_____

89 What is the origin of the pilosebaceous units? _____
_____

90 From outer to inner, what are the three regions of a hair?
1 _____ 3 _____
2 _____

91 Where is the hair shaft and hair root located? _____
_____

92 What is the function of the hair matrix? _____
_____

93 What dermal structure is located by the matrix and contains blood vessels? _____
_____

Hair growth

94 What happens at the matrix during the active stage of hair growth? _____
_____
_____

95 What stage follows the active stage? _____
_____ What happens during this stage? _____
_____

Hair Follicle

96 What is the name of the lower expanded portion of the hair follicle? _____

97 What is the function of the root hair plexus? _____
_____

98 What are the two sheaths of the wall of the hair follicle?
1 _____ 2 _____

99 Where does the external root sheath begin? _____
_____

100 Where does hair growth occur? _____
_____

101 What is the name of the dermal projection that brings blood into close proximity of the hair matrix? _____

102 Where is the arrector pili muscle located? _____
_____

103 What is the result of the contraction of the arrector pili muscle? _____
_____

Sebaceous (Oil) Glands

104 Where are sebaceous glands located? _____
_____

105 What is the name and function of the substance produced by sebaceous glands? _____
_____

Sweat (Sudoriferous) Glands

106 Where are sweat glands located? _____
_____

107 What are the two types of sweat glands?
1 _____ 2 _____

ECCRINE, OR MEROCRINE, SWEAT GLANDS

108 Which type of sweat gland is the most abundant? _____
_____

109 Where do the ducts of eccrine sweat glands open? _____
_____

110 What is the function of eccrine sweat glands? _____
_____

APOCRINE SWEAT GLANDS

111 Where are apocrine sweat glands located? _____
_____

112 Where do the ducts of apocrine sweat glands open? _____
_____

113 When do apocrine sweat glands become active? _____

114 What is the speculated function of apocrine sweat glands?
_____
_____

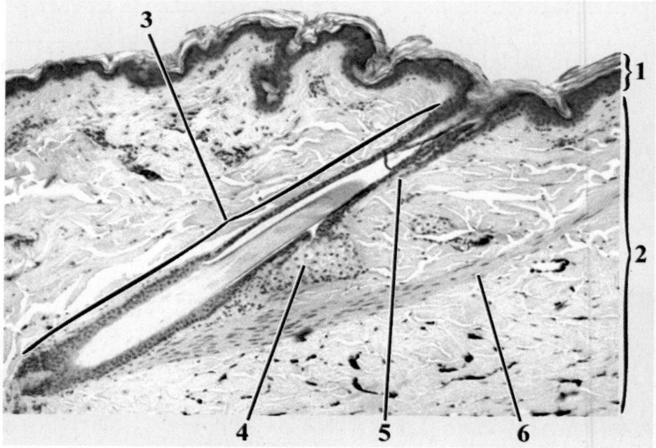

Figure 9.15

115 In reference to **Figure 9.15**, identify #1 - #6.
1 _____ 4 _____
2 _____ 5 _____
3 _____ 6 _____

Chapter 9 - Integumentary System

116 In reference to **Figure 9.15**, what is the function of #1? _____

117 In reference to **Figure 9.15**, what is the function of #2? _____

118 In reference to **Figure 9.15**, what is the function of #4? _____

119 In reference to **Figure 9.15**, what is the function of #6? _____

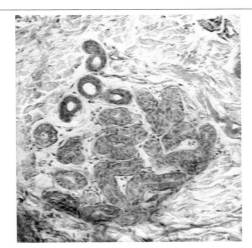

Figure 9.16
120 What accessory structure is shown in **Figure 9.16**? _____

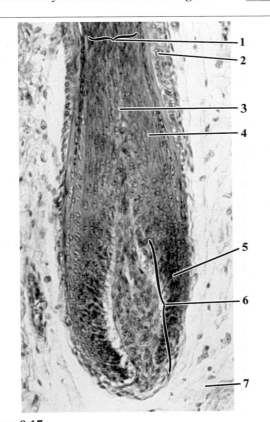

Figure 9.17
121 In reference to **Figure 9.17**, identify #1 - #7.
1 _____   5 _____
2 _____   6 _____
3 _____   7 _____
4 _____

122 In reference to **Figure 9.17**, what is the function of #5? _____

123 In reference to **Figure 9.17**, what is the function of #6? _____

SKIN INNERVATION

124 What are some of the sensory functions associated with skin? _____

125 What are some of the motor functions associated with skin? _____

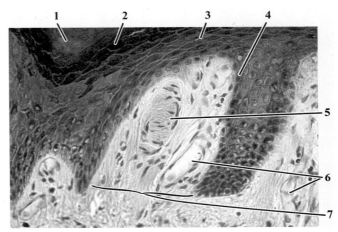

Figure 9.18
126 In reference to **Figure 9.18**, identify #1 - #7.
1 _____   5 _____
2 _____   6 _____
3 _____   7 _____
4 _____

127 In reference to **Figure 9.18**, what is the function of #5? _____

Chapter 9 - Integumentary System

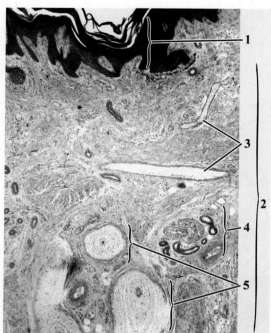

Figure 9.19

128 In reference to **Figure 9.19**, identify #1 - #5.

1 _____   4 _____
2 _____   5 _____
3 _____

129 In reference to **Figure 9.19**, what is the function of #5? __
_____
_____

NAILS

130 What are two functions of nails? _____
_____
_____

131 What are the three parts of a nail?

1 _____   3 _____
2 _____

132 Why is the nail bed pinkish? _____
_____ How would a decrease of blood oxygenation influence the color of the nail beds? _____
_____

133 Where is the root of the nail located? _____
_____

134 What is the function of the nail matrix? _____
_____

135 What is the origin of the cuticle, or eponychium? _____
_____

136 Where is the hyponychium located? _____
_____
_____

Bone Tissue and Bones - Worksheets

BONE STRUCTURE AND CLASSIFICATION

1. What do the bones form? _____

2. What are five functions of bones?
   (1) _____
   (2) _____
   (3) _____
   (4) _____
   (5) _____

3. What are the four classifications of bones?
   1 _____ 3 _____
   2 _____ 4 _____

4. Describe long bones. _____

5. Describe short bones. _____

6. Describe flat bones. _____

7. Describe irregular bones. _____

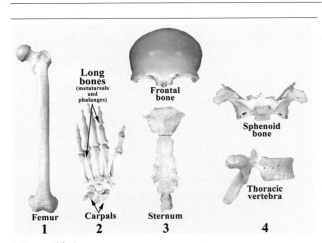

Figure 10.1

8. In reference to **Figure 10.**, identify #1 - #5.
   1 _____ 4 _____
   2 _____ 5 _____
   3 _____

GROSS ANATOMY OF A LONG BONE

9. Where is the diaphysis located? _____

10. What type of bone tissue produces the diaphysis? _____

11. Where is the medullary cavity located? _____

12. Where are the epiphyses located? _____

13. What type of bone tissue mostly forms the epiphyses? __

14. What are trabeculae? _____

15. What is the epiphyseal line? _____

16. Where is the periosteum located? _____

17. Name the two layers of the periosteum and give their location. 1 _____
    2 _____

18. What types of cells are found in the cellular layer of the periosteum? _____

19. What type of tissue forms the fibrous layer of the periosteum? _____

20. What is the function of the fibrous layer of the periosteum? _____

21. Where is the endosteum located? _____

22. What types of cells are found in the endosteum? _____

23. Where is articular cartilage located? _____

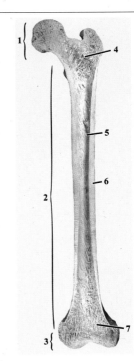

Figure 10.2

24. In reference to **Figure 10.2**, identify #1 - #7.
    1 _____ 5 _____
    2 _____ 6 _____
    3 _____ 7 _____
    4 _____

132 Chapter 10 - Bone Tissue and Bones

Name _____
Class _____

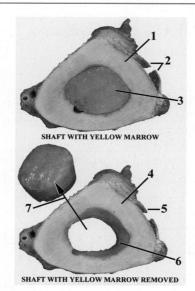

Figure 10.3

25 In reference to **Figure 10.3**, identify #1 -#6.
1 _____   4 _____
2 _____   5 _____
3 _____   6 _____
                              7 _____

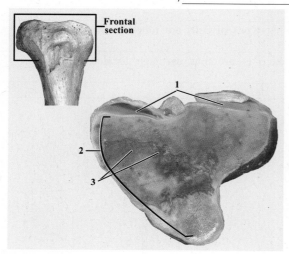

Figure 10.4

26 In reference to **Figure 10.4**, identify #1 -#3.
1 _____   3 _____
2 _____

27 **Match Terms with Descriptions**
   Terms
   A   Articular cartilage
   B   Compact bone
   C   Diaphysis (shaft)
   D   Endosteum
   E   Epiphyseal line
   F   Epiphyses
   G   Medullary (marrow) cavity
   H   Periosteum
   I   Red bone marrow
   J   Spongy bone
   K   Trabeculae
   L   Yellow bone marrow

Descriptions

_____ Composed mostly of osteoprogenitor cells, osteoblasts, and osteoclasts.
_____ Consists mostly of adipocytes and functions as a nutrient (fat) storage site.
_____ Portion of bone that consists mostly of strong, dense bone called compact bone.
_____ Consists of an inner cellular layer and an outer fibrous layer.
_____ Described as proximal or distal according to their respective closer or farther placement from the body.
_____ Expanded ends of the long bone.
_____ Forms an inner framework of spongy bone that is strong, light, and contains red bone marrow.
_____ Forms the longitudinal axis of the bone.
_____ Found within the medullary (marrow) cavity of long bones
_____ Hyaline cartilage that coves the end surfaces of the epiphyses where the bones form a joint.
_____ In adults, it contains mostly fatty (yellow) bone marrow.
_____ Internally, these regions are composed of spongy bone.
_____ Its fibrous layer is mostly dense irregular connective tissue that provides attachment sites for tendons, ligaments, and into the bone itself by perforating fibers (Sharpey's fibers) which penetrate the cellular layer into the matrix of compact bone.
_____ Its cellular layer is composed mostly of osteoprogenitor cells (stem cells), bone-producing cells (osteoblasts), and bone-removing cells (osteoclasts).
_____ Large cavity within the diaphysis of a long bone and partially extends into the epiphyses.
_____ Line of bone formed by trabeculae at the site where a cartilage growth area, the epiphyseal plate, was located.
_____ Located in the spongy bone tissue of the epiphyses of long bones and within the spongy bone tissue of all other bones.
_____ Membrane that covers the outer surface of the diaphysis and the epiphyses (except their ends)
_____ Membrane that lines the medullary cavity, the trabeculae of spongy bone (mostly in the epiphyses), and extends into the central canals of the osteons (Haversian systems).
_____ Strong, dense bone that forms the diaphysis (shaft) of long bones.
_____ Strong, dense bone that forms the exterior surfaces of all bones.
_____ Thin plates of bone that form the internal framework of the epiphyses of long bones and most other bones.
_____ Tissue where the formed elements of blood are produced.
_____ Type of bone that is composed of thin plates called trabeculae.

GROSS ANATOMY OF A FLAT BONE

28 Describe the characteristics of a flat bone. _____
_____

29 Where is the spongy bone tissue of flat bones located? ___
_____

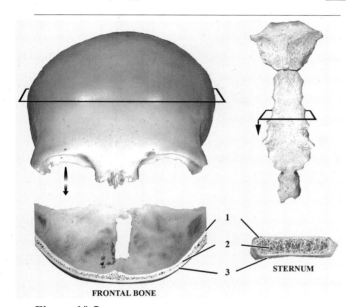

Figure 10.5

30 In reference to **Figure 10.5**, identify #1 -#3.
1 _____ 3 _____
2 _____

MEMBRANES OF BONE

31 Where is the periosteum located? _____
_____
_____

32 What are three functions of the periosteum?
1 _____
_____
2 _____
_____
3 _____
_____

33 What type of tissue is found in the outer layer of the periosteum? _____

34 What are perforating fibers? _____
_____

35 What types of cells are found in the inner cellular layer of the periosteum? _____
_____

36 What is the function of the inner cellular layer of the periosteum? _____
_____

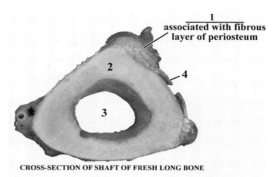

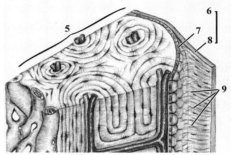

Figure 10.6

37 In reference to **Figure 10.6**, identify #1 - #9.
1 _____ 6 _____
2 _____ 7 _____
3 _____ 8 _____
4 _____ 9 _____
5 _____

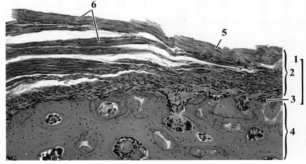

Figure 10.7

38 In reference to **Figure 10.7**, identify #1 - #6.
1 _____ 4 _____
2 _____ 5 _____
3 _____ 6 _____

39 In reference to **Figure 10.7**, what is the function of #2?
_____
_____

40 In reference to **Figure 10.7**, what is the function of #3?
_____
_____

41 In reference to **Figure 10.7**, explain how the collagen fibers at #5 connect into the matrix of bone. _____
_____
_____

42 Give three locations of the endosteum.
1 _____
2 _____
3 _____

Chapter 10 - Bone Tissue and Bones

43. What are three types of cells of the endosteum?
    1 _____ 3 _____
    2 _____
44. What is the function of the endosteum? _____
    _____

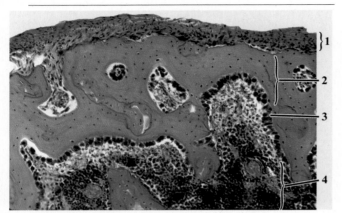

Figure 10.8
45. In reference to **Figure 10.8**, identify #1 - #4.
    1 _____ 3 _____
    2 _____ 4 _____
46. In reference to **Figure 10.8**, what is the function of #3?
    _____
    _____

BONE TISSUE AND STRUCTURE
47. What are the two types of bone tissue?
    1 _____ 2 _____

MATRIX OF BONE TISSUE
48. What is bone matrix? _____
    _____
49. What forms the inorganic portion of the bone matrix? _____
    _____
50. What is the function of the mineral salts of bone? _____
    _____
51. What forms the organic portion of the bone matrix? _____
    _____
52. What is the function of collagenous fibers? _____
    _____

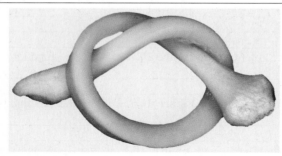

Figure 10.9
53. In reference to **Figure 10.9**, what portion of the bone's matrix is shown? _____ How was this bone processed? _____
    _____

Figure 10.10
54. In reference to **Figure 10.10**, what portion of the bone's matrix is shown? _____ How was this bone processed? _____
    _____

COMPACT BONE
55. Where is compact bone tissue located? _____
    _____
56. What are the structural units of compact bone? _____
    _____
57. How are the lamellae of an osteon organized? _____
    _____
58. Where are the osteocytes of an osteon located? _____
    _____
59. What are interstitial lamellae? _____
    _____
60. What is the location of perforating canals? _____

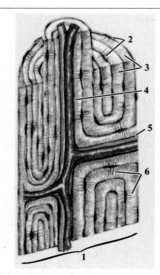

Figure 10.11
61. In reference to **Figure 10.11**, identify #1 - #6.
    1 _____ 4 _____
    2 _____ 5 _____
    3 _____ 6 _____
62. What do central canals contain? _____
    _____
63. Canaliculi are the pathways for what structures? _____
    _____

64 What are lacunae? _____
   _____
65 What are lacunae interconnected with? _____
   _____
66 How do osteocytes stay in communication with the blood vessels in the central canals? _____
   _____
   _____

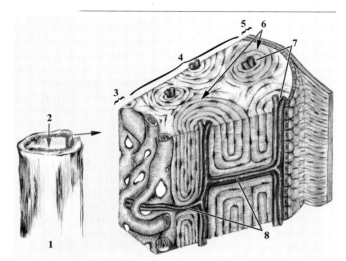

Figure 10.12
67 In reference to **Figure 10.12**, identify #1 - #8.
   1 _____ 5 _____
   2 _____ 6 _____
   3 _____ 7 _____
   4 _____ 8 _____

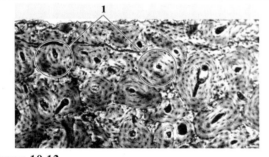

Figure 10.13
68 In reference to **Figure 10.13**, identify #1.
   1 _____

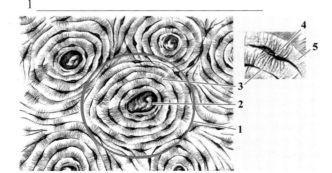

Figure 10.14
69 In reference to **Figure 10.14**, identify #1 - #5.
   1 _____ 4 _____
   2 _____ 5 _____
   3 _____

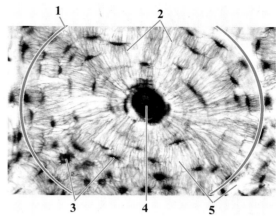

Figure 10.15
70 In reference to **Figure 10.15**, identify #1 - #5.
   1 _____ 4 _____
   2 _____ 5 _____
   3 _____

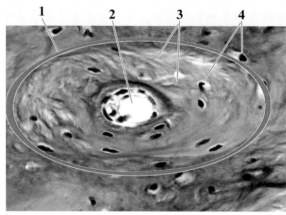

Figure 10.16
71 In reference to **Figure 10.16**, identify #1 - #4.
   1 _____ 3 _____
   2 _____ 4 _____
72 In reference to **Figure 10.16**, what is the organizational unit of the compact bone? _____
73 In reference to **Figure 10.16**, what is the function of #2?
   _____
   _____
74 In reference to **Figure 10.16**, what is the function of #3?
   _____
   _____
75 In reference to **Figure 10.16**, what is the function of #4?
   _____
   _____

Chapter 10 - Bone Tissue and Bones

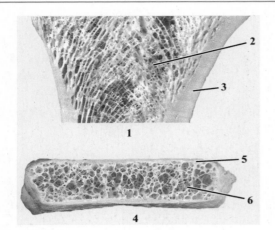

Figure 10.17

76. In reference to **Figure 10.17**, identify #1 - #6.
   1. _____ 4. _____
   2. _____ 5. _____
   3. _____ 6. _____

CELLS OF BONE TISSUE

77. What are four types of cells found in bone tissue?
   1. _____ 3. _____
   2. _____ 4. _____

78. What is the function of osteoprogenitor cells? _____

79. Where are osteoprogenitor cells found? _____

82. What is the function of osteoblasts? _____

81. Where are osteoblasts found? _____

82. What is osteoid? _____

83. What type of cells do osteoblasts differentiate into? _____

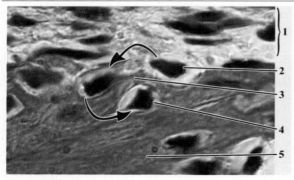

Figure 10.18

84. In reference to **Figure 10.18**, identify #1 - #5.
   1. _____ 4. _____
   2. _____ 5. _____
   3. _____

85. In reference to **Figure 10.18**, explain the events that occur at #6 - #7. _____

86. Where are osteocytes found? _____

87. What is the function of osteocytes? _____

88. What are canaliculi? _____

89. What is found in canaliculi? _____

90. What type of junctions interconnect the cell branches? _____

91. Where is interstitial fluid located? _____

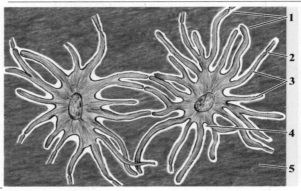

Figure 10.19

92. In reference to **Figure 10.19**, identify #1 - #5.
   1. _____ 4. _____
   2. _____ 5. _____
   3. _____

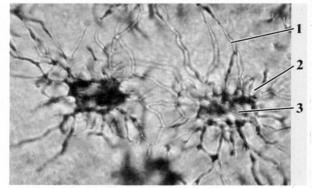

Figure 10.20

93. In reference to **Figure 10.20**, identify #1 - #3
   1. _____ 3. _____
   2. _____

94. In reference to **Figure 10.20**, what is the function of #2? _____

95. In reference to **Figure 10.20**, what is the function of #3? _____

96. Where are osteoclasts located? _____

97. What are two reasons why bone matrix would be destroyed? _____

Chapter 10 - Bone Tissue and Bones **137**

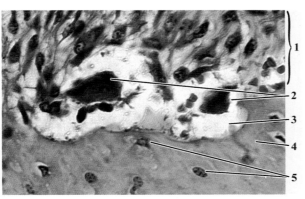

Figure 10.21

98 In reference to **Figure 10.21**, identify #1 - #5.
1 _____ 4 _____
2 _____ 5 _____
3 _____

99 In reference to **Figure 10.21**, what is the function of #2? _____

100 What happens to the minerals that are released from the erosion of the matrix? _____

BONE DEVELOPMENT and GROWTH

101 What is ossification? _____

102 What is calcification? _____

103 What portion of bone tissue is calcified? _____

104 What are the two process of bone formation?
1 _____
2 _____

105 Where does intramembranous ossification occur and what does it produce? _____

106 Where does endochondral ossification occur and what does it produce? _____

INTRAMEMBRANOUS OSSIFICATION

107 When does intramembranous ossification begin? _____

108 What is the function of osteoprogenitor cells? _____

109 What do osteoblasts produce first? _____

110 What happens to ostoid? _____

111 When are osteocytes formed? _____

112 What type of bone tissue is first formed by intramembranous ossification? _____

113 What membrane forms on the surfaces of the trabecular network? _____

114 What is the function of the periosteum of flat bones?____

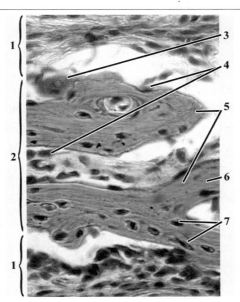

Figure 10.22

115 In reference to **Figure 10.22**, identify #1 - #7.
1 _____ 5 _____
2 _____ 6 _____
3 _____ 7 _____
4 _____

116 In reference to **Figure 10.22**, what is the function of #3? _____

117 In reference to **Figure 10.22**, what is the function of #4? _____

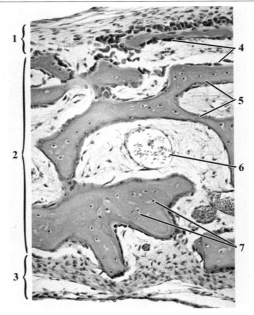

Figure 10.23

118 In reference to **Figure 10.23**, identify #1 - #6.
1 _____ 4 _____
2 _____ 5 _____
3 _____ 6 _____

138 Chapter 10 - Bone Tissue and Bones

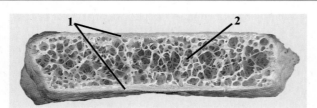

Figure 10.24

119 In reference to **Figure 10.24**, identify #1 - #3.
   1 _____ 2 _____

INTRAMEMBRANOUS OSSIFICATION
FETAL SKULL

120 What are three ways intramembranous ossification is observed in the fetal skull?
   1 _____
   _____
   2 _____
   _____
   3 _____
   _____

121 What are fontanels? _____
   _____

122 Name and give the location of the fontanels.
   1 _____
   _____
   2 _____
   _____
   3 _____
   _____
   4 _____
   _____

123 What is the function of the fontanels? _____
   _____
   _____

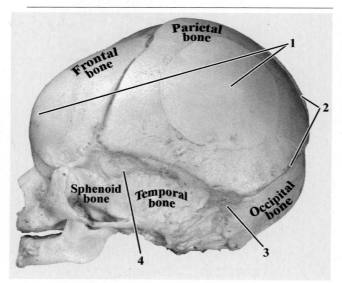

Figure 10.25

124 In reference to **Figure 10.25**, identify #1 - #4.
   1 _____ 3 _____
   2 _____ 4 _____

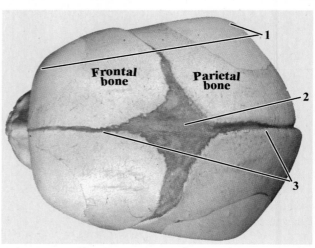

Figure 10.26

125 In reference to **Figure 10.26**, identify #1 - #3.
   1 _____ 3 _____
   2 _____

ENDOCHONDRAL OSSIFICATION

126 What serves as the site and model for endochondral ossification? _____
   _____

127 What is the name of the area of the cartilage diaphysis where endochondral ossification begins? _____
   _____

128 Before hyaline cartilage of the diaphysis undergoes ossification, what happens at the perichondrium? _____
   _____

129 As a bony collar is being produced, what happens to the older cartilage cells? _____
   _____

130 What happens to the cartilage cells when their matrix calcifies? _____
   _____

131 When does ossification begin? _____
   _____
   _____

132 **Match Terms with Descriptions**
   Terms
   A   Diaphysis
   B   Epiphyses
   C   Hypertrophic cartilage
   D   Metaphysis
   E   Perichondrium
   F   Periosteum
   G   Primary ossification center
   H   Trabeculae
   Descriptions
   _____ Area in the center of the hyaline cartilage model (diaphysis) that exhibits cartilage enlargement (hypertrophy), degeneration (cavitation), invasion by blood vessels and osteoprogenitor cells, and osteoblasts form bony trabeculae.
   _____ Area of centrally located chondrocytes that have undergone enlargement (hypertrophy). Also, min-

eralization (calcification) of the cartilage matrix occurs at this time.
_____ Bone plates formed by osteoblasts.
_____ Fibrous connective tissue on the surface of the cartilage.
_____ Hyaline cartilage ends of the diaphysis.
_____ In early development, a thin bone collar surrounds its internal framework of trabeculae.
_____ Its cellular layer produces appositional growth (growth in diameter).
_____ Its outer fibrous layer connects to tendons, ligaments, and by inward penetrating fibers into the bone matrix.
_____ Layer on the surface of bone that consists of an outer fibrous layer and an inner cellular layer.
_____ Region between an epiphysis and the diaphysis.
_____ Region where longitudinal growth occurs.
_____ Replaced by the periosteum which produces an initial bony collar around the cartilage shaft.
_____ Shaft of the bone.

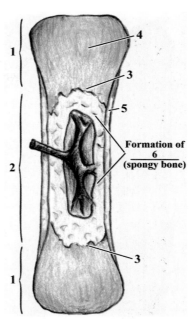

Figure 10.28

135 In reference to **Figure 10.28**, identify #1 - #6.
1 _____ 4 _____
2 _____ 5 _____
3 _____ 6 _____

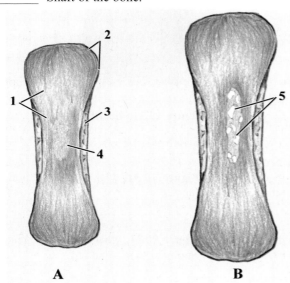

Figure 10.27

133 In reference to **Figure 10.27**, identify #1 - #5.
1 _____ 4 _____
2 _____ 5 _____
3 _____

134 In reference to **Figure 10.27**, what are the next sequential events in the development of #5? _____

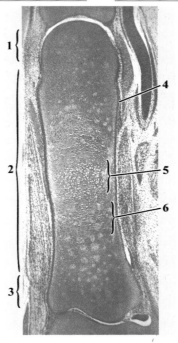

Figure 10.29

136 In reference to **Figure 10.29**, identify #1 - #6.
1 _____ 4 _____
2 _____ 5 _____
3 _____ 6 _____

137 In reference to **Figure 10.29**, what develops at #4? _____

138 In reference to **Figure 10.29**, what is the next sequential event in the development of #5? _____

140   Chapter 10 - Bone Tissue and Bones

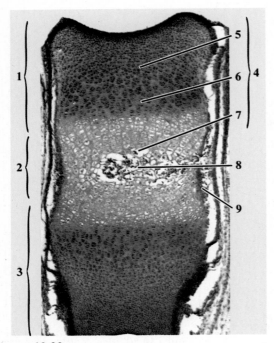

Figure 10.30

139 In reference to **Figure 10.30**, identify #1 - #9.
   1 _____   6 _____
   2 _____   7 _____
   3 _____   8 _____
   4 _____   9 _____
   5 _____

140 In reference to **Figure 10.30**, what is the function of area #5? _____
_____

141 In reference to **Figure 10.30**, what characterizes area #6?
_____
_____

142 In reference to **Figure 10.30**, what characterizes area #7?
_____
_____

143 In reference to **Figure 10.30**, what characterizes area #8?
_____
_____

144 In reference to **Figure 10.30**, what is the next sequential event in the development of #8? _____
_____

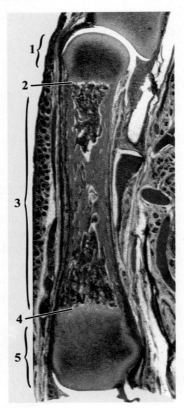

Figure 10.31

145 In reference to **Figure 10.31**, identify #1 - #5.
   1 _____   4 _____
   2 _____   5 _____
   3 _____

146 In reference to **Figure 10.31**, what is the function of #2 (and #4)? _____
_____

147 In reference to **Figure 10.31**, what is the next sequential event in the development of #1 (and #5)? _____
_____

Secondary Ossification Center and Epiphyseal Plate

148 Where does the secondary ossification center form? _____
_____

149 List the sequential events that occur at the secondary (and the primary) ossification centers:
   1 _____
   2 _____
   3 _____
   4 _____
   5 _____
   6 _____
   7 _____
   8 _____
   9 _____
   10 _____
   11 _____

150 What type of tissue forms the epiphyseal plate? _____
_____

Name _____
Class _____

Chapter 10 - Bone Tissue and Bones   141

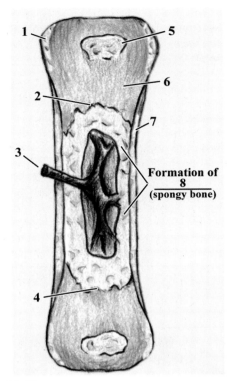

Figure 10.32

151 In reference to **Figure 10.32**, identify #1 - #8.
  1 _____ 5 _____
  2 _____ 6 _____
  3 _____ 7 _____
  4 _____ 8 _____

152 In reference to **Figure 10.32**, what are the next sequential events in the development of #4 ? _____
  _____
  _____

153 What is a metaphysis? _____
  _____

154 What hyaline cartilage region is located near the epiphyseal surface of the epiphyseal plate? _____
  _____

155 The region where cartilage cells undergo growth and maturation is called the _____

156 What happens to the matrix surrounding the hypertrophic cartilage? _____
  _____

157 What happens to the cartilage cells when their surrounding matrix calcifies? _____

158 What happens at the diaphyseal surface (the metaphysis) of the epiphyseal plate? _____
  _____
  _____

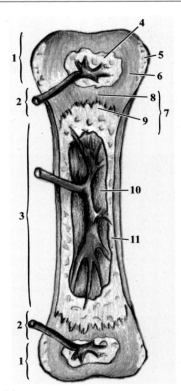

Figure 10.33

159 In reference to **Figure 10.33**, identify #1 - #11.
  1 _____ 7 _____
  2 _____ 8 _____
  3 _____ 9 _____
  4 _____ 10 _____
  5 _____ 11 _____
  6 _____

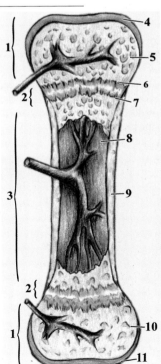

Figure 10.34

160 In reference to **Figure 10.34**, identify #1 - #11.

142 Chapter 10 - Bone Tissue and Bones

Name _____
Class _____

1 _____ 7 _____
2 _____ 8 _____
3 _____ 9 _____
4 _____ 10 _____
5 _____ 11 _____
6 _____

161 In reference to **Figure 10.34**, what is the function of #2?
_____
_____

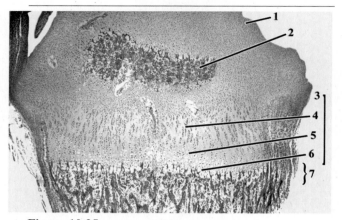

Figure 10.35

162 In reference to **Figure 10.35**, identify #1 - #7.
1 _____ 5 _____
2 _____ 6 _____
3 _____ 7 _____
4 _____

163 In reference to **Figure 10.35**, what is the function of #4?
_____
_____

164 In reference to **Figure 10.35**, what is occurring at #5?
_____
_____

165 In reference to **Figure 10.35**, why do cavities form at #6?
_____
_____

166 In reference to **Figure 10.35**, bone growth (ossification) at #7 produces an increase in bone _____ (length or diameter).

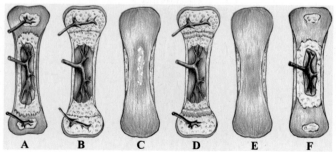

Figure 10.36

167 In reference to **Figure 10.36**, list the bone development sketches (A - E) in the order of increasing maturity.
_____

GROWTH OF THE DIAPHYSIS
LONGITUDINAL GROWTH (GROWTH IN LENGTH)

168 What are the two types of bone growth:
1 _____
2 _____

169 Where does longitudinal growth occur? _____
_____

170 Where does appositional growth occur? _____
_____

171 What is the metaphysis? _____
_____
_____

172 What are three events that occur at the epiphyseal surface of the metaphysis?
1 _____
2 _____
3 _____

173 What are two events that occur at the diaphyseal surface of the metaphysis?
1 _____
2 _____

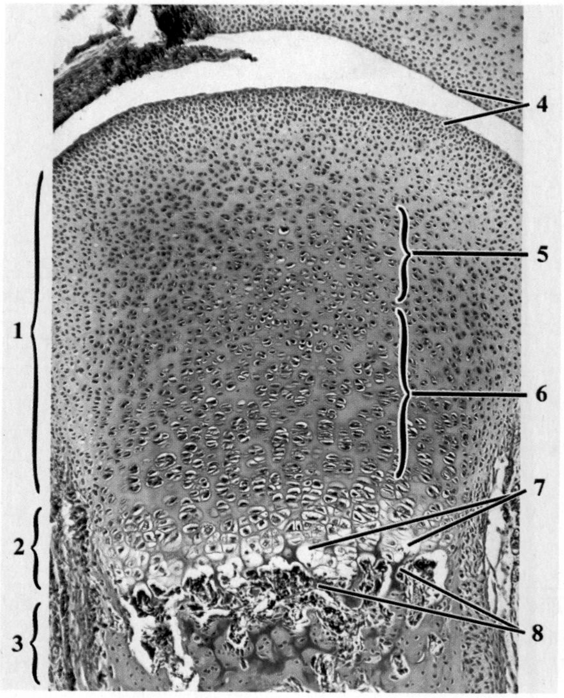

Figure 10.37

174 In reference to **Figure 10.37**, identify #1 - #6.
1 _____ 5 _____
2 _____ 6 _____
3 _____ 7 _____
4 _____ 8 _____

175 At puberty the epiphyseal plates gradually narrow as cartilage growth activity decreases, and osteoblast growth activity increases mostly due to _____.

176 What is an epiphyseal plate? _____
_____

177 What is an epiphyseal line? _____
_____

Chapter 10 - Bone Tissue and Bones   143

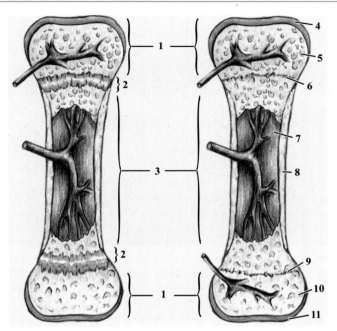

Figure 10.38
178 In reference to **Figure 10.38**, identify #1 - #13.
1 _____   7 _____
2 _____   8 _____
3 _____   9 _____
4 _____   10 _____
5 _____   11 _____
6 _____

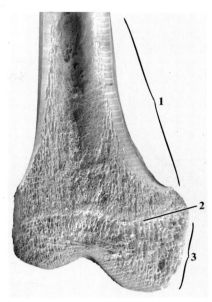

Figure 10.40
180 In reference to **Figure 10.40**, identify #1 - #3.
1 _____   3 _____
2 _____

APPOSITIONAL GROWTH (GROWTH IN DIAMETER)
181 Where does growth that increases bone diameter occur?
_____
_____

182 What happens to periosteal blood vessels? _____
_____

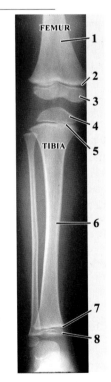

Figure 10.39
179 In reference to **Figure 10.39**, identify #1 - #8.
1 _____   5 _____
2 _____   6 _____
3 _____   7 _____
4 _____   8 _____

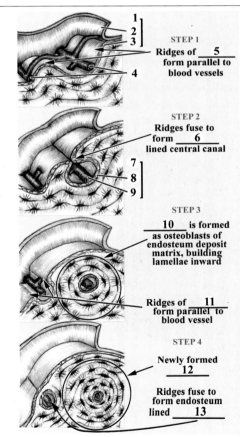

Figure 10.41
183 In reference to **Figure 10.41**, identify #1 - #13.

144 Chapter 10 - Bone Tissue and Bones

Name _____
Class _____

1 _____ 8 _____
2 _____ 9 _____
3 _____ 10 _____
4 _____ 11 _____
5 _____ 12 _____
6 _____ 13 _____
7 _____

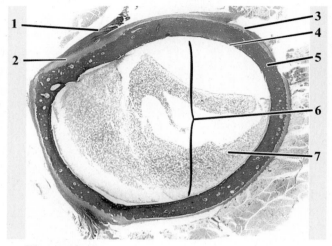

Figure 10.42

184 In reference to **Figure 10.42**, identify #1 - #7.
1 _____ 4 _____
2 _____ 5 _____
3 _____ 6 _____
                           7 _____

185 In reference to **Figure 10.42**, what is the function of #2?
_____
_____

186 In reference to **Figure 10.42**, what is the function of #3?
_____
_____

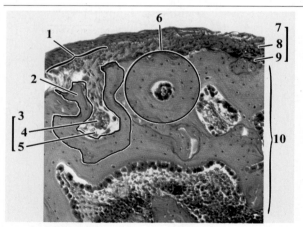

Figure 10.43

187 In reference to **Figure 10.43**, identify #1 - #9.
1 _____ 6 _____
2 _____ 7 _____
3 _____ 8 _____
4 _____ 9 _____
5 _____ 10 _____

Figure 10.44

188 In reference to **Figure 10.44**, identify #1 - #9.
1 _____ 6 _____
2 _____ 7 _____
3 _____ 8 _____
4 _____ 9 _____
5 _____

189 In reference to **Figure 10.44**, what is occurring at #1?
_____
_____

190 In reference to **Figure 10.44**, the growth produced at #5 increases the bone's _____ (length or diameter)?

191 In reference to **Figure 10.44**, what is the function of #4?
_____

192 In reference to **Figure 10.44**, what is the function of #7?
_____
_____

BONE DYNAMICS
(GROWTH, REMODELING, AND MAINTENANCE)

193 Give three reasons why bone remodeling is a lifelong process. _____
_____
_____

194 What roles do dietary protein and mineral salts play in the formation of bone matrix? _____
_____

195 What are the functions of vitamin C and vitamin D? ____
_____
_____

196 What effects do lifestyle have on bone structure? _____
_____

197 What is the function of growth hormone? _____
_____

198 What is the function of thyroxine? _____
_____

199 What two hormones regulate blood ionic calcium level? _____
_____

200 What two hormones are produced by the thyroid gland?
_____

201 What hormone is produced by the parathyroid gland?
_____

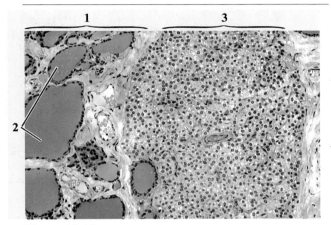

Figure 10.45

202 In reference to **Figure 10.45**, identify #1 - #3.
  1 _____ 3 _____
  2 _____

Calcitonin

203 When is calcitonin produced? _____
_____

204 What is the function of calcitonin? _____
_____

205 What are the two targets of calcitonin?
  1 _____ 2 _____

206 What is the effect of calcitonin on osteobasts? _____
_____

207 What is the effect of calcitonin on the kidney? _____
_____

208 When calcium levels are high, intestinal absorption of calcium is _____ (low or high) due to _____ (increased or decreased) levels of parathyroid hormone.

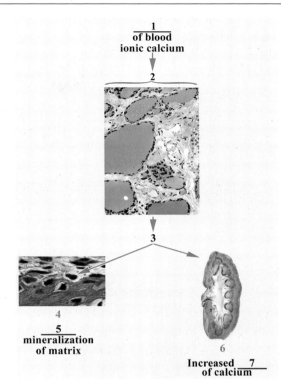

Figure 10.46

209 In reference to **Figure 10.46**, complete #1 - #7.
  1 _____
  2 _____
  3 _____
  4 _____
  5 _____
  6 _____
  7 _____

146 Chapter 10 - Bone Tissue and Bones

Parathyroid hormone

210 When is parathyroid hormone produced? _____

211 What is the function of parathyroid hormone? _____

212 What are three targets of parathyroid hormone?
1 _____ 3 _____
2 _____

213 What is the effect of parathyroid hormone on osteoclasts? _____

214 What is the effect of parathyroid hormone on the kidney? _____

215 What is the effect of parathyroid hormone on the intestines? _____

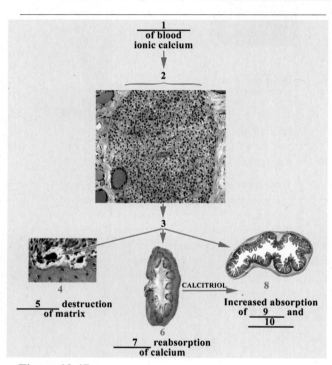

Figure 10.47

216 In reference to **Figure 10.47**, complete #1 - #7.
1 _____
2 _____
3 _____
4 _____
5 _____
6 _____
7 _____
8 _____
9 _____
10 _____

BONE REMODELING

217 What two processes are involved in bone remodeling?
1 _____ 2 _____

218 What two layers of bone are involved in remodeling?
1 _____ 2 _____

219 How are interstitial lamellae formed? _____

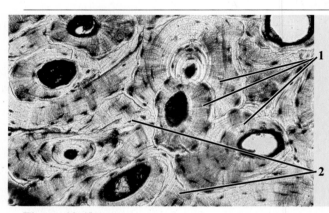

Figure 10.48

220 In reference to **Figure 10.48**, identify #1 and #2.
1 _____ 2 _____

The Skeleton - Worksheets

1. What forms the skeletal system? _____
   _____
2. What are six functions of bones?
   1 _____
   2 _____
   3 _____
   4 _____
   5 _____
   6 _____
3. What is the function of the skeletal cartilages? _____
   _____
4. What is the function of the skeletal ligaments? _____
   _____
5. What is the function of the skeletal joints? _____
   _____
6. What are the three types of structural joints?
   1 _____ 3 _____
   2 _____
7. How is a fibrous joint formed? _____
   _____
8. Functionally, fibrous joints provide for _____
   _____ with _____ movement.
9. How is a cartilaginous joint formed? _____
   _____
10. Functionally, cartilaginous joints provide for _____
    _____ with _____ movement.
11. How is a synovial joint formed? _____
    _____
12. Functionally, synovial joints are _____
    _____ reinforced by _____.

BONE TERMINOLOGY

13. **Match Terms with Descriptions**
    Terms
    A Canal   H Head
    B Condyle   I Line
    C Crest   J Process
    D Epicondyle   K Shaft
    E Facet   L Spine
    F Foramen   M Trochanter
    G Fossa   N Tubecle
                        O Tuberosity

    Descriptions
    _____ Enlarged end of a bone, often rounded.
    _____ Large, rounded, roughened process.
    _____ Long, middle portion of a long bone.
    _____ Narrow, sharp projection.
    _____ Narrow ridge of bone.
    _____ Passageway through the bone
    _____ Process located above a condyle that functions as an attachment site for tendons or ligaments.
    _____ Projection from the surface of the bone.
    _____ Ridge of bone that is less prominent than a crest.
    _____ Round or oval hole in a bone that functions as a passageway for nerves, blood vessesl, etc.
    _____ Rounded projection that has a smooth surface for articulation to another bone
    _____ Shallow depression
    _____ Small, rounded, roughened process.
    _____ Smooth, flat surface that functions as a site for articulation.
    _____ Very large, blunt process.

DIVISIONS OF THE SKELETON

14. What are the two major divisions of the skeleton?
    1 _____ 2 _____
15. What are four divisions of the axial skeleton?
    1 _____ 3 _____
    2 _____ 4 _____
16. What are four divisions of the appendicular skeleton?
    1 _____ 3 _____
    2 _____ 4 _____

AXIAL SKELETON

17. What are the two major regions of the skull?
    1 _____ 2 _____
18. What are the location and function of the cranial bones? __
    _____
19. Name the cranial bones.
    1 _____ 4 _____
    2 _____ 5 _____
    3 _____ 6 _____
20. What are sutures? _____
21. Name and give the location of the sutures associated with the parietal bones.
    1 _____
    _____
    2 _____
    _____
    3 _____
    _____
    4 _____
    _____
22. Name the superficial facial bones.
    1 _____ 4 _____
    2 _____ 5 _____
    3 _____

148 Chapter 11 - The Skeleton

Name _____
Class _____

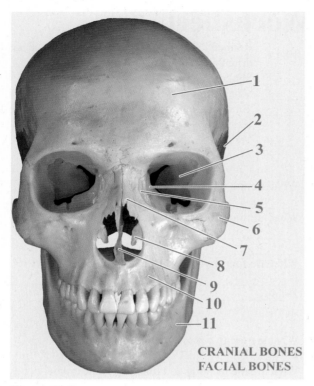

CRANIAL BONES
FACIAL BONES

23 **Figure 11.1**
In reference to **Figure 11.1**, identify #1 - #9.

1 _____   6 _____
2 _____   7 _____
3 _____   8 _____
4 _____   9 _____
5 _____

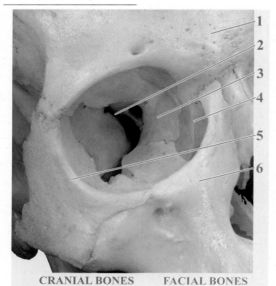

CRANIAL BONES   FACIAL BONES

Figure 11.2
24 In reference to **Figure 11.2**, identify #1 - #6.

1 _____   5 _____
2 _____   6 _____
3 _____
4 _____

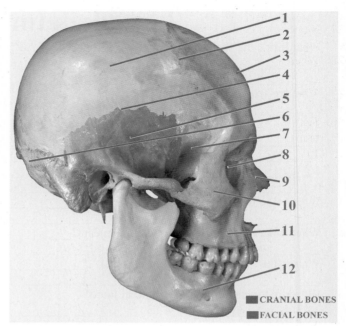

■ CRANIAL BONES
■ FACIAL BONES

Figure 11.3
25 In reference to **Figure 11.3**, identify #1 - #12.

1 _____   7 _____
2 _____   8 _____
3 _____   9 _____
4 _____   10 _____
5 _____   11 _____
6 _____   12 _____

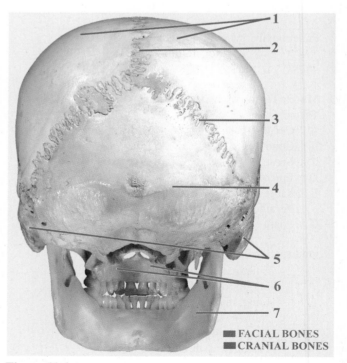

■ FACIAL BONES
■ CRANIAL BONES

Figure 11.4
26 In reference to **Figure 11.4**, identify #1 - #7.

1 _____   5 _____
2 _____   6 _____
3 _____   7 _____
4 _____

SKULL'S ASSOCIATED BONES

27 Name the bones that are associated with the skull.
    1 _____ 2 _____
28 Where is the hyoid bone located? _____
    _____
29 Where are the ossicles located? _____
    _____
30 Starting with the ossicle attached to the tympanic membrane, list sequentially the bones in a set of ossicles.
    1 _____ 2 _____ 3 _____

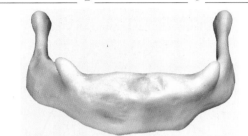

Figure 11.5
31 In reference to **Figure 11.5**, identify the bone. _____

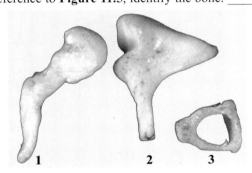

Figure 11.6
32 In reference to **Figure 11.6**, identify #1 - #3.
    1 _____ 2 _____ 3 _____

VERTEBRAL COLUMN

33 How many vertebrae are found in the vertebral column?
    _____
34 Name the five regions of the vertebral column and give the number of vertebrae associated with each
    1 _____, # _____   4 _____, # _____
    2 _____, # _____   5 _____, # _____
    3 _____, # _____
35 The intervertebral joints are formed from _____ and are located _____
    _____
36 The joint between C_1 and C_2 is a synovial _____ joint that provides _____ movement.

Cervical division
37 How many vertebrae are in the cervical division? _____
38 The cervical vertebrae are the bones of the _____.
39 The first cervical vertebra, C_1, is named the _____.
40 The articulation between the atlas and the occipital bone forms a synovial _____ joint that allows _____
    _____

Thoracic division
41 How many vertebrae are in the thoracic division? _____
42 Each thoracic vertebra articulates with a _____.
43 The thoracic vertebrae form the posterior aspect of the
    _____.

Lumbar division
44 How many vertebrae are in the lumbar division? _____
45 What bone is located inferior to the lumbar vertebrae?
    _____

Sacrum
46 Where is the sacrum located? _____
    _____
47 The sacrum is formed from _____ fused sacral vertebrae.
48 The sacrum has _____ articulations.
49 Superiorly, the sacrum articulates with the _____ sacral vertebra and inferiorly with the _____; each articulation forms a _____ joint, called a _____ _____ joint.
50 Laterally, the sacrum articulates with each _____ bone and forms a synovial _____ joint. The joint is called the _____ joint and allows _____ movements.

Coccyx
51 The coccyx is located _____
    _____
52 The coccyx is formed from _____ fused coccygeal vertebrae.
53 The coccyx articulates with the _____ at a fibrocartilage _____ joint.

150 Chapter 11 - The Skeleton

Name _____
Class _____

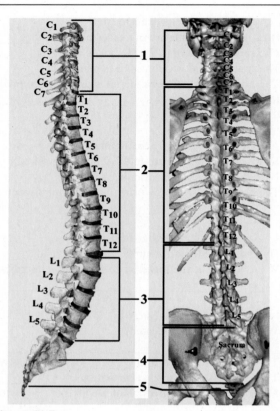

Figure 11.7

54 In reference to **Figure 11.7**, identify #1 - #5.
  1 _____   4 _____
  2 _____   5 _____
  3 _____

THORAX (THORACIC CAGE)
55 What forms the thorax (the thoracic cage)?
  1 _____   3 _____
  2 _____

Thoracic vertebrae
56 What forms the posterior boundary of the thorax? _____
_____
57 Each thoracic vertebra articulates with a _____ of ribs.

Rib cage
58 What forms the rib cage? _____
_____
59 Each rib articulates posteriorly with a _____.
60 Each rib has _____ vertebral articulation sites, except the floating ribs, each has _____.
60 The rib-vertebra articulations are synovial _____ joints.
62 Rib pairs #1 - #7 are commonly called the _____, or named by their attachment the _____.
63 The ribs attach to the sternum by the _____.
64 The costal cartilages function as _____ joints, called the _____ joints.
65 Rib pairs 8 - 12 are commonly called the _____.
66 The first three pairs of false ribs are joined by _____. Named by their articulation sites, they are called the _____ ribs.
67 The last two pairs of false ribs are commonly called the _____. Named by their articulation sites, they are called the _____ ribs.

Sternum
68 The anterior boundary of the thorax is formed by the _____.
69 The sternum articulates with the _____, and the _____.
70 The sternal articulation with the clavicle is a synovial ____ _____ that allows limited movement is most directions.
71 The sternal articulation with the ribs is by the _____ cartilages.
72 The costal cartilages function as _____ joints, the _____ joints, which allow slight movements between the sternum and the _____.

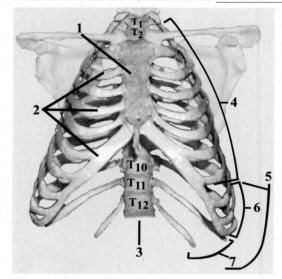

Figure 11.8
73 In reference to **Figure 11.8**, identify #1 - #7.
  1 _____   5 _____
  2 _____   6 _____
  3 _____   7 _____
  4 _____

APPENDICULAR SKELETON
74 What are the four divisions of the appendicular skeleton?
  1 _____   3 _____
  2 _____   4 _____

PECTORAL (SHOULDER) GIRDLE

75 What bones form the pectoral girdle?
1 _____ 2 _____

76 The clavicle articulates medially with the _____ and laterally with the _____.

77 The clavicle's articulation with the sternum is a synovial _____ joint that allows _____ in most directions.

78 The clavicle's articulation with the scapula is a synovial _____ joint.

79 The scapula articulates medially with the _____ and laterally with the _____.

80 The scapula's articulation with the head of the humerus forms a synovial _____ joint, commonly called the _____ joint.

81 The shoulder joint allows a _____ range of movement in _____ directions.

82 The scapula's articulation with the clavicle is a synovial _____ joint.

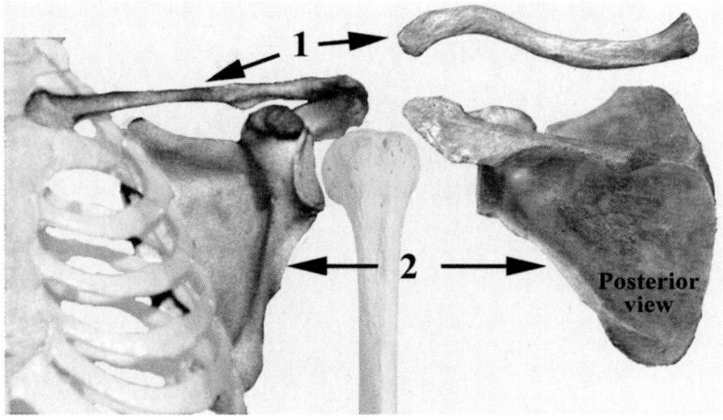

Figure 11.9

83 In reference to **Figure 11.9**, identify bones #1 - #2.
1 _____ 2 _____

UPPER LIMB

84 What bones form the upper limb?
1 _____ 4 _____
2 _____ 5 _____
3 _____ 6 _____

Arm

85 The bone of the arm is the _____.

86 Proximally, the humerus articulates with the _____ and forms a synovial _____ joint, commonly called the _____.

87 The shoulder joint allows a _____ range of movement in _____ directions.

88 Distally, the humerus articulates with the _____ and the _____ and forms a synovial _____ joint, commonly called the _____ joint.

89 The distal medial surface of the humerus articulates with the _____ and the distal lateral surface articulates with the _____.

Forearm

90 The two bones of the forearm are the medial _____ and the lateral _____.

Radius

91 Proximally, the radius articulates with both the _____ and the _____.

92 The articulation of the head of the radius with the distal _____ surface of the humerus forms a portion of the synovial _____ joint, commonly called the _____ joint.

93 The _____ articulation of the head of the radius with the ulna forms a synovial _____ joint allowing _____ of the forearm and hand.

94 Rotation of the forearm and hand resulting in the palm facing backward or downward is _____.

95 Rotation of the forearm and hand resulting in the palm facing forward or upward is _____.

96 Distally, the radius articulates with _____ and forms a synovial _____ joint, commonly called the _____ joint.

Ulna

97 Proximally, the ulna articulates with the _____ and forms a portion of the synovial _____ joint, commonly called the _____ joint.

98 Proximally and _____ the ulna articulates with the radius and forms a synovial _____ joint that allows _____ of the forearm and hand.

99 Rotation of the forearm and hand resulting in the palm facing backward or downward is _____.

100 Rotation of the forearm and hand resulting in the palm facing forward or upward is _____.

101 Distally, the ulna articulates with an articular _____, which separates the ulna from the _____.

Hand

102 The three groups of bones of the hand are the
1 _____ 3 _____
2 _____

103 Proximally, two carpal bones articulate with the _____ to form a synovial _____ joint, commonly called the _____ joint.

152   Chapter 11 - The Skeleton

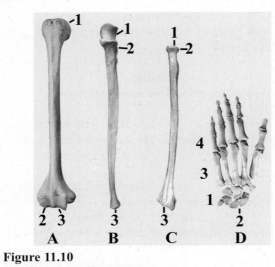

Figure 11.10

104 In reference to **Figure 11.10 - A**
   • identify the bone at A _____
   • #1 articulates with the _____ and forms a synovial _____ joint, commonly called the _____ joint.
   • #2 articulates with the _____, and forms a portion of the synovial _____ joint, commonly called the _____ joint
   • #3 articulates with the _____, and forms a portion of the synovial _____ joint, commonly called the _____ joint.

105 In reference to **Figure 11.10 - B**
   • identify the bone at B _____
   • #1 articulates with _____ and forms a synovial _____ joint, a portion of the joint commonly called the _____ joint.
   • #2 articulates with the _____, and forms a synovial _____ joint that allows a _____ of the forearm and hand.
   • Rotation of the forearm and hand resulting in the palm facing backward or downward is _____.
   • Rotation of the forearm and hand resulting in the palm facing forward or upward is _____.
   • #3 articulates with an _____, which _____ the carpals from the ulna.

106 In reference to **Figure 11.10 - C**
   • identify the bone at C _____
   • #1 articulates with the _____ and forms a synovial _____ joint, a portion of the joint commonly called the _____ joint.
   • #2 articulates with the _____, and forms a synovial _____ joint that allows _____ of the forearm and hand.
   • Rotation of the forearm and hand resulting in the palm facing backward or downward is _____.
   • Rotation of the forearm and hand resulting in the palm facing forward or upward is _____.
   • #3 articulates with _____, and forms a synovial _____ joint, commonly called the _____ joint.

107 In reference to **Figure 11.10 - D**
   • identify the bones at #1 _____
   • the two bones at #2 articulate with the _____, and form a synovial _____ joint, commonly called the _____ joint.
   • identify the group of bones at #3 _____
   • identify the group of bones at #4 _____

PELVIC (HIP) GIRDLE

108 What bones form the pelvic girdle? _____
   _____

119 Anteriorly, the two coxal bones articulate at a _____ joint commonly called the _____.

110 The pubis symphysis is formed from _____ cartilage and is _____ moveable.

111 Posteriorly, the two coxal bones articulate with the _____.

112 Each coxa articulates with the sacrum and forms a joint commonly called the _____.

113 The sacroiliac joint is a synovial _____ joint and has _____ movement.

114 Laterally, the coxal bones (coxae) articulate with the _____, and each forms a synovial _____ joint commonly called the _____.

115 The ball-and-socket joints, the hip joints, allow _____ _____ of movements in all directions.

Name ______________________
Class ______________________

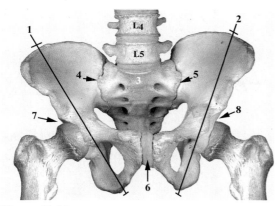

Figure 11.11

116 In reference to **Figure 11.11**, identify #1 - #8.

1 ______________ 5 ______________
2 ______________ 6 ______________
3 ______________ 7 ______________
4 ______________ 8 ______________

Coxal bones (coxae)

117 A coxal bone (coxa) is formed from the fusion of three bones, the:

1 ______________ 3 ______________
2 ______________

118 Posteriorly, the ilium of a coxa articulates with the __________ at the synovial __________ joint commonly called the ______________.

119 Anteriorly, the pubis of a coxa articulates with the ______ ______________, at the ______________ joint commonly called the ______________.

120 Laterally, a large socket of the coxa (the acetabulum) articulates with the ______________ and forms a synovial ______________ joint, commonly called the ________.

Right coxa, medial view   Right coxa, lateral view

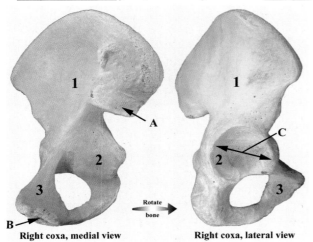

Figure 11.12

121 In reference to **Figure 11.12**, identify #1 - #3.

1 ______________ 3 ______________
2 ______________

122 In reference to **Figure 11.12**, "A" articulates with the ____ at the synovial ______________ joint commonly called the ____________.

122 In reference to **Figure 11.12**, "B" articulates with _______ ______________ at the ______________ joint, commonly called the ______________.

124 In reference to **Figure 11.12**, "C" articulates with the ___ ______________ at the synovial ______________ joint, commonly called the _______ joint.

Pelvis

125 What bones form the pelvis?

1 ______________ 3 ______________
2 ______________

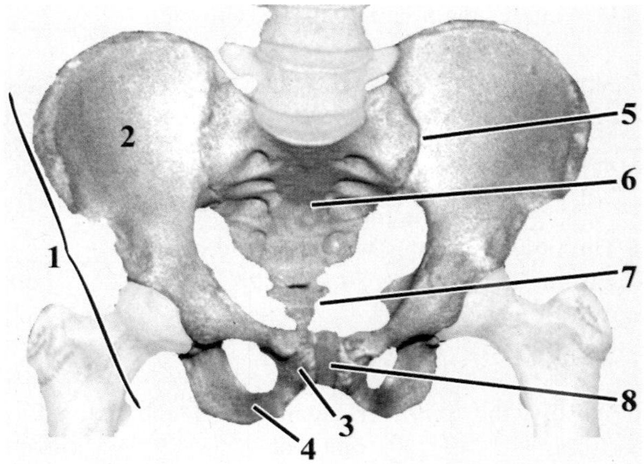

Figure 11.13

126 In reference to **Figure 11.13**, identify #1 - #8.

1 ______________ 5 ______________
2 ______________ 6 ______________
3 ______________ 7 ______________
4 ______________ 8 ______________

LOWER LIMB

127 The bones of the lower limb are the:

1 ______________ 5 ______________
2 ______________ 6 ______________
3 ______________ 7 ______________
4 ______________

Thigh

128 The bone of the thigh is the ______________.

129 Proximally, the medial _____ shaped head of the femur articulates with a lateral _________ of the coxa forming a synovial ______________ joint, commonly called the _____ joint.

130 Distally, the femur articulates with the __________ and the __________.

131 Distally, the femur's smooth _________ surface articulates with the __________, and forms a synovial __________ joint.

132 Distally, the femur's medial and lateral ______________ articulate with the _________ medial and lateral condyles, and form a synovial __________ joint.

133 The articulations formed by the femur-patella and femur-tibia form a complex synovial joint, commonly called the ________ joint.

154 Chapter 11 - The Skeleton

Patella
134 The patella articulates with the smooth anterior distal surface of the _____, and forms a synovial _____ joint. The patella-femur articulation is the anterior joint of the _____ joint.

Leg
135 The bones of the leg are the medial _____ and the lateral _____.

Tibia
136 Proximally, the tibia articulates with the _____ and the _____.
137 Superiorly, the tibia's proximal condyles (medial and lateral condyles) articulate with the femur's distal _____, and forms a synovial _____ joint; together with the femur-patella articulation, the _____ joint is formed.
138 The tibia articulates with the head of the _____ at the inferior surface of the tibia's _____ condyle, and forms a synovial _____ joint.
139 Distally, the tibia articulates with two bones: the _____ and the _____.
140 Distally, the tibia's articulation with the fibula forms a strong _____ joint that _____ movement.
141 Distally, the tibia forms a synovial articulation with a tarsal bone, the _____.
142 The tibia-talus articulation is a synovial _____ joint and forms a portion of the joint commonly called the _____.
143 The ankle joint is formed by the distal synovial articulations of the tibia and the _____ with the talus.

Fibula
144 Proximally, the head of the fibula articulates at the inferior surface of the tibia's _____, and forms a synovial _____ joint.
145 Distally, the fibula articulates with two bones: the _____ and the carpal bone, the _____.
146 Distally, the fibula articulates with the lateral surface of the carpal bone, the _____.
147 Distally, the fibula articulates with the_____ and forms a strong _____ joint that _____ movement.
148 The articulations of the fibula-talus and tibia-talus form a synovial _____ joint, commonly called the _____ joint.

Foot
149 The three regions of foot bones are the:
1 _____ 3 _____
2 _____
150 The tarsal bones are a group of _____ bones.
151 The talus articulates with two bones: the _____ and the _____, and forms a synovial _____ joint commonly called the _____ joint.
152 The metatarsal bones are a group of _____ bones.
153 The phalanges are the bones of the _____. Each toe has _____ phalanges except the great (big) toe which has _____ phalanges.

In reference to **Figure 11.14**, identify #1 - #32.

1 _____ 17 _____
2 _____ 18 _____
3 _____ 19 _____
4 _____ 20 _____
5 _____ 21 _____
6 _____ 22 _____
7 _____ 23 _____
8 _____ 24 _____
9 _____ 25 _____
10 _____ 26 _____
11 _____ 27 _____
12 _____ 28 _____
13 _____ 29 _____
14 _____ 30 _____
15 _____ 31 _____
16 _____ 32 _____

Name _____
Class _____

Chapter 12 - Axial Skeleton   155

Axial Skeleton - Worksheets

1. The axial skeleton forms the _____ axis of the body and includes the bones of the
   1 _____   3 _____
   2 _____   4 _____

   ### SKULL
2. The two major region of the skull are the
   1 _____   2 _____

   ### Cranial Bones
3. The cranial bones are the bones that _____
   _____.
4. The cranial bones are the
   1 _____   4 _____
   2 _____   5 _____
   3 _____   6 _____

   ### Facial Bones
5. The facial bones are the bones that _____.
6. The superficial facial bones are the
   1 _____   4 _____
   2 _____   5 _____
   3 _____
7. The deep facial bones are the
   1 _____   2 _____

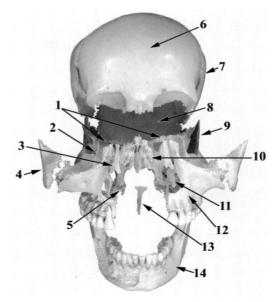

Figure 12.2

9. In reference to **Figure 12.2**, identify #1 - #14.
   1 _____   8 _____
   2 _____   9 _____
   3 _____   10 _____
   4 _____   11 _____
   5 _____   12 _____
   6 _____   13 _____
   7 _____   14 _____

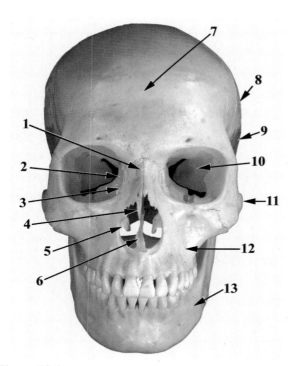

Figure 12.1

8. In reference to **Figure 12.1**, identify #1 - #13.
   1 _____   8 _____
   2 _____   9 _____
   3 _____   10 _____
   4 _____   11 _____
   5 _____   12 _____
   6 _____   13 _____
   7 _____

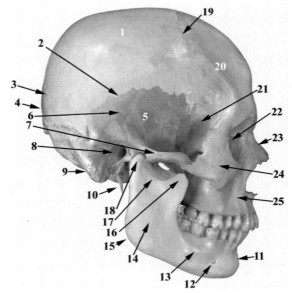

Figure 12.3

10. In reference to **Figure 12.3**, identify #1 - #25.
    1 _____   14 _____
    2 _____   15 _____
    3 _____   16 _____
    4 _____   17 _____
    5 _____   18 _____
    6 _____   19 _____
    7 _____   20 _____

156 Chapter 12 - Axial Skeleton

Name _____
Class _____

8 _____   21 _____
9 _____   22 _____
10 _____   23 _____
11 _____   24 _____
12 _____   25 _____
13 _____

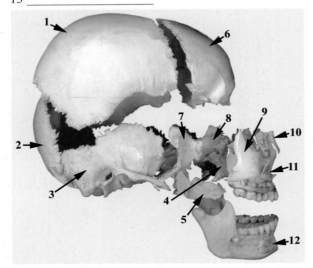

Figure 12.4

11 In reference to **Figure 12.4**, identify #1 - #12.

1 _____   7 _____
2 _____   8 _____
3 _____   9 _____
4 _____   10 _____
5 _____   11 _____
6 _____   12 _____

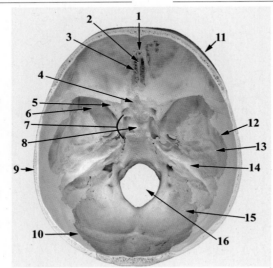

Figure 12.5

12 In reference to **Figure 12.5**, identify #1 - #16.

1 _____   9 _____
2 _____   10 _____
3 _____   11 _____
4 _____   12 _____
5 _____   13 _____
6 _____   14 _____
7 _____   15 _____
8 _____   16 _____

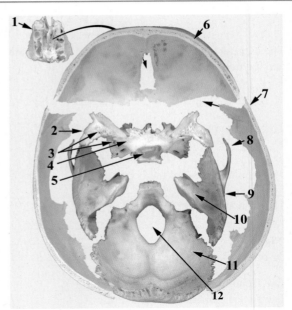

Figure 12.6

13 In reference to **Figure 12.6**, identify #1 - #12.

1 _____   7 _____
2 _____   8 _____
3 _____   9 _____
4 _____   10 _____
5 _____   11 _____
6 _____   12 _____

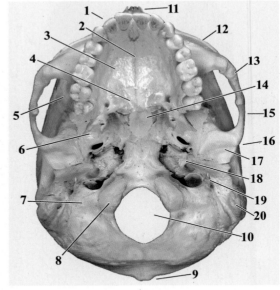

Figure 12.7

14 In reference to **Figure 12.7**, identify #1 - #19.

1 _____   10 _____
2 _____   11 _____
3 _____   12 _____
4 _____   13 _____
5 _____   14 _____
6 _____   15 _____
7 _____   16 _____
8 _____   17 _____
9 _____   18 _____
                             19 _____

Chapter 12 - Axial Skeleton 157

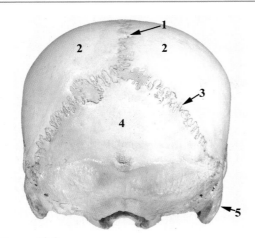

Figure 12.8
15 In reference to **Figure 12.8**, identify #1 - #5.
   1 _____   4 _____
   2 _____   5 _____
   3 _____

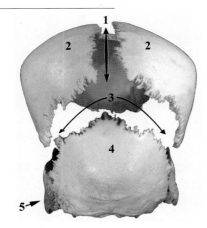

Figure 12.9
16 In reference to **Figure 12.9**, identify #1 - #5.
   1 _____   4 _____
   2 _____   5 _____
   3 _____

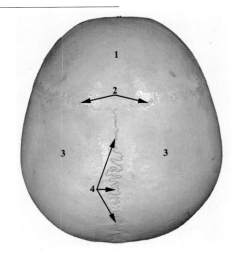

Figure 12.10
17 In reference to **Figure 12.10**, identify #1 - #4.
   1 _____   3 _____
   2 _____   4 _____

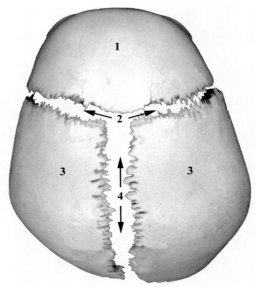

Figure 12.11
18 In reference to **Figure 12.11**, identify #1 - #4.
   1 _____   3 _____
   2 _____   4 _____

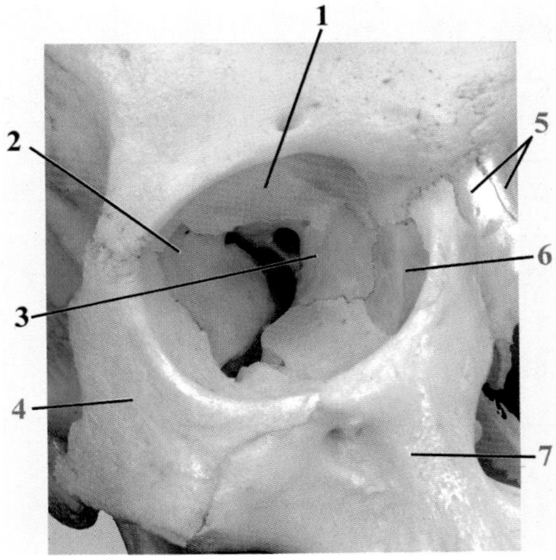

Figure 12.12
19 In reference to **Figure 12.12**, identify #1 - #7.
   1 _____   5 _____
   2 _____   6 _____
   3 _____   7 _____
   4 _____

BONES OF THE CRANIUM
FRONTAL BONE

20 The two regions of the frontal bone are the
   1 _____   2 _____
21 The squamous region forms the flat anterior portion of the bone, commonly called the _____.
22 The orbital region forms the _____ of the orbits and a portion of the roofs of the _____ cavities.
23 Internally, the frontal bones forms the _____ boundary and _____ of the cranial cavity.

158 Chapter 12 - Axial Skeleton

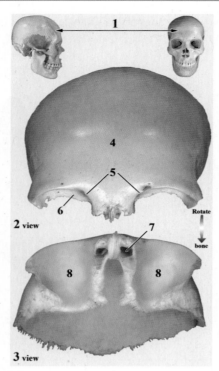

Figure 12.13

24 In reference to **Figure 12.13**, identify #1 - #8.
1 _____ 5 _____
2 _____ 6 _____
3 _____ 7 _____
4 _____ 8 _____

PARIETAL BONES

25 Externally, the paired parietal bones form the _____ and _____ portions of the cranium.

26 Internally, the paired parietal bones form the _____ and _____ portions of the cranial cavity.

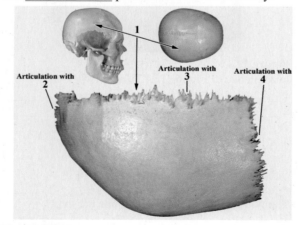

Figure 12.14

27 In reference to **Figure 12.14**, identify #1 - #4.
1 _____ 3 _____
2 _____ 4 _____

28 Sutures are _____ joints between the flat bones of the skull.

29 Sutures associated with the parietal bones are the:
1 _____ 3 _____
2 _____ 4 _____

Match Terms with Descriptions
30 **Terms**
   A Coronal suture
   B Lambdoidal suture
   C Sagittal suture
   D Squamosal suture

Descriptions
   _____ Formed at the articulation of each parietal bone with each temporal bone.
   _____ Formed at the articulation of the parietal bones with the frontal bone.
   _____ Formed at the articulation of the parietal bone with the occipital bone.
   _____ Formed at the midline articulation of the two parietal bones.

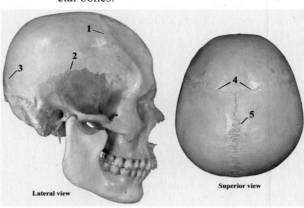

Figure 12.15

31 In reference to **Figure 12.15**, identify #1 - #5.
1 _____ 4 _____
2 _____ 5 _____
3 _____

OCCIPITAL BONES

32 Externally, the occipital bones form the _____ and _____ portions of the cranium.

33 Internally, the occipital bones form the _____ boundary and _____ floor of the cranial cavity.

34 **Match Terms with Descriptions**
Terms
   A External occipital protuberance
   B Foramen magnum
   C Occipital condyles

Descriptions
   _____ Articulate with the superior facets of the first cervical vertebra, the atlas, and form a synovial condyloid joint.
   _____ Large foramen (hole) in the floor of cranium that serves as a passageway for the spinal cord to the brain stem (medulla oblongata)
   _____ Medial protuberance posterior and superior to the foramen magnum.
   _____ Their condyloid joint allows a rocking motion of the head, especially as seen in the "yes" head movement.
   _____ Two rounded articular projections lateral to the foramen magnum.

Chapter 12 - Axial Skeleton 159

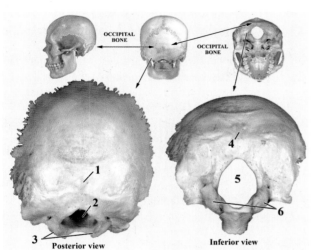

Figure 12.16

35 In reference to **Figure 12.16**, identify #1 - #6.
1 _____   4 _____
2 _____   5 _____
3 _____   6 _____

TEMPORAL BONES

36 Externally, the paired temporal bones form the _____ portions of the cranium.

37 Internally, the paired temporal bones form the _____ boundary and the _____ portion of the cranial cavity.

38 Five major portions of the temporal bone are:
1 _____   4 _____
2 _____   5 _____
3 _____

39 **Match Terms with Descriptions**
   Terms
   A External auditory (acoustic) meatus
   B Internal auditory (acoustic) meatus
   C Mandibular fossa
   D Mastoid portion
   E Mastoid process
   F Petrous portion
   G Squamous portion
   H Styloid process
   I Tympanic portion
   J Zygomatic process

   Descriptions
   _____ Depression on the inferior surface that articulates with the mandibular condyle forming the temporomandibular joint (TMJ).
   _____ Inferior conical process extending from the mastoid portion of the temporal bone.
   _____ Large bony canal that allows sound to reach the tympanic membrane.
   _____ Lateral flattened portion that contacts the parietal bone at the squamous suture.
   _____ Located posterior and inferior to the external auditory meatus.
   _____ Portion of the temporal bone that is located anterior to the mastoid portion and forms a portion of the outer auditory meatus.
   _____ Posterior portion of the temporal bone. Extending downward, forms a conical projection, the mastoid process.
   _____ Pyramidal shaped portion that extends inward from the lateral boundary of the cranial cavity (squamous region).
   _____ Rounded process projecting anteriorly from the squamous region and articulating with the temporal process of the zygomatic bone.
   _____ Separates the sphenoid bone and the posterior occipital bone.
   _____ Sharp, long process inferior to the external auditory meatus
   _____ Small canal located posteriorly and centrally in the petrous portion of the temporal bone. The meatus serves as a passageway for several nerves and a blood vessel.
   _____ This process and the temporal process of the zygomatic bone form the zygomatic arch.

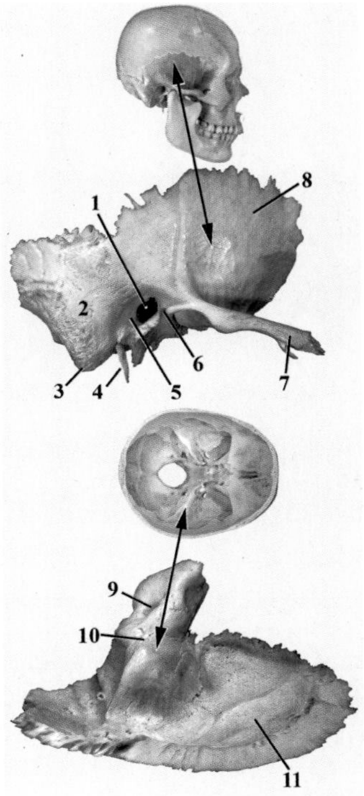

Figure 12.17

40 In reference to **Figure 12.17**, identify #1 - #11.
1 _____   7 _____
2 _____   8 _____
3 _____   9 _____
4 _____   10 _____
5 _____   11 _____
6 _____

SPHENOID BONE

41 Externally, the sphenoid bone forms a part of the _____ cranial walls, and the _____ boundary of the orbits.

42 Internally, the sphenoid bone is _____ located in the cranium and forms the _____ cranial floor, the _____ boundary of the cranial cavity, and the _____ walls of the nasopharynx.

160 Chapter 12 - Axial Skeleton

43 The sphenoid consists of a centrally located _____, from which three processes project: the (1) _____ _____, the (2) _____ _____, and the (3) _____ processes.

Figure 12.18

44 In reference to **Figure 12.18**, identify #1 - #13.
1 _____
2 _____
3 _____
4 _____
5 _____
6 _____
7 _____
8 _____
9 _____
10 _____
11 _____
12 _____
13 _____

45 **Match Terms with Descriptions**
Terms
A Body
B Greater wings
C Hypophyseal fossa
D Lesser wings
E Pterygoid processes
F Sella turcica
Descriptions
_____ Deep depression on the superior surface of the body.
_____ Deepest portion of the sella turcica.
_____ Forms a portion of the lateral walls of the nasopharynx.
_____ Large, wing-like processes which project laterally from the body.
_____ Location of the pituitary (hypophysis).
_____ Project inferiorly from the body.
_____ Rectangular portion located in the central aspect of cranial cavity.
_____ Small, wing-like processes that project laterally from the anterior portion of the body.
_____ Superior surface houses the sella turcica.

ETHMOID BONE

46 The ethmoid is located _____ to the sphenoid bone and forms most of the boundary around the _____ _____.

47 **Match Terms with Descriptions**
Terms
A Cribriform (horizontal) plate
B Crista galli
C Olfactory foramina
D Orbital plates (two)
E Perpendicular plate
F Superior and middle nasal conchae (2 each)
Descriptions
_____ Each forms one of the medial walls of the orbits.
_____ Houses the olfactory foramina.
_____ Inward projections into the nasal cavity.
_____ Small openings in the cribriform plate that serve as passageways for the olfactory nerves.
_____ Small, triangular projection located in the middle of the cribriform plate.
_____ Superior portion of the ethmoid bone which forms an anterior, medial portion of the cranial floor.
_____ Thin, bony plate that forms the upper portion of the nasal septum.

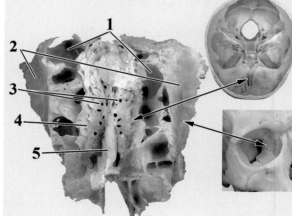

48 **Figure 12.19**
In reference to **Figure 12.19**, identify #1 - #5.
1 _____   4 _____
2 _____   5 _____
3 _____

MANDIBLE

49 The mandible forms the _____ _____ and is the only _____ bone of the skull.
50 The major regions of the mandible are the (1) _____ and two vertical (2) _____.
51 The mandible's condyles articulate with the _____ fossae of the _____ bones.

Chapter 12 - Axial Skeleton   161

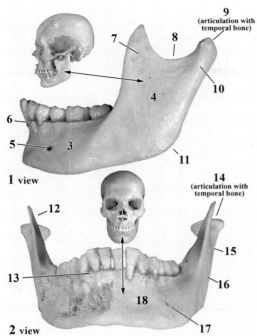

Figure 12.20

52 In reference to **Figure 12.20**, identify #1 - #18.

1 _____   10 _____
2 _____   11 _____
3 _____   12 _____
4 _____   13 _____
5 _____   14 _____
6 _____   15 _____
7 _____   16 _____
8 _____   17 _____
9 _____   18 _____

53 **Match Terms with Descriptions**
   Terms
   A   Alveolar process (margin)
   B   Angle
   C   Body
   D   Condyle
   E   Condyloid process
   F   Coronoid process
   G   Mandibular notch
   H   Mental foramen
   I   Rami
   Descriptions
   _____ Angular portion where the body meets the ramus.
   _____ Anterior process of the ramus.
   _____ Articulates with the articular disc of the temporo-mandibular joint.
   _____ Depression between the mandibular condyle and the coronoid process
   _____ Each is located inferior to the second premolar near the midline of the body.
   _____ Each consists of a condyle and a neck.
   _____ Each terminates to form two processes, the condyloid process and the coronoid process.
   _____ Horizontal portion of the mandible.
   _____ Portion of the condyloid process that articulates with the articular disc of the temporomandibular joint.
   _____ Portion of the body that contains the teeth.
   _____ Superior and posterior processes of the mandible.
   _____ Vertical portions of the mandible.

MAXILLARY BONES (MAXILLAE)

54 The maxillary bones form the (1) _____, (2) the anterior portion of the _____ of the mouth, (3) a portion of the floor of the _____, and (4) a portion of the _____ walls of the nasal cavity.

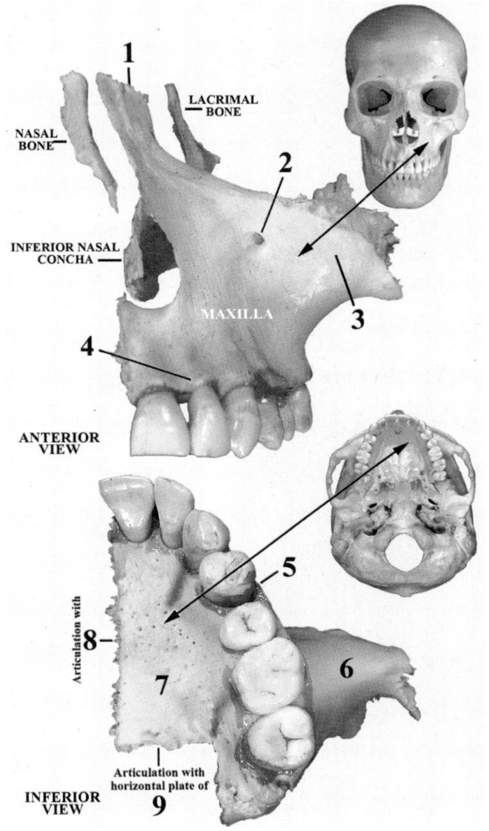

Figure 12.21

55 In reference to **Figure 12.21**, identify #1 - #9.

1 _____   6 _____
2 _____   7 _____
3 _____   8 _____
4 _____   9 _____
5 _____

56 **Match Terms with Descriptions**
   Terms
   A   Alveolar processes (margins)
   B   Palatine processes
   Descriptions
   _____ Horizontal extensions of the maxillae that form most of the roof (hard palate) of the mouth.
   _____ Posteriorly, articulate with the horizontal plates of the palatine bones.
   _____ Portion of the maxilla that contains the teeth.

NASAL BONES

57 The nasal bones are the bones of the _____.
58 Together, the nasal bones form the _____ of the nose.

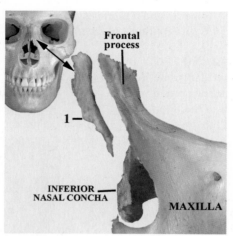

Figure 12.22

59 In reference to **Figure 12.22**, identify #1.
    1 _____

LACRIMAL BONES

60 The lacrimal bones form the _____ walls of the orbits.

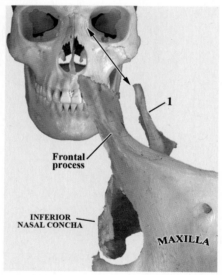

Figure 12.23

61 In reference to **Figure 12.23**, identify #1.
    1 _____

INFERIOR NASAL CONCHAE

62 The inferior nasal conchae are located _____ to the middle nasal conchae.
63 The inferior nasal conchae articulate with the medial walls of the _____, which form the inferior boundaries of the nasal cavities.

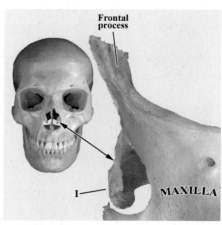

Figure 12.24

64 In reference to **Figure 12.24**, identify #1.
    1 _____

ZYGOMATIC BONES

65 The zygomatic bones form the _____ and the lateral, inferior portion of each _____.

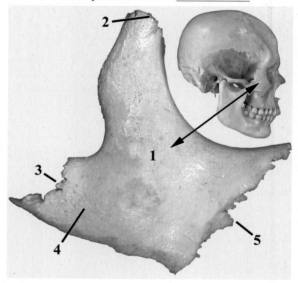

Figure 12.25

66 In reference to **Figure 12.25**, identify #1 - #5.
    1 _____   4 _____
    2 _____   5 _____
    3 _____

67 **Match Terms with Descriptions**
    Terms
    A Frontal surface
    B Maxillary surface
    C Temporal process
    D Temporal surface
    Descriptions
    _____ Posterior projection that joins with the zygomatic process of the temporal bone.
    _____ Surface that articulates with the frontal bone.
    _____ Surface that articulates with the maxillary bone.
    _____ Surface that articulates with the temporal bone.
    _____ With the temporal process of the zygomatic bone both form the zygomatic arch.

PALATINE BONES

68 The palatine bones are located between the _____ and the _____ bone.

69 The horizontal plates of the palatine bones form the posterior portion of the _____.

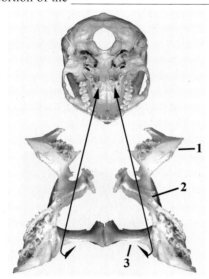

Figure 12.26

70 In reference to **Figure 12.26**, identify #1 - #3.
    1 _____   3 _____
    2 _____

71 **Match Terms with Descriptions**
    Terms
    A   Horizontal plate
    B   Orbital process
    C   Perpendicular plate
    Descriptions
    _____ Forms a portion of the posterior wall of the orbit.
    _____ Fuse to form the posterior portion of the bony roof of the mouth, the hard palate.
    _____ Thin lateral upright portion of the palatine bone.

VOMER

72 The vomer forms the inferior portion of the _____ _____.

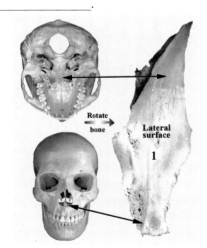

Figure 12.27

73 In reference to **Figure 12.27**, identify bone #1.
    1 _____

FORAMINA, CANALS, AND FISSURES

74 **Match Terms with Descriptions**
    Terms
    A   Canal
    B   Fissure
    C   Foramen (pl., foramina)
    D   Meatus
    Descriptions
    _____ Body passage (a body opening).
    _____ Deep groove or furrow, usually involving a wall.
    _____ Opening through which blood vessels and/or nerves pass.
    _____ Passageway for blood vessels and/or nerves.

75 **Match Terms with Descriptions**
    Terms
    A   Carotid foramen
    B   Condyloid canal (posterior)
    C   Foramen lacerum
    D   Foramen magnum
    E   Foramen ovale
    F   Foramen rotundum
    G   Foramen spinosum
    H   Greater palatine foramen
    I   Hypoglossal canal (anterior condyloid foramen)
    J   Incisive foramen
    K   Inferior orbital fissure
    L   Infraorbital foramen
    M   Internal auditory (acoustic) meatus
    N   Jugular foramen
    O   Lesser palatine foramen
    P   Mental foramen
    Q   Olfactory foramina
    R   Optic foramen
    S   Stylomastoid foramen
    T   Superior orbital fissure
    U   Supraorbital foramen
    Descriptions
    _____ Best observed in the inferior surface of the petrous portion of each temporal bone.
    _____ Best observed inferior to the orbit in the superior portion of each maxilla.
    _____ Best observed in the lateral, inferior portion of each orbit, between the greater wing of the sphenoid and each maxilla.
    _____ Function as passageways for the olfactory nerves (cranial nerve I) that originate from the nasal mucosa.
    _____ Functions as the passageway for an oculomotor nerve (cranial nerve III), a trochlear nerve (cranial nerve IV), ophthalmic nerve, and the abducens nerve (cranial nerve VI).
    _____ Functions as the passageway for each infraorbital nerve and blood vessels.
    _____ Functions as a passageway for a small vein associated with a large venous transverse sinus.
    _____ Functions as the passageway for each maxillary and zygomatic nerve, and the infraorbital vessels.

Chapter 12 - Axial Skeleton

_____ Functions as an exit for the facial nerve and entrance of the stylomastoid artery.

_____ Functions as the passageway for the medulla oblongata, the spinal part of the accessory nerve (cranial nerve XI), and vertebral arteries.

_____ Functions to transmit the facial and auditory nerves, and blood vessels (an auditory artery).

_____ Functions as a passageway for the hypoglossal nerve (cranial nerve twelve) and the ascending pharyngeal artery.

_____ Functions as the passageway for the optic nerve (cranial nerve II) and the ophthalmic artery into the orbital cavity.

_____ Functions as the passageway for the mandibular nerve.

_____ Functions as a passageway for each jugular vein, glossopharyngeal nerve (cranial nerve IX), vagus nerve (cranial nerve X), and the cranial part of the accessory nerve (cranial nerve XI).

_____ Functions as the passageway for the anterior palatine nerve and palatine blood vessels.

_____ Functions as the passageway for the internal carotid artery and several blood vessels.

_____ Functions as the passageway for the maxillary nerve.

_____ Functions as the passageway for the middle and posterior palatine nerves.

_____ Functions as the passageway for the middle meningeal vessels.

_____ Functions as a passageway for the mental nerve and blood vessels.

_____ Located in the posterior angle of each of the greater wings of the sphenoid bone.

_____ Located in the mandible inferior to each of the second premolars.

_____ Located at the mid-lateral surface of the horizontal plate of each palatine bone.

_____ Located in the cribriform plate of the ethmoid bone.

_____ Located in the medial part of each of the greater wings of the sphenoid.

_____ Located at the posterior-lateral surface of the horizontal plate of each palatine bone.

_____ Located lateral to the base of each of the occipital condyles.

_____ Located in each of the medial aspects of the small wings of the sphenoid bone

_____ Located in the posterior part of each of the greater wings of the sphenoid bone.

_____ Located between the petrous portion of each temporal bone and the occipital bone.

_____ Located at the anterior medial boundary of the palatine processes of maxillae.

_____ Located near the base of the styloid process of each temporal bone.

_____ Located at the anterior, inferior portion of the occipital bone.

_____ Located in the superior central aspect of the petrous portion of each temporal bone.

_____ Located posterior to each occipital condyle.

_____ Located between the roof and the lateral wall of each orbit.

_____ Located between the sphenoid and each lateral surface of the basal portion of the occipital bone.

_____ Located superior to the margin of each orbit. Functions as the passageway for the supraorbital nerve and blood vessels.

_____ Opens into the incisive canals, which transmits the nasopalatine nerves and branches of the palatine blood vessels.

_____ Passageway for each carotid artery.

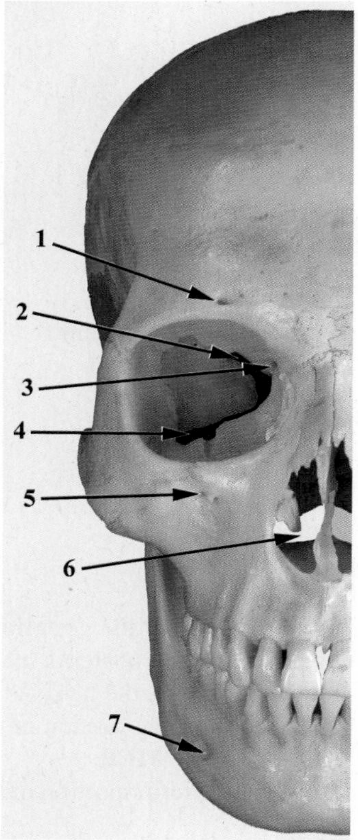

Figure 12.28

76 In reference to **Figure 12.28**, identify #1 - #7.

1 _____   5 _____
2 _____   6 _____
3 _____   7 _____
4 _____

Chapter 12 - Axial Skeleton

PARANASAL SINUSES

79 What lines the paranasal sinuses? _____

80 What are the four bones that house the paranasal sinuses?
    1 _____   3 _____
    2 _____   4 _____

81 Where do the paranasal sinuses drain? _____

82 What are four functions of the paranasal sinuses?
    1 _____
    2 _____
    3 _____
    4 _____

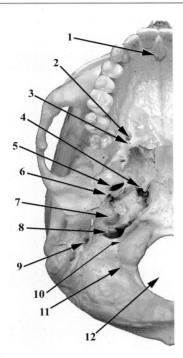

Figure 12.29

77 In reference to **Figure 12.29**, identify #1 - #12.
    1 _____   7 _____
    2 _____   8 _____
    3 _____   9 _____
    4 _____   10 _____
    5 _____   11 _____
    6 _____   12 _____

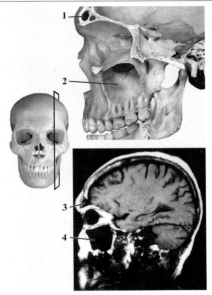

Figure 12.31

83 In reference to **Figure 12.31**, identify #1 - #4.
    1 _____   3 _____
    2 _____   4 _____

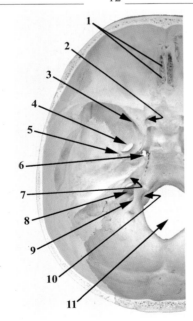

Figure 12.30

78 In reference to **Figure 12.30**, identify #1 - #12.
    1 _____   7 _____
    2 _____   8 _____
    3 _____   9 _____
    4 _____   10 _____
    5 _____   11 _____
    6 _____

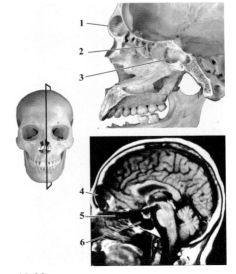

Figure 12.32

84 In reference to **Figure 12.32**, identify #1 - #6.
    1 _____   4 _____
    2 _____   5 _____
    3 _____   6 _____

166 Chapter 12 - Axial Skeleton

SKULL'S ASSOCIATED BONES

85 The bones associated with the skull are the _____ and the _____.

HYOID BONE

86 The hyoid bone is a horseshoe shaped boned inferior to the _____.

87 The hyoid bone is attached by ligaments to the _____ _____ of each temporal bone.

88 The hyoid bone functions to attach _____ of the tongue, larynx, and pharynx.

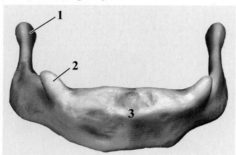

Figure 12.33

89 In reference to **Figure 12.33**, identify #1 - #3.
1 _____ 3 _____
2 _____

90 **Match Terms with Descriptions**
   Terms
   A Body
   B Greater horns
   C Lesser horns
   Descriptions
   _____ Anterior central part.
   _____ Attaches the hyoid to the styloid processes of the temporal bones (by a stylohyoid ligament).
   _____ Serves as attachment sites for several muscles of the tongue.
   _____ Serves as an attachment site for several muscles of the tongue, pharynx (throat), and larynx (voice box).
   _____ Supports the larynx (voice box).

OSSICLES

91 The ossicles conduct sound vibrations across the _____ _____, a bony cavity within the _____ bone.

92 Starting from the eardrum, the three ossicles are the:
1 _____ 3 _____
2 _____

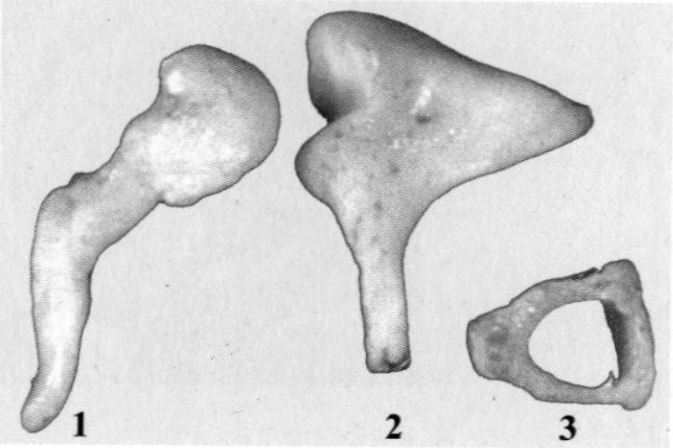

Figure 12.34

93 In reference to **Figure 12.34**, identify #1 - #3.
1 _____ 3 _____
2 _____

VERTEBRAL COLUMN

94 The vertebral column consists of 33 _____, divided into five divisions.

95 The five divisions of the vertebral column and their number of vertebrae are:
1 _____ # _____
2 _____ # _____
3 _____ # _____
4 _____ # _____ fused
5 _____ # _____ fused

96 Cartilaginous joints, the _____, are found between each individual component except between _____ _____.

97 Cartilaginous joints are made of _____ and are strong, _____ moveable joints.

98 The articulation between C_1 and C_2 is a _____ joint, and allows _____ of the head.

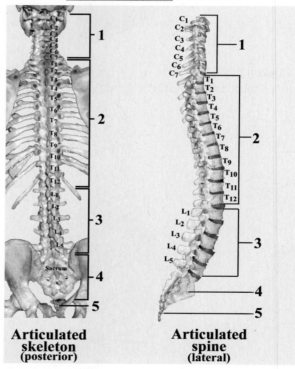

Articulated skeleton (posterior) **Articulated spine (lateral)**

Figure 12.35

99 In reference to **Figure 12.35**, identify #1 - #5.
1 _____ 4 _____
2 _____ 5 _____
3 _____

TYPICAL VERTEBRAL STRUCTURE

100 **Match Terms with Descriptions**
   Terms
   A Body (centrum)
   B Inferior articular processes
   C Laminae
   D Pedicles
   E Spinous process
   F Superior articular processes
   G Transverse processes
   H Vertebral arch (neural arch)
   I Vertebral foramen (neural foramen)

Name _____
Class _____

Descriptions

_____ Articulate with the vertebra above, at its inferior articular processes.
_____ Articulate with the vertebra below, at its superior articular processes.
_____ Consists of two portions: (1) a pair of anterior processes called pedicles and (2) a pair of posterior plates called laminae.
_____ Fuse to form a posterior midline process, the spinous process.
_____ Opening formed by boundaries of the body and the vertebral arch.
_____ Pair of short processes that attach the vertebral arch to the body.
_____ Pair of flattened processes that extend from the pedicles and fuse to form the posterior, midline spinous process.
_____ Paired processes that extend laterally from the vertebral arch.
_____ Paired processes that extend inferiorly from the vertebral arch; each with a smooth-surfaced articular facet.
_____ Paired processes that extend superiorly from the vertebral arch; each with a smooth-surfaced articular facet.
_____ Passageway for the spinal cord and its surrounding meninges.
_____ Single, midline posterior process formed from the union of the laminae.
_____ Vertebra's anterior weight-bearing region.
_____ Vertebra's posterior region that is attached to the body.

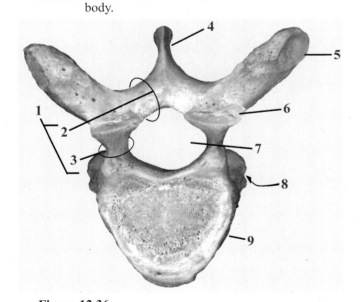

Figure 12.36
101 In reference to **Figure 12.36**, identify #1 - #9.
1 _____   6 _____
2 _____   7 _____
3 _____   8 _____
4 _____   9 _____
5 _____

REGIONAL CHARACTERISTICS OF VERTEBRAE
CERVICAL DIVISION

102 The cervical division consists of _____ vertebrae.
103 The first two cervical vertebrae, the _____ and the _____ are modified for movements of the _____.

Regional Feature of Cervical Vertebrae

104 When compared to the other vertebral regions:
    The bodies of the cervical vertebrae are _____
    The vertebral arches are _____
    The spinous process are _____ (except C_7)
    The transverse processes have a _____

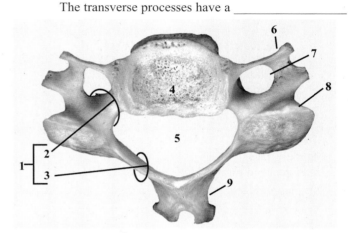

Figure 12.37
105 In reference to **Figure 12.37**, identify #1 - #9.
1 _____   6 _____
2 _____   7 _____
3 _____   8 _____
4 _____   9 _____
5 _____

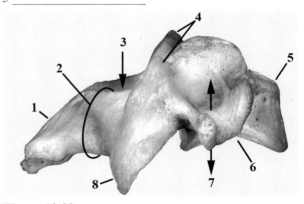

Figure 12.38
106 In reference to **Figure 12.38**, identify #1 - #8.
1 _____   5 _____
2 _____   6 _____
3 _____   7 _____
4 _____   8 _____

ATLAS and AXIS

107 The first cervical vertebra is called the _____.
108 The atlas articulates superiorly with the _____ _____ and inferiorly with the _____.
109 The second cervical vertebra is called the _____.
110 The axis articulates superiorly with the _____ and inferiorly with the _____.

Chapter 12 - Axial Skeleton

Atlas

111 Four unique features of the atlas are:
1 _____
2 _____
3 _____
4 _____

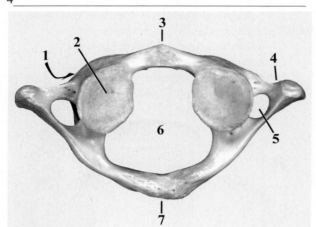

Figure 12.39

112 In reference to **Figure 12.39**, identify #1 - #7.
1 _____   5 _____
2 _____   6 _____
3 _____   7 _____
4 _____

Axis

113 Superiorly, the articulation of the _____ with the inner boundary of the anterior arch of the atlas forms a synovial _____ joint that allows _____ of the head.

114 Inferiorly, the axis forms a _____ joint with the _____ cervical vertebra.

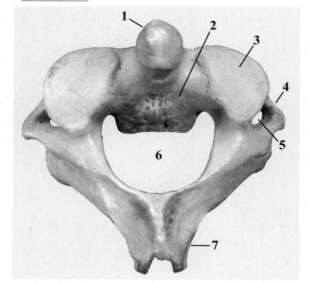

Figure 12.40

115 In reference to **Figure 12.40**, identify #1 - #7.
1 _____   5 _____
2 _____   6 _____
3 _____   7 _____
4 _____

Articulation of Cervical Vertebrae

116 The articulation of the superior articular processes of the atlas with the _____ forms a synovial _____ joint.

117 The _____ of the axis fits within the inner boundary of the anterior arch of the _____ forming a synovial _____ joint.

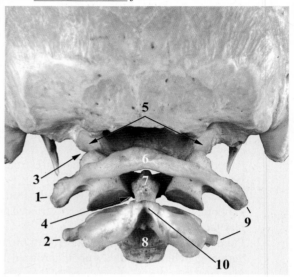

Figure 12.41

118 In reference to **Figure 12.41**, identify #1 - #10.
1 _____   6 _____
2 _____   7 _____
3 _____   8 _____
4 _____   9 _____
5 _____   10 _____

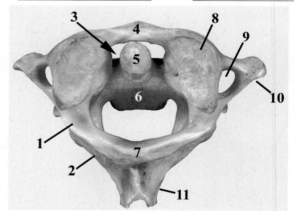

Figure 12.42

119 In reference to **Figure 12.42**, identify #1 - #11.
1 _____   7 _____
2 _____   8 _____
3 _____   9 _____
4 _____   10 _____
5 _____   11 _____
6 _____

120 Located between the bodies of C3 - C7 are the _____ _____.

121 Intervertebral discs are cartilaginous (fibrocartilage) joints that provide for _____ movements and absorb _____.

122 The _____ foramina form a vertical passageway for the vertebral blood vessels and nerves.
123 The _____ foramina are formed between the vertebrae and allow the passage of spinal nerves.

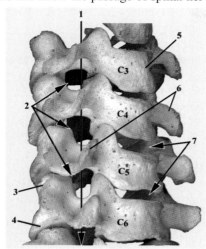

Figure 12.43
124 In reference to **Figure 12.43**, identify #1 - #7.
1 _____   5 _____
2 _____   6 _____
3 _____   7 _____
4 _____

THORACIC DIVISION
125 The thoracic division is the _____ division of the vertebral column and consists of _____ vertebrae.

Regional Feature of Thoracic Vertebrae
126 When compared to the other vertebral regions:
The thoracic vertebrae are _____ than the cervical and _____ than the lumbar vertebrae.
The spinous process are _____.
The bodies all have _____ for articulation with the heads of the _____.
The transverse processes have _____ for articulation with the tubercles of the _____ (except T_{11} - T_{12}).

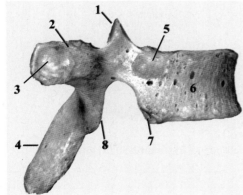

127 **Figure 12.44**
In reference to **Figure 12.44**, identify #1 - #8.
1 _____   5 _____
2 _____   6 _____
3 _____   7 _____
4 _____   8 _____

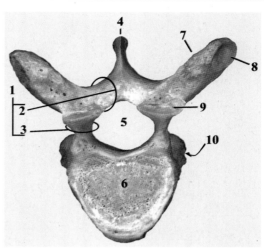

Figure 12.45
128 In reference to **Figure 12.45**, identify #1 - #10.
1 _____   6 _____
2 _____   7 _____
3 _____   8 _____
4 _____   9 _____
5 _____   10 _____

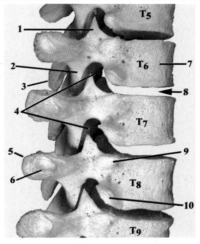

Figure 12.46
129 In reference to **Figure 12.46**, identify #1 - #10.
1 _____   6 _____
2 _____   7 _____
3 _____   8 _____
4 _____   9 _____
5 _____   10 _____

130 The thoracic vertebrae, T_2 - T_8 have _____ on each side of their bodies for articulation with the _____ of the ribs.
131 The thoracic vertebra, T_1, has one superior _____ facet and one inferior _____ on each side of its body.
132 The thoracic vertebrae, T_9 - T_{12} have one _____ _____ on each side of their body.
133 All thoracic vertebrae, except for T_{11} and T_{12}, have facets on their _____ processes.
134 The thoracic vertebrae T_{11} and T_{12}, articulate with the _____ ribs.
135 The lack of articulation with the transverse processes give the floating ribs greater _____.

170 Chapter 12 - Axial Skeleton

Name _____
Class _____

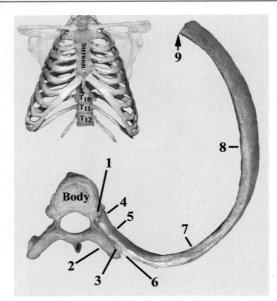

Figure 12.47

136 In reference to **Figure 12.47**, identify #1 - #9.

1 _____ 6 _____
2 _____ 7 _____
3 _____ 8 _____
4 _____ 9 _____
5 _____

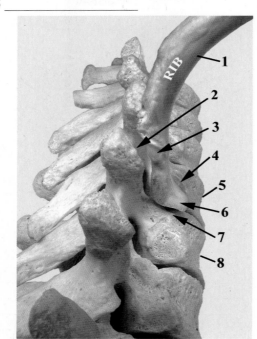

Figure 12.48

137 In reference to **Figure 12.48**, identify #1 - #8.

1 _____ 5 _____
2 _____ 6 _____
3 _____ 7 _____
4 _____ 8 _____

LUMBAR DIVISION (L$_1$ - L$_5$)

138 The lumbar division is the _____ division of the spinal column.

Regional Feature of Thoracic Vertebrae

139 When compared to the other vertebral regions:
The lumbar vertebrae are _____ than the cervical and the thoracic vertebrae.
The lumbar vertebrae have spinous processes that are _____.
The lumbar vertebrae have _____ that are large and project medially.

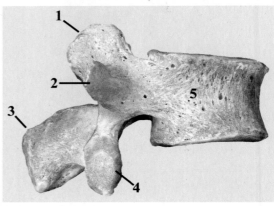

Figure 12.49

140 In reference to **Figure 12.49**, identify #1 - #5.

1 _____ 4 _____
2 _____ 5 _____
3 _____

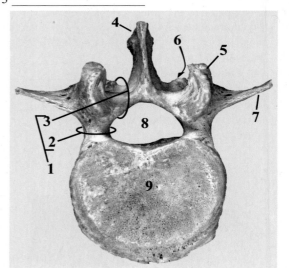

Figure 12.50

141 In reference to **Figure 12.50**, identify #1 - #9.

1 _____ 6 _____
2 _____ 7 _____
3 _____ 8 _____
4 _____ 9 _____
5 _____

Chapter 12 - Axial Skeleton 171

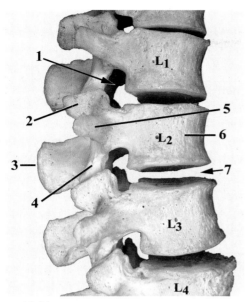

Figure 12.51

142 In reference to **Figure 12.51**, identify #1 - #7.

1 _____   5 _____
2 _____   6 _____
3 _____   7 _____
4 _____

SACRAL DIVISION

143 **Match Terms with Descriptions**

Terms

A	Alae	F	Sacral canal
B	Auricular surface	G	Sacral foramina
C	Body	H	Sacral hiatus
D	Lateral sacral crest	I	Sacral promontory
E	Median sacral crest	J	Superior articular processes

Descriptions

_____ Articulation between this surface and the sacrum forms the sacroiliac joint.

_____ Enlarged, inferior opening of sacral canal.

_____ Formed by the intervertebral foramina of the fused, sacral vertebrae.

_____ Formed by the posterior spinous processes of the fused, sacral vertebrae.

_____ Formed by a series of tubercles that represent the transverse processes of the fused sacral vertebrae.

_____ Formed by the vertebral foramina of the fused, sacral vertebrae.

_____ Four pairs are observed on the anterior and posterior surfaces.

_____ Lateral portion of each ala that articulates with the ilium of the coxal bone.

_____ Midline region formed by the bodies of fused, sacral vertebrae (S_1 - S_5).

_____ Superior processes that articulate with inferior articular processes of L_5.

_____ Superior anterior (pelvic) portion of the body of the sacrum (S_1).

_____ Wing-like portions of the superior, lateral aspects of sacrum.

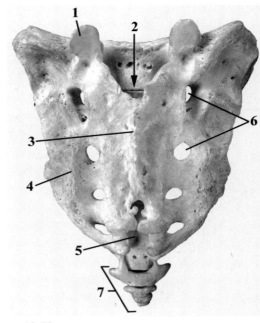

144 **Figure 12.52**

In reference to **Figure 12.52**, identify #1 - #7.

1 _____   5 _____
2 _____   6 _____
3 _____   7 _____
4 _____

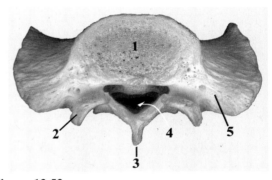

Figure 12.53

145 In reference to **Figure 12.53**, identify #1 - #5.

1 _____   4 _____
2 _____   5 _____
3 _____

172 Chapter 12 - Axial Skeleton

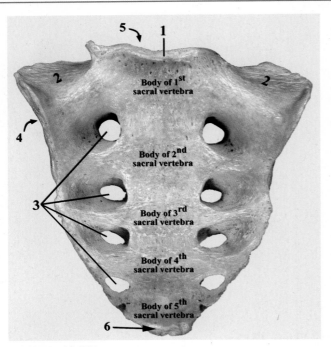

Figure 12.54

146 In reference to **Figure 12.54**, identify #1 - #3.
1 _____   3 _____
2 _____

147 In reference to **Figure 12.54**, identify articulations #4 - #6.
4 _____
5 _____
6 _____

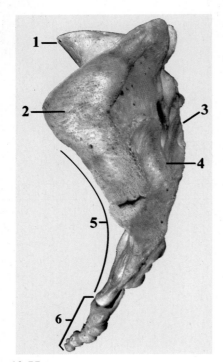

Figure 12.55

148 In reference to **Figure 12.55**, identify #1 - #6.
1 _____   4 _____
2 _____   5 _____
3 _____   6 _____

COCCYX

149 The coccyx is formed from _____ fused rudimentary vertebrae

150 The coccygeal vertebrae lack vertebral _____ and _____.

151 Superiorly the coccyx articulates with the _____.

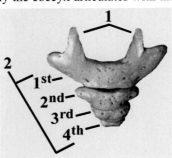

Figure 12.56

152 In reference to **Figure 12.56**, identify #1 - #2.
1 _____   2 _____

THORAX (THORACIC CAGE)

153 The thoracic cage consists of the (1) _____, the (2) _____ (ribs and the costal cartilages), and the (3) _____.

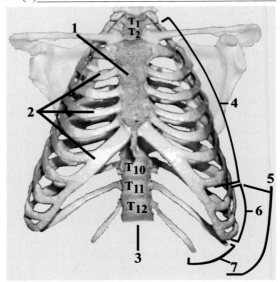

Figure 12.57

154 In reference to **Figure 12.57**, identify #1 - #7.
1 _____   5 _____
2 _____   6 _____
3 _____   7 _____
4 _____

Thoracic vertebrae

155 The thoracic vertebrae form the _____ of the thorax.

156 All thoracic vertebrae have _____ (complete and\or demi) on the sides of their _____ for articulation with the _____.

157 All thoracic vertebrae, except T_{11} - T_{12}, have facets on their _____ processes for articulation with the tubercles of the ribs.

Name _____
Class _____

Chapter 12 - Axial Skeleton

Rib cage and Ribs

158 The rib cage consists of _____ pairs of ribs and their associated _____.
159 The pairs of ribs 1 - 7 are called the _____ ribs or _____ ribs.
160 The vertebro-sternal ribs articulate posteriorly with the _____.
161 The vertebro-sternal ribs articulate anteriorly with the _____ by the _____ cartilages.
162 The pairs of ribs 8 - 12 are called the _____.
163 The first three pairs of false ribs are called the _____ _____ ribs.
164 The last two pairs of false ribs are called the _____ ribs, or _____ ribs.

RIBS

165 Three divisions of a rib are:
  1 _____
  2 _____
  3 _____

166 **Match Terms with Descriptions**
  Terms
    A Anterior (Sternal) Extremity
    B Body
    C Head
    D Neck
    E Posterior (Vertebral) Extremity
    F Tubercle
  Descriptions
    _____ Constricted region adjacent to the head (not found on floating, or vertebral, rib pairs #11 - #12).
    _____ Divided into three regions: (1) head, (2) neck, and (3) tubercle (except ribs #11 - #12 which lack necks and tubercles).
    _____ Flattened and merges with the costal cartilages at its concave surface.
    _____ Major portion of rib located between the two extremities.
    _____ Posterior portion of the rib that articulates with the body of the vertebra.
    _____ Projection from posterior portion of the rib that articulates with the transverse process of the vertebra (not found on floating, or vertebral, rib pairs #11 - #12).
    _____ Projects anteriorly and slightly downward.

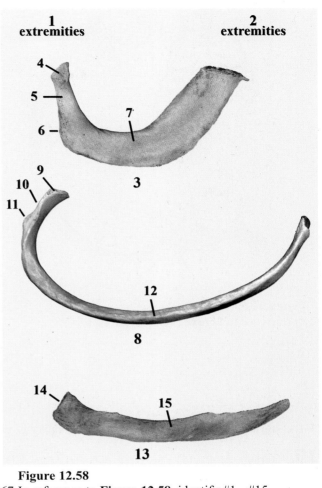

Figure 12.58

167 In reference to **Figure 12.58**, identify #1 - #15.
  1 _____ 8 _____
  2 _____ 9 _____
  3 _____ 10 _____
  4 _____ 11 _____
  5 _____ 12 _____
  6 _____ 13 _____
  7 _____ 14 _____
                     15 _____

Articulations of the Ribs

168 Posteriorly, the ribs articulate with the _____ _____.
169 All ribs articulate with the _____ of the thoracic vertebrae, and except for the _____ ribs, the ribs also articulate with the _____ processes of the vertebrae.
170 The ribs-vertebrae articulations are synovial _____ joints.
171 Anteriorly, the ribs articulate with the _____ by way of the _____ cartilages.

Chapter 12 - Axial Skeleton

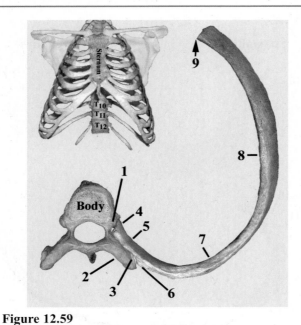

Figure 12.59
172 In reference to **Figure 12.59**, identify #1 - #9.
1 _____ 6 _____
2 _____ 7 _____
3 _____ 8 _____
4 _____ 9 _____
5 _____

STERNUM

173 The **sternum** forms the _____ boundary of the _____.

174 The sternum is divided into three major portions: **(1)** a superior portion called the _____, **(2)** a middle portion called the _____, and **(3)** an inferior portion called the _____.

175 The sternum articulates with each (1) _____ of each pectoral girdle and the (2) _____ of rib pairs #1 - #7.

176 **Match Terms with Descriptions**
   Terms
   A   Body
   B   Clavicular notches
   C   Manubrium
   D   Xiphoid process
   Descriptions
   _____ Articulates with each respective clavicle of each pectoral girdle at the clavicular notches.
   _____ Articulates with the costal cartilages of a portion of rib pair #2 and completely with the costal cartilages of rib pairs #3 - #7.
   _____ Each articulates with the sternal end of its respective clavicle.
   _____ Inferior portion of sternum.
   _____ Middle portion of the sternum.
   _____ Superior, lateral depressions of the manubrium.
   _____ Superior portion of sternum.

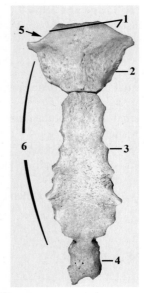

Figure 12.60
177 In reference to **Figure 12.60**, identify #1 - #4.
1 _____ 3 _____
2 _____ 4 _____
178 In reference to **Figure 12.60**, identify articulations #5 - #6.
5 _____
6 _____

Articulations of the Sternum

179 Each lateral surface of the sternum articulates with the _____ associated with ribs #1 - #7, the _____ ribs.

180 The sternocostal joints are _____ joints, which function as strong _____ moveable joints.

181 A _____ joint is formed at each sternal articulation with each clavicle.

182 A sternoclavicular joint is a synovial _____ joint.

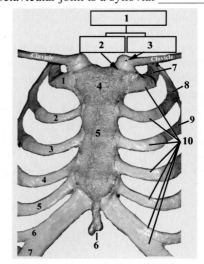

Figure 12.61
183 In reference to **Figure 12.61**, identify #1 - #9.
1 _____ 6 _____
2 _____ 7 _____
3 _____ 8 _____
4 _____ 9 _____
5 _____ 10 _____

The Appendicular Skeleton - Worksheets

1. What are the four divisions of the appendicular skeleton?
   1. _____ 3. _____
   2. _____ 4. _____

   ### PECTORAL GIRDLE

2. The two bones of a pectoral girdle are the
   1. _____ 2. _____

   ### CLAVICLES

3. Where are the clavicles located? _____
   _____
   What two bones form articulation sites with a clavicle?
   1. _____ 2. _____

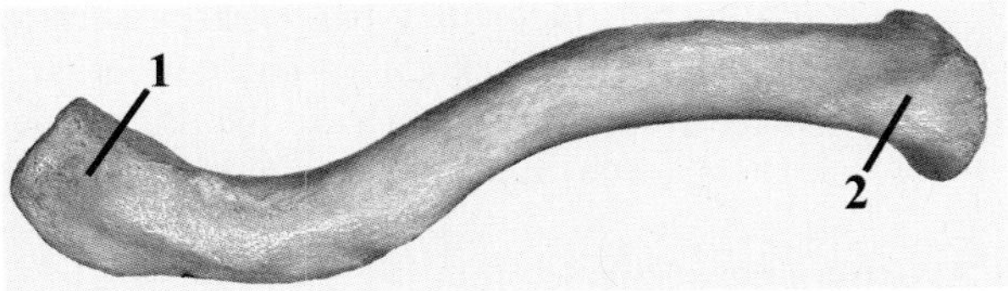

Figure 13.1

4. In reference to **Figure 13.1**, identify #1 and #2.
   1. _____ 2. _____

5. **Match Terms with Descriptions**
   Terms
   A   Acromial end
   B   Sternal end
   Descriptions
   _____ Articulates with acromion process of the scapula.
   _____ Articulates with clavicular notch of the sternum's manubrium.
   _____ Lateral, flattened (superiorly-inferiorly) end.
   _____ Medial, broad, triangular-shaped end.

   Articulations of the Clavicles

6. The sternal end of the clavicle articulates with the _____ _____ of the _____ and forms the synovial _____ joint, the _____ joint.

7. The acromial end of the clavicle articulates with the _____ of the _____ and forms the synovial _____ joint, the _____ joint.

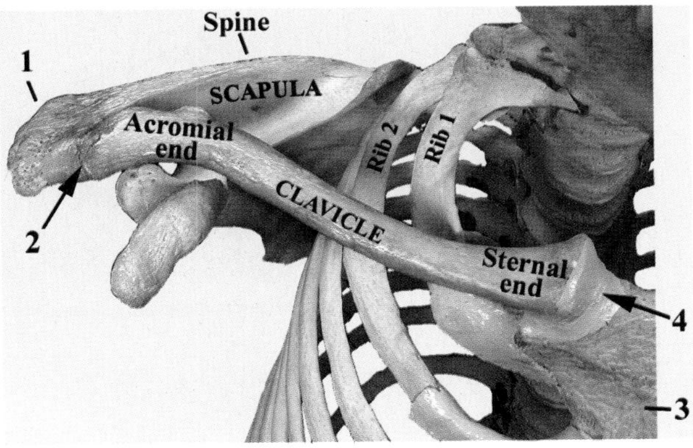

Figure 13.2

8. In reference to **Figure 13.2**, identify #1- #4.
   1. _____ 3. _____
   2. _____ 4. _____

SCAPULAS

9. Where are the scapulas located? _____
   _____
   _____

10. What two bones articulate with a scapula?
    1. _____ 2. _____

11. What are the three borders of a scapula?
    1. _____ 2. _____ 3. _____

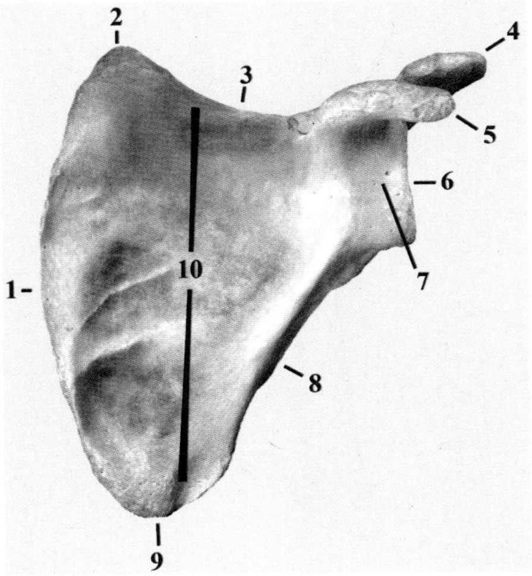

Figure 13.3

12. In reference to **Figure 13.3**, identify #1 - #10.
    1. _____ 6. _____
    2. _____ 7. _____
    3. _____ 8. _____
    4. _____ 9. _____
    5. _____ 10. _____

13. In reference to **Figure 13.3**, the bone that articulates at #6 is the _____ and at #4 is the _____.

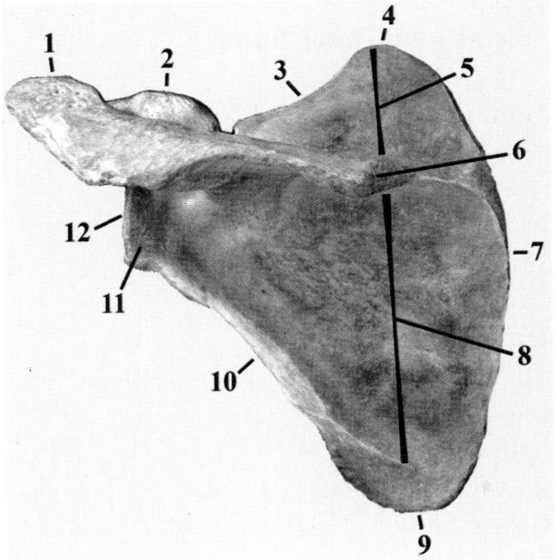

Figure 13.4

14. In reference to **Figure 13.4**, identify #1 - #12.
    1. _____ 7. _____
    2. _____ 8. _____
    3. _____ 9. _____
    4. _____ 10. _____
    5. _____ 11. _____
    6. _____ 12. _____

Chapter 13 - Appendicular Skeleton

15 In reference to **Figure 13.4**, the bone that articulates at #1 is the _____ and at #12 is the _____.

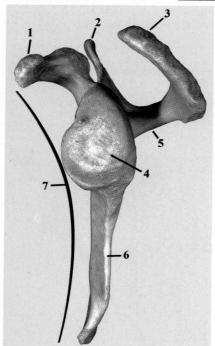

Figure 13.5

16 In reference to **Figure 13.5**, identify #1 - #7.
1 _____   4 _____
2 _____   5 _____
3 _____   6 _____
                    7 _____

17 In reference to **Figure 13.5**, surface #7 is anterior, posterior, or lateral? _____
In reference to **Figure 13.5**, the bone that articulates at #3 is the _____ and at #4 is the _____.

18 **Match Terms with Descriptions**
   Terms
   A Acromion process
   B Axillary border
   C Coracoid process
   D Glenoid cavity
   E Inferior angle
   F Infraspinatous fossa
   G Lateral angle
   H Medial angle
   I Spine
   J Superior border
   K Supraspinatous fossa
   L Vertebral border
   Descriptions
   _____ Articulates with the acromial end of clavicle to form the acromioclavicular joint.
   _____ Articulation with the head of the humerus forms the shoulder (glenohumeral) joint.
   _____ Concavity superior to the spine that serves as the origin of the supraspinatous muscle.
   _____ Formed by the fusion of the superior border and the vertebral (medial) border.
   _____ Formed by the union of the axillary (lateral) and vertebral (medial) borders.
   _____ Formed by the union of the axillary (lateral) border and the superior border.
   _____ Forms the superior boundary of the scapula.
   _____ Large lateral shallow depression that is the articulation site for the head of the humerus
   _____ Lateral boundary of the scapula.
   _____ Lateral termination of the spine
   _____ Medial boundary of the scapula.
   _____ Projects anteriorly from the superior border and functions as a site of muscle attachment.
   _____ Prominent ridge along the posterior surface that terminates in the lateral acromion process.
   _____ Shallow concavity inferior the the spine that serves as the origin of the infraspinatous muscle.

Articulations of the Scapula

19 The scapula articulates with two bones, the
   1 _____   2 _____

20 The scapula's _____ process articulates with the acromial end of its respective clavicle forming the _____. This joint is a freely moveable synovial _____ joint that allows slight movements.

21 The scapula's _____ articulates with its respective humerus and forms the shoulder, or _____ joint.

22 The glenohumeral joint is a synovial _____ joint that allows a wide range of movements.

23 The scapula's posterior attachment to the thorax is _____.

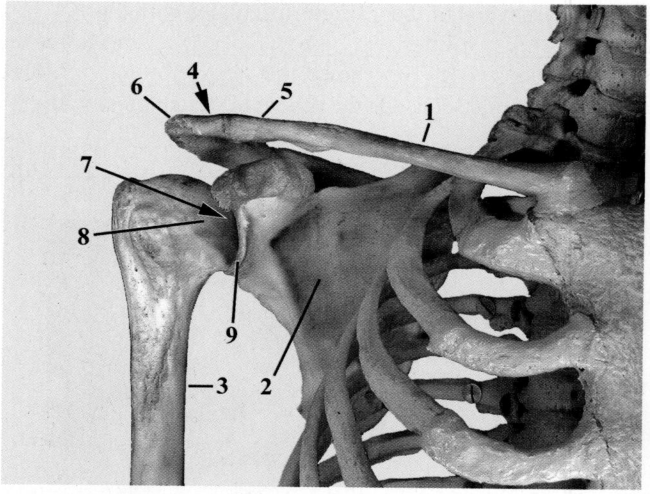

Figure 13.6

24 In reference to **Figure 13.6**, identify #1 - #9.
1 _____   6 _____
2 _____   7 _____
3 _____   8 _____
4 _____   9 _____
5 _____

Name _____
Class _____

Chapter 13 - Appendicular Skeleton **177**

UPPER LIMB
25 The three division of the upper limb are the: **(1)** _____,
    (2) _____, and **(3)** _____.

ARM
HUMERUS
Proximal Views of the Humerus
26 Proximally, the _____ of the humerus articulates with
    the _____ cavity of its respective _____,
    forming the _____ joint.

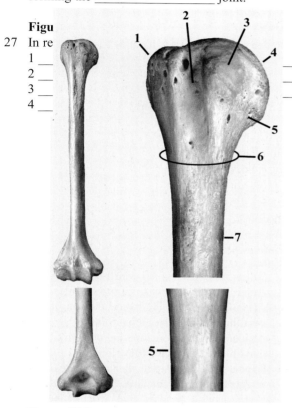
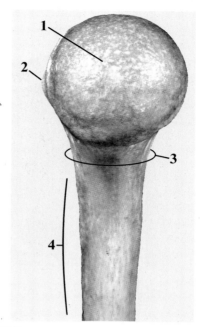

Figure 13.8
28 In reference to **Figure 13.8**, identify #1 - #5.
    1 _____ 4 _____
    2 _____ 5 _____
    3 _____

27 In reference to Figure 13.8, identify #1 - #7 (?)
    1 ___
    2 ___
    3 ___
    4 ___

Figure 13.9
29 In reference to **Figure 13.9**, identify #1 - #4.
    1 _____ 3 _____
    2 _____ 4 (surface) _____

30 **Match Terms with Descriptions**
    Terms
    A Anatomical neck
    B Body (shaft)
    C Greater tubercle
    D Head
    E Intertubercular groove
    F Lesser tubercle
    G Surgical neck
    Descriptions
    _____ Articulates with the glenoid cavity of the scapula.
    _____ Large, rounded projection lateral and inferior to the head.
    _____ Major cylindrical portion of the bone.
    _____ Portion of humerus that forms synovial freely moveable ball-and-socket joint.
    _____ Portion of humerus that forms the shoulder (glenohumeral) joint.
    _____ Proximal, medial hemispherical portion of humerus.
    _____ Slight constriction at the circumference of the smooth, articular surface of the head.
    _____ Slightly constricted portion inferior to the greater and lesser tubercles.
    _____ Small rounded projection anterior and inferior to the head.
    _____ Small groove between the greater and lesser tubercles.

Proximal Articulations of the Humerus
31 Proximally, the _____ of the humerus articulates
    with the _____ cavity of the _____.
32 The articulation of the humerus and the scapula form the
    shoulder, or _____ joint.

178 Chapter 13 - Appendicular Skeleton

33 The shoulder joint is a freely moveable synovial _____ joint.

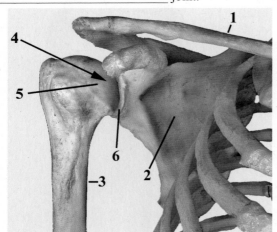

Figure 13.10

34 In reference to **Figure 13.10**, identify #1 - #6.
1 _____ 4 _____
2 _____ 5 _____
3 _____ 6 _____

Distal Views of the Humerus

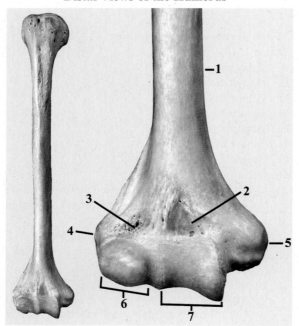

Figure 13.11

35 In reference to **Figure 13.11**, identify #1 - #7.
1 _____ 5 _____
2 _____ 6 _____
3 _____ 7 _____
4 _____

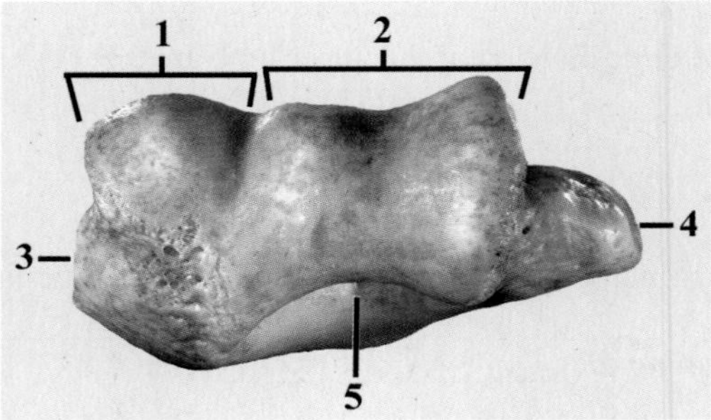

Figure 13.12

36 In reference to **Figure 13.12**, identify #1 - #5.
1 _____ 4 _____
2 _____ 5 _____
3 _____

37 **Match Terms with Descriptions**
Terms
A Body (shaft)
B Capitulum
C Coronoid fossa
D Lateral epicondyle
E Medial epicondyle
F Olecranon fossa
G Radial fossa
H Trochlea

Descriptions
_____ Anterior depression superior to capitulum.
_____ Anterior depression superior to trochlea.
_____ Articulation forms the lateral portion of the synovial hinge joint, the elbow joint.
_____ Articulation forms the medial portion of the synovial hinge joint, the elbow joint.
_____ Distal, medial condyle that articulates with the trochlear notch of the ulna.
_____ Distal, lateral condyle that articulates with the head of the radius.
_____ Large process on the distal, medial aspect of humerus.
_____ Large depression located on the distal and posterior aspect of the humerus.
_____ Major cylindrical portion of the bone.
_____ Process on distal, lateral aspect of humerus.
_____ Receives the coronoid process of the ulna when the forearm is flexed.
_____ Receives the head of the radius when the forearm is flexed.
_____ Receives the olecranon process of ulna when the forearm is extended.

Distal Articulations of the Humerus

38 The _____ of the humerus articulates with the _____ of the radius.
39 The articulation between the humerus and the radius forms the _____ joint.
40 The _____ of the humerus articulates with the _____ of the ulna.
41 The articulation between the humerus and the ulna forms the _____ joint.

Chapter 13 - Appendicular Skeleton **179**

42 The two joints, the humeroradial and the humeroulnar joints, form the synovial hinge joint, the _____ joint.

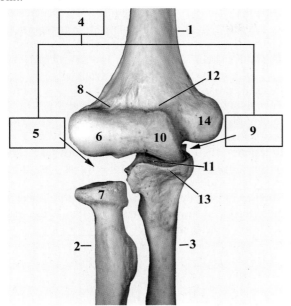

Figure 13.13

43 In reference to **Figure 13.13**, identify #1 - #13.
1 _____   8 _____
2 _____   9 _____
3 _____   10 _____
4 _____   11 _____
5 _____   12 _____
6 _____   13 _____
7 _____

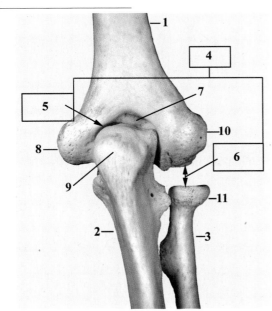

Figure 13.14

44 In reference to **Figure 13.14**, identify #1 - #10.
1 _____   6 _____
2 _____   7 _____
3 _____   8 _____
4 _____   9 _____
5 _____   10 _____

FOREARM

45 The lateral bone of the forearm is the _____, and the medial bone of the forearm is the _____.
46 Each paired radius and ulna form a proximal articulation with the humerus, the _____ joint.
47 The radius forms two _____ joints, the proximal and distal _____ joints, that allow _____ of the forearm.

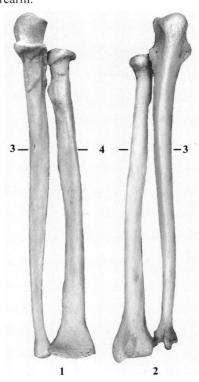

Figure 13.15

48 In reference to **Figure 13.15**, identify #1 - #4.
1 (view) _____   3 _____
2 (view) _____   4 _____

ULNA
Views of Proximal Structures of the Ulna

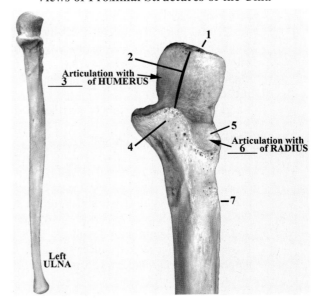

Figure 13.16

180 Chapter 13 - Appendicular Skeleton

49 In reference to **Figure 13.16**, identify #1 - #7.
    1 _____ 5 _____
    2 _____ 6 _____
    3 _____ 7 _____
    4 _____

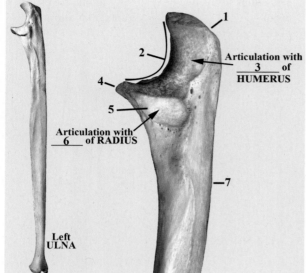

Figure 13.17

50 In reference to **Figure 13.17**, identify #1 - #7.
    1 _____ 5 _____
    2 _____ 6 _____
    3 _____ 7 _____
    4 _____

51 **Match Terms with Descriptions**
    Terms
    A   Coronoid process
    B   Olecranon process
    C   Radial notch
    D   Shaft (body)
    E   Trochlear notch
    Descriptions
    _____ Articulates with the head of the radius forming a proximal synovial pivot joint.
    _____ Articulates with the trochlea of the humerus forming a portion of the synovial hinge joint, the elbow joint.
    _____ Depression located laterally to the coronoid process.
    _____ Large concavity formed between the olecranon and coronoid processes.
    _____ Major cylindrical portion of the bone.
    _____ Process that moves into the coronoid fossa of the humerus when the forearm is flexed.
    _____ Process that is a large process located superiorly and posteriorly on ulna's proximal portion.
    _____ Process that is located superiorly and anteriorly on the proximal portion of the ulna.
    _____ Process that moves into the olecranon fossa of the humerus when the forearm is extended.

Proximal Articulations of the Ulna

52 The ulna forms two proximal articulations, one with the _____ and the other with the _____.
53 The ulna's _____ notch articulates with the _____ of the humerus to form a portion of the synovial _____ joint, the _____ joint.
54 The ulna's lateral _____ notch articulates with the _____ of the radius to form a synovial _____ joint, the proximal _____ joint.
55 Two pivot joints, the proximal and distal _____ joints allow rotation of the distal _____ over the ulna producing _____ and _____.
56 Rotation of the forearm so that the palms face backwards (posteriorly) or downward (inferiorly) and the radius _____ over the ulna is _____.
57 Rotation of the forearm so that the palms face forward (anteriorly) or upward (superiorly) and the radius is _____ to the ulna is _____.

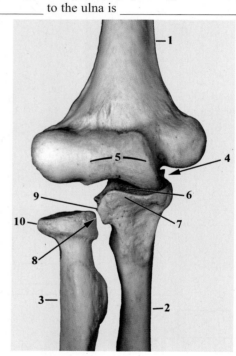

Figure 13.18

58 In reference to **Figure 13.18**, identify #1 - #10.
    1 _____ 6 _____
    2 _____ 7 _____
    3 _____ 8 _____
    4 _____ 9 _____
    5 _____ 10 _____

Name _____
Class _____

Chapter 13 - Appendicular Skeleton   **181**

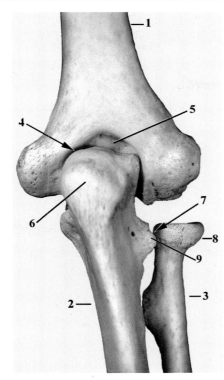

Figure 13.19
59 In reference to **Figure 13.19**, identify #1 - #9.
1 _____ 6 _____
2 _____ 7 _____
3 _____ 8 _____
4 _____ 9 _____
5 _____

Views of Distal Structures of the Ulna

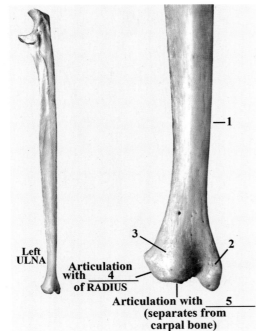

Figure 13.20
60 In reference to **Figure 13.20**, identify #1 - #5.
1 _____ 4 _____
2 _____ 5 _____
3 _____

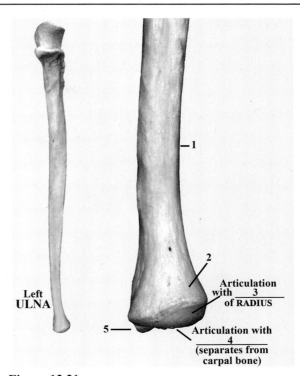

Figure 13.21
61 In reference to **Figure 13.21**, identify #1 - #5.
1 _____ 4 _____
2 _____ 5 _____
3 _____

62 **Match Terms with Descriptions**
   Terms
   A Head
   B Shaft (body)
   C Styloid process

   Descriptions
   _____ Articulates laterally with the ulnar notch of the radius forming a distal synovial pivot joint, the distal radioulnar joint.
   _____ Distal knob-like terminus of ulna.
   _____ Major cylindrical portion of the bone.
   _____ Small, sharp process located on the posterior and medial aspect of the ulna, inferior to its head.
   _____ Terminates at an articular disc that separates it from the wrist (carpal) bones.

   Distal Articulations of the Ulna
63 The head of the ulna articulates with an _____ disc that separates it from the _____ bones, and the head articulates with the _____ notch of the radius.
64 The inferior surface of the _____ disc, which separates the ulna from the carpal bones, and the articulation of the radius with two carpal bones forms the _____ joint.
65 The head of the ulna articulates with the _____ notch of the radius forming the synovial pivot joint, the distal _____ joint.

182 Chapter 13 - Appendicular Skeleton

66 Two pivot joints, the proximal and distal _____ joints, allow _____ of the forearm.

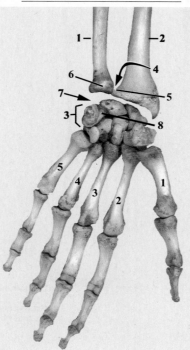

Figure 13.22

67 In reference to **Figure 13.22**, identify #1 - #8.
1 _____   5 _____
2 _____   6 _____
3 _____   7 _____
4 _____   8 _____

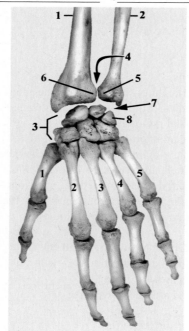

Figure 13.23

68 In reference to **Figure 13.23**, identify #1 - #8.
1 _____   5 _____
2 _____   6 _____
3 _____   7 _____
4 _____   8 _____

RADIUS

69 The radius is the _____ bone of the forearm.

Proximal views of Radius

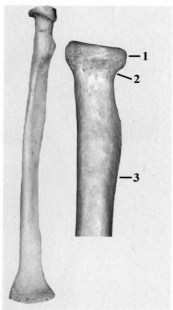

Figure 13.24

70 In reference to **Figure 13.24**, identify #1 - #3.
1 _____   3 _____
2 _____

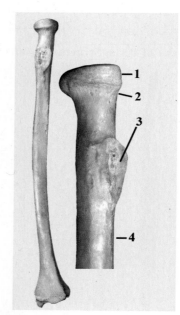

Figure 13.25

71 In reference to **Figure 13.25**, identify #1 - #4.
1 _____   3 _____
2 _____   4 _____

72 **Match Terms with Descriptions**
  Terms
  A Head
  B Neck
  C Radial tuberosity
  D Shaft (body)

Name _____
Class _____

Chapter 13 - Appendicular Skeleton

Descriptions

_____ Constricted portion just inferior to and continuous with the head.
_____ Major cylindrical portion of the bone.
_____ Medially articulates with the radial notch of the ulna forming the synovial pivot joint, the proximal radioulnar joint.
_____ Proximal portion shaped like a knob.
_____ Small rough projection located medial and inferior to the neck.
_____ Superiorly articulates with the capitulum of the humerus forming the synovial hinge joint, the humeroradial joint.

Proximal Articulations of Radius

73 The _____ of the radius articulates with the _____ of the humerus and the _____ notch of the ulna.
74 The articulation of the capitulum of the humerus with the _____ of the radius forms a synovial _____ joint, the _____ joint.
75 The humeroradial joint is the _____ portion of the synovial _____ joint, the _____ joint.
76 The articulation of the _____ of the radius with the radial notch of the ulna forms the synovial _____ joint, the _____ joint.
77 The two pivot joints, the proximal and distal radioulnar joints allow _____ of the forearm.
78 Rotation of the forearm so that the palms face backwards (posteriorly) or downward (inferiorly) and the radius _____ over the ulna is _____.
79 Rotation of the forearm so that the palms face forward (anteriorly) or upward (superiorly) and the radius is _____ to the ulna is _____.

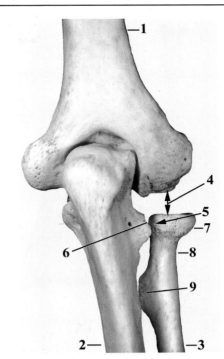

Figure 13.27

81 In reference to **Figure 13.27**, identify #1 - #9.
1 _____ 6 _____
2 _____ 7 _____
3 _____ 8 _____
4 _____ 9 _____
5 _____

Distal Views of the Radius

Posterior View

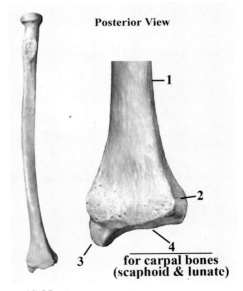

Figure 13.28

82 In reference to **Figure 13.28**, identify #1 - #4.
1 _____ 3 _____
2 _____ 4 _____

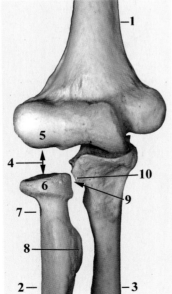

Figure 13.26

80 In reference to **Figure 13.26**, identify #1 - #10.
1 _____ 6 _____
2 _____ 7 _____
3 _____ 8 _____
4 _____ 9 _____
5 _____ 10 _____

184 Chapter 13 - Appendicular Skeleton

Name _____
Class _____

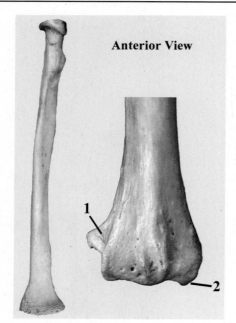

Figure 13.29

83 In reference to **Figure 13.29**, identify #1 - #4.
    1 _____ 2_____

84 **Match Terms with Descriptions**
    Terms
    A Shaft (body)
    B Styloid process
    C Ulnar notch
    Descriptions
    _____ Articulates with the head of the ulna forming the synovial pivot joint, the distal radioulnar joint.
    _____ Major cylindrical portion of the bone.
    _____ Small depression on the distal, medial end of the radius.
    _____ Small projection on the distal, lateral end of the radius.

    Distal Articulations of the Radius

85 The _____ notch of the radius articulates with the head of the ulna forming the synovial _____ joint, the distal _____ joint.

86 The distal and proximal _____ joints allow _____ of the radius for pronation and supination.

87 The most distal surface of the radius articulates with two _____ bones, the lunate and scaphoid bones, forming a major portion of the _____ joint.

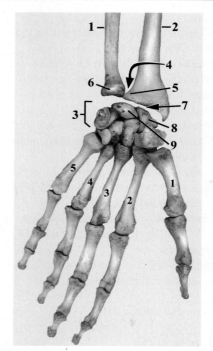

Figure 13.30

88 In reference to **Figure 13.30**, identify #1 - #9.
    1 _____ 6_____
    2 _____ 7_____
    3 _____ 8_____
    4 _____ 9_____
    5 _____

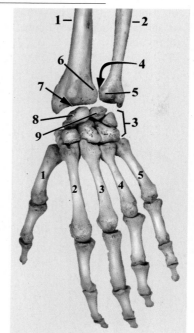

Figure 13.31

89 In reference to **Figure 13.31**, identify #1 - #9.
    1 _____ 6_____
    2 _____ 7_____
    3 _____ 8_____
    4 _____ 9_____
    5 _____

Chapter 13 - Appendicular Skeleton

Proximal Articulations of the Ulna and Radius

90 Proximally, the ulna and the radius form two articulations, one with the _____ and the other with each other.

91 The ulna and the radius articulate with the two distal condyles of the humerus to form a synovial _____ joint, the _____ joint.

92 The trochlea of the humerus articulates with the _____ notch of the ulna forming the _____ joint.

93 The capitulum of the humerus articulates with the _____ of the radius forming the _____ joint.

94 The humeroulnar and the humeroradial joints form the _____ joint.

95 The ulna and the radius articulate with each other to form the synovial _____ joint, the proximal _____ joint.

96 The proximal radioulnar joint allows _____ of the forearm (and palm).

Distal Articulations of the Ulna and Radius

97 Distally, the ulna and radius form three articulations: (1) the radius articulates with two _____ bones, (2) the ulna articulates with an _____ disc, and (3) the radius and ulna articulate with _____.

98 The distal articular surface of the radius articulates with two _____ bones, the _____ and the _____.

99 The articulation of the radius with two carpal bones forms a major portion of the _____ joint.

100 The _____ of the ulna articulates with an _____ disc.

101 The inferior surface of the articular disc articulates with a _____ bone, the _____ bone.

102 The articulation of the articular disc with the triquetrum bone forms the medial portion of the _____ joint.

103 The ulna and the radius articulate with each other to form the synovial _____ joint, the distal _____ joint.

104 The distal and proximal radioulnar joints allow _____ of the forearm (and palm).

Rotation of the Forearm - Pronation and Supination

105 The proximal and distal radioulnar joints allow _____ of the distal _____ over the ulna.

106 Pronation is _____ of the forearm so that the palm faces _____ or _____ and the radius _____ over the ulna.

107 Supination is _____ of the forearm so that the palm faces _____ or _____ and the radius is _____ to the ulna.

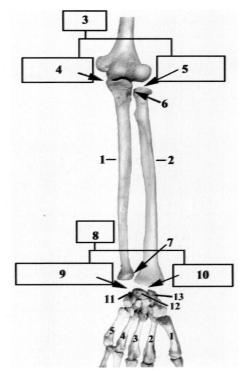

108 Figure 13.32
In reference to Figure 13.32, identify #1 - #12.

1 _____ 7 _____
2 _____ 8 _____
3 _____ 9 _____
4 _____ 10 _____
5 _____ 11 _____
6 _____ 12 _____

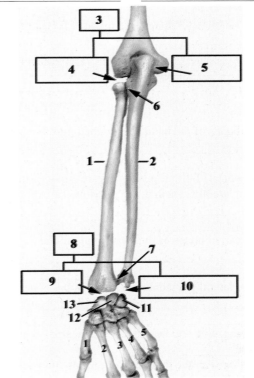

Figure 13.33
109 In reference to Figure 13.33, identify #1 - #12.

186 Chapter 13 - Appendicular Skeleton

1 _____   7 _____
2 _____   8 _____
3 _____   9 _____
4 _____   10 _____
5 _____   11 _____
6 _____   12 _____

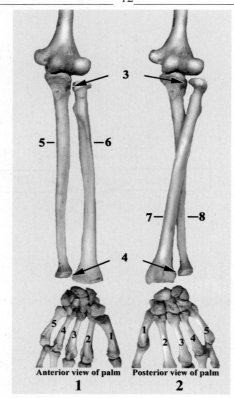

Figure 13.34

110 In reference to **Figure 13.34**, identify #1 - #8.

1 _____   5 _____
2 _____   6 _____
3 _____   7 _____
4 _____   8 _____

HAND

111 The bones of the hand are grouped into three divisions:
  1 _____   3 _____
  2 _____

112 The bones of the carpus (wrist) are arranged in _____ rows.

113 From lateral to medial, the proximal carpal bones are:
  1 _____   3 _____
  2 _____   4 _____

114 The carpal bones of the wrist joint are the
  1 _____   3 _____
  2 _____

115 From lateral to medial, the distal carpal bones are the:
  1 _____   3 _____
  2 _____   4 _____

116 Starting with the thumb, the metacarpal bones are numbered _____.

117 Starting with the thumb, the digits are numbered ____.

118 Each digit has _____ phalanges except the _____, which has only _____ phalanges.

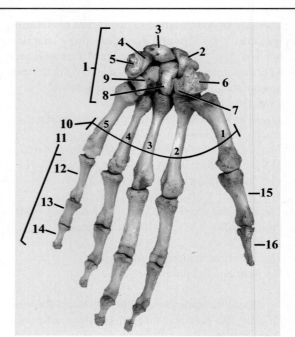

Figure 13.35

119 In reference to **Figure 13.35**, identify #1 - #14.

1 _____   8 _____
2 _____   9 _____
3 _____ 10 _____
4 _____ 11 _____
5 _____ 12 _____
6 _____ 13 _____
7 _____ 14 _____

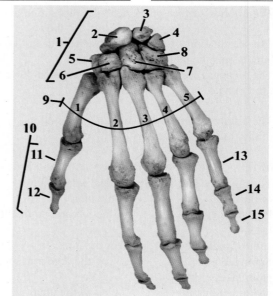

Figure 13.36

120 In reference to **Figure 13.36**, identify #1 - #13.

1 _____   8 _____
2 _____   9 _____
3 _____ 10 _____
4 _____ 11 _____
5 _____ 12 _____
6 _____ 13 _____
7 _____

Chapter 13 - Appendicular Skeleton 187

Proximal Articulations of the Carpus, the Wrist Joint

121 The synovial _____ wrist joint is formed by two articulations: (**1**) the _____ articulation, the articulation between the two _____ bones, the scaphoid and lunate bones with the distal articular surface of the _____, and (**2**) the _____ bone articulates with the inferior surface of an articular disc.

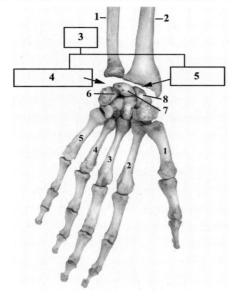

Figure 13.37
122 In reference to **Figure 13.37**, identify #1 - #8.
1 _____ 5 _____
2 _____ 6 _____
3 _____ 7 _____
4 _____ 8 _____

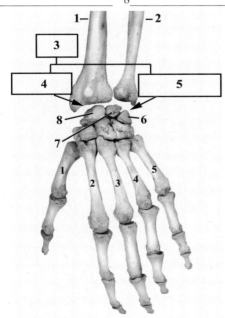

Figure 13.38
123 In reference to **Figure 13.38**, identify #1 - #8.
1 _____ 5 _____
2 _____ 6 _____
3 _____ 7 _____
4 _____ 8 _____

PELVIC GIRDLE
COXAL BONES (OSSA COXAE)

124 The pelvic girdle is formed by the paired _____ bones. Each coxal bone consists of three regions: (1) _____, (2) _____, and (3) _____.

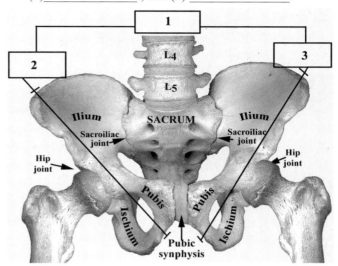

Figure 13.39
125 In reference to **Figure 13.39**, identify #1 - #3.
1 _____ 3 _____
2 _____

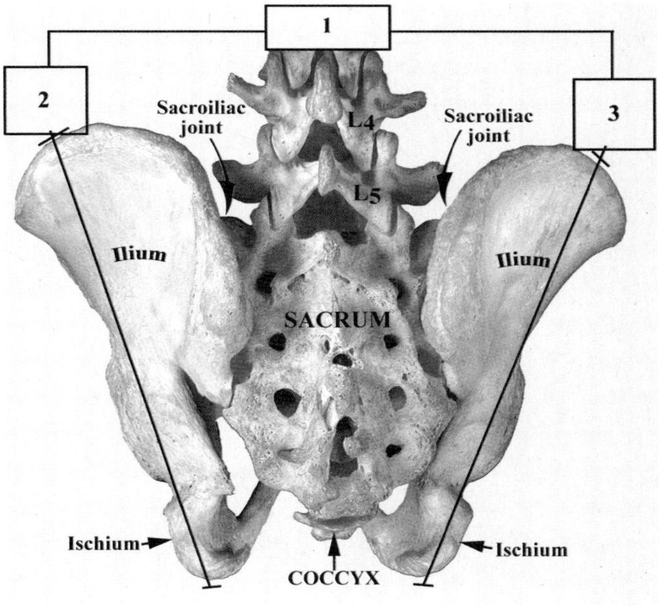

Figure 13.40
126 In reference to **Figure 13.40**, identify #1 - #3.
1 _____ 3 _____
2 _____

Chapter 13 - Appendicular Skeleton

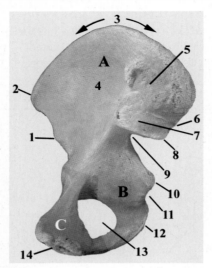

Figure 13.41

127 In reference to **Figure 13.41**:
Identify A - C.
A _____   C _____
B _____
Identify #1 - #14.
1 _____   8 _____
2 _____   9 _____
3 _____   10 _____
4 _____   11 _____
5 _____   12 _____
6 _____   13 _____
7 _____   14 _____

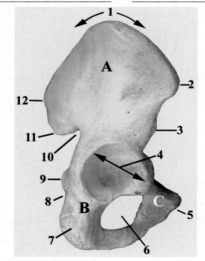

Figure 13.42

128 In reference to **Figure 13.42**:
Identify A - C.
A _____   C _____
B _____
Identify #1 - #14.
1 _____   7 _____
2 _____   8 _____
3 _____   9 _____
4 _____   10 _____
5 _____   11 _____
6 _____   12 _____

129 Match Terms with Descriptions
Coxal Bone
Terms
A   Acetabulum
B   Ilium
C   Ischium
D   Obturator foramen
E   Pubis

Descriptions
_____ Anterior, inferior portion of coxal bone.
_____ Inferior, posterior portion of coxal bone.
_____ Large socket on the lateral central portion of coxal bone.
_____ Large foramen inferior to the acetabulum.
_____ Superior, lateral portion of coxal bone.

Ilium
Terms
A   Ala
B   Anterior inferior iliac spine
C   Anterior superior iliac spine
D   Auricular surface
E   Greater sciatic notch
F   Iliac crest
G   Ilium
H   Posterior superior iliac spine
I   Posterior inferior iliac spine

Descriptions
_____ Articulates with the sacrum to form the synovial gliding sacroiliac joint.
_____ Blunt process that forms the posterior end of the iliac crest.
_____ Blunt process that forms the anterior end of the iliac crest.
_____ Large indentation inferior to the posterior inferior iliac spine.
_____ Large superior, lateral, wing-like portion of the ilium.
_____ Largest portion of the coxal bone.
_____ Rough articular surface located on the posterior, medial, iliac surface.
_____ Small blunt process inferior to the posterior superior iliac spine.
_____ Small blunt process inferior to the anterior superior iliac spine.
_____ Superior border of the ilium.

Ischium
Terms
A   Ischium
B   Ischial spine
C   Ischial tuberosity
D   Lesser sciatic notch

Descriptions
_____ Inferior, posterior portion of the coxal bone.
_____ Thin, pointed process located on the superior, posterior border of the ischium.
_____ Rough, enlarged inferior portion of the ischium.
_____ Small indentation inferior to the ischial spine.

Pubis (Pubic Bone)

Terms
A Pubic symphysis
B Pubis (pubic bone)

Descriptions
_____ Anterior, inferior portion of the coxal bone.
_____ Cartilaginous joint that forms anterior articulation

Articulations of the Coxal Bones

130 The medial and slightly superior _____ surface of each coxal bone articulates with each lateral _____ surface of the sacrum.
131 Each coxal bone articulation with the sacrum forms a synovial _____ joint, the _____ joint.
132 Anteriorly, the pubis of each coxal bone articulate at a _____ joint, the pubic _____.
133 Laterally, each socket, called the _____, provides articulation with the _____ of each femur.
134 The acetabulum and head of the femur form a synovial _____ joint, the _____ joint.

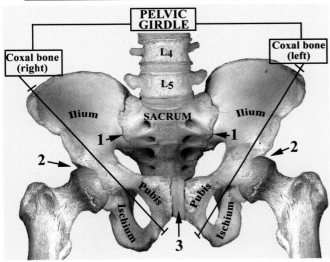

Figure 13.43
135 In reference to **Figure 13.43**, identify #1 - #3.
1 _____ 3 _____
2 _____

PELVIS

136 The pelvis is formed from three components: (1) the _____ girdle, (2) _____, and (3) _____.
137 As a portion of the appendicular skeleton, the pelvis is formed from the _____.
138 As a portion of the axial skeleton, the pelvis is formed from the _____ and the _____.

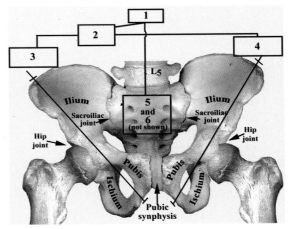

Figure 13.44
139 In reference to **Figure 13.44**, identify #1 - #6.
1 _____ 4 _____
2 _____ 5 _____
3 _____ 6 _____

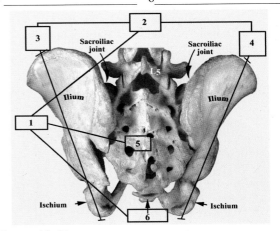

Figure 13.45
140 In reference to **Figure 13.45**, identify #1 - #6.
1 _____ 4 _____
2 _____ 5 _____
3 _____ 6 _____

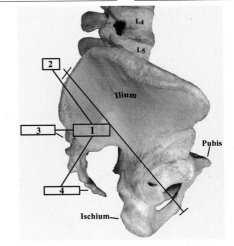

Figure 13.46
141 In reference to **Figure 13.46**, identify #1 - #4.
1 _____ 3 _____
2 _____ 4 _____

190 Chapter 13 - Appendicular Skeleton

142 Regions of the Pelvis
Terms
- A   Anal region
- B   False pelvis
- C   Pelvic inlet
- D   Pelvic outlet
- E   Pelvic brim
- F   Perineum
- G   True pelvis
- H   Urogenital region

Descriptions
_____ Anterior area of perineum
_____ Area of tissue that forms boundary of pelvic outlet
_____ Area contains anal canal and anus
_____ Area contains external urogenital organs
_____ Forms the pelvic cavity
_____ Inferior opening of the true pelvis
_____ Posterior area of perineum
_____ Region inferior to the pelvic brim
_____ Region superior to the pelvic brim
_____ Region bounded by flaring portions of iliac bones.
_____ Superior margin of the true pelvis
_____ Superior opening into the true pelvis

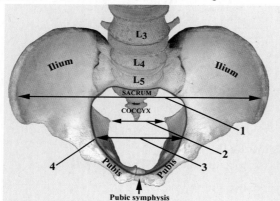

Figure 13.47

143 In reference to **Figure 13.47**, identify #1 - #4.
1 _____ 3 _____
2 _____ 4 _____

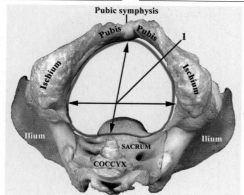

Figure 13.48

144 In reference to **Figure 13.48**, identify #1.
1 _____

Articulations of the Pelvis

145 Superiorly, the sacrum articulates at a _____ joint, the _____ disc, with the _____.

146 Laterally, each articular socket (acetabulum) forms a synovial _____ joint, a _____ joint, with the _____ of each respective femur.

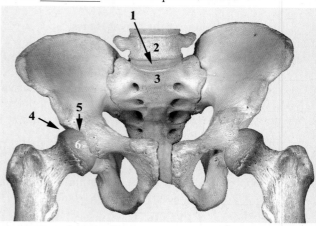

Figure 13.49

147 In reference to **Figure 13.49**, identify #1 - #6.
1 _____ 4 _____
2 _____ 5 _____
3 _____ 6 _____

MAJOR DIFFERENCES BETWEEN MALE AND FEMALE PELVIC STRUCTURE

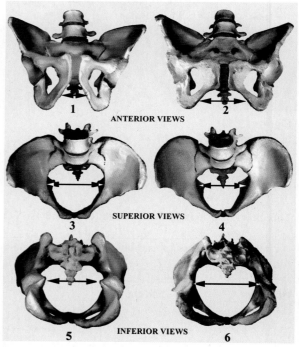

Figure 13.50

148 In reference to **Figure 13.50**, identify #1 - #6 as to **male** or **female**.
1 _____ 4 _____
2 _____ 5 _____
3 _____ 6 _____

Name _____
Class _____

Chapter 13 - Appendicular Skeleton   191

149 Match Terms with Descriptions
   Terms
   A   Female pelvis
   B   Male pelvis

   Descriptions
   _____ Acetabula are larger, directed more posteriorly, and are closer together.
   _____ Iliac bones are more flaring, and the distance between the iliac bones are wider (greater distance between the anterior iliac spines).
   _____ Ischial spines are blunter and less inward curved.
   _____ Pelvic brim is narrow and heart shaped.
   _____ Pelvic outlet is wide.
   _____ Pelvis is narrow, with heaver, thicker, and rougher bones.
   _____ Pubic arch forms an angle less than 90°.
   _____ Pubic arch forms an angle of over 90°.
   _____ Sacrum and coccyx pelvis are less inward curved.

THE LOWER LIMB
150 The lower limb is divided into: **(1)** the _____, **(2)** the _____, **(3)** the _____, and **(4)** the _____.

THIGH

FEMUR
Proximal Views of the Femur

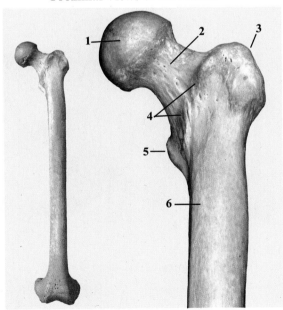

Figure 13.51
151 In reference to **Figure 13.51**, identify #1 - #6.
   1 _____ 4 _____
   2 _____ 5 _____
   3 _____ 6 _____

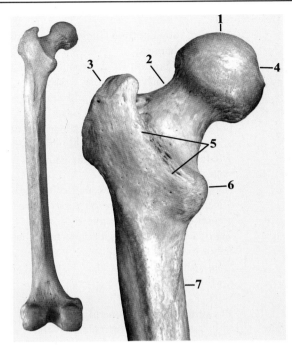

Figure 13.52
152 In reference to **Figure 13.52**, identify #1 - #7.
   1 _____ 5 _____
   2 _____ 6 _____
   3 _____ 7 _____
   4 _____

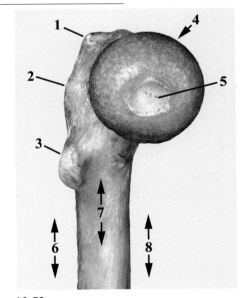

Figure 13.53
153 In reference to **Figure 13.53**, identify #1 - #8.
   1 _____ 5 _____
   2 _____ 6 _____
   3 _____ 7 _____
   4 _____ 8 _____

154 Match Terms with Descriptions
   Terms
   A   Fovea capitis   E   Intertrochanteric crest
   B   Greater trochanter   F   Lesser trochanter
   C   Head   G   Neck
   D   Intertrochanteric line H   Shaft (body)

192   Chapter 13 - Appendicular Skeleton

Name _____
Class _____

Descriptions
_____ Anterior, rough portion connecting the greater and lesser trochanters.
_____ Cylindrical portion of the femur that is slightly bowed anteriorly-posteriorly.
_____ Large, rounded, proximal end that articulates at the acetabulum of the coxal bone.
_____ Large, blunt process located superiorly and laterally at the junction of the head with the neck.
_____ Medially angled portion that connects the head with the shaft (body).
_____ Ovoid depression at the center of the head.
_____ Posterior, narrow ridge of bone connecting the greater and lesser trochanters.
_____ Site of a short tendon that connects the head of the femur to the coxal bone.
_____ Small, blunt process located posteriorly and inferiorly to the neck.

Proximal Articulations of Femur
155 The _____ of each femur articulates with it respective _____ of each _____ bone, each forming a _____ joint, the _____ joint.

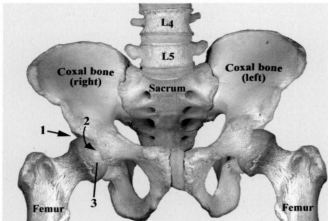

Figure 13.54
156 In reference to **Figure 13.54**, identify #1 - #3.
1 _____ 3 _____
2 _____

Distal Views of Femur

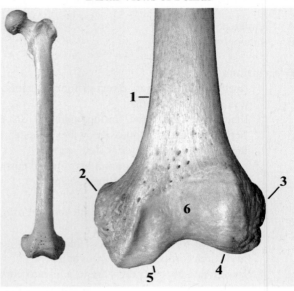

Figure 13.55
157 In reference to **Figure 13.55**, identify #1 - #6.
1 _____ 4 _____
2 _____ 5 _____
3 _____ 6 _____

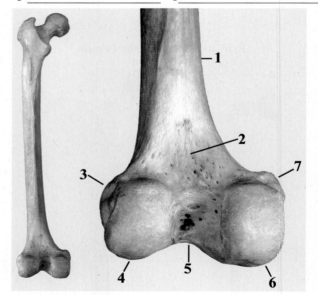

Figure 13.56
158 In reference to **Figure 13.56**, identify #1 - #7.
1 _____ 5 _____
2 _____ 6 _____
3 _____ 7 _____
4 _____

Chapter 13 - Appendicular Skeleton   193

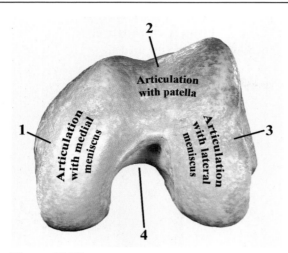

Figure 13.57

159 In reference to **Figure 13.57**, identify #1 - #4.
   1 _____   3 _____
   2 _____   4 _____

160 **Match Terms with Descriptions**
   Terms
   A   Intercondylar fossa
   B   Lateral epicondyle
   C   Lateral condyle
   D   Medial epicondyle
   E   Medial condyle
   F   Patellar surface
   B   Shaft (body)
   Descriptions
   _____ Anterior, shallow depression between the medial and lateral condyles that articulates with the patella.
   _____ Cylindrical portion of the femur that is slightly bowed anteriorly-posteriorly.
   _____ Houses two smooth articular surfaces, the medial and lateral facets, that articulate with the respective patellar surfaces.
   _____ Lateral, rounded projection superior to the lateral condyle.
   _____ Lateral, large, rounded projection that articulates with the lateral condyle of the tibia.
   _____ Medial, rounded projection superior to the medial condyle.
   _____ Medial, large, rounded projection that articulates with the medial condyle of the tibia.
   _____ Posterior, deep depression between the medial and lateral condyles.

   Distal Articulations of the Femur

161 The femur's medial and lateral condyles articulate with _____ pads, the medial and lateral menisci.
162 The medial and lateral menisci are located between the medial and lateral _____ of the femur and the medial and lateral condyles of the _____.
163 The femur-tibia articulation forms one portion of the synovial _____ joint, the _____ joint.
164 The anterior patellar surface of the femur articulates with the _____.

165 The femur-patella articulation forms a synovial _____ joint, the other portion of the _____ joint.

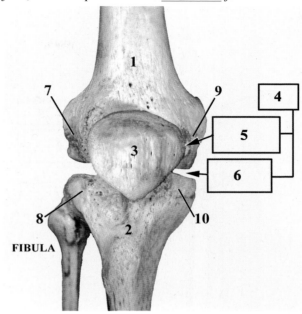

Figure 13.58

166 In reference to **Figure 13.58**, identify #1 - #10.
   1 _____   6 _____
   2 _____   7 _____
   3 _____   8 _____
   4 _____   9 _____
   5 _____ 10 _____

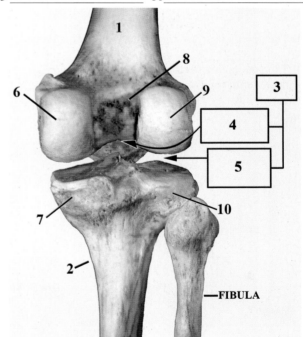

Figure 13.59

167 In reference to **Figure 13.59**, identify #1 - #10.
   1 _____   6 _____
   2 _____   7 _____
   3 _____   8 _____
   4 _____   9 _____
   5 _____ 10 _____

194   Chapter 13 - Appendicular Skeleton

LEG
TIBIA AND FIBULA

172 The bones of the leg are the medial _____ and the lateral _____.

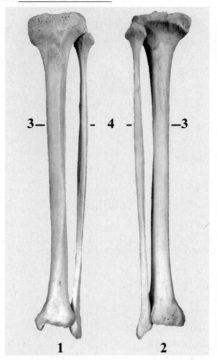

Figure 13.62

173 In reference to **Figure 13.62**, identify #1 - #4.

1 _____   3 _____
2 _____   4 _____

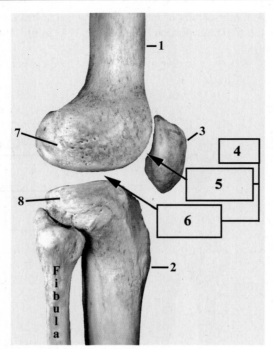

Figure 13.60

168 In reference to **Figure 13.60**, identify #1 - #8.

1 _____   5 _____
2 _____   6 _____
3 _____   7 _____
4 _____   8 _____

KNEE
PATELLA

169 The patella is a flat, triangular bone that is embedded in a _____.

170 The patella functions in increasing the _____ of the anterior thigh muscles and in _____ of the knee joint.

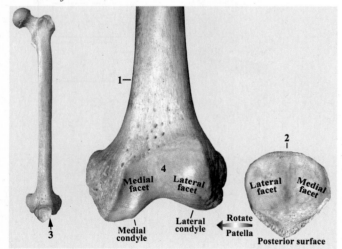

Figure 13.61

171 In reference to **Figure 13.61**, identify #1 - #4.

1 _____   3 _____
2 _____   4 _____

TIBIA
Proximal Views of the Tibia

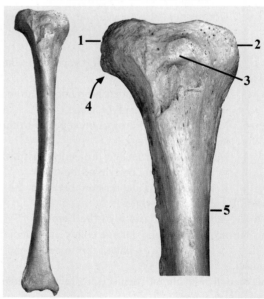

Figure 13.63

174 In reference to **Figure 13.63**, identify #1 - #5.

1 _____   4 _____
2 _____   5 _____
3 _____

Name _____
Class _____

Chapter 13 - Appendicular Skeleton **195**

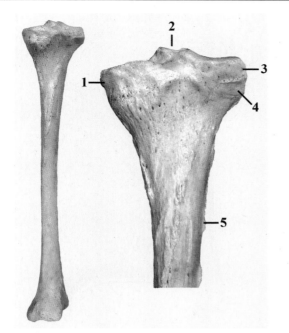

Figure 13.64
175 In reference to **Figure 13.64**, identify #1 - #5.
1 _____ 4 _____
2 _____ 5 _____
3 _____

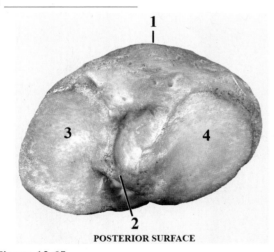

Figure 13.65
176 In reference to **Figure 13.65**, identify #1 - #5.
1 _____ 3 _____
2 _____ 4 _____

177 **Match Terms with Descriptions**
   Terms
   A Anterior crest
   B Lateral condyle
   C Medial condyle
   D Shaft (body)
   E Tibial tuberosity
   Descriptions
   _____ Central portion of the bone.
   _____ Distally terminates at the ankle in an expanded, concave surface that articulates with the superior portion of the talus (a tarsal bone).
   _____ Inferiorly and posteriorly structure has a circular surface that articulates with the head of the fibula.
   _____ Its triangular form produces three distinctive borders.
   _____ Lateral projection with a superior, ovoid, concave surface that articulates with the lateral condyle of the femur.
   _____ Medial projection with an ovoid, concave surface that articulates with the medial condyle of the femur.
   _____ Roughened, rounded projection anterior and inferior to the condyles.
   _____ The anterior, sharp border of shaft.

Proximal Articulations of the Tibia
178 Proximally the tibia articulates with two bones, the _____ and the _____.
179 The medial and lateral _____ of the tibia articulate with their respective medial and lateral _____ of the femur, forming the _____ joint, a portion of the _____ joint.
180 The tibia's lateral condyle articulates at its posterior inferior surface with the _____ of the fibula.
181 The proximal articulation of the head of the fibula with the tibia's posterior inferior lateral condyle forms the synovial _____ joint, the proximal _____ joint.

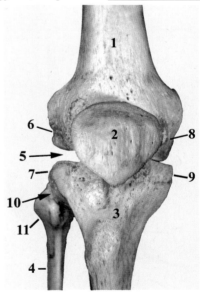

Figure 13.66
182 In reference to **Figure 13.66**, identify #1 - #11.
1 _____ 7 _____
2 _____ 8 _____
3 _____ 9 _____
4 _____ 10 _____
5 _____ 11 _____
6 _____

196 Chapter 13 - Appendicular Skeleton

Distal Views of Tibia

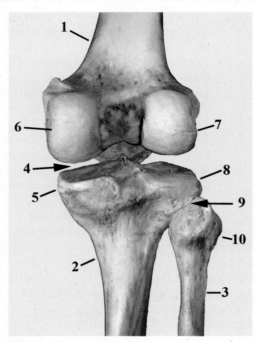

Figure 13.67
183 In reference to **Figure 13.67**, identify #1 - #10.
1 _____   6 _____
2 _____   7 _____
3 _____   8 _____
4 _____   9 _____
5 _____   10 _____

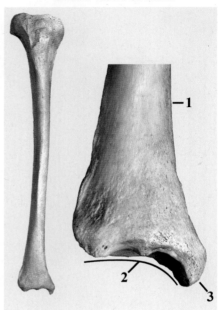

Figure 13.69
185 In reference to **Figure 13.69**, identify #1 - #5.
1 _____   3 _____
2 _____

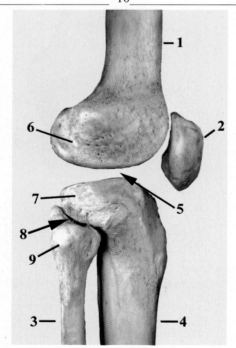

Figure 13.68
184 In reference to **Figure 13.68**, identify #1 - #9.
1 _____   6 _____
2 _____   7 _____
3 _____   8 _____
4 _____   9 _____
5 _____

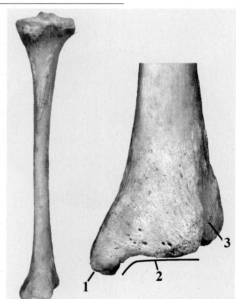

Figure 13.70
186 In reference to **Figure 13.70**, identify #1 - #5.
1 _____   3 _____
2 _____

187 **Match Terms with Descriptions**
   Terms
   A   Inferior Articular Surface
   B   Lateral Articular Surface
   C   Medial malleolus
   Descriptions
   _____ Articulates with the distal fibula and forms the fibrous distal tibiofibular joint.
   _____ Articulates with the medial surface of the talus (a tarsal bone).
   _____ Articulates with the superior surface of the tarsal bone, the talus.
   _____ Medial, inferior process that forms the medial portion of the ankle.

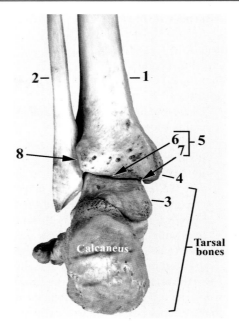

Figure 13.72
189 In reference to **Figure 13.72**, identify #1 - #9.
   1 _____   5 _____
   2 _____   6 _____
   3 _____   7 _____
   4 _____   8 _____

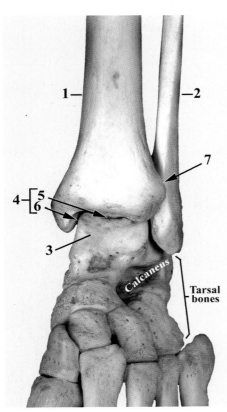

Figure 13.71
188 In reference to **Figure 13.71**, identify #1 - #7.
   1 _____   5 _____
   2 _____   6 _____
   3 _____   7 _____
   4 _____

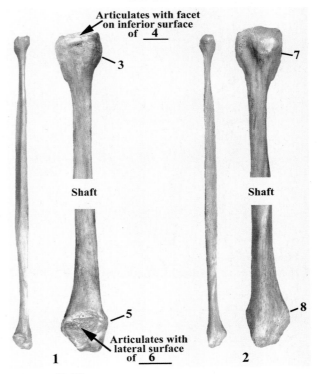

Figure 13.73
190 In reference to **Figure 13.73**, identify #1 - #8.
   1 _____   5 _____
   2 _____   6 _____
   3 _____   7 _____
   4 _____   8 _____

198 Chapter 13 - Appendicular Skeleton

191 Match Terms with Descriptions
Terms
A Head
B Lateral malleolus
C Shaft (body)

Descriptions
_____ Distal, pyramid-shaped portion that forms the lateral portion of the ankle.
_____ Forms a synovial gliding joint, the proximal tibiofibular joint.
_____ Its smooth, inner, triangular surface articulates with the lateral portion of the talus (tarsal bone).
_____ Proximal, expanded portion of the fibula.
_____ Superior, medial surface is flattened and articulates with the lateral condyle of the tibia.
_____ Thin middle portion of the fibula, extending from the head distally to the lateral malleolus.

Proximal Articulations of the Fibula

192 Proximally, the _____ of the fibula articulates with a facet on the inferior posterior surface of the tibia's _____.

193 The proximal fibula-tibia articulation forms a synovial _____ joint, the proximal _____ joint.

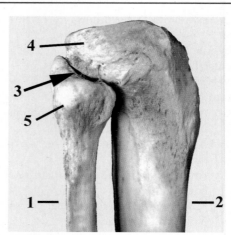

Figure 13.75
195 In reference to **Figure 13.75**, identify #1 - #5.
1 _____ 4 _____
2 _____ 5 _____
3 _____

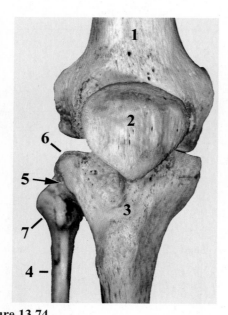

Figure 13.74
194 In reference to **Figure 13.74**, identify #1 - #7.
1 _____ 5 _____
2 _____ 6 _____
3 _____ 7 _____
4 _____

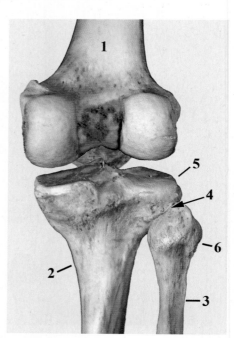

Figure 13.76
196 In reference to **Figure 13.76**, identify #1 - #6.
1 _____ 4 _____
2 _____ 5 _____
3 _____ 6 _____

Distal Articulations of the Fibula

197 Distally, the fibula's medial surface articulates with the lateral surface of the tibia forming the fibrous distal _____ joint.

198 The fibula's lateral malleolus articulates with the lateral surface of the _____ forming the lateral portion of the synovial _____ joint, the _____ joint.

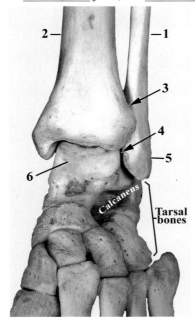

Figure 13.77
199 In reference to **Figure 13.77**, identify #1 - #5.

1 _____ 4 _____
2 _____ 5 _____
3 _____ 6 _____

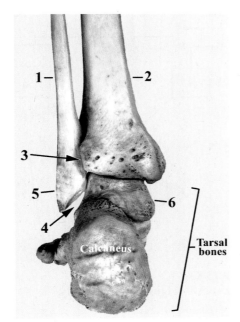

Figure 13.78
200 In reference to **Figure 13.78**, identify #1 - #6.

1 _____ 4 _____
2 _____ 5 _____
3 _____ 6 _____

FOOT

201 The bones of the foot are grouped into the (1) _____, (2) _____, and (3) _____.

Views of the Foot

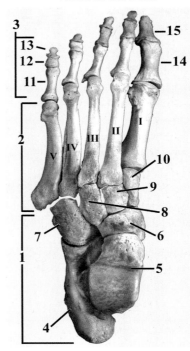

Figure 13.79
202 In reference to **Figure 13.79**, identify #1 - #15.

1 _____ 8 _____
2 _____ 9 _____
3 _____ 10 _____
4 _____ 11 _____
5 _____ 12 _____
6 _____ 13 _____
7 _____ 14 _____
                 15 _____

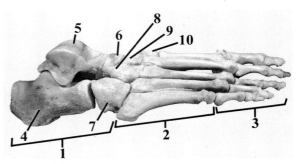

Figure 13.80
203 In reference to **Figure 13.80**, identify #1 - #10.

1 _____ 6 _____
2 _____ 7 _____
3 _____ 8 _____
4 _____ 9 _____
5 _____ 10 _____

200 Chapter 13 - Appendicular Skeleton

Name _____
Class _____

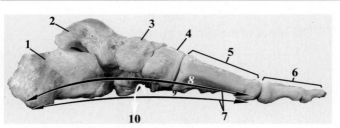

Figure 13.81
204 In reference to **Figure 13.81**, identify #1 - #19.
1 _____ 6 _____
2 _____ 7 _____
3 _____ 8 _____
4 _____ 9 _____
5 _____ 10 _____

Structure of the Foot
Tarsus
205 The tarsus is composed of _____ tarsal bones.
206 The major weight bearing bones of the tarsus are the _____ and the _____.
207 Weight is transferred to the superior portion of the _____ bone by its articulation with the distal articular surface of the _____.
208 The medial _____ of the _____ and the lateral _____ of the _____ provide for medial and lateral ankle support.
209 The calcaneus bone receives weight from the inferior portion of the _____ and transfers weight to the _____.
Metatarsus
210 The metatarsus is composed of _____ metatarsal bones.
211 Beginning _____ the metatarsal bones are numbered _____.
Phalanges
212 There are _____ phalanges.
213 Starting _____, the toes are numbered _____.
214 The toes have _____ phalanges, except the _____, which has only _____ phalanges.

Articulations of the Foot
215 Proximally, the articulation of the foot is the synovial _____ joint, the _____ joint, or _____ joint.
216 The ankle joint is formed by two _____ articulations and one _____ articulation.
217 The tibia-talus articulations are formed (1) by the superior surface of the _____ articulating with the inferior articular surface of the _____, and (2) the medial surface of the _____ articulates with the medial _____.
218 The fibula-talus articulation is formed by the lateral surface of the _____ articulating with the lateral _____ of the _____.

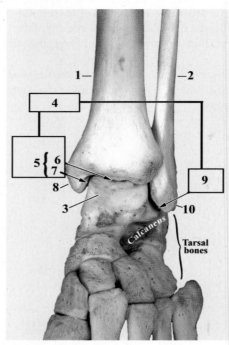

Figure 13.82
219 In reference to **Figure 13.82**, identify #1 - #19.
1 _____ 6 _____
2 _____ 7 _____
3 _____ 8 _____
4 _____ 9 _____
5 _____ 10 _____

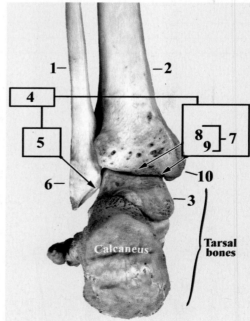

220 **Figure 13.83**
In reference to **Figure 13.83**, identify #1 - #19.
1 _____ 6 _____
2 _____ 7 _____
3 _____ 8 _____
4 _____ 9 _____
5 _____ 10 _____

Articulations - Worksheet

1. What are articulations? _____
   _____
   _____

2. What are the three functional types of joints?
   1 _____ 3 _____
   2 _____

3. Synarthroses are _____ joints.
4. Amphiarthroses are _____ joints.
5. Diarthroses are _____ joints.
6. Structurally, joints are classified in respect to the presence or absence of a _____.
7. If a joint lacks a joint cavity, the joint is classified in respect to the type of _____ material between adjacent bones.
8. The three structural classifications of joints are:
   1 _____ 3 _____
   2 _____
9. A fibrous joint is characterized by the _____ of a joint cavity.
10. A fibrous joint is joined by _____ connective tissue.
11. A cartilaginous joint is characterized by the _____ of a joint cavity.
12. A cartilaginous joint is joined by _____.
13. A synovial joint is characterized by the _____ of a joint cavity.

FIBROUS JOINTS

14. Three types of fibrous joints are
    1 _____ 3 _____
    2 _____

SUTURES

15. What are sutures located? _____
    _____

16. How are sutures formed? _____
    _____

17. Sutures are classified functionally as _____ joints.

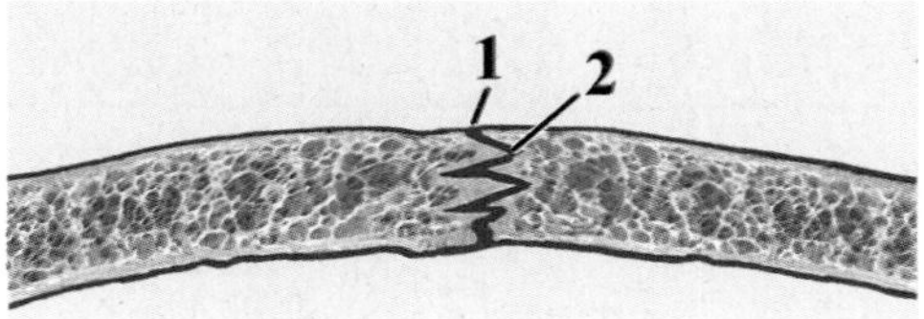

Figure 14.1

18. In reference to **Figure 14.1**, what structural type of joint is shown at #1? _____
19. In reference to **Figure 14.1**, identify #2. _____
    _____
20. In reference to **Figure 14.1**, what functional type of joint is shown? _____
21. In reference to **Figure 14.1**, are movements allowed? ___

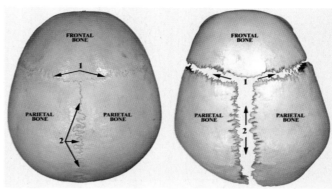

Figure 14.2

22. In reference to **Figure 14.2**, what structural and functional types of joints are shown? _____
    _____
23. In reference to **Figure 14.2**, identify #1 and #2.
    1 _____ 2 _____

GOMPHOSIS

24. What is a gomphosis? _____
    _____
25. What is the name of the connective tissue ligament of a gomphosis? _____
    _____
26. Functionally, a gomphosis is classified as _____, a joint that allows _____ movement.

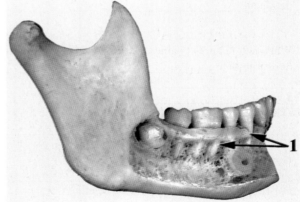

Figure 14.3

27. In reference to **Figure 14.3**, what structural and functional types of joints are shown at #1? _____
    _____
28. In reference to **Figure 14.3**, the joints allow _____ movements.

SYNDESMOSES

29. How are syndesmoses formed? _____
    _____
    _____

30. Functionally, syndesmoses are classified as _____, a joint that allows movement _____.

202 Chapter 14 - Articulations

Name _____
Class _____

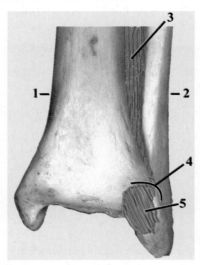

Figure 14.4

31 In reference to **Figure 14.4**, identify #1 - #5.
    1_____ 4_____
    2_____ 5_____
    3_____

32 In reference to **Figure 14.4**, structurally #4 is what type of joint? _____

33 In reference to **Figure 14.4**, functionally, #4 is a _____ _____ joint, a joint that allows movement _____.

34 In reference to **Figure 14.4**, the name of the joint at #4 is the distal _____.

CARTILAGINOUS JOINTS

35 What two features characterize the formation of cartilaginous joints? 1 _____
    2 _____

36 What are two types of cartilaginous joints?
    1_____ 2_____

SYNCHONDROSES

37 How are synchondroses formed? _____
    _____

38 Functionally, synchondroses may be either _____ or _____.

39 A synarthrotic joint allows _____ movement, and a amphiarthrotic joint is _____ moveable.

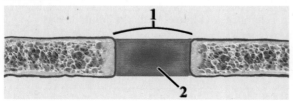

Figure 14.5

40 In reference to **Figure 14.5**, identify #1 - #2.
    1_____ 2_____

Sternocostal Joints

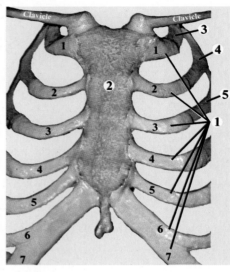

Figure 14.6

41 In reference to **Figure 14.6**, identify #1 - #5.
    1_____ 4_____
    2_____ 5_____
    3_____

42 In reference to **Figure 14.6**, the joints formed at #1 structurally are _____.

43 In reference to **Figure 14.6**, the joints formed at #1 are functionally _____ moveable.

SYMPHYSES

44 What type of binding material forms symphyses? _____
    _____

45 What are two examples of symphyses?
    1_____ 2_____

46 Functionally, symphyses are classified as _____, joints, joints that are _____ moveable.

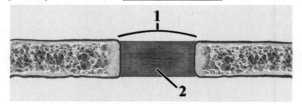

Figure 14.7

47 In reference to **Figure 14.7**, identify #1 - #2.
    1_____ 2_____

Intervertebral joints and Pubic Symphysis

48 What is the binding material for the intervertebral joints?
    _____

49 Where are intervertebral joints located? _____
    _____

50 Functionally, the intervertebral joints are _____, joints that provide for _____ movements.

51 Where is the pubic symphysis located? _____
    _____

52 Functionally, the pubic symphysis is _____, a joint that provides for _____ movements.

Name _____
Class _____

Chapter 14 - Articulations   203

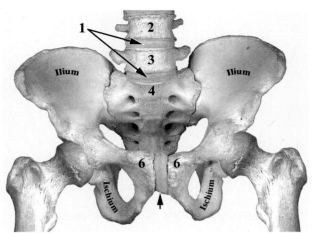

Figure 14.8

53 In reference to **Figure 14.8**, identify #1 - #6.
    1_____ 4_____
    2_____ 5_____
    3_____ 6_____
54 In reference to **Figure 14.8**, the joints formed at #1 and #5 structurally are _____.
55 In reference to **Figure 14.8**, functionally the joints formed at #1 and #5 are _____, joints that allow _____ movements.

SYNOVIAL JOINTS

56 Synovial joints are not directly joined by a binding type of _____ material. Instead, synovial joints have a _____ joint cavity and are diarthrotic, or _____ moveable.
57 Synovial joints are distinguished by
    1_____ 4_____
    2_____ 5_____
    3_____ 6_____
58 The ends of the articulating bones of a synovial joint are covered by _____.
59 The articulating bones of a synovial joint are separated by a _____, filled with _____ fluid.
60 An _____ capsule encloses a synovial joint cavity.
61 The two layers of the articular capsule are the
    1_____ 2_____
62 The outer layer of the articular capsule is the _____ layer.
63 The inner layer of the articular capsule is the _____ membrane, which produces _____ fluid.
64 What is the function of synovial fluid? _____
    _____
65 What reinforces the articular capsule? _____
    _____
66 An articular disc is formed from _____ and extend _____ from the articular capsule to either partially or complete separate the joint.
67 Synovial joints are classified according to the type of ____ _____ they allow.

68 What are three classifications of synovial joints according to the number of axes in which the synovial joints move?
    1_____ 3_____
    2_____
69 Synovial joints that do not move in any one axis are described as _____.
70 Synovial joints that have only one axis of movement and move only in one plane are called _____.
71 Synovial joints that have two axes of movement and move in two planes are called _____.
72 Synovial joints that have more than two axes of movement and move in more than two planes are called _____.

TYPES OF MOVEMENTS
ANGULAR MOVEMENTS

73 What are angular movements? _____
    _____
74 What are four angular movements in diarthrotic joints?
    1_____ 3_____
    2_____ 4_____

FLEXION AND EXTENSION

75 What movement is produced by flexion? _____
    _____
76 What movement is produced by extension? _____
    _____
77 What is hyperextension? _____
    _____

Figure 14.9

78 In reference to **Figure 14.9**, identify #1 - #8.
    1_____ 5_____
    2_____ 6_____
    3_____ 7_____
    4_____ 8_____

Chapter 14 - Articulations

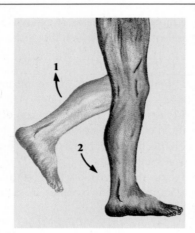

Figure 14.10

79 In reference to **Figure 14.10**, identify #1 - #2.
    1_____ 2_____

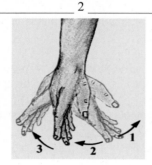

Figure 14.11

80 In reference to **Figure 14.11**, identify #1 - #3.
    1_____ 3_____
    2_____

ABDUCTION AND ADDUCTION

81 What movement is produced by abduction? _____
    _____

82 What movement is produced by adduction? _____
    _____

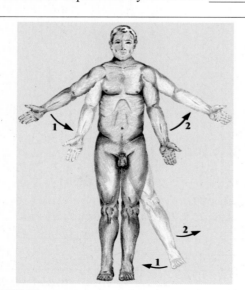

Figure 14.12

83 In reference to **Figure 14.12**, identify #1 - #4.
    1_____ 3_____
    2_____ 4_____

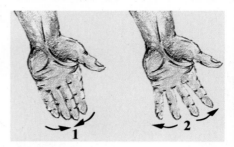

Figure 14.13

84 In reference to **Figure 14.13**, identify #1 - #2.
    1_____ 2_____

CIRCUMDUCTION

85 What movement is produced by circumduction? _____
    _____

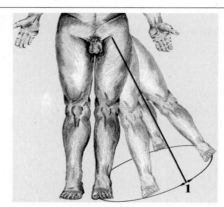

Figure 14.14

86 In reference to **Figure 14.14**, identify #1.
    1_____

CIRCULAR MOVEMENTS

87 What do circular movements allow? _____
    _____

88 What are three circular movements?
    1_____ 3_____
    2_____

Rotation

89 What movement is produced by rotation? _____
    _____

90 What movement is produced by pronation? _____
    _____
    _____

91 What movement is produced by supination? _____
    _____
    _____

92 The radius crosses over the ulna when _____
    (pronation or supination) occurs.

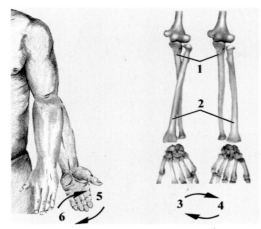

Figure 14.15

93 In reference to **Figure 14.15**, identify #1 - #6.
   1_____ 4_____
   2_____ 5_____
   3_____ 6_____

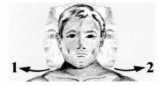

Figure 14.16

94 In reference to **Figure 14.16**, identify #1 - #2.
   1_____ 2_____

SPECIAL MOVEMENTS
Elevation - Depression, Retraction - Protraction

95 What type of movement is produced by elevation? _____

96 What type of movement is produced by depression? _____

97 What type of movement is produced by retraction? _____

98 What type of movement is produced by protraction? _____

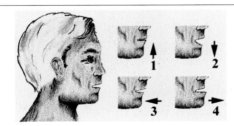

Figure 14.17

99 In reference to **Figure 14.17**, identify #1 - #4.
   1_____ 3_____
   2_____ 4_____

Dorsiflexion and Plantar flexion

100 What movement is produced by dorsiflexion? _____

101 What movement is produced by plantar flexion? _____

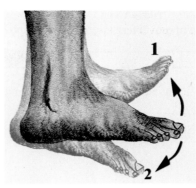

Figure 14.18

102 In reference to **Figure 14.18**, identify #1 - #2.
   1_____ 2_____

Inversion and Eversion

103 What movement is produced by inversion? _____

104 What movement is produced by eversion? _____

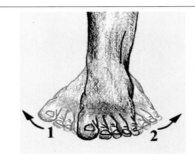

Figure 14.19

105 In reference to **Figure 14.19**, identify #1 - #2.
   1_____ 2_____

Opposition

106 What movement is produced by opposition? _____

SYNOVIAL JOINT TYPES

107 The classification of synovial joints is according to
   1 _____
   2 _____

108 According to the classification of synovial joints by type of movement, synovial joints are all _____.

109 According to the axes of movement synovial joints are classified as
   1_____ 3_____
   2_____ 4_____

110 Nonaxial joints have _____ axis of movement and are described as _____ joints.

111 Uniaxial joints have _____ axis of movement and are described as _____ and _____ joints.

112 Biaxial joints have _____ axes of movement and are described as _____ and _____ joints.

113 Multiaxial joints have _____ axes of movement and are described as _____ joints.

206 Chapter 14 - Articulations

NONAXIAL JOINTS
114 What type of movement is allowed by nonaxial joints? ____

GLIDING (PLANE) JOINTS
115 Describe the structure of gliding (plane) joints. ____

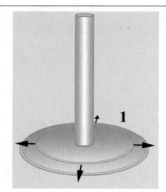

Figure 14.20
116 In reference to **Figure 14.20**, identify the type of joint. ____

Intervertebral joints

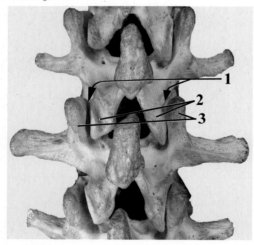

Figure 14.21
117 In reference to **Figure 14.21**, identify #1 - #3.
  1 _____ 3 _____
  2 _____
118 In reference to **Figure 14.21**, what are the names of the joints shown at #1? ____
119 In reference to **Figure 14.21**, what structural type of joint is shown at #1? ____
120 In reference to **Figure 14.21**, what functional type of joint is shown at #1? ____
121 In reference to **Figure 14.21**, what movement is allowed at the joint shown at #1? ____

Proximal tibiofibular joint

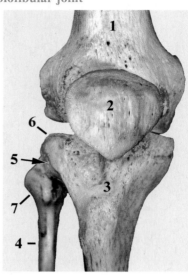

Figure 14.22
122 In reference to **Figure 14.22**, identify #1 - #7.
  1 _____ 5 _____
  2 _____ 6 _____
  3 _____ 7 _____
  4 _____
123 In reference to **Figure 14.22**, what is the name of the joint shown at #5? ____
124 In reference to **Figure 14.22**, what structural type of joint is shown at #5? ____
125 In reference to **Figure 14.22**, what functional type of joint is shown at #5? ____
126 In reference to **Figure 14.22**, what movement is allowed at the joint shown at #5? ____

Sacroiliac joint

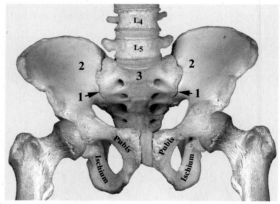

Figure 14.23
127 In reference to **Figure 14.23**, identify #1 - #3.
  1 _____ 3 _____
  2 _____
128 In reference to **Figure 14.23**, what is the name of the joint shown at #1? ____
129 In reference to **Figure 14.23**, what structural type of joint is shown at #1? ____
130 In reference to **Figure 14.23**, what functional type of joint is shown at #1? ____
131 In reference to **Figure 14.23**, what movement is allowed at the joint shown at #1? ____

UNIAXIAL JOINTS

132 Uniaxial joints have only one _____ of movement and move in only one _____.

133 Two types of uniaxial joints are
1 _____ 2 _____

HINGE JOINTS

134 Hinge joint only allow the movements of
1 _____ 2 _____

135 Three examples of hinge joints are the
1 _____ 3 _____
2 _____

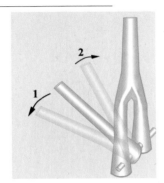

Figure 14.24

136 In reference to **Figure 14.24**, identify the type of joint. ___
_____

137 In reference to **Figure 14.24**, identify the movements shown at #1 and #2.
1 _____ 2 _____

Elbow - Humeroulnar joint and Humeroradial Joints

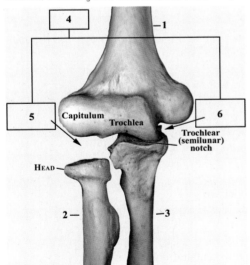

Figure 14.25

138 In reference to **Figure 14.25**, identify #1 - #6.
1 _____ 4 _____
2 _____ 5 _____
3 _____ 6 _____

139 In reference to **Figure 14.25**, what is the name of the joint shown at #4? _____

140 In reference to **Figure 14.25**, what structural type of joint is shown at #4? _____

141 In reference to **Figure 14.25**, what functional type of joint is shown at #4? _____

142 In reference to **Figure 14.25**, what movements are allowed at the joint shown at #4? _____

Knee - Tibiofemoral and Femoropatellar Joints

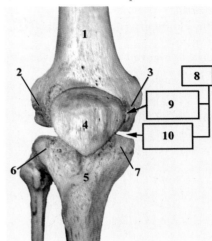

Figure 14.26

143 In reference to **Figure 14.26**, identify #1 - #6.
1 _____ 4 _____
2 _____ 5 _____
3 _____ 6 _____

144 In reference to **Figure 14.26**, what is the name of the joint shown at #4? _____

145 In reference to **Figure 14.26**, what structural type of joint is shown at #4? _____

146 In reference to **Figure 14.26**, what functional type of joint is shown at #4? _____

147 In reference to **Figure 14.26**, what movement is allowed at the joint shown at #4? _____

Ankle Joint - Tibia-Talus and Fibula-Talus Articulations

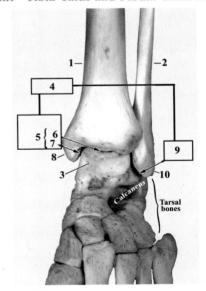

Figure 14.27

148 In reference to **Figure 14.27**, identify #1 - #10.
1 _____ 6 _____
2 _____ 7 _____
3 _____ 8 _____
4 _____ 9 _____
5 _____ 10 _____

208   Chapter 14 - Articulations

149 In reference to **Figure 14.27**, what is the name of the joint shown at #4? _____
150 In reference to **Figure 14.27**, what structural type of joint is shown at #4? _____
151 In reference to **Figure 14.27**, what functional type of joint is shown at #4? _____
152 In reference to **Figure 14.27**, what movement is allowed at the joint shown at #4? _____

PIVOT JOINTS
153 What movement is allowed by pivot joints? _____
154 What are two examples of pivot joints?
   1_____ 2 _____

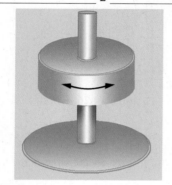

Figure 14.28
155 In reference to **Figure 14.28**, identify the type of joint. ___
_____

ATLAS (C₁) AND AXIS (C₂) - ATLANTO-AXIAL JOINT

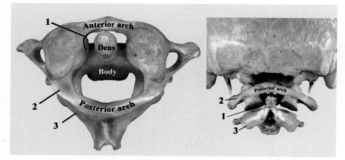

Figure 14.29
156 In reference to **Figure 14.29**, identify #1 - #3.
   1_____ 3 _____
   2_____
157 In reference to **Figure 14.29**, what is the name of the joint shown at #1? _____
158 In reference to **Figure 14.29**, what structural type of joint is shown at #1? _____
159 In reference to **Figure 14.29**, what functional type of joint is shown at #1? _____
160 In reference to **Figure 14.29**, what movement is allowed at the joint shown at #1? _____

Radius and Ulna - Proximal and Distal Radioulnar Joints

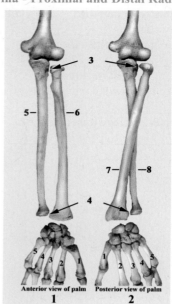

Figure 14.30
161 In reference to **Figure 14.30**, identify #1 - #8.
   1_____ 5 _____
   2_____ 6 _____
   3_____ 7 _____
   4_____ 8 _____
162 What is pronation? _____
_____
163 What is supination? _____
_____

164 In reference to **Figure 14.30**, what are the names of the joints shown at #3 and #4? _____
165 In reference to **Figure 14.30**, what structural type of joint is shown at #3 and #4? _____
166 In reference to **Figure 14.30**, what functional type of joint is shown at #3 and #4? _____
167 In reference to **Figure 14.30**, what movement is allowed at the joints shown at #3 and #4? _____

BIAXIAL JOINTS
168 Biaxial joints have _____ axes of movement and move in two _____.
169 What are two examples of biaxial joints?
   1_____ 2 _____

CONDYLOID JOINTS
170 How are condyloid joints formed? _____
_____

171 What are five movements allowed by condyloid joints?
   1_____ 4 _____
   2_____ 5 _____
   3_____
172 What are three examples of condyloid joints?
   1_____
   2_____
   3_____

Name _____
Class _____

Chapter 14 - Articulations   209

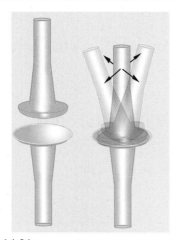

Figure 14.31
173 In reference to **Figure 14.31**, identify the type of joint.
_____

Atlas and Occipital Condyles - Atlanto-Occipital Joint

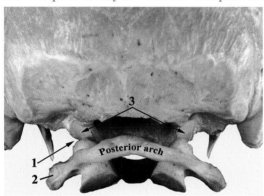

Figure 14.32
174 In reference to **Figure 14.32**, identify #1 - #3.
　1 _____　3 _____
　2 _____
175 In reference to **Figure 14.32**, what is the name of the joint shown at #1? _____
176 In reference to **Figure 14.32**, what structural type of joint is shown at #1? _____
177 In reference to **Figure 14.32**, what functional type of joint is shown at #1? _____
178 In reference to **Figure 14.32**, what movements are allowed at the joint shown at #1?
　1 _____　4 _____
　2 _____　5 _____
　3 _____
179 Which joint allows rotation of the head? _____
_____

Wrist - Radiocarpal Articulation

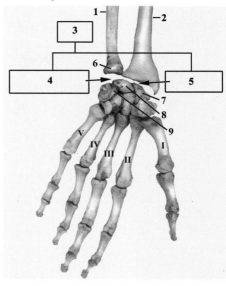

Figure 14.33
180 In reference to **Figure 14.33**, identify #1 - #9.
　1 _____　6 _____
　2 _____　7 _____
　3 _____　8 _____
　4 _____　9 _____
　5 _____
181 In reference to **Figure 14.33**, what is the name of the joint shown at #5? _____
182 In reference to **Figure 14.33**, what structural type of joint is shown at #5? _____
183 In reference to **Figure 14.33**, what functional type of joint is shown at #5? _____
184 In reference to **Figure 14.33**, what movements are allowed at the joint shown at #5?
　1 _____　4 _____
　2 _____　5 _____
　3 _____

Knuckles - Metacarpophalangeal Joints

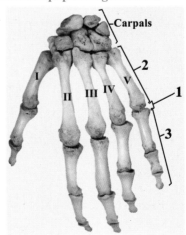

Figure 14.34
185 In reference to **Figure 14.34**, identify #1 - #3.
　1 _____
　2 _____
　3 _____

210 Chapter 14 - Articulations

186 In reference to **Figure 14.34**, what are the names of the joints shown at #1? _____
187 In reference to **Figure 14.34**, what structural type of joint is shown at #1? _____
188 In reference to **Figure 14.34**, what functional type of joint is shown at #1? _____
189 In reference to **Figure 14.34**, what movements are allowed at the joints shown at #1?
  1 _____ 4 _____
  2 _____ 5 _____
  3 _____

SADDLE JOINTS

190 What types of surfaces are present on both articulations of saddle joints? _____
_____
191 What is an example of a saddle joint? _____
_____

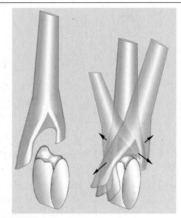

Figure 14.35
192 In reference to **Figure 14.35**, identify the type of joint.
_____

Saddle Joint - Carpometacarpal Joint

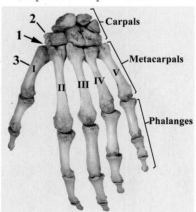

Figure 14.36
193 In reference to **Figure 14.36**, identify #1 - #3.
  1 _____
  2 _____
  3 _____
194 In reference to **Figure 14.36**, what is the name of the joint shown at #1? _____
195 In reference to **Figure 14.36**, what structural type of joint is shown at #1? _____

196 In reference to **Figure 14.36**, what functional type of joint is shown at #1? _____
197 In reference to **Figure 14.36**, what movements are allowed at the joint shown at #1?
  1 _____ 4 _____
  2 _____ 5 _____
  3 _____

MULTIAXIAL JOINTS

198 Multiaxial joints have _____ axes of movement and move in _____ planes.
199 Multiaxial joints are structurally _____ joints.
200 What movements are allowed by ball-and-socket joints?
_____
_____

BALL-AND-SOCKET JOINTS

201 What are two examples of ball-and-socket joints?
  1 _____
  2 _____

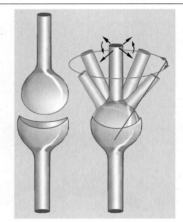

Figure 14.37
202 In reference to **Figure 14.37**, identify the type of joint.
_____

Hip (Coxal) Joint

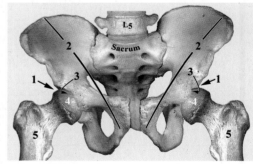

Figure 14.38
203 In reference to **Figure 14.38**, identify #1 - #5.
  1 _____ 4 _____
  2 _____ 5 _____
  3 _____
204 In reference to **Figure 14.38**, what is the name of the joint shown at #1? _____
205 In reference to **Figure 14.38**, what structural type of joint is shown at #1? _____
206 In reference to **Figure 14.38**, what functional type of joint is shown at #1? _____

Name _____
Class _____

Chapter 14 - Articulations 211

207 In reference to **Figure 14.38**, what movements are allowed at the joint shown at #1?
1 _____ 5 _____
2 _____ 6 _____
3 _____ 7 _____
4 _____

Shoulder (Glenohumeral) Joint

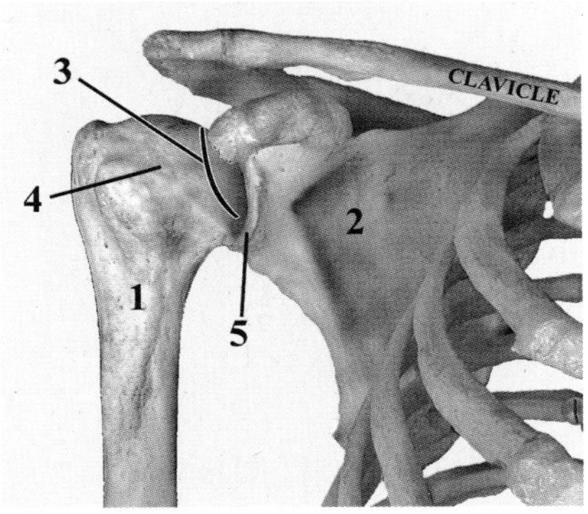

Figure 14.39

208 In reference to **Figure 14.39**, identify #1 - #5.
1 _____ 4 _____
2 _____ 5 _____
3 _____

209 In reference to **Figure 14.39**, what is the name of the joint shown at #3? _____

210 In reference to **Figure 14.39**, what structural type of joint is shown at #3? _____

211 In reference to **Figure 14.39**, what functional type of joint is shown at #3? _____

In reference to **Figure 14.39**, what movements are allowed at the joint shown at #3?
212 1 _____ 4 _____
2 _____ 5 _____
3 _____ 6 _____

Representative Synovial Joints
Shoulder Joint

213 The shoulder joint is also called the _____ joint, and structurally is a _____ joint..

214 The shoulder joint is formed between the _____ of the humerus and the _____ of the scapula.

215 The movements allowed by the shoulder joint are
1 _____ 4 _____
2 _____ 5 _____
3 _____ 6 _____

LAB ACTIVITY 19

Shoulder Joint and Skeleton
Medial Structures of a Scapula

216 Match Terms with Descriptions
Terms
A Acromion process
B Axillary border
C Coracoid process
D Glenoid cavity

Descriptions
_____ Articulates with acromial end of clavicle to form acromioclavicular joint.
_____ Articulation with head of humerus forms the shoulder, or glenohumeral joint
_____ Lateral boundary of the scapula.
_____ Large lateral shallow depression that is articulation for the head of humerus.
_____ Lateral termination of the spine of scapula.
_____ Projects anteriorly from superior border and functions as site of muscle attachment.

Humerus - Structures of Proximal Views
217 Match Terms with Descriptions
Terms
A Anatomical neck
B Body
C Greater tubercle
D Head
E Intertubercular groove
F Lesser tubercle

Descriptions
_____ Forms the shoulder (glenohumeral) joint, a synovial freely moveable ball-and-socket joint.
_____ Large, rounded projection lateral and inferior to the head.
_____ Proximal, medial hemispherical portion of humerus that articulates with the glenoid cavity of the scapula.
_____ Slight constriction at the circumference of the smooth, articular surface of the head.
_____ Small groove between the greater and lesser tubercles.
_____ Small, rounded projection anterior and inferior to the head.

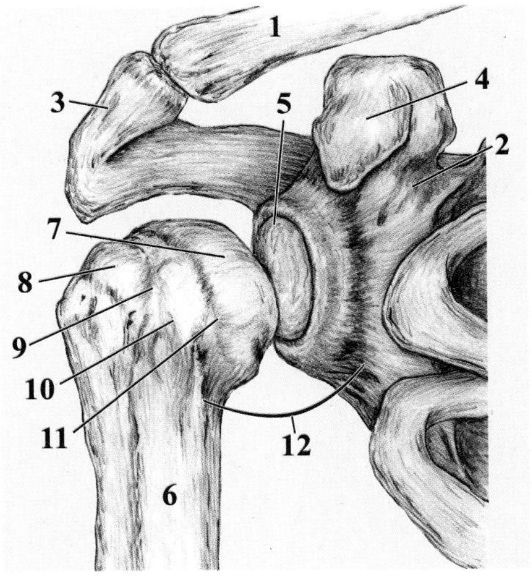

Figure 14.40

218 In reference to **Figure 14.40**, identify #1 - #12.
1 _____ 7 _____
2 _____ 8 _____
3 _____ 9 _____
4 _____ 10 _____
5 _____ 11 _____
6 _____ 12 _____

212 Chapter 14 - Articulations

SHOULDER JOINT AND LIGAMENTS

219 The ligaments of the shoulder joint are:
1 _____ 4 _____
2 _____ 5 _____
3 _____

220 **Match Terms with Descriptions**
  Terms
   A Articular capsule
   B Coracohumeral ligament
   C Glenohumeral ligaments
   D Glenoid labrum
   E Transverse humeral ligament
  Descriptions
   _____ Loose fibrous capsule of the shoulder joint.
   _____ Supports the superior and anterior regions of articular capsule and aids in anchoring the head of humerus to the glenoid cavity.
   _____ Located between the coracoid process of the scapula and the greater tubercle of the humerus.
   _____ Rim of fibrocartilage around the margin of the glenoid cavity.
   _____ Attaches to long head of biceps brachii and supplies most of the strength to keep the head of the humerus against the glenoid cavity.
   _____ Located between the greater and lesser tubercles of the humerus and closes the intertubercular groove.

221 The rotator cuff muscles are positioned between the _____ and the proximal humerus.

222 In addition to providing for arm movements, the rotator cuff muscles structurally _____ the shoulder joint and function to _____ and hold the _____ of the humerus in the glenoid cavity.

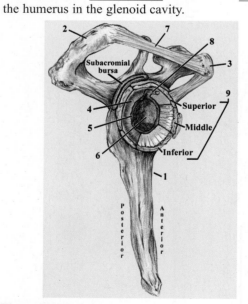

Figure 14.41
223 In reference to **Figure 14.41**, identify #1 - #9.
1 _____ 6 _____
2 _____ 7 _____
3 _____ 8 _____
4 _____ 9 _____
5 _____

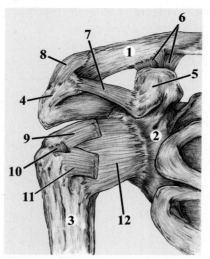

Figure 14.42
224 In reference to **Figure 14.42**, identify #1 - #12.
1 _____
2 _____
3 _____
4 _____
5 _____
6 _____
7 _____
8 _____
9 _____
10 _____
11 _____
12 _____

ELBOW JOINT
Humeroulnar joint and Humeroradial Joints

225 The elbow joint is a _____ joint formed between the humerus and the _____ and the humerus and the _____.

226 The portion of the elbow joint between the humerus and the ulna is called the _____ joint.

227 The portion of the elbow joint between the humerus and the radius is called the _____ joint.

228 The elbow joint allows the movements of _____ and _____.

ELBOW JOINT AND SKELETAL ARTICULATIONS
Distal Structures of the Humerus

229 **Match Terms with Descriptions**
  Terms
   A Capitulum
   B Coronoid fossa
   C Lateral epicondyle
   D Medial epicondyle
   E Olecranon fossa
   F Radial fossa
   G Trochlea
  Descriptions
   _____ Anterior depression superior to capitulum
   _____ Anterior depression superior to trochlea that receives the coronoid process of the ulna when the forearm is flexed.

_____ Distal, lateral condyle that articulates with the head of the radius.
_____ Distal, medial condyle that articulates with the trochlear notch of the ulna
_____ Forms the lateral portion of the synovial hinge joint, the elbow joint.
_____ Forms the medial portion of the synovial hinge joint, the elbow joint
_____ Large depression located on the distal and posterior aspect of the humerus.
_____ Large process on the distal, medial aspect of humerus.
_____ Process on distal, lateral aspect of humerus.
_____ Receives the head of the radius when the forearm is flexed.
_____ Receives the olecranon process of ulna when the forearm is extended
_____ Serves as an attachment site for the radial collateral ligament of the elbow joint and some extensor muscles of the forearm.
_____ Serves as an attachment site for the ulnar collateral ligament of the elbow joint and some flexor muscles of the forearm.

Proximal Structures of the Ulna

230 **Match Terms with Descriptions**

Terms
A Coronoid process
B Olecranon process
C Radial notch
D Trochlear notch

Descriptions
_____ Articulates with the head of the radius forming a proximal synovial pivot joint.
_____ Articulates with the trochlea of the humerus forming a portion of the synovial hinge joint, the elbow joint.
_____ Depression located laterally to the coronoid process.
_____ Large concavity formed between the olecranon and coronoid processes.
_____ Large process located superiorly and posteriorly on the proximal portion of the ulna
_____ Located superiorly and anteriorly on the proximal portion of the ulna
_____ Moves into the olecranon fossa of the humerus when the forearm is extended.
_____ Moves into the coronoid fossa of the humerus when the forearm is flexed.

Proximal Structures of the Radius

231 **Match Terms with Descriptions**

Terms
A Head
B Neck
C Radial tuberosity

Descriptions
_____ Proximal portion shaped like a knob.
_____ Articulates with the capitulum of the humerus forming the synovial hinge joint, the humeroradial joint.
_____ Constricted portion just inferior to and continuous with the head.
_____ Small rough projection located medial and inferior to the neck.

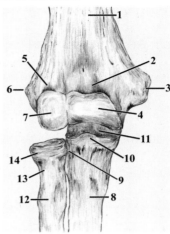

Figure 14.43
232 In reference to **Figure 14.43**, identify #1 - #14.
1 _____ 8 _____
2 _____ 9 _____
3 _____ 10 _____
4 _____ 11 _____
5 _____ 12 _____
6 _____ 13 _____
7 _____ 14 _____

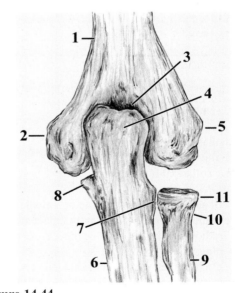

Figure 14.44
233 In reference to **Figure 14.44**, identify #1 - #11.
1 _____ 7 _____
2 _____ 8 _____
3 _____ 9 _____
4 _____ 10 _____
5 _____ 11 _____
6 _____

214 Chapter 14 - Articulations

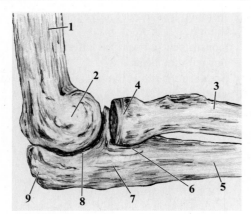

Figure 14.45

234 In reference to **Figure 14.45**, identify #1 - #9.
1 _____ 6 _____
2 _____ 7 _____
3 _____ 8 _____
4 _____ 9 _____
5 _____

ELBOW JOINT AND LIGAMENTS

LAB ACTIVITY 22

235 What are the ligaments of the elbow joint?
1 _____
2 _____
3 _____

236 What forms the primary joint of the elbow joint? _____
_____
_____

237 In addition to the humeroulnar joint, what other joint forms the elbow joint? _____
_____

238 What form the humeroradial joint? _____
_____
_____

239 **Match Terms with Descriptions**
   Terms
   A Articular capsule
   B Radial collateral ligament
   C Ulnar collateral ligament
   Descriptions
   _____ Located between the lateral epicondyle of the humerus and the annular ligament (of proximal radioulnar joint) and the lateral margin of the ulna.
   _____ Located between the medial epicondyle of the humerus and the medial margins of the coronoid and olecranon processes of the ulna.
   _____ Provides lateral stability to the elbow joint.
   _____ Provides medial stability to the elbow joint.
   _____ Reinforced medially and laterally by the radial and ulnar collateral ligaments.
   _____ Surrounds the elbow joint and is relative thin in its anterior and posterior aspects.

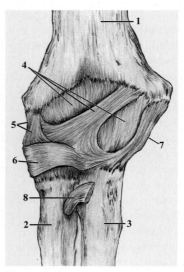

Figure 14.46

240 In reference to **Figure 14.46**, identify #1 - #8.
1 _____ 5 _____
2 _____ 6 _____
3 _____ 7 _____
4 _____ 8 _____

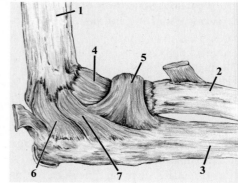

Figure 14.47

241 In reference to **Figure 14.47**, identify #1 - #7.
1 _____ 5 _____
2 _____ 6 _____
3 _____ 7 _____
4 _____

HIP (COXAL) JOINT

242 What forms the hip joint? _____
_____
_____

243 What movements are allowed at the hip joint?
1 _____ 5 _____
2 _____ 6 _____
3 _____ 7 _____
4 _____

HIP (COXAL) JOINT AND SKELETAL ARTICULATIONS
Regions and Structures of Coxal Bone

244 **Match Terms with Descriptions**
   Terms
   A Acetabulum
   B Ilium
   C Ischium
   D Pubis

Name _____
Class _____

Chapter 14 - Articulations   215

Descriptions

_____ Anterior, inferior portion of coxal bone.
_____ Inferior, posterior portion of coxal bone.
_____ Large socket on the lateral central portion of coxal bone.
_____ Superior, lateral portion of coxal bone.

Proximal Structures of the Femur

245 **Match Terms with Descriptions**

Terms
A Fovea capitis
B Greater trochanter
C Head
D Intertrochanteric crest
E Intertrochanteric line
F Lesser trochanter
G Neck

Descriptions

_____ Anterior, rough portion connecting the greater and lesser trochanters.
_____ Large, blunt process located superiorly and laterally at the junction of the head with the neck.
_____ Large, rounded, proximal end that articulates at the acetabulum of the coxal bone.
_____ Medially angled portion that connects the head with the shaft (body).
_____ Ovoid depression at the center of the head
_____ Posterior, narrow ridge of bone connecting the greater and lesser trochanters.
_____ Site of a short tendon that connects the head of the femur to the coxal bone.
_____ Small, blunt process located posteriorly and inferiorly to the neck.

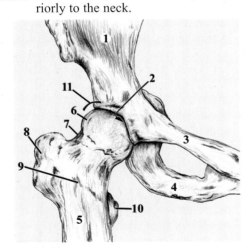

246 **Figure 14.48**
In reference to **Figure 14.48**, identify #1 - #14.
1 _____   7 _____
2 _____   8 _____
3 _____   9 _____
4 _____ 10 _____
5 _____ 11 _____
6 _____

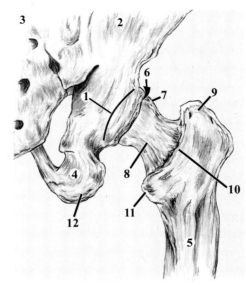

Figure 14.49
247 In reference to **Figure 14.49**, identify #1 - #14.
1 _____   7 _____
2 _____   8 _____
3 _____   9 _____
4 _____ 10 _____
5 _____ 11 _____
6 _____ 12 _____

HIP JOINT AND LIGAMENTS

248 The ligaments of the hip are the
1 _____   5 _____
2 _____   6 _____
3 _____   7 _____
4 _____

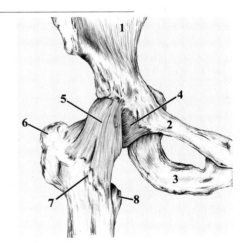

Figure 14.50
249 In reference to **Figure 14.50**, identify #1 - #8.
1 _____   5 _____
2 _____   6 _____
3 _____   7 _____
4 _____   8 _____

Chapter 14 - Articulations

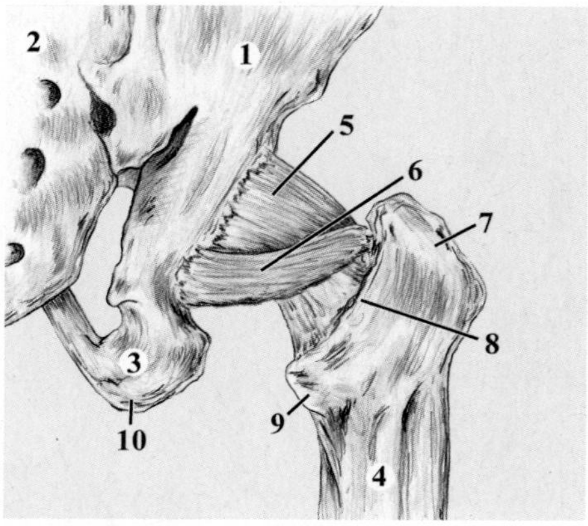

Figure 14.51

250 In reference to **Figure 14.51**, identify #1 - #10.

1 _____ 6 _____
2 _____ 7 _____
3 _____ 8 _____
4 _____ 9 _____
5 _____ 10 _____

251 Match Terms with Descriptions

Terms
- A Acetabular labrum
- B Articular capsule
- C Iliofemoral ligament
- D Ischiocapsular ligament
- E Ligament of femoral head (Ligamentum teres)
- F Pubocapsular Ligament
- G Transverse acetabular ligament

Descriptions

_____ Anterior ligament of the hip joint located between the anterior iliac spine of the ilium and the intertrochanteric line of the femur.

_____ Extends across the acetabular notch and completes the protective rim of the acetabular labrum.

_____ Functions to deepen the acetabulum, and it protects the margin of the acetabulum.

_____ Inferior ligament of the hip joint.

_____ Located between the fovea capitus of the femur and the base of the acetabular notch and the transverse ligament.

_____ Posterior ligament of the hip joint.

_____ Rim of fibrocartilage located at the margin of the acetabulum.

_____ Strong fibrous capsule reinforced by several ligaments (especially the iliofemoral, pubocapsular, and ischiocapsular ligaments).

KNEE JOINT
Tibiofemoral and Femoropatellar Joints

252 What forms the knee joint? _____

253 What forms the tibiofemoral joint? _____

254 What forms the femoropatellar joint? _____

255 What movements are allowed at the knee joint? _____

Distal Structures of the Femur

256 Match Terms with Descriptions

Terms
- A Intercondylar fossa
- B Lateral condyle
- C Lateral epicondyle
- D Medial condyle
- E Medial epicondyle
- F Patellar surface

Descriptions

_____ Anterior, shallow depression between the medial and lateral condyles that articulates with the patella.

_____ Lateral, large, rounded projection that articulates with the lateral condyle of the tibia.

_____ Lateral, rounded projection superior to the lateral condyle.

_____ Medial, large, rounded projection that articulates with the medial condyle of the tibia.

_____ Medial, rounded projection superior to the medial condyle.

_____ Posterior, deep depression between the medial and lateral condyles.

Proximal Structures of the Tibia

257 Match Terms with Descriptions

Terms
- A Intercondyloid eminence (spine)
- B Lateral condyle
- C Medial condyle
- D Tibial tuberosity

Descriptions

_____ Anterior and posterior to this structure are rough depressions, the anterior and posterior intercondylar fossae, that serve as attachment sites for (1) the anterior and posterior cruciate ligaments and (2) the menisci of the knee joint.

_____ Inferiorly and posteriorly, this structure has a small, smooth, circular surface that articulates with the head of the fibula.

_____ Proximal, lateral projection with a superior, ovoid, concave surface that articulates with the lateral condyle of the femur.

_____ Proximal, medial projection with an ovoid, concave surface that articulates with the medial condyle of the femur.

_____ Roughened, rounded projection anterior and inferior to the condyles.

_____ Superior projection located between the medial and lateral facets of the condyles

258 The triangular bone of the knee that is embedded in a tendon is the _____.

259 The patella functions to increase the _____ of the anterior thigh muscles and to _____ the knee joint.

Name _____
Class _____

Chapter 14 - Articulations 217

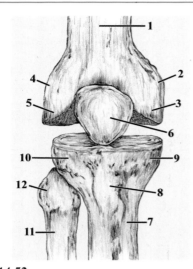

Figure 14.52

260 In reference to **Figure 14.52**, identify #1 - #12.

1 _____ 7 _____
2 _____ 8 _____
3 _____ 9 _____
4 _____ 10 _____
5 _____ 11 _____
6 _____ 12 _____

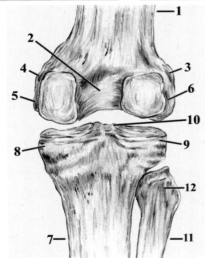

Figure 14.53

261 In reference to **Figure 14.53**, identify #1 - #12.

1 _____ 7 _____
2 _____ 8 _____
3 _____ 9 _____
4 _____ 10 _____
5 _____ 11 _____
6 _____ 12 _____

KNEE JOINT AND LIGAMENTS

262 The major ligaments of the knee joint include the:

1 _____
2 _____
3 _____
4 _____
5 _____

263 **Match Terms with Descriptions**

Terms
A   Articular capsule
B   Collateral ligaments (tibial and fibular)
C   Cruciate ligaments (anterior and posterior)
D   Menisci (medial and lateral)
E   Patellar ligament

Descriptions

_____ Broad flat ligament that is located between the medial epicondyle of the femur and the medial epicondyle and medial surface of the tibia.

_____ Called medial and lateral as to their location between the medial and lateral condyles of the femur and the tibia.

_____ Consists of a thin fibrous membrane that is strengthened by its intimate association with connective tissues (tendons, ligaments, and extensions from the periostea).

_____ Continuation of the tendon of the quadriceps femoris muscle.

_____ Crescent, or semilunar, shaped fibrocartilage disks located between the condyles of the femur and the tibia

_____ Fibrous cord that is located between the lateral epicondyle of the femur and head of the fibula.

_____ Function to (1) deepen the articular surface of the tibia (2) act as cushions (shock absorption), and (3) provide stability to the joint.

_____ Functions in the prevention of knee rotation during flexion.

_____ Fuses with the medial meniscus.

_____ Inserts on the tibial tuberosity.

_____ Positions names "anterior" and "posterior" are in reference to their attachment to the tibia.

_____ Provide considerable strength to the knee joint and are located in the middle of the knee joint at the intercondylar fossa.

_____ Reinforced by (1) the ligaments of the Vasti (vastus lateralis, vastus intermedius, vastus medialis) to form the medial and lateral patellar retinacula and by the (2) patellar ligament

218 Chapter 14 - Articulations

Name _____
Class _____

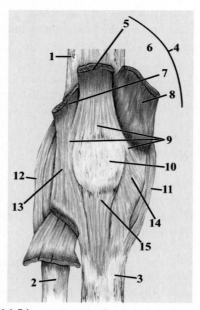

Figure 14.54
264 In reference to **Figure 14.54**, identify #1 - #15.

1 _____
2 _____
3 _____
4 _____
5 _____
6 _____
7 _____
8 _____
9 _____
10 _____
11 _____
12 _____
3 _____
4 _____
5 _____
6 _____
7 _____
8 _____
9 _____
10 _____
11 _____
12 _____
13 _____

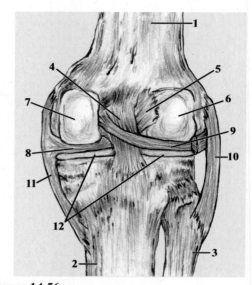

Figure 14.56
In reference to **Figure 14.56**, identify #1 - #12.

1 _____
2 _____
3 _____
4 _____
5 _____
6 _____
7 _____
8 _____
9 _____
10 _____
11 _____
12 _____

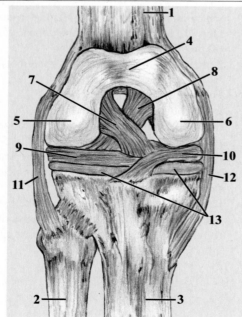

Figure 14.55
265 In reference to **Figure 14.55**, identify #1 - #13.

1 _____
2 _____

Muscular System - Worksheets

FUNCTIONAL GROUPS

1. What are four functional groups of muscles?
   1. _____
   2. _____
   3. _____
   4. _____

2. What is the function of a prime mover? _____

3. What is the function of an antagonist? _____

4. What is the function of a synergist? _____

5. What is the function of a fixator? _____

NAMING SKELETAL MUSCLES

6. What are seven structural and/or functional characteristics used to name muscles?
   1. _____   5. _____
   2. _____   6. _____
   3. _____   7. _____
   4. _____

TYPES OF MOVEMENTS
Angular Movements

7. What are angular movements? _____

8. What are four angular movements that can occur in diarthrotic (moveable) joints?
   1. _____   3. _____
   2. _____   4. _____

Flexion and Extension

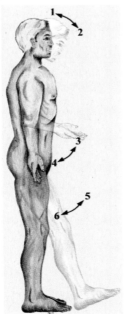

Figure 15.1

9. In reference to **Figure 15.1**, identify movements #1-#6.
   1. _____   4. _____
   2. _____   5. _____
   3. _____   6. _____

10. What type of movement is produced by a flexor? _____

11. The biceps brachii muscle is an example of a _____ muscle. The biceps brachii muscle flexes the _____ and _____.

12. What type of movement is produced by an extensor? ___

13. The triceps brachii muscle is an example of a _____ muscle. The triceps brachii muscle extends the _____ and _____.

Adductor and Abductor

14. What type of movement is produced by an abductor? ___

15. The deltoid muscle is an example of an _____ muscle. The deltoid muscle _____ the arm.

16. What type of movement is produced by an adductor? ___

17. The pectoralis major and the latissimus dorsi muscles are examples of _____ muscles. The pectoralis major and the latissimus dorsi muscle _____ the arm.

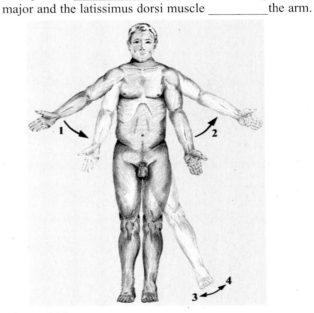

Figure 15.2

18. In reference to **Figure 15.2**, identify movements #1-#4.
    1. _____   3. _____
    2. _____   4. _____

220 Chapter 15 - Muscular System

CIRCULAR MOVEMENTS

19 What are circular movements? _____
_____

20 What are three circular movements?
1 _____ 3 _____
2 _____

Rotator, Supinator, and Pronator Muscles

21 What type of movement is produced by a rotator? _____
_____
_____

22 What type of movement is produced by a pronator? ___
_____
_____
_____

23 What type of movement is produced by a supinator? ___
_____
_____
_____

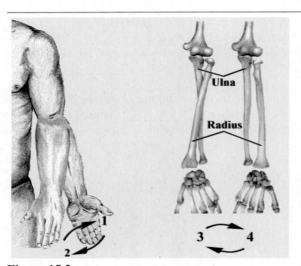

Figure 15.3

24 In reference to **Figure 15.3**, identify movements #1-#4.
1 _____ 3 _____
2 _____ 4 _____

SPECIAL MOVEMENTS
Levator and Depressor Muscles

25 What type of movement is produced by a levator? _____
_____

26 The masseter muscle is an example of a _____ muscle.

27 The masseter muscle is the prime mover in the closure of the mouth, which results from the _____ of the mandible

28 What type of movement is produced by a depressor? ___
_____
_____

29 The platysma and the digastric muscles are examples of _____ muscles. The platysma and the digastric muscles function in _____ of the mandible, thus, opening the mouth.

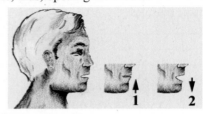

Figure 15.4

30 In reference to **Figure 15.4**, identify movements #1-#2.
1 _____ 2 _____

31 What type of movement is produced by a sphincter? ____
_____

32 Two examples of sphincter muscles are the _____
_____ and the _____
_____

33 What type of movement is produced by a tensor? _____
_____
_____

34 An example of a tensor muscle is the _____
_____, which steadies the trunk-thigh region of the body.

Name _____
Class _____

Chapter 15 - Muscular System 221

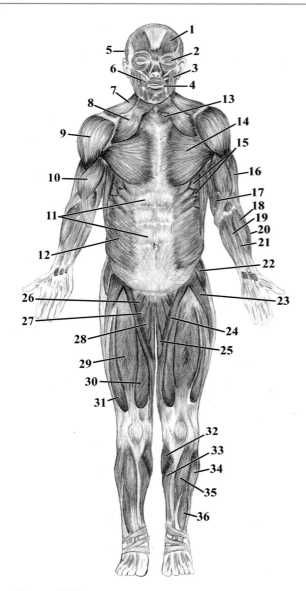

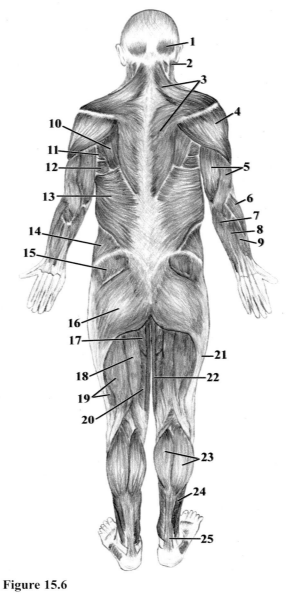

Figure 15.5
35 In reference to **Figure 15.5**, identify muscles #1-#36.

1 _____	19 _____
2 _____	20 _____
3 _____	21 _____
4 _____	22 _____
5 _____	23 _____
6 _____	24 _____
7 _____	25 _____
8 _____	26 _____
9 _____	27 _____
10 _____	28 _____
11 _____	29 _____
12 _____	30 _____
13 _____	31 _____
14 _____	32 _____
15 _____	33 _____
16 _____	34 _____
17 _____	35 _____
18 _____	36 _____

Figure 15.6
36 In reference to **Figure 15.6**, identify muscles #1-#25.

1 _____	14 _____
2 _____	15 _____
3 _____	16 _____
4 _____	17 _____
5 _____	18 _____
6 _____	19 _____
7 _____	20 _____
8 _____	21 _____
9 _____	22 _____
10 _____	23 _____
11 _____	24 _____
12 _____	25 _____
13 _____	

222 Chapter 15 - Muscular System

Name _____
Class _____

PRINCIPAL SUPERFICIAL MUSCLES
MUSCLES OF FACIAL EXPRESSION AND MASTICATION
Muscles of Facial Expression

37 **Frontalis**
Location - _____
Action - _____
Origin - _____
Insertion - _____

38 **Occipitalis**
Location - _____
Action - _____
Origin - _____
Insertion - _____

39 **Zygomaticus (major and minor)**
Location - _____
Action - _____
Origin - _____
Insertion - _____

40 **Orbicularis oris**
Location - _____
Action - _____
Origin - _____
Insertion - _____

41 **Platysma**
Location - _____
Action - _____
Origin - _____
Insertion - _____

Muscles of Mastication

42 **Temporalis**
Location - _____
Action - _____
Origin - _____
Insertion - _____

43 **Masseter**
Location - _____
Action - _____
Origin - _____
Insertion - _____

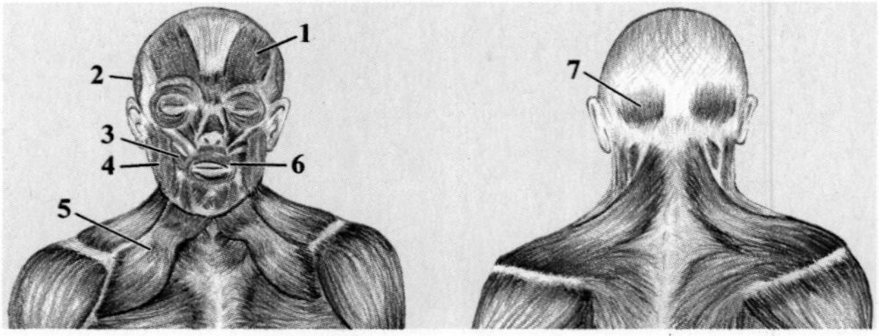

Figure 15.7

44 In reference to **Figure 15.7**, identify muscles #1-#6.
1 _____ 5 _____
2 _____ 6 _____
3 _____ 7 _____
4 _____

MUSCLE THAT MOVES THE HEAD

45 **Sternocleidomastoid**
Location - _____
Action - _____
Origin - _____
Insertion - _____

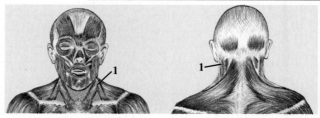

Figure 15.8

46 In reference to **Figure 15.8**, identify muscle #1.
1 _____

Name _____
Class _____

Chapter 15 - Muscular System   223

MUSCLES OF THE ABDOMINAL WALL

47 **External oblique**
    Location - _____
    Action   - _____
    Origin   - _____
    Insertion - _____

48 **Rectus abdominis**
    Location - _____
    Action   - _____
    Origin   - _____
    Insertion - _____

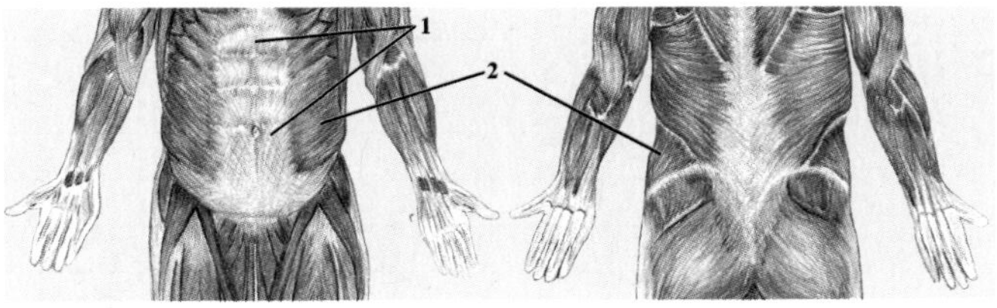

49 **Figure 15.9**
    In reference to **Figure 15.9**, identify muscles #1-#2.
    1 _____ 2 _____

MUSCLES THAT MOVE THE SHOULDER

50 **Serratus anterior**
    Location - _____
    Action   - _____
    Origin   - _____
    Insertion - _____

51 **Trapezius**
    Location - _____
    Action   - _____
    Origin   - _____
    Insertion - _____

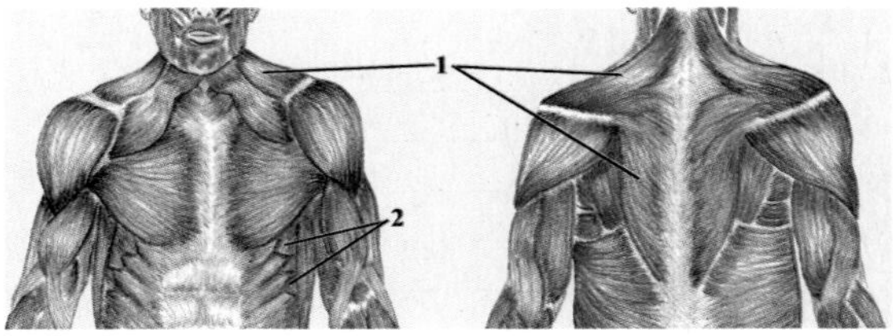

52 **Figure 15.10**
    In reference to **Figure 15.10**, identify muscles #1-#2.
    1 _____ 2 _____

MUSCLES THAT MOVE THE ARM

53 **Pectoralis major**
    Location - _____
    Action   - _____
    Origin   - _____
    Insertion - _____

54 **Latissimus dorsi**
    Location - _____
    Action   - _____
    Origin   - _____
    Insertion - _____

55 **Deltoid**
    Location - _____
    Action   - _____
    Origin   - _____
    Insertion - _____

224   Chapter 15 - Muscular System

Name _____
Class _____

56 **Infraspinatus**
   Location - _____
   Action   - _____
   Origin   - _____
   Insertion - _____

57 **Teres major**
   Location - _____
   Action   - _____
   Origin   - _____
   Insertion - _____

58 **Teres minor**
   Location - _____
   Action   - _____
   Origin   - _____
   Insertion - _____

59 **Figure 15.11**
   In reference to **Figure 15.11**, identify muscles #1-#6.
   1 _____   4 _____
   2 _____   5 _____
   3 _____   6 _____

MUSCLES THAT MOVE THE FOREARM

60 **Triceps brachii**
   Location - _____
   Action   - _____
   Origin   - _____
   Insertion - _____
   Insertion - Olecranon process of the ulna

61 **Biceps brachii**
   Location - _____
   Action   - _____
   Origin   - _____
   Insertion - _____

62 **Brachialis**
   Location - _____
   Action   - _____
   Origin   - _____
   Insertion - _____

63 **Brachioradialis**
   Location - _____
   Action   - _____
   Origin   - _____
   Insertion - _____

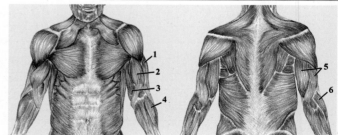

Figure 15.12
64 In reference to **Figure 15.12**, identify muscles #1-#5.
   1 _____   4 _____
   2 _____   5 _____
   3 _____

Chapter 15 - Muscular System

MUSCLES THAT MOVE THE WRIST, HAND, AND FINGERS

65 **Flexor carpi radialis**
   Location -
   Action -
   Origin -
   Insertion -

66 **Palmaris longus**
   Location -
   Action -
   Origin -
   Insertion -

67 **Flexor carpi ulnaris**
   Location -
   Action -
   Origin -
   Insertion -

68 **Extensor carpi radialis longus**
   Location -
   Action -
   Origin -
   Insertion -

69 **Extensor carpi ulnaris**
   Location -
   Action -
   Origin -
   Insertion -

70 **Extensor digitorum**
   Location -
   Action -
   Origin -
   Insertion -

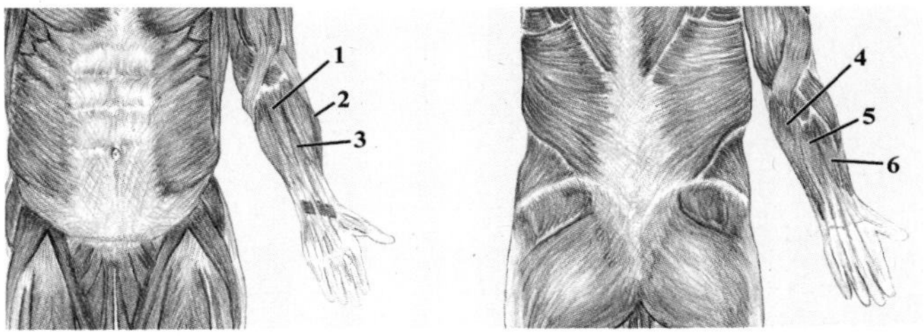

Figure 15.13

71 In reference to **Figure 15.13**, identify muscles #1-#6.
   1 _____ 4 _____
   2 _____ 5 _____
   3 _____ 6 _____

MUSCLES THAT MOVE THE THIGH

72 **Gluteus maximus**
   Location -
   Action -
   Origin -
   Insertion -

73 **Gluteus medius**
   Location -
   Action -
   Origin -
   Insertion -

Chapter 15 - Muscular System

74 **Iliopsoas**
   Location -
   Action   -
   Origin   -
   Insertion -

75 **Pectineus**
   Location -
   Action   -
   Origin   -
   Insertion -

76 **Tensor fasciae latae**
   Location -
   Action   -
   Origin   -
   Insertion -

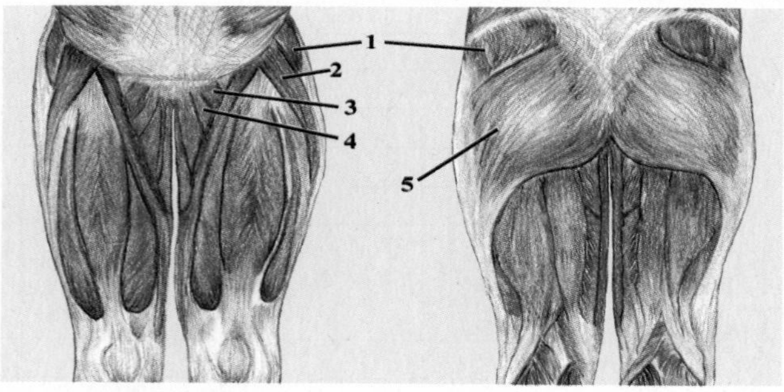

77 **Figure 15.14**
   In reference to **Figure 15.14**, identify muscles #1-#5.
   1 _____   4 _____
   2 _____   5 _____
   3 _____

Muscles that Move the Leg

78 **Sartorius**
   Location -
   Action   -
   Origin   -
   Insertion -

79 **Adductor longus**
   Location -
   Action   -
   Origin   -
   Insertion -

80 **Gracilis**
   Location -
   Action   -
   Origin   -
   Insertion -

81 **Adductor magnus**
   Location -
   Action   -
   Origin   -
   Insertion -

Name _____
Class _____

Chapter 15 - Muscular System   227

QUADRICEPS FEMORIS

The quadriceps femoris is a muscle formed by the muscle group which includes the rectus femoris, vastus intermedius, vastus lateralis, and vastus medialis.

82 **Rectus femoris**
    Location - _____
    Action   - _____
             _____
             _____
    Origin   - _____
    Insertion - _____

83 **Vastus lateralis**
    Location - _____
    Action   - _____
             _____
             _____
    Origin   - _____
    Insertion - _____

84 **Vastus medialis**
    Location - _____
    Action   - _____
             _____
             _____
    Origin   - _____
    Insertion - _____

HAMSTRINGS

The hamstrings are the biceps femoris, semitendinosus, and the semimembranosus. They have a common origin and extend and flex the thigh.

85 **Biceps femoris**
    Location - _____
    Action   - _____
             _____
             _____
    Origin   - _____
    Insertion - _____

86 **Semitendinosus**
    Location - _____
    Action   - _____
             _____
    Origin   - _____
    Insertion - _____

87 **Semimembranosus**
    Location - _____
    Action   - _____
             _____
             _____
    Origin   - _____
             _____
    Insertion - _____

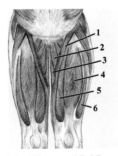 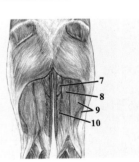

88 **Figure 15.15**
    In reference to **Figure 15.15**, identify muscles #1-#10.
    1 _____   6 _____
    2 _____   7 _____
    3 _____   8 _____
    4 _____   9 _____
    5 _____   10 _____

228 Chapter 15 - Muscular System

MUSCLES THAT MOVE THE FOOT AND TOES

89 **Peroneus longus**
    Location - _____
    Action   - _____
                 _____
    Origin   - _____
    Insertion - _____

90 **Extensor digitorum longus**
    Location - _____
    Action   - _____
                 _____
    Origin   - _____
    Insertion - _____

91 **Tibialis anterior**
    Location - _____
    Action   - _____
                 _____
    Origin   - _____
    Insertion - _____

92 **Gastrocnemius**
    Location - _____
    Action   - _____
                 _____
    Origin   - _____
    Insertion - _____

93 **Soleus**
    Location - _____
    Action   - _____
                 _____
    Origin   - _____
    Insertion - _____

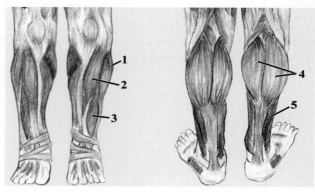

Figure 15.16

94 In reference to **Figure 15.16**, identify muscles #1-#5.
    1 _____   4 _____
    2 _____   5 _____
    3 _____

Muscles and Contraction - Worksheets

1. What are the three types of muscle tissue?
   1 _____ 2 _____ 3 _____

TYPES AND FUNCTION OF MUSCLE TISSUE

SKELETAL MUSCLE TISSUE

2. What are three functions of skeletal muscle tissue?
   1 _____
   2 _____
   3 _____

CARDIAC MUSCLE TISSUE

3. What are two functions of cardiac muscle tissue?
   1 _____
   2 _____

4. What chambers of the heart are responsible for delivery of blood into its two primary circuits? _____

5. What does the pulmonary circuit involve? _____

6. What does the systemic circuit involve? _____

SMOOTH MUSCLE TISSUE

7. What are two functions of smooth muscle tissue?
   1 _____
   2 _____

8. Where does the regulation of the diameter of hollow organs play a significant role? _____

9. Where does regulation of movement of materials through hollow organs play a significant role? _____

CHARACTERISTICS OF MUSCLE TISSUE

10. What are four characteristics of muscle tissue?
    1 _____ 3 _____
    2 _____ 4 _____

Excitability

11. What is muscle tissue excitability? _____

12. What are two ways muscle tissue can be excited? _____

13. What does excitation of the fiber lead to? _____

Contractility

14. What is muscle contractility? _____

15. What is muscle tension? _____

Extensibility

16. What is muscle extensibility? _____

17. Where are two examples that demonstrate muscle extensibility? _____

Elasticity

18. What is muscle elasticity? _____

SKELETAL MUSCLES

19. What forms skeletal muscles? _____

20. What is the function of blood vessels that are associated with muscles? _____

21. What two types of nerves are associated with muscles? _____

22. What do sensory nerves monitor? _____

23. What is the function of motor fibers? _____

24. What is the neurotransmitter released at the neuromuscular junctions of skeletal muscle? _____

Skeletal Muscle Fiber

25. What is the functional unit of a skeletal muscle? _____

26. Describe the structure of a skeletal muscle fiber. _____

27. What is the name of the connection (synapse) between a motor neuron and a muscle fiber? _____

28. What is a fascicle? _____

Connective Tissues of a Muscle

29. The connective tissue that surrounds each individual muscle fiber is called the _____.

30. The endomysium is continuous with the _____, the connective tissue that surrounds a group of muscle fibers, a _____.

31. The perimysium is continuous with the _____, the connective tissue that surrounds the entire muscle.

32. In most muscles the epimysium continues as a _____, attaching the muscle to a _____.

230 Chapter 16 - Muscles and Contraction

Name _____
Class _____

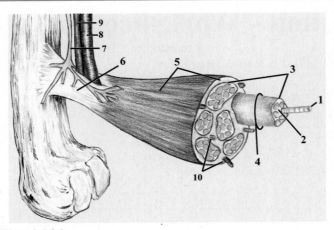

Figure 16.1

33 In reference to **Figure 16.1**, identify #1-10.

1 _____ 6 _____
2 _____ 7 _____
3 _____ 8 _____
4 _____ 9 _____
5 _____ 10 _____

34 Place the following in sequence by numbering from 1 - 8. Start with the generation of tension to the transfer of force to the matrix of bone:

_____ Sharpey's fibers   _____ Tendon
_____ Muscle fiber   _____ Periosteum
_____ Epimysium   _____ Perimysium
_____ Endomysium   _____ Matrix of bone

LAB ACTIVITY 1

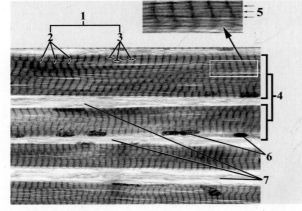

Figure 16.2

35 In reference to **Figure 16.2**, identify #1 - #7.

1 _____ 5 _____
2 _____ 6 _____
3 _____ 7 _____
4 _____

36 What are myofibrils? _____
_____

37 Striations of the fiber are produced by the alignment of the striations of the _____.

38 Striations are formed by the alternating pattern of thin and thick _____ along the length of the myofibrils.

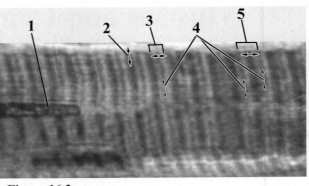

Figure 16.3

39 In reference to **Figure 16.3**, identify #1 - #5.

1 _____ 4 _____
2 _____ 5 _____
3 _____

40 The dark bands of the fiber are called the _____ and are composed of _____.

41 What are two regions where the A bands are associated with other proteins?

1 _____
_____

2 _____
_____

42 The light bands of the fiber are called the _____ and are composed of _____.

43 The thin filaments have a central region called _____.

44 What is the function of the Z lines (discs)? _____

45 A sarcomere is the region between two adjacent _____ and is defined as the functional unit of _____.

LAB ACTIVITY 2

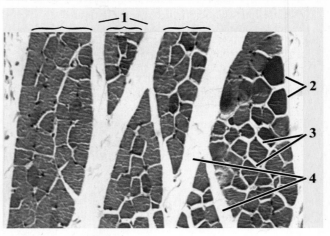

Figure 16.4

46 In reference to **Figure 16.4**, identify #1 - #4.

1 _____ 3 _____
2 _____ 4 _____

Name _____
Class _____

Chapter 16 - Muscles and Contraction   231

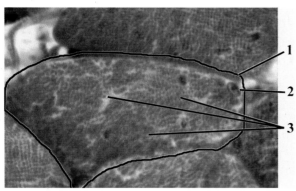

Figure 16.5

47 In reference to **Figure 16.5**, identify #1 - #3.
    1 _____ 3 _____
    2 _____

48 What forms the myofibrils? _____

49 What contractile protein is found in the thin filaments? ___

50 What contractile protein is found in the thick filaments? _____

Attachments of a Muscle

51 How is a direct muscle attachment formed? _____
    _____
    _____

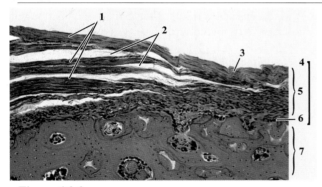

Figure 16.6

52 In reference to **Figure 16.6**, identify #1 - #6.
    1 _____ 4 _____
    2 _____ 5 _____
    3 _____ 6 _____

53 How is an indirect muscle attachment formed? _____
    _____

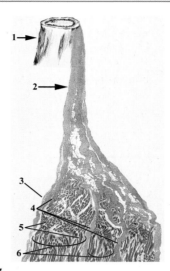

Figure 16.7

54 In reference to **Figure 16.7**, identify #1 - #6.
    1 _____ 4 _____
    2 _____ 5 _____
    3 _____ 6 _____

55 What is a muscle's insertion? _____
    _____

56 What is a muscle's origin? _____
    _____

MICROANATOMY OF A SKELETAL MUSCLE FIBER

57 What are four characteristics of a skeletal muscle fiber?
    1 _____ 3 _____
    2 _____ 4 _____

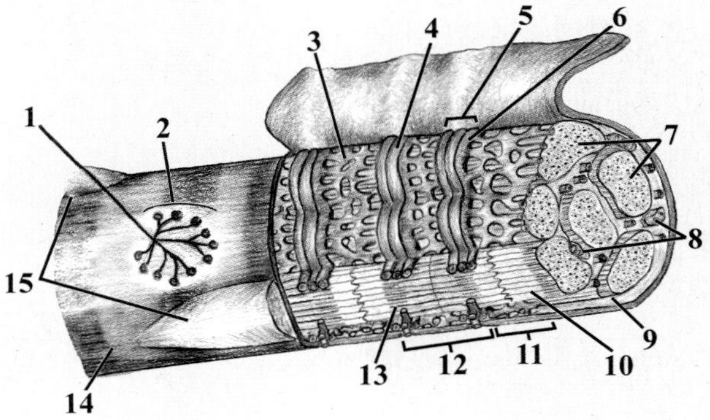

Figure 16.8

58 In reference to **Figure 16.8**, identify #1 - #6.
    1 _____ 9 _____
    2 _____ 10 _____
    3 _____ 11 _____
    4 _____ 12 _____
    5 _____ 13 _____
    6 _____ 14 _____
    7 _____ 15 _____
    8 _____

59 What is a neuromuscular junction? _____
    _____

232 Chapter 16 - Muscles and Contraction

Name _____
Class _____

Neuromuscular Junction

60 What is the name of the region of the sarcolemma (plasma membrane) that is responsive to a neurotransmitter? _____

61 What begins at the postsynaptic membrane? _____

62 Where does the action potential spread? _____

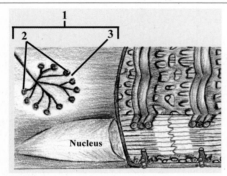

Figure 16.9

63 In reference to **Figure 16.9**, identify #1 - #3.
    1 _____
    2 _____
    3 _____

64 In reference to **Figure 16.9**, what is the function of #2? _____

65 In reference to **Figure 16.9**, what is the function of #3? _____

Sarcolemma, Sarcoplasm, and T tubules

66 What is the sarcolemma? _____

67 In addition to functioning as the boundary of the cell, what is another function of the sarcolemma? _____

68 What does excitability mean? _____

69 What is sarcoplasm? _____

70 Where are T tubules located? _____

71 What is the function of T tubules? _____

72 What is a triad? _____

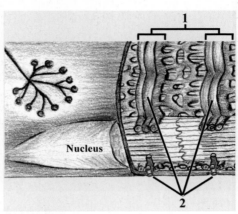

Figure 16.10

73 In reference to **Figure 16.10**, identify #1 - #2.
    1 _____ 2 _____

Sarcoplasmic reticulum

74 What forms the sarcoplasmic reticulum? _____

75 Where is the sarcoplasmic reticulum located? _____

76 Where are the terminal cisternae located? _____

77 What is the function of the calcium ion pumps of the sarcoplasmic reticulum? _____

78 What happens as a result of the arrival of an action potential at the terminal cisternae? _____

79 What do the calcium ions initiate? _____

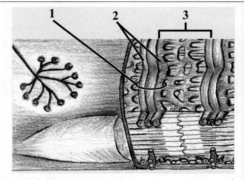

Figure 16.11

80 In reference to **Figure 16.11**, identify #1 - #3.
    1 _____ 3 _____
    2 _____

Myofibrils

81 What two types of filaments are found in the myofibrils? _____

82 The arrangement of the filaments produces cross-bands called _____. The dark band is the _____, and the light band is the _____.

83 The thick filaments are composed of the contractile protein _____.

Name _____
Class _____

Chapter 16 - Muscles and Contraction   233

84 The heads of myosin molecules bind to the contractile proteins of the thin filament called _____. Once bound, the heads _____ to produce the force of contraction.
85 In addition to thick filaments, the A band contains _____ filaments in a region called the zone of _____.
86 The region of the A band where thin filaments are not located is called the _____ zone.

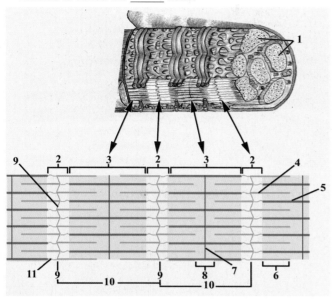

Figure 16.12
87 In reference to **Figure 16.12**, identify #1 - #11.
    1 _____ 7 _____
    2 _____ 8 _____
    3 _____ 9 _____
    4 _____ 10 _____
    5 _____ 11 _____
    6 _____

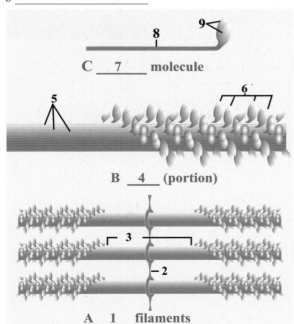

Figure 16.13
88 In reference to **Figure 16.13**, identify #1 - #9.

1 _____ 6 _____
2 _____ 7 _____
3 _____ 8 _____
4 _____ 9 _____
5 _____

I Bands
89 What are six components of the I bands?
    1 _____ 4 _____
    2 _____ 5 _____
    3 _____ 6 _____
90 What is the contractile protein of the I bands? _____
    _____
91 In a resting muscle fiber the myosin binding sites (of actin) are covered by the _____ complex.
92 In a resting muscle fiber, the subunit of troponin that binds _____ is open because calcium is stored in the _____.

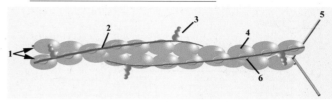

Figure 16.14
93 In reference to **Figure 16.14**, identify #1 - #6.
    1 _____ 4 _____
    2 _____ 5 _____
    3 _____ 6 _____
94 In reference to **Figure 16.14**, are the myosin binding sites (active sites) of actin blocked or unblocked? _____
95 In reference to **Figure 16.14**, what binds calcium ions? ___
    _____
96 When are calcium ions released into the sarcoplasm? ____
    _____
97 What happens when calcium binds with troponin? _____
    _____
    _____
98 When does cross-bridge interaction begin? _____
    _____

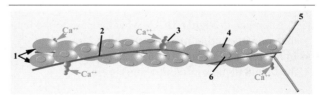

Figure 16.15
99 In reference to **Figure 16.15**, identify #1 - #6.
    1 _____ 4 _____
    2 _____ 5 _____
    3 _____ 6 _____
100 In reference to **Figure 16.15**, are the myosin binding sites (active sites) of actin blocked or unblocked? _____

234 Chapter 16 - Muscles and Contraction

Sarcomere

101 What is the functional unit of contraction? _____

102 A sarcomere is described as the region between adjacent _____.

103 When a muscle fiber contracts, the _____ pulls the _____ inward, resulting in a _____ in length of the sarcomere as the I bands are pulled _____.

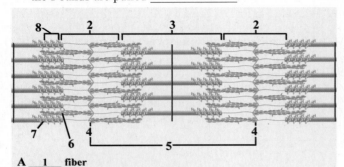

A __1__ fiber

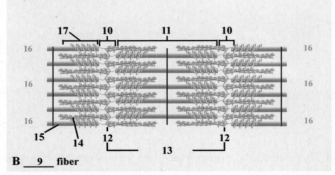

B __9__ fiber

Figure 16.16

104 In reference to **Figure 16.16**, identify #1 - #16.
1 _____ 9 _____
2 _____ 10 _____
3 _____ 11 _____
4 _____ 12 _____
5 _____ 13 _____
6 _____ 14 _____
7 _____ 15 _____
8 _____ 16 _____
                17 _____

105 In reference to **Figure 16.16**, what happens to the length of the zone of overlap in the contracting fiber? _____

106 In reference to **Figure 16.16**, what happens to the length of the I band in the contracting muscle fiber? _____

INNERVATION OF THE MUSCLE FIBER
NEUROMUSCULAR JUNCTIONS AND MUSCLE SPINDLES
Muscle Spindles - Proprioceptors

107 What controls skeletal muscle contraction? _____

108 What are proprioceptors? _____

109 Where are proprioceptors commonly located? _____

110 In addition to monitoring muscle contraction, what is another function of muscle spindles? _____

LAB ACTIVITY 3

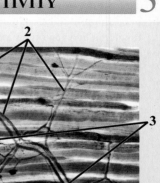

Figure 16.17

111 In reference to **Figure 16.17**, identify #1 - #3.
1 _____ 3 _____
2 _____

Figure 16.18

112 In reference to **Figure 16.18**, identify #1 - #3.
1 _____ 3 _____
2 _____

113 In reference to **Figure 16.18**, what is the function of #1? _____

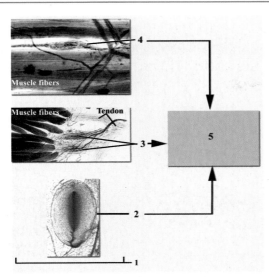

Figure 16.19

114 In reference to **Figure 16.19**, identify #1 - #5.
   1 _____ 4 _____
   2 _____ 5 _____
   3 _____

Neuromuscular Junction

115 After the central nervous system integrates information from proprioceptors, what type of neuron sends information to the muscle fibers?

116 What is the name of the union (synapse) between an axon of a motor neuron and the muscle fiber? _____

117 What is the name of the neurotransmitter released from the presynaptic membrane? _____

118 Where does acetylcholine bind? _____

LAB ACTIVITY 4

119 What is a synapse? _____

120 What are three components of a chemical synapse?
   1 _____
   2 _____
   3 _____

121 What ions move as a result of the binding of acetylcholine to its receptors on the postsynaptic membrane? _____

122 What is the result of the movement of sodium ions into the fiber? _____

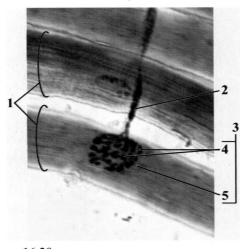

Figure 16.20

123 In reference to **Figure 16.20**, identify #1 - #5.
   1 _____
   2 _____
   3 _____
   4 _____
   5 _____

124 In reference to **Figure 16.20**, where does #2 originate? ___

125 In reference to **Figure 16.20**, what chemical is released at #4? _____

126 In reference to **Figure 16.20**, the binding of the neurotransmitter at # _____ first produces depolarization, then an action potential spreads across the _____.

CONTRACTION OF SKELETAL MUSCLE FIBER

127 What two coupled events are involved in the contraction of a muscle fiber? _____

128 Where does excitation of the muscle fiber begin? _____

129 When does the excitation phase end? _____

130 When does contraction begin? _____

131 What does the sliding filament theory explain? _____

Neuromuscular Junction

132 The arrival of an action potential at the axon terminals opens their _____ channels, which results in the exocytosis of _____.

133 Acetylcholine diffuses across the synaptic cleft and binds to its receptors located on the _____.

134 The binding of acetylcholine to its receptors results in the opening of _____ channels in the postsynaptic membrane.

135 Sodium moves into the sarcoplasm resulting in the _____ _____ of the postsynaptic membrane, and the generation of an _____ that spreads across the sarcolemma.

Chapter 16 - Muscles and Contraction

136 The depolarization of the sarcolemma spreads inward by the _____. From the T tubules the terminal cisternae depolarize and release _____.

137 Acetylcholine is rapidly degraded by the enzyme _____ _____, returning the synapse to its initial state, ready to respond to the arrival of another _____.

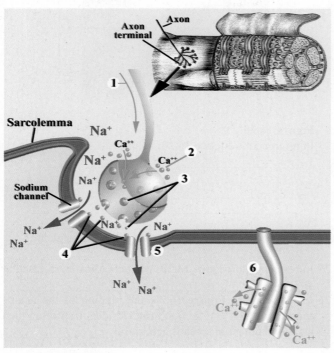

Figure 16.21

138 In reference to **Figure 16.21**, match the following descriptions with their appropriate numbers.
____ Action potential is generated at postsynaptic membrane and propagated along sarcolemma
____ Action potential arrives at axon terminal
____ Vesicles release acetylcholine into synaptic cleft
____ Acetylcholine binds to sodium channel receptors
____ Action potential is propagated along T tubules
____ Calcium ions enter axon terminal and promote exocytosis of acetylcholine

139 The depolarization of the T tubules stimulates the _____ _____ to release _____.

140 Calcium ions bond to _____, and results in the movement of _____ complex away from its blocking position on actin.

141 The interaction of myosin with actin allows the sliding of the _____ inward.

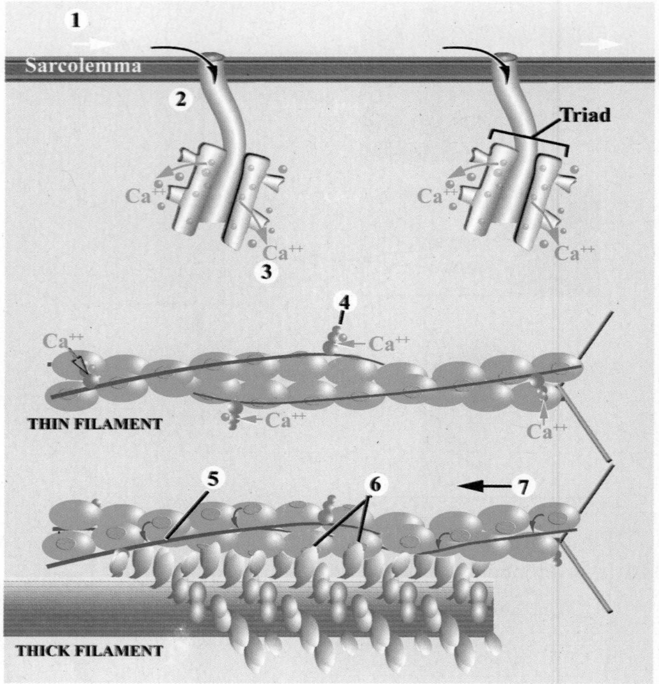

Figure 16.22

142 In reference to **Figure 16.22**, match the following descriptions with their appropriate numbers.
____ Action potential is propagated along T tubules
____ Terminal cisternae release calcium ions
____ Action potential is propagated along sarcolemma
____ Troponin-tropomyosin complex moves to expose myosin binding sites on actin
____ Thin filaments slide inward
____ Calcium ions bond to troponin
____ Myosin heads (cross-bridges) bind to actin

CONTRACTION OF THE MUSCLE FIBER

143 In the high energy state myosin has bound to the high energy molecule _____.

144 In the high energy state, the head of myosin faces away from the center of the ____ band.

145 In the low energy state, myosin has reacted with _____, and the its head is now positioned _____ of the A band.

146 The _____ of the head is an expression of mechanical energy (derived from the chemical energy in the breakdown of ATP).

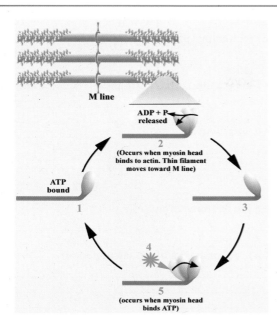

Figure 16.23

147 In reference to **Figure 16.23**, identify #1 - #5.

1 _____ 4 _____
2 _____ 5 _____
3 _____

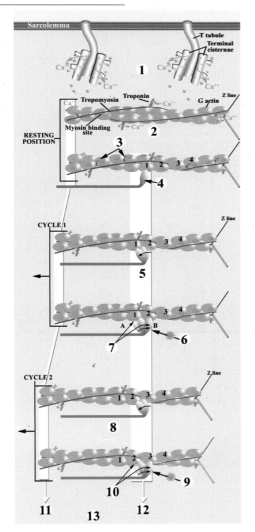

Figure 16.24

148 In reference to **Figure 16.24**, match the following descriptions with their appropriate numbers.

____ Binding of calcium ions to troponin
____ High energy configuration and attachment of myosin to G actin (#1)
____ Power stroke moves G actin (#1) inward
____ Power stroke moves G actin (#2) inward
____ Movement of troponin-tropomyosin
____ Myosin attaches to binding site on G actin (#1)
____ ATP binds to myosin head (cross-bridge)
____ ATP binding results in myosin detachment from G actin (#1)
____ ATP binding results in myosin high energy configuration and attachment to G actin (#2)
____ ATP binding results in myosin detachment from G actin (#2)
____ ATP binding results in myosin high energy configuration and attachment to G actin (#3)
____ Exposure of myosin binding sites
____ Terminal cisternae release calcium ions
____ Thick filaments do not move
____ Thin filaments slide inward toward M line
____ Contraction continues until calcium ion removal restores blocking position of troponin-tropomyosin

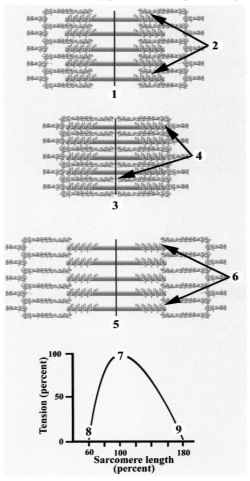

Figure 16.25

149 In reference to **Figure 16.25**, identify #1 - #9.

Chapter 16 - Muscles and Contraction

1 _____
2 _____
3 _____
4 _____
5 _____
6 _____
7 _____
8 _____
9 _____

150 What are two factors that influence the tension produced by a contracting fiber? _____

151 When is maximal tension produced? _____

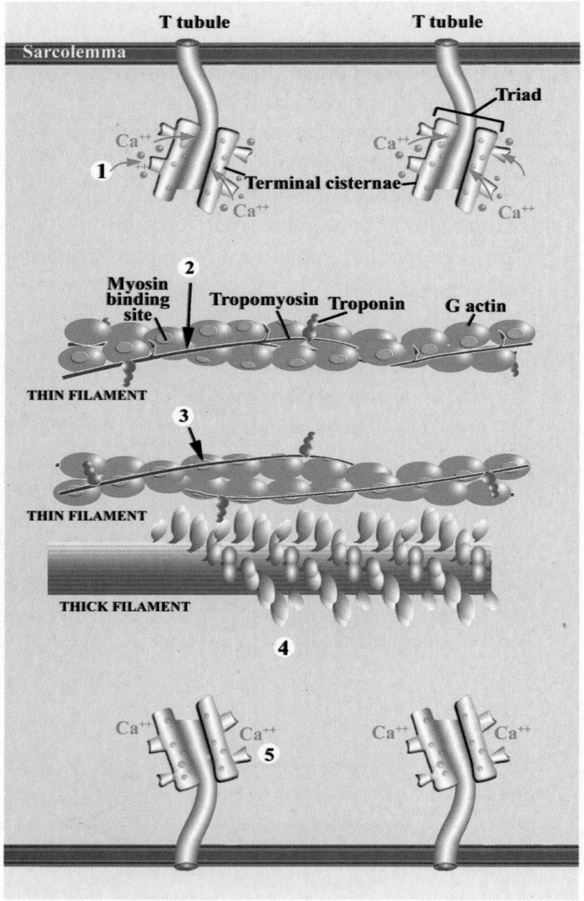

Figure 16.26

152 In reference to **Figure 16.26**, identify #1 - #5.
1 _____
2 _____
3 _____
4 _____
5 _____

TWITCH

153 The regulation of the sarcoplasmic concentration of _____ is essential in the control of muscle contraction.

154 Why does a single brief stimulus produce a brief single contraction? _____

155 What is a twitch? _____

156 What is a myogram? _____

157 What are the three events of a twitch?
1 _____   2 _____
3 _____

158 What are four events of the latent period?
1 _____
2 _____
3 _____
4 _____

159 What are three events of the contraction phase?
1 _____
2 _____
3 _____

160 When does the relaxation phase begin? _____

161 When does the relaxation phase end? _____

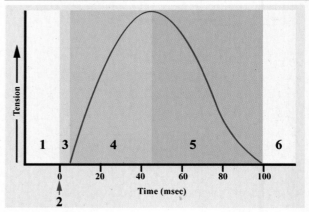

Figure 16.27

162 In reference to **Figure 16.27**, identify #1 - #6.
1 _____   4 _____
2 _____   5 _____
3 _____   6 _____

163 In reference to **Figure 16.27**, why does a delay exist between the time of the stimulus and the production of tension? _____

Name _____
Class _____

Chapter 16 - Muscles and Contraction 239

164 In reference to **Figure 16.27**, what causes the steady increase in tension? _____
_____

165 In reference to **Figure 16.27**, what causes the steady decrease in tension? _____
_____
_____

TWITCH VELOCITY

166 What is one factor that accounts for the different velocities of twitches? _____
_____

167 Fibers that primary derive energy from glycolysis are _____ twitch fibers.

168 Fibers that primary derive energy from aerobic metabolism are _____ twitch fibers.

169 Give an example of a fast twitch muscle and a slow twitch muscle. _____
_____

MUSCLE RESPONSES
VARYING THE CONTRACTION BY MOTOR UNIT RECRUITMENT

170 What is the function of a graded muscle response? _____
_____
_____

171 What is a motor unit? _____
_____

172 What is the effect of increasing the number of contracting motor units? _____
_____

173 What is threshold stimulus for a muscle? _____
_____

174 What happens to a muscle as voltage is gradually increased above threshold? _____

175 What is a maximal stimulus? _____
_____

176 Why doesn't increasing the voltage above the maximal stimulus increase muscle tension? _____
_____

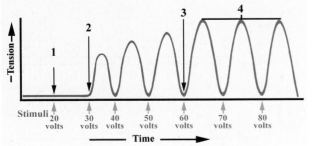

Figure 16.28

177 In reference to **Figure 16.28**, identify #1 - #4.
1 _____ 3 _____
2 _____ 4 _____

178 In reference to **Figure 16.28**, what is the voltage for the threshold stimulus? _____

179 In reference to **Figure 16.28**, what is voltage is subthreshold? _____

180 In reference to **Figure 16.28**, what voltage is the maximal stimulus? _____

LAB ACTIVITY 5

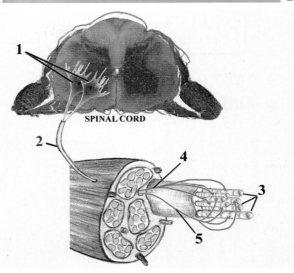

Figure 16.29

181 In reference to **Figure 16.29**, identify #1 - #5.
1 _____ 4 _____
2 _____ 5 _____
3 _____

182 In reference to **Figure 16.29**, which motor unit is the largest? _____

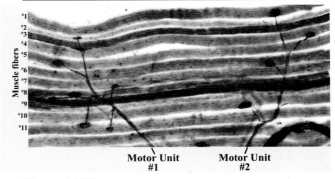

Figure 16.30

183 In reference to **Figure 16.30,** which motor unit involves the most fibers? _____

184 Which fibers (list numbers) are located in motor unit #2?
_____

TREPPE- STAIRCASE EFFECT OR STAIRCASE PHENOMENON

185 What is treppe? _____
_____
_____

186 What are three factors that contribute to treppe?
1 _____
_____
2 _____
_____
3 _____
_____

240 Chapter 16 - Muscles and Contraction

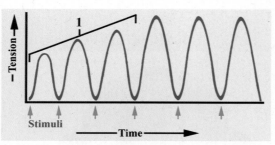

Figure 16.31

187 In reference to **Figure 16.31**, identify #1. _____
_____

188 In reference to **Figure 16.31,** was the intensity of the stimulus changed? _____

ISOTONIC AND ISOMETRIC CONTRACTIONS
Isotonic Contraction

189 What is an isotonic contraction? _____
_____
_____

190 What are the two types of isotonic contraction? _____

191 What is a concentric isotonic contraction? _____
_____

192 In a concentric isotonic contraction does the muscle develop enough tension to overcome the load? _____

193 What is an eccentric isotonic contraction? _____
_____

194 In an eccentric isotonic contraction does the muscle develop enough tension to overcome the load? _____

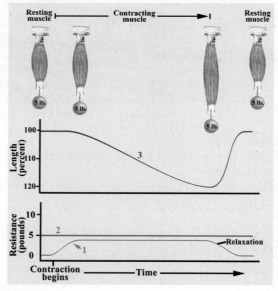

Figure 16.32

195 In reference to **Figure 16.32**, is a concentric or eccentric isotonic contraction shown? _____

196 In reference to **Figure 16.32**, identify #1 - #3.
1 _____ 3 _____
2 _____

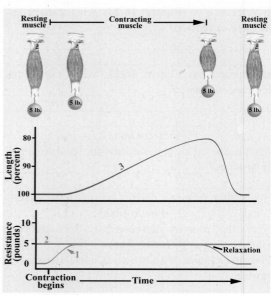

Figure 16.33

197 In reference to **Figure 16.33**, is a concentric or eccentric isotonic contraction shown? _____

198 In reference to **Figure 16.33**, identify #1 - #3.
1 _____ 3 _____
2 _____

Isometric Contraction

199 What is an isometric contraction? _____
_____

200 In an isometric contraction does the muscle develop enough tension to overcome the load? _____

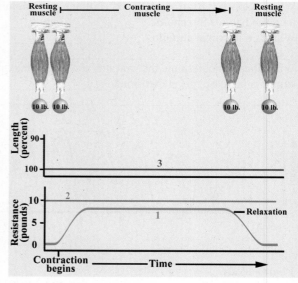

Figure 16.34

201 In reference to **Figure 16.34**, identify #1 - #3.
1 _____ 3 _____
2 _____

Varying the Contraction by Frequency of Stimulation

202 Besides motor unit recruitment, what is another way muscle contraction can be graded (varied)? _____

Wave Summation (summation of twitches) and Tetanus

203 One way to modify twitches to produce smooth sustained muscle contractions is to increase the _____ of the stimulation.

204 What is wave (temporal) summation? _____

205 How is incomplete tetanus produced? _____

206 What characterizes a muscle in incomplete tetanus? _____

207 If the frequency of successive stimuli is increased, what happens to the phases of relaxation? _____

208 What is complete tetanus? _____

209 Why does complete tetanus occur? _____

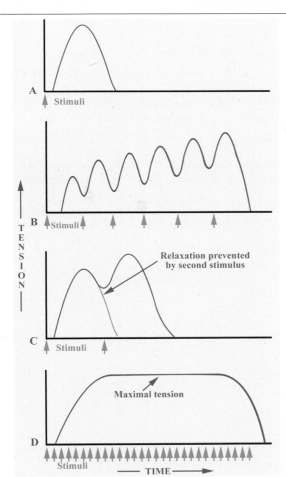

Figure 16.35

210 In reference to **Figure 16.35**, identify A - D.
A _____ C _____
B _____ D _____

211 In reference to **Figure 16.35**, which myogram shows a muscle that quivers as it rapidly undergoes contraction and relaxation? _____

212 In reference to **Figure 16.35**, which myogram shows a muscle in a complete state of contraction? _____

LAB ACTIVITY 6

Frog Gastrocnemius Muscle

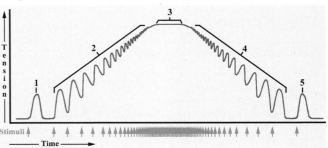

Figure 16.36

213 In reference to **Figure 16.36**, identify #1 - #5.
1 _____ 4 _____
2 _____ 5 _____
3 _____

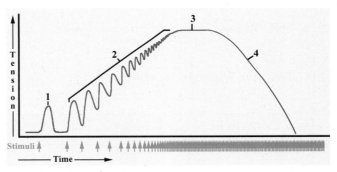

Figure 16.37

214 In reference to **Figure 16.37**, identify #1 - #4.
1 _____ 3 _____
2 _____ 4 _____

215 What causes fatigue in an isolated muscle? _____

MUSCLE ENERGETICS
Muscle Fiber - Resting

216 What are two primary sources of energy for resting muscles? _____

Fatty Acids

217 What source supplies about 95% of the energy for the resting muscle? _____

218 Where are fatty acids catabolized? _____

219 What is aerobic metabolism? _____

220 In the resting muscle fiber, what waste products are produced by the catabolism of fatty acids? _____

221 What captures the chemical energy from the catabolism of fatty acids? _____

Chapter 16 - Muscles and Contraction

222 In a resting muscle fiber, some ATP is "stored," and some is used to create another energy rich molecule called _____ phosphate.

223 The reaction for the production of creatine phosphate is
ATP + _____ –> _____ + _____
                    Glucose

224 In the resting muscle fiber, how is glucose mostly utilized? _____
_____

225 What is glycogen? _____

226 When is glycogen mostly used by muscles? _____

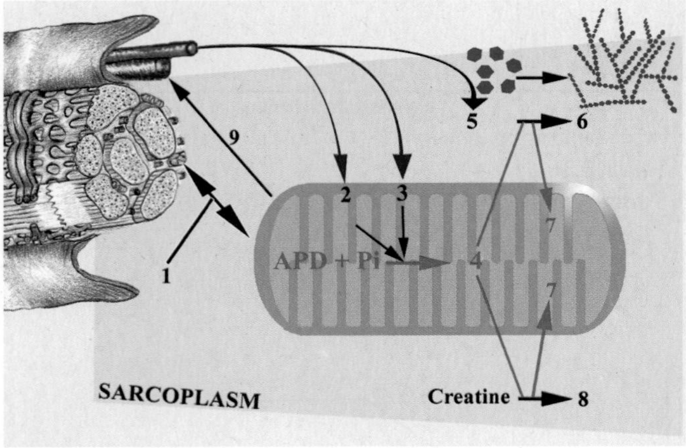

Figure 16.38

227 In reference to **Figure 16.38**, identify #1 - #9.
1 _____   6 _____
2 _____   7 _____
3 _____   8 _____
4 _____   9 _____
5 _____

MUSCLE FIBER - MODERATE PROLONGED ACTIVITY

228 What two sources of energy are catabolized to meet higher energy requirements?? _____

229 The reaction for the catabolism of creatine phosphate is:
_____ + _____ –> ATP + _____

230 What energy source provides for prolonged ATP production? _____

231 What are two sources of glucose?
1 _____
2 _____

232 What are two pathways, and where does each occur, for the catabolism of glucose?
1 _____
_____
2 _____
_____

233 For each molecule of glucose, which pathway provides the most ATP? _____

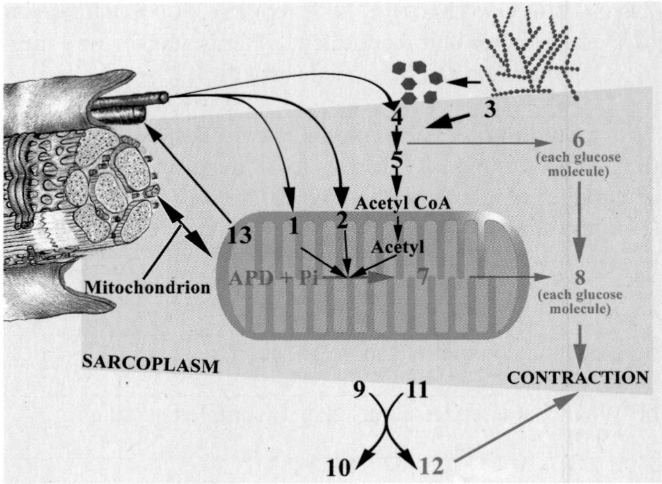

Figure 16.39

234 In reference to **Figure 16.39**, identify #1 - #13.
1 _____   8 _____
2 _____   9 _____
3 _____   10 _____
4 _____   11 _____
5 _____   12 _____
6 _____   13 _____
7 _____

MUSCLE FIBER - PEAK ACTIVITY

235 What continues as the primary source of energy in a muscle in peak activity? _____

236 What is the difference in the utilization of glucose between a muscle in moderate activity and peak activity? _____
_____

237 As energy demands increase about how many times faster can glycolysis increase above aerobic oxidation? _____

238 What happens to pyruvic acid that is not destined to enter the mitochondria for oxidation? _____
_____

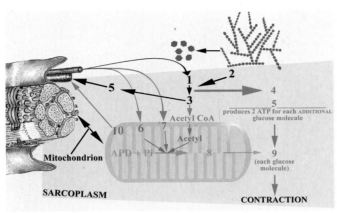

Figure 16.40

239 In reference to **Figure 16.40**, identify #1 - #10.

1 _____ 6 _____
2 _____ 7 _____
3 _____ 8 _____
4 _____ 9 _____
5 _____ 10 _____

MUSCLE ENERGETICS and TYPES OF FIBERS

240 What are two general categories of muscle fibers?
   1 _____ 2 _____
241 What is the primary source of fuel in moderate and peak activities for both Type I and Type II muscle fibers? _____
242 Type I fibers are also called _____ twitch _____ fibers.
243 Type I fibers have abundant mitochondria and depend mostly upon the _____ pathway of glucose oxidation.
244 Type I fibers are rich in the oxygen binding molecule, _____, and thus are also called _____ fibers.
245 Type I fibers are _____ to fatigue.
246 Type II fibers are also called _____ twitch _____ fibers.
247 Type II fibers have relatively few _____ and depend mostly upon the _____ pathway of glucose oxidation.
248 Type II fibers are rich in stored fuel called _____.
249 Type II fibers are low in oxygen binding _____, and are called _____ fibers.
251 Type II fibers fatigue _____.

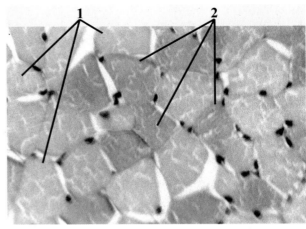

Figure 16.41

251 In reference to **Figure 16.41**, identify #1 & #2.
   1 _____ 2 _____
252 In reference to **Figure 16.41**, which fibers show stained glycogen? _____
253 In reference to **Figure 16.41**, are the fibers at #2 called slow-twitch red fibers or fast-twitch white fibers? _____

SMOOTH MUSCLE TISSUE

254 Describe the appearance of smooth muscle. _____
255 Smooth muscle is contains both _____ and _____ filaments. However, the filaments are not organized into _____.
256 The thin filaments are associated with _____ filaments, not Z lines (discs).
257 Dense bodies are found at the association of intermediate filaments with the _____.
258 What gives smooth muscle the ability to be increasingly stretched without greatly reducing the ability to generate tension? _____

Chapter 16 - Muscles and Contraction

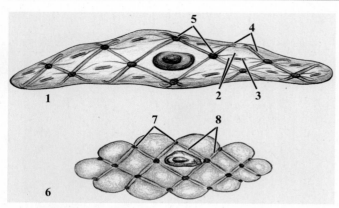

Figure 16.42

259 In reference to **Figure 16.42**, identify #1 - #8.
1 _____   5 _____
2 _____   6 _____
3 _____   7 _____
4 _____   8 _____

Single-unit (Visceral) Smooth Muscle

260 How is single-unit smooth muscle organized? _____

261 Where is single-unit smooth muscle mostly located? _____

262 What are the names of the two layers of single-unit smooth muscle found in the walls of most of the digestive, respiratory, urinary, and reproductive tracts? _____

263 What effect does contraction of the circular layer have upon an organ, such as the intestine? _____

264 What effect does contraction of the longitudinal layer have upon an organ, such as the intestine? _____

265 What type of cell junction allows the spread of an action potential from cell-to-cell? _____

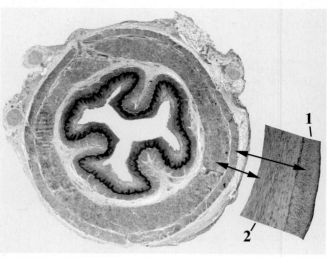

Figure 16.43

266 In reference to **Figure 16.43**, identify #1 & #2.
1 _____   2 _____

LAB ACTIVITY 7

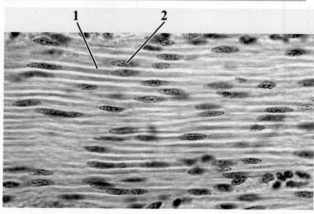

Figure 16.44

267 In reference to **Figure 16.44**, identify #1 & #2.
1 _____   2 _____

268 Wavelike contractions of single-unit smooth muscle in the digestive tract is called _____ waves.

269 In reference to **Figure 16.44**, single-unit smooth muscle from a cross section of the small intestine, the layer is called the _____ layer.

Contraction of Smooth Muscle

270 What are four similarities between the contraction of smooth and skeletal muscle?
1 _____
2 _____
3 _____
4 _____

271 What are three ways smooth muscle contraction can be initiated?
1 _____
2 _____
3 _____

272 What do all three initiators of contraction lead to? _____

273 Where does most of the calcium ions originate? _____

274 What do the calcium ions bond to? _____

275 What is the function of the calcium-calmodulin complex? _____

276 What is the function of myosin light chain kinase (MLCK)? _____

277 When does contraction of smooth muscle begin? _____

278 When does contraction of smooth muscle end? _____

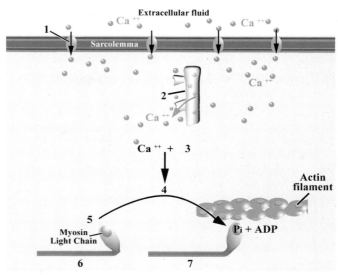

Figure 16.45

279 In reference to **Figure 16.45**, identify #1 - #7.
1 _____ 5 _____
2 _____ 6 _____
3 _____ 7 _____
4 _____

FEATURES OF SMOOTH MUSCLE
Stress-relaxation Response, Length-Tension Changes

280 What is the function of the stress-relaxation response of smooth muscle? _____

281 What features of smooth muscle allow the stress-relaxation response? _____

282 What feature of smooth muscle allows the muscle to stretch and still produce considerable tension? _____

246 Chapter 16 - Muscles and Contraction

CARDIAC MUSCLE

283 Describe the structure of cardiac muscle fibers. _____
_____
_____

284 How are the T tubules of cardiac muscle different from skeletal muscle?
_____
_____

285 Like skeletal and smooth muscle, contraction is dependent upon the presence of _____ ions.

286 Like smooth muscle, most of the calcium ions are delivered from the _____.

287 What structures are found at intercalated discs?
_____
_____

288 What is the function of gap junctions? _____
_____

289 What is the function of desmosomes? _____
_____

290 What are pacemaker cells? _____
_____

291 What controls the pacemaker cells of the heart? _____
_____

LAB ACTIVITY 8

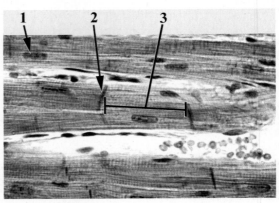

Figure 16.47

293 In reference to **Figure 16.47**, identify #1 - #3.
1 _____   3 _____
2 _____

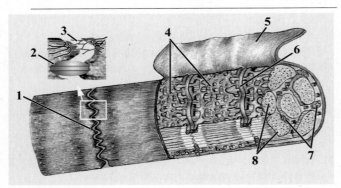

Figure 16.46

292 In reference to **Figure 16.46**, identify #1 - #8.
1 _____   5 _____
2 _____   6 _____
3 _____   7 _____
4 _____   8 _____

Nervous System - Worksheets

1. What are the two major divisions of the nervous system?
   1 _____ 2 _____
2. What are the components of the central nervous system?
   1 _____ 2 _____
3. What are some of the functions of the central nervous system? _____
4. What forms the peripheral nervous system? _____
5. What are the two divisions of the peripheral nervous system? 1 _____
   2 _____
6. What is the function of the afferent (sensory) division? _____
7. What is the function of the efferent (motor) division? _____
8. What does the somatic division involve? _____
9. What are two components of the somatic division? _____
10. What does the visceral division involve? _____
11. What is the function of the visceral sensory component? _____
12. What is the function of the motor component? _____
13. What are the two divisions of the autonomic nervous system? _____
14. What does the autonomic division control? _____
15. What division directs information flow from the special senses to the central nervous system? _____

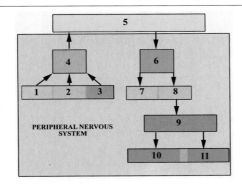

Figure 17.1

16. In reference to **Figure 17.1**, identify #1 - #11.
    1 _____ 7 _____
    2 _____ 8 _____
    3 _____ 9 _____
    4 _____ 10 _____
    5 _____ 11 _____
    6 _____

NEURONS

17. What are three functions of neurons? _____
18. Where are the cell bodies of neurons of the central nervous system located? _____
19. Where are the cell bodies of neurons of the peripheral nervous system located? _____

NEURON STRUCTURE

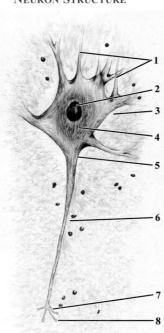

Figure 17.2

20. In reference to **Figure 17.2**, identify #1 - #8.
    1 _____ 5 _____
    2 _____ 6 _____
    3 _____ 7 _____
    4 _____ 8 _____
21. Dendrites traditionally are defined as processes that conduct impulses _____ the cell body.
22. Functionally, dendrites function as the _____ portion of the neuron.
23. Axons traditionally are defined as processes that conduct impulses _____ the cell body.
24. Where is the axon hillock located? _____
25. What are telodendria? _____

248 Chapter 17 - Nervous System

26 What is located at the end of each telodendron? _____
_____

27 What is a synapse? _____
_____
_____

28 Functionally, the axon is the process that _____ and _____ the nerve impulse, and releases a _____ at its terminus.

CLASSIFICATION OF NEURONS

29 What are three classifications of neurons according to function? _____
_____

30 What is the function of sensory neurons? _____

31 What is the function of motor neurons? _____

32 What is the function of association neurons (interneurons)? _____
_____

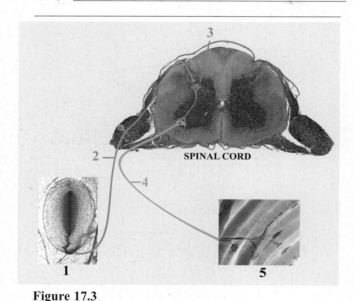

Figure 17.3

33 In reference to **Figure 17.3**, identify #1 - #5.
1 _____ 4 _____
2 _____ 5 _____
3 _____

34 What are three classifications of neurons according to structure? _____
_____

35 A unipolar neuron has a single fibrous process that terminates with _____
_____

36 A bipolar neuron has _____ fibrous process, each arises from _____ sides of the cell's body.

37 Multipolar neurons have _____ associated with the cell's body.

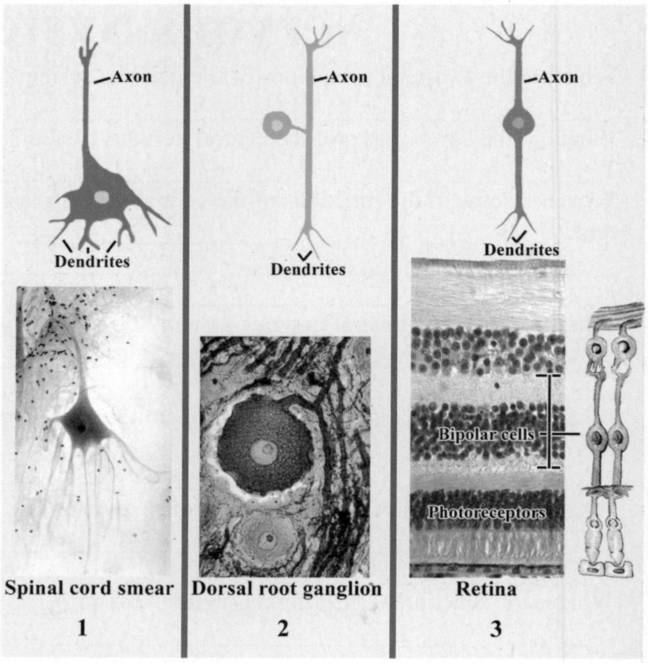

Figure 17.4

38 In reference to **Figure 17.4**, identify #1 - #3.
1 _____ 3 _____
2 _____

NEUROGLIA

39 What are neuroglia? _____
_____

40 What are four varieties of neuroglia found in the central nervous system?
1 _____ 3 _____
2 _____ 4 _____

41 What are two types of neuroglia found in the peripheral nervous system?
1 _____ 2 _____

Neuroglia of the Central Nervous System

42 Where are ependymal cells located? _____

43 What is the function of ependymal cells? _____
_____

44 What are two functions of astrocytes? _____
_____

45 What is a function of oligodendrocytes? _____
_____

46 What is a function of microglia? _____
_____

Neuroglia of the Peripherial Nervous System

47 Where are satellite cells located? _____
_____

48 Where are Schwann cells located? _____
_____

49 Schwann cells tightly wrap axons to produced _____ axons or remain in close association to produce _____ axons.

Name _____
Class _____

Chapter 17 - Nervous System   **249**

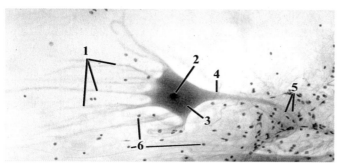

Figure 17.5

50 In reference to **Figure 17.5**, identify #1 - #6.
    1 _____ 4 _____
    2 _____ 5 _____
    3 _____ 6 _____

51 In reference to **Figure 17.5**, what is the function of #1?
    _____
    _____

52 In reference to **Figure 17.5**, what is the function of #4?
    _____
    _____

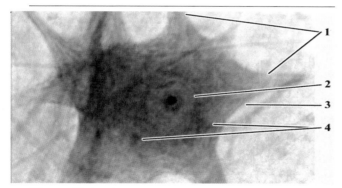

Figure 17.6

53 In reference to **Figure 17.6**, identify #1 - #4.
    1 _____ 3 _____
    2 _____ 4 _____

MYELINATED and UNMYELINATED AXONS
Peripheral Nervous System

54 What is myelin? _____
    _____

55 How are myelinated axons formed? _____
    _____
    _____

56 How are unmyelinated axons formed? _____
    _____
    _____

57 What are nodes of Ranvier? _____
    _____

58 What are nerves? _____
    _____

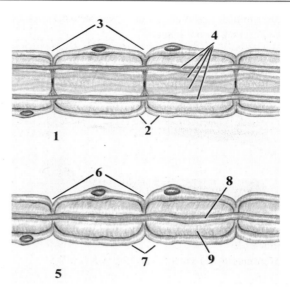

Figure 17.7

59 In reference to **Figure 17.7**, identify #1 - #9.
    1 _____ 6 _____
    2 _____ 7 _____
    3 _____ 8 _____
    4 _____ 9 _____
    5 _____

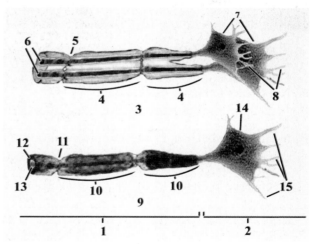

Figure 17.8

60 In reference to **Figure 17.8**, identify #1 - #15.
    1 _____ 9 _____
    2 _____ 10 _____
    3 _____ 11 _____
    4 _____ 12 _____
    5 _____ 13 _____
    6 _____ 14 _____
    7 _____ 15 _____
    8 _____

Central Nervous System

61 In the central nervous system, what is the name of the neurolgia that myelinate axons? _____

62 In the central nervous system, where are most of the myelinated axons located? _____
    _____

250 Chapter 17 - Nervous System

63 In the central nervous system, where are most of the unmyelinated axons located? _____

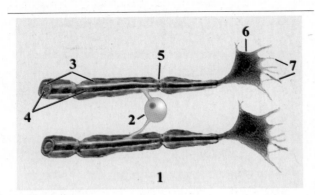

Figure 17.9

64 In reference to **Figure 17.9**, identify #1 - #7.
1 _____   5 _____
2 _____   6 _____
3 _____   7 _____
4 _____

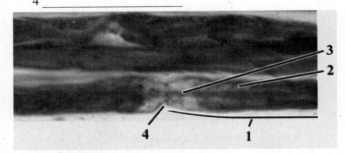

Figure 17.10

65 In reference to **Figure 17.10**, identify #1 - #4.
1 _____   3 _____
2 _____   4 _____

66 In reference to **Figure 17.10**, what is the function of #1? _____

67 In reference to **Figure 17.10**, what is the function of #4? _____

68 In reference to **Figure 17.10**, what is the function of #3? _____

NERVE

69 What is a nerve? _____

70 What are three classifications of nerves according to the direction of impulse conduction? _____

Organization of a Nerve

71 Where is the endoneurium located? _____

72 Where is the perineurium located? _____

73 Where is the epineurium located? _____

74 What are fascicles? _____

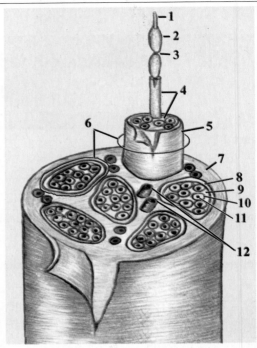

Figure 17.11

75 In reference to **Figure 17.11**, identify #1 - #12.
1 _____   7 _____
2 _____   8 _____
3 _____   9 _____
4 _____   10 _____
5 _____   11 _____
6 _____   12 _____

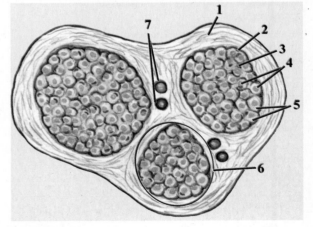

Figure 17.12

76 In reference to **Figure 17.12**, identify #1 - #6.
1 _____   4 _____
2 _____   5 _____
3 _____   6 _____

Name _____
Class _____

Chapter 17 - Nervous System   **251**

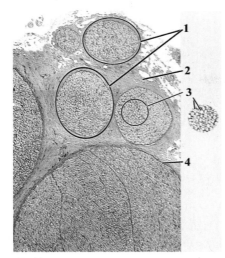

Figure 17.13

77 In reference to **Figure 17.13**, identify #1 - #4.
   1 _____   3 _____
   2 _____   4 _____

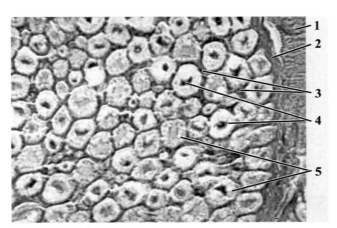

Figure 17.14

78 In reference to **Figure 17.14**, identify #1 - #5.
   1 _____   4 _____
   2 _____   5 _____
   3 _____

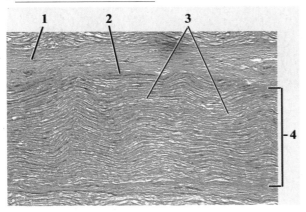

Figure 17.15

79 In reference to **Figure 17.15**, identify #1 - #4.
   1 _____   3 _____
   2 _____   4 _____

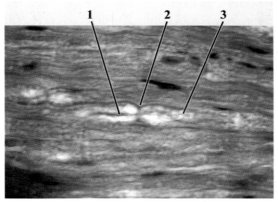

Figure 17.16

80 In reference to **Figure 17.16**, identify #1 - #3
   1 _____   3 _____
   2 _____

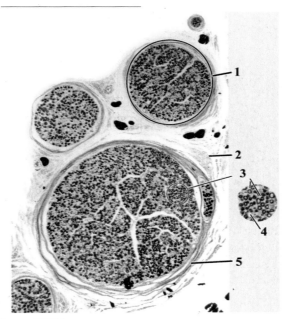

Figure 17.17

81 In reference to **Figure 17.17**, identify #1 - #5.
   1 _____   4 _____
   2 _____   5 _____
   3 _____

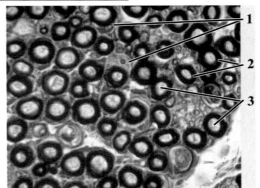

Figure 17.18

82 In reference to **Figure 17.18**, identify #1 - #3.
   1 _____   3 _____
   2 _____

252 Chapter 17 - Nervous System

SPECIALIZED NEURON ENDINGS

83 Where are specialized neuron endings of the PNS located? _____

84 What are receptors? _____

85 What is a stimulus? _____

86 What are effectors? _____

87 What are neuromuscular synapses? _____

88 What are neuroglandular synapses? _____

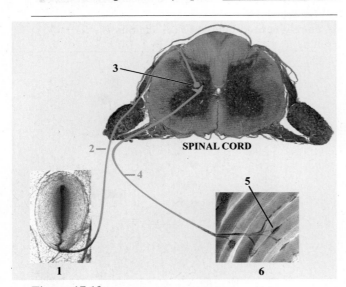

Figure 17.19

89 In reference to **Figure 17.19**, identify #1 - #6.
1 _____ 4 _____
2 _____ 5 _____
3 _____ 6 _____

Pacinian Corpuscle

90 What is the function of a Pacinian corpuscle? _____

91 Where are Pacinian corpuscles located? _____

92 What is a graded potential? _____

93 What happens when a graded potential reaches threshold? _____

94 What happens when the action potential reaches the axon terminals? _____

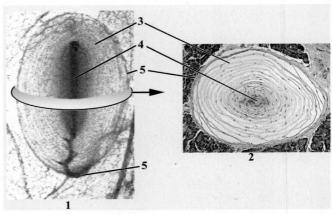

Figure 17.20

95 In reference to **Figure 17.20**, identify #1 - #5.
1 _____ 4 _____
2 _____ 5 _____
3 _____

96 In reference to **Figure 17.20**, what is the function of #4? _____

Neuromuscular Junctions

97 What is a neuromuscular junction? _____

98 Where is the synaptic cleft located? _____

99 Where is the motor end plate located? _____

100 What does the motor end plate contain? _____

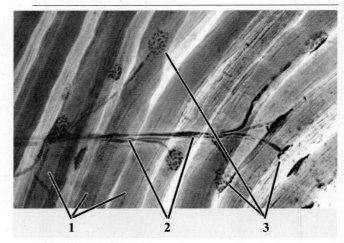

Figure 17.21

101 In reference to **Figure 17.21**, identify #1 - #3.
1 _____ 3 _____
2 _____

Name _____
Class _____

Chapter 17 - Nervous System   253

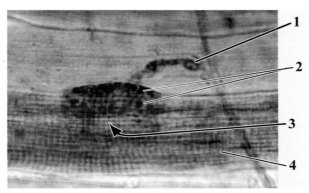

Figure 17.22
102 In reference to **Figure 17.22**, identify #1 - #4.
   1 _____ 3 _____
   2 _____ 4 _____
103 In reference to **Figure 17.22**, what is the function of #1?
   _____
   _____
104 In reference to **Figure 17.22**, what is the function of #2?
   _____
   _____
105 In reference to **Figure 17.22**, what is the function of #3?
   _____
   _____

HUMAN BRAIN

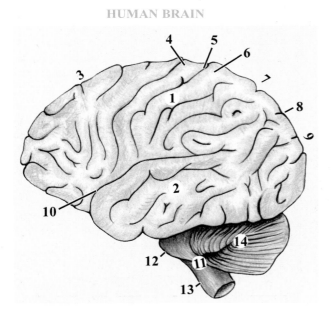

Figure 17.23
106 In reference to **Figure 17.23**, identify #1 - #14.
   1 _____ 8 _____
   2 _____ 9 _____
   3 _____ 10 _____
   4 _____ 11 _____
   5 _____ 12 _____
   6 _____ 13 _____
   7 _____ 14 _____

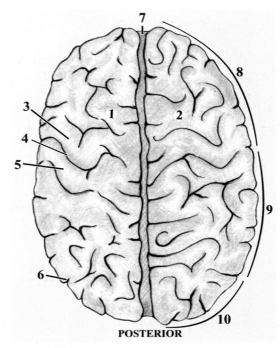

Figure 17.24
107 In reference to **Figure 17.24**, identify #1 - #10.
   1 _____ 6 _____
   2 _____ 7 _____
   3 _____ 8 _____
   4 _____ 9 _____
   5 _____ 10 _____

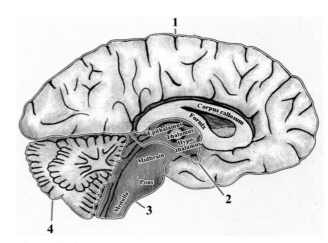

Figure 17.25
108 In reference to **Figure 17.25**, identify #1 - #5.
   1 _____ 3 _____
   2 _____ 4 _____

254 Chapter 17 - Nervous System

Name _____
Class _____

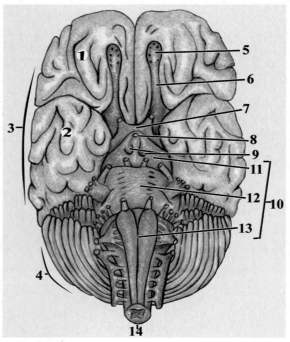

Figure 17.26

109 In reference to **Figure 17.26**, identify #1 - #14.

1 _____ 8 _____
2 _____ 9 _____
3 _____ 10 _____
4 _____ 11 _____
5 _____ 12 _____
6 _____ 13 _____
7 _____ 14 _____

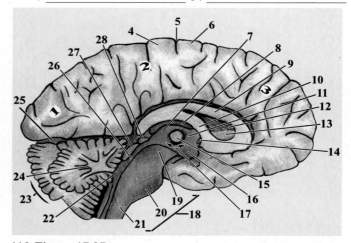

110 **Figure 17.27**

In reference to **Figure 17.27**, identify #1 - #28.

1 _____ 15 _____
2 _____ 16 _____
3 _____ 17 _____
4 _____ 18 _____
5 _____ 19 _____
6 _____ 20 _____
7 _____ 21 _____
8 _____ 22 _____
9 _____ 23 _____
10 _____ 24 _____
11 _____ 25 _____
12 _____ 26 _____
13 _____ 27 _____
14 _____ 28 _____

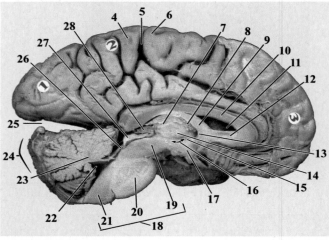

111 **Figure 17.28**

In reference to **Figure 17.28**, identify #1 - #28.

1 _____ 15 _____
2 _____ 16 _____
3 _____ 17 _____
4 _____ 18 _____
5 _____ 19 _____
6 _____ 20 _____
7 _____ 21 _____
8 _____ 22 _____
9 _____ 23 _____
10 _____ 24 _____
11 _____ 25 _____
12 _____ 26 _____
13 _____ 27 _____
14 _____ 28 _____

CEREBRUM

112 What connects the right and left cerebral hemispheres? _____

113 What are some of the functions of the cerebrum? _____

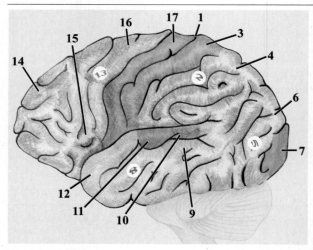

Figure 17.29

Name _____
Class _____

Chapter 17 - Nervous System   255

114 In reference to **Figure 17.29**, identify #1 - #17.
   1 _____   10 _____
   2 _____   11 _____
   3 _____   12 _____
   4 _____   13 _____
   5 _____   14 _____
   6 _____   15 _____
   7 _____   16 _____
   8 _____   17 _____
   9 _____

115 What are the four lobes of the cerebrum? _____
_____

116 What are gyri? _____
_____

117 What are sulci? _____

118 What is a fissure? _____

119 What is the name of the fissure that separates the cerebrum into its right and left hemispheres? _____

120 What is the name of the fissure that separates the cerebrum from the cerebellum? _____
_____

121 What divides each hemisphere's frontal lobe from its parietal lobe? _____
_____

122 What are five functional regions of the frontal lobes?
   1 _____
   _____
   2 _____
   _____
   3 _____
   _____
   4 _____
   _____
   5 _____
   _____

123 What is the function of the precentral gyrus? _____
_____

124 Where is each parietal lobe located? _____
_____

125 What is the primary function of the parietal lobe? _____
_____
_____

126 Where is the primary sensory (somatosensory) area located? _____
_____

127 Where is the somatosensory association area located and what is its function? _____
_____
_____

128 The primary sensory (somatosensory) area houses neurons that receive relayed information from receptors in the _____ and from receptors in muscles, tendons, and joints, the _____.

129 What is the function of the occipital lobes? _____
_____

130 What information does the primary visual cortex receive? _____
_____

131 What is the function of the visual association area? _____
_____

132 The lateral sulcus separates the temporal lobe from the _____
_____

133 What are four functions of the temporal lobes?
   1 _____
   2 _____
   3 _____
   4 _____

134 What is the function of the primary auditory cortex? _____
_____

135 What is the function of the auditory association areas? _____
_____

136 Where is the cerebral cortex located and of what does it mostly consist? _____
_____

137 Where is the cerebral white matter located and of what does it mostly consist? _____
_____

138 What are the lateral ventricles? _____
_____

139 Which ventricle communicates with the lateral ventricles? _____
_____

140 What is the function of the choroid plexus? _____
_____

141 Where is the corpus callosum located? _____
_____

142 What is the function of the corpus callosum? _____
_____

143 Where is the fornix located? _____
_____

144 What is the function of the fornix? _____
_____

145 Where is the septum pellucidum located? _____
_____

Cerebrum, silver impregnation, section

146 The surface of the cerebrum is modified into numerous rounded ridges called _____.

147 Separating the ridges are shallow grooves called _____.

148 The substance of the cerebrum is organized into an outer _____ and the inner _____.

149 The cerebral cortex consists of _____ matter, and is divided into two regions, the outer _____ layer and the inner _____ layer.

150 The cerebral white matter consists mostly of _____ _____ and supporting cells (neuroglia).

256 Chapter 17 - Nervous System

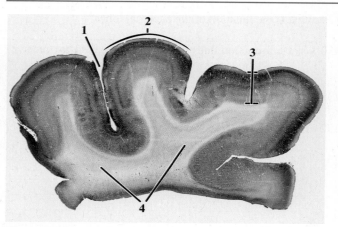

Figure 17.30
151 In reference to **Figure 17.30**, identify #1 - #4.
1 _____ 3 _____
2 _____ 4 _____

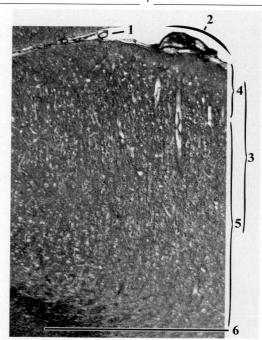

Figure 17.31
152 In reference to **Figure 17.31**, identify #1 - #6.
1 _____ 4 _____
2 _____ 5 _____
3 _____ 6 _____

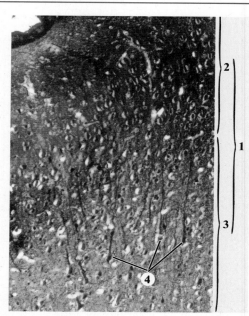

Figure 17.32
153 In reference to **Figure 17.32**, identify #1 - #4.
1 _____ 3 _____
2 _____ 4 _____

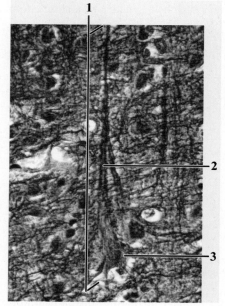

Figure 17.33
154 In reference to **Figure 17.33**, identify #1 - #3.
1 _____ 3 _____
2 _____

Cerebral cortex

155 What are the two most common types of neurons of the granular layer of the cerebral cortex? _____
_____

156 What is the function of granule cells? _____
_____

157 What is the function of pyramidal cells? _____
_____

Diencephalon

158 What are the three regions of the diencephalon?
1 _____ 3 _____
2 _____

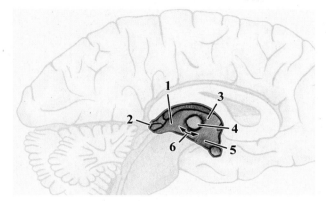

Figure 17.34

159 In reference to **Figure 17.34**, identify #1 - #6.
1 _____ 4 _____
2 _____ 5 _____
3 _____ 6 _____

160 What are the thalami? _____

161 What structure connects the thalami across the third ventricle? _____

162 What is the primary function of the thalami? _____

163 Where is the epithalamus located? _____

164 Where is the pineal gland located? _____

165 What is the function of the pineal gland (body)? _____

166 Where is the hypothalamus located? _____

167 What are five functions of the hypothalamus?
1 _____
2 _____
3 _____
4 _____
5 _____

168 Where is the infundibulum located? _____

169 The infundibulum functions as a pathway for _____ and _____ that leave the hypothalamus and enter the pituitary gland.

170 Where is the third ventricle located? _____

171 The third ventricle receives CSF from the _____ ventricles and its choroid plexus.

172 CSF from the third ventricle drains into the _____

Brain Stem

173 Where is the brain stem located? _____

174 What are the three regions of the brain stem?
1 _____ 3 _____
2 _____

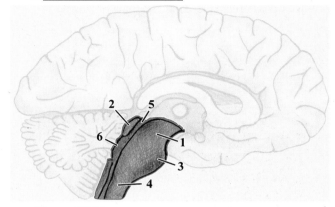

Figure 17.35

175 In reference to **Figure 17.35**, identify #1 - #6.
1 _____ 4 _____
2 _____ 5 _____
3 _____ 6 _____

176 Where is the midbrain located? _____

177 What are five functions of the midbrain?
1 _____
2 _____
3 _____
4 _____
5 _____

178 What forms the ventral-lateral surface of the midbrain? _____

179 What forms the posterior surface of the midbrain? _____

180 What are the corpora quadrigemina? _____

181 What is the function of the superior colliculi? _____

182 What is the function of the inferior colliculi? _____

183 Where is the pons located? _____

258 Chapter 17 - Nervous System

184 What are three functions of the pons?
1 _____
2 _____
3 _____

185 Where is the medulla oblongata located? _____
_____
_____

186 What are five functions of the medulla oblongata?
1 _____
2 _____
3 _____
4 _____
5 _____

187 Where is the cerebral aqueduct located? _____
_____

188 What is the function of the cerebral aqueduct? ____
_____

189 Where is the fourth ventricle located? _____
_____

190 In addition to receiving CSF produced by its _____
_____, the fourth ventricle receives CSF from the _____ ventricle by way of the cerebral aqueduct.

191 From the fourth ventricle, CSF moves into the _____ _____, a space formed by under the _____ mininx, a membrane covering that surrounds the brain and the spinal cord.

CEREBELLUM

192 Where is the cerebellum located? _____
_____
_____

193 What do the functions of the cerebellum include? _____
_____
_____

194 The cerebellum has _____ hemispheres connected by a region called the _____.

195 What are cerebellar folia? _____
_____

196 What are the arbor vitae? _____
_____

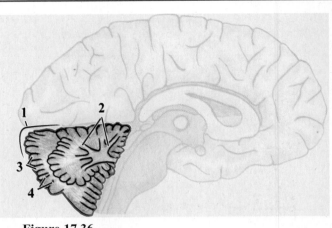

Figure 17.36

197 In reference to **Figure 17.36**, identify #1 - #4.
1 _____ 3 _____
2 _____ 4 _____

Cerebellum, silver impregnation, section

198 The substance of the cerebellum is organized into the outer _____ and the inner _____.

199 From outer to inner, what are the three layers of the cerebellar cortex? _____
_____

200 Where is the cerebellar white matter located? _____
_____

201 What is the function of the molecular layer? _____
_____

202 What forms the Purkinje layer? _____
_____

203 Where do the Purkinje neuron axons extend? _____

204 What types of cells are found in the granular layer? _____
_____

205 Granular cells receive _____ from neurons that enter the cerebellum, and their axons extend _____
_____

206 Cerebellar white matter consists mostly of _____
_____

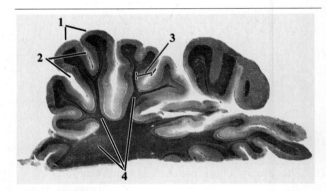

Figure 17.37

In reference to **Figure 17.37**, identify #1 - #4.
_____ 3 _____
_____ 4 _____

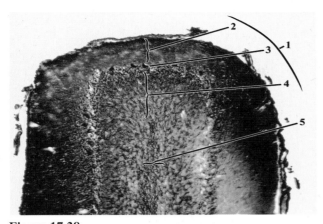

Figure 17.38
208 In reference to **Figure 17.38**, identify #1 - #5.
1 _____ 4 _____
2 _____ 5 _____
3 _____

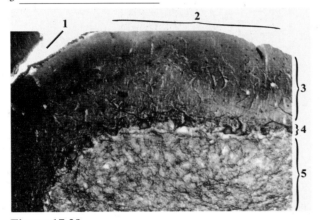

Figure 17.39
209 In reference to **Figure 17.39**, identify #1 - #5.
1 _____ 4 _____
2 _____ 5 _____
3 _____

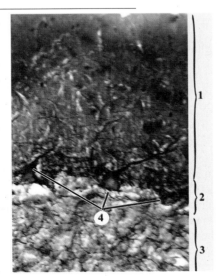

Figure 17.40
210 In reference to **Figure 17.40**, identify #1 - #4.
1 _____ 3 _____
2 _____ 4 _____

VENTRICLES

211 What are ventricles of the brain? _____
_____

212 What do the ventricles contain? _____
_____

213 What medially separates the two lateral ventricles? _____

214 What does an interventricular foramen connect? _____
_____

215 What is the function of the choroid plexus? _____
_____

216 CSF from the lateral ventricles drains into the _____
_____ by way of the _____.

217 Where is the third ventricle located? _____
_____

218 Where does the third ventricle receive its CSF? _____
_____

219 CSF from the third ventricle drains into the _____
_____.

220 Where does the fourth ventricle receive its CSF? _____
_____

221 CSF from the fourth ventricle drains into the _____
_____
_____

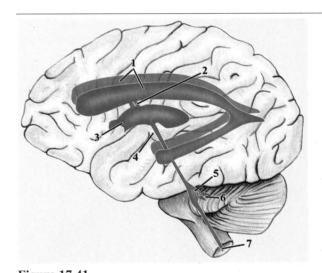

Figure 17.41
222 In reference to **Figure 17.41**, identify #1 - #7.
1 _____ 5 _____
2 _____ 6 _____
3 _____ 7 _____
4 _____

MENINGES

223 What are the meninges? _____
_____

224 What are functions of the meninges? _____
_____
_____

225 From outer to inner, what are the three meninges
1 _____ 3 _____
2 _____

226 Where does the outer surface of the brain's dura mater attach? _____

260 Chapter 17 - Nervous System

227 What does the inner surface of the brain's dura contain? _____

228 What is the superior sagittal sinus? _____

229 Where is the superior sagittal sinus located? _____

230 What does the superior sagittal sinus receive? _____

231 Where is the arachnoid meninx located? _____

232 Where is cerebrospinal fluid located? _____

233 What are functions of cerebrospinal fluid? _____

234 Where is the pia mater located? _____

CEREBROSPINAL FLUID

235 What structure functions as the site for the production of cerebrospinal fluid? _____

236 Where are the choroid plexuses located? _____

237 CSF from the fourth ventricle moves into the _____, where it circulates around the _____ and the _____.

238 Where are the arachnoid granulations located? _____

239 What is the function of arachnoid granulations? _____

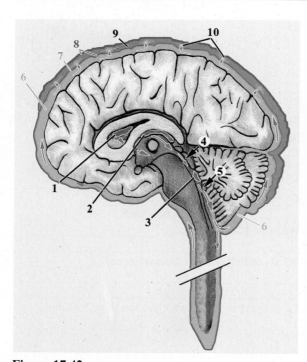

Figure 17.42
240 In reference to **Figure 17.42**, identify #1 - #10.

1 _____
2 _____
3 _____
4 _____   8 _____
5 _____   9 _____
6 _____   10 _____
7 _____

CRANIAL NERVES

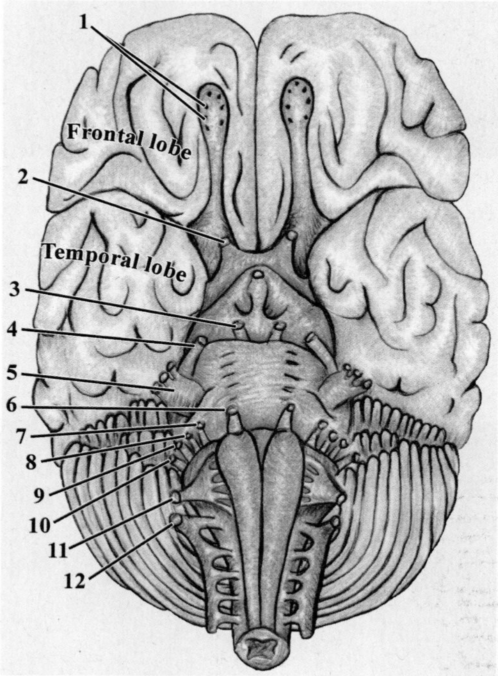

Figure 17.43
241 In reference to **Figure 17.43**, identify #1 - #12.
1 _____ 7 _____
2 _____ 8 _____
3 _____ 9 _____
4 _____ 10 _____
5 _____ 11 _____
6 _____ 12 _____

242 Match the following cranial nerves with their function.
A Olfactory nerves (I)   G Facial nerve (VII)
B Optic nerve (II)   H Vestibulocochlear nerve (VIII)
C Oculomotor nerve (III)   I Glossopharyngeal nerve (IX)
D Trochlear nerve (IV)   J Vagus nerve (X)
E Trigeminal nerve (V)   K Accessory nerve (XI)
F Abducens nerve (VI)   L Hypoglossal nerve (XII)

_____ Efferent to parotid salivary gland (parasympathetic) and motor fibers to muscles of the pharynx

_____ Five major branches: (1) temporal, (2) zygomatic, (3) buccal, (4) mandibular, and (5) cervical.

_____ Meet at the optic chiasma

_____ Motor fibers from the cranial root diverge to the larynx, pharynx, and soft palate

_____ Motor fibers to muscles of the larynx (voice box) and pharynx (swallowing) and organs of the thoracic and abdominal cavities (parasympathetic)

_____ Motor fibers to some of the muscles of the eyeball and

Name _____
Class _____

Chapter 17 - Nervous System **261**

_____ to the muscle which raises the eyelid
_____ Motor fibers to the lateral rectus muscle of the eyeball
_____ Motor fibers to the superior oblique muscle
_____ Motor nerve of the face
_____ Sensory input from taste receptors of the posterior tongue and input from thoracic and abdominal organs for regulation of digestive activity, breathing, and heart rate
_____ Sensory input of taste and general sensation from the posterior one-third of the tongue, input from the pharynx
_____ Sensory nerve of hearing and equilibrium
_____ Sensory nerves of smell
_____ Sensory nerve of the face with some motor fibers to a few muscles of mastication
_____ Sensory nerve of vision
_____ Spinal root supplies motor fibers to the trapezius and the sternocleidomastoid muscles.
_____ Supplies motor fibers to some muscles of the tongue
_____ Three divisions are (1) the opthalmic division, (2) the maxillary division, and (3) the mandibular division.
_____ Two divisions, the (1) cochlear division and the (2) vestibular division.

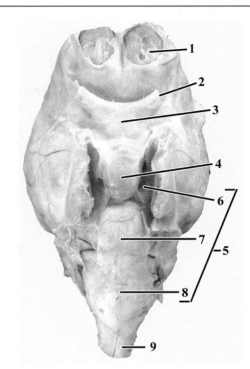

Figure 17.45
244 In reference to **Figure 17.45**, identify #1 - #9.
1 _____ 6 _____
2 _____ 7 _____
3 _____ 8 _____
4 _____ 9 _____
5 _____

DISSECTION OF THE SHEEP BRAIN

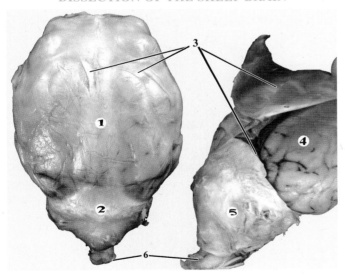

Figure 17.44
243 In reference to **Figure 17.44**, identify #1 - #6.
1 _____ 4 _____
2 _____ 5 _____
3 _____ 3 _____

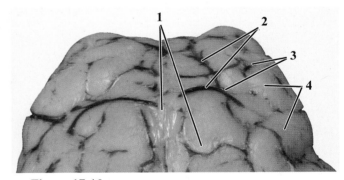

Figure 17.46
245 In reference to **Figure 17.46**, identify #1 - #4.
1 _____ 3 _____
2 _____ 4 _____

262 Chapter 17 - Nervous System

Name _____
Class _____

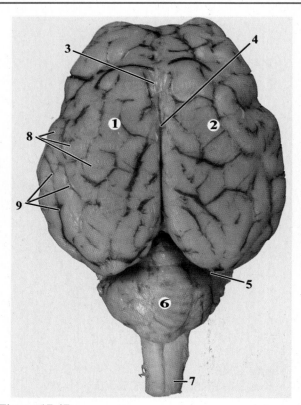

Figure 17.47
246 In reference to **Figure 17.47**, identify #1 - #9.
1 _____   6 _____
2 _____   7 _____
3 _____   8 _____
4 _____   9 _____
5 _____

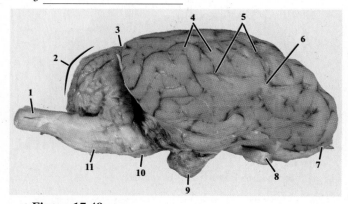

Figure 17.48
247 In reference to **Figure 17.48**, identify #1 - #11.
1 _____   7 _____
2 _____   8 _____
3 _____   9 _____
4 _____   10 _____
5 _____   11 _____
6 _____

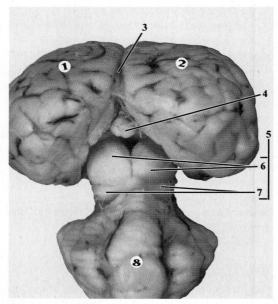

Figure 17.49
248 In reference to **Figure 17.49**, identify #1 - #8.
1 _____   5 _____
2 _____   6 _____
3 _____   7 _____
4 _____   8 _____

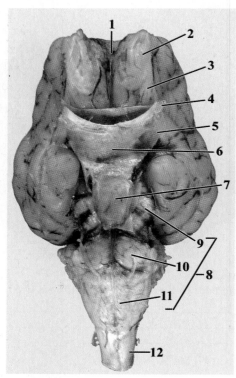

Figure 17.50
249 In reference to **Figure 17.50**, identify #1 - #12.
1 _____   7 _____
2 _____   8 _____
3 _____   9 _____
4 _____   10 _____
5 _____   11 _____
6 _____   12 _____

Chapter 17 - Nervous System 263

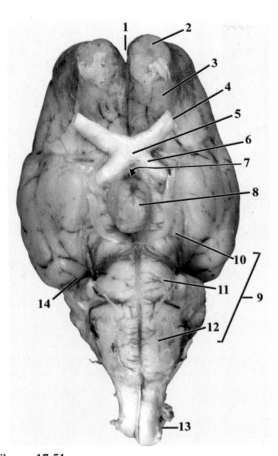

Figure 17.51
250 In reference to **Figure 17.51**, identify #1 - #14.

1 _____ 8 _____
2 _____ 9 _____
3 _____ 10 _____
4 _____ 11 _____
5 _____ 12 _____
6 _____ 13 _____
7 _____ 14 _____

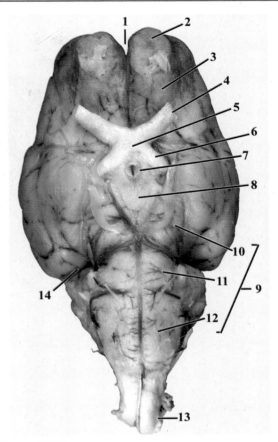

Figure 17.52
251 In reference to **Figure 17.52**, identify #1 - #14.

1 _____ 7 _____
2 _____ 8 _____
3 _____ 10 _____
4 _____ 11 _____
5 _____ 12 _____
6 _____ 13 _____
7 _____ 14 _____

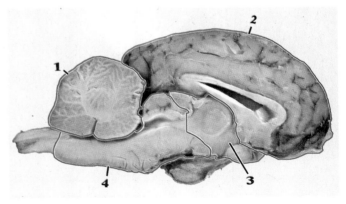

Figure 17.53
252 In reference to **Figure 17.53**, identify #1 - #4.

1 _____ 3 _____
2 _____ 4 _____

264 Chapter 17 - Nervous System

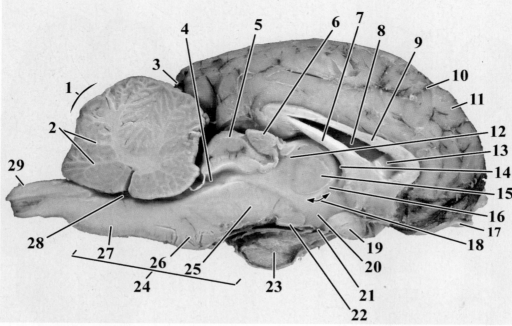

253 In reference to **Figure 17.54**, identify #1 - #29.

1 _____	11 _____	21 _____
2 _____	12 _____	22 _____
3 _____	13 _____	23 _____
4 _____	14 _____	24 _____
5 _____	15 _____	25 _____
6 _____	16 _____	26 _____
7 _____	17 _____	27 _____
8 _____	18 _____	28 _____
9 _____	19 _____	29 _____
10 _____	20 _____	

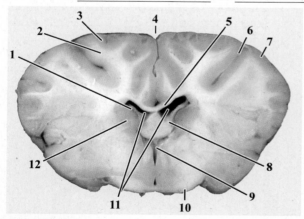

Figure 17.55

254 In reference to **Figure 17.55**, identify #1 - #12.

1 _____	7 _____
2 _____	8 _____
3 _____	9 _____
4 _____	10 _____
5 _____	11 _____
6 _____	12 _____

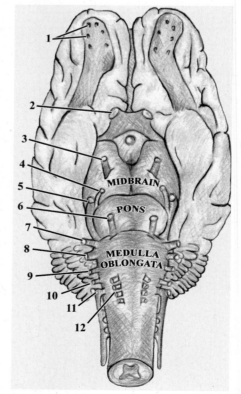

Figure 17.56

255 In reference to **Figure 17.56**, identify #1 - #12.
1 _____ 7 _____
2 _____ 8 _____
3 _____ 9 _____
4 _____ 10 _____
5 _____ 11 _____
6 _____ 12 _____

SPINAL CORD

256 What is the function of the spinal cord? _____

257 Where is the spinal cord located? _____

258 Where does the spinal cord begin and end? _____

259 In the adult, the tip of the conus medullaris is at what vertebral level? _____

260 What is the function of the filum terminale? _____

261 Where do the spinal nerves (except C_1) exit the vertebral column? _____

262 What is the basis for naming the spinal nerves? _____

Meninges

263 From outer to inner, what are the names of the meninges? _____

264 In the spinal cord, what separates the dura mater from the surrounding vertebrae? _____

265 What does the subarachnoid space contain? _____

266 Where is the pia mater located? _____

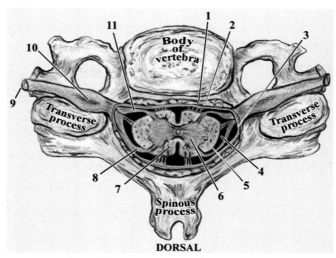

Figure 17.57

267 In reference to **Figure 17.57**, identify #1 - #11.
1 _____ 7 _____
2 _____ 8 _____
3 _____ 9 _____
4 _____ 10 _____
5 _____ 11 _____
6 _____

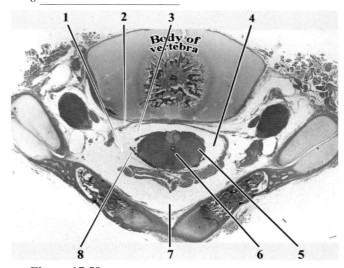

Figure 17.58

268 In reference to **Figure 17.58**, identify #1 - #8.
1 _____ 5 _____
2 _____ 6 _____
3 _____ 7 _____
4 _____ 8 _____

Chapter 17 - Nervous System

Spinal Cord Structure

269 What is the name of the neural tissue located to the inside the spinal cord? _____

270 What is the name of the neural tissue located to the outside of the spinal cord? _____

271 What are four regions of the gray matter?
1 _____ 3 _____
2 _____ 4 _____

272 What is the organization of the white matter? _____

274 What axons are found in the dorsal root? _____

275 What part of the sensory neuron is housed in the dorsal root ganglion? _____

276 What axons are found in the ventral root? _____

277 What do the dorsal and ventral roots unite to form? _____

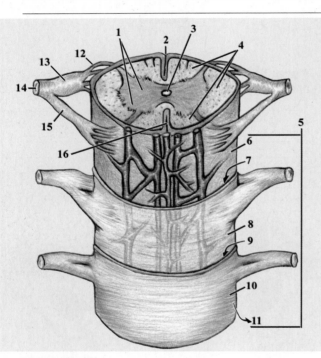

Figure 17.59

273 In reference to **Figure 17.59**, identify #1 - #16.
1 _____ 9 _____
2 _____ 10 _____
3 _____ 11 _____
4 _____ 12 _____
5 _____ 13 _____
6 _____ 14 _____
7 _____ 15 _____
8 _____ 16 _____

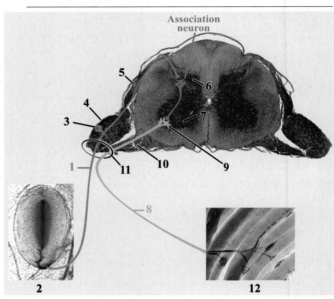

Figure 17.60

278 In reference to **Figure 17.60**, identify #1 - #12.
1 _____ 7 _____
2 _____ 8 _____
3 _____ 9 _____
4 _____ 10 _____
5 _____ 11 _____
6 _____ 12 _____

Chapter 17 - Nervous System   267

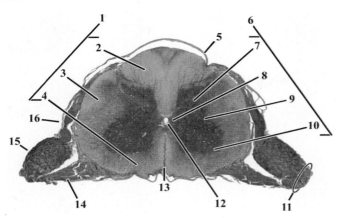

Figure 17.61

279 In reference to **Figure 17.61**, identify #1 - #16.
1 _____   9 _____
2 _____   10 _____
3 _____   11 _____
4 _____   12 _____
5 _____   13 _____
6 _____   14 _____
7 _____   15 _____
8 _____   16 _____

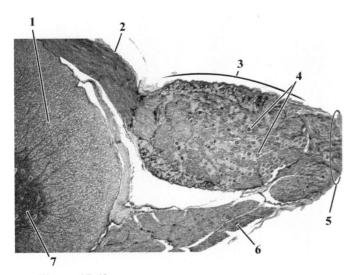

Figure 17.62

280 In reference to **Figure 17.62**, identify #1 - #7.
1 _____   5 _____
2 _____   6 _____
3 _____   7 _____
4 _____

281 In reference to **Figure 17.62**, what axons are found at #2? _____

282 In reference to **Figure 17.62**, what axons are found at #6? _____

283 In reference to **Figure 17.62**, what is the function of the tissue found at #1? _____

284 In reference to **Figure 17.62**, what is the function of the tissue found at #7? _____

285 In reference to **Figure 17.62**, is #5 part of the CNS or PNS? _____

286 In reference to **Figure 17.62**, is #5 sensory, motor, or mixed (sensory and motor) in function? _____

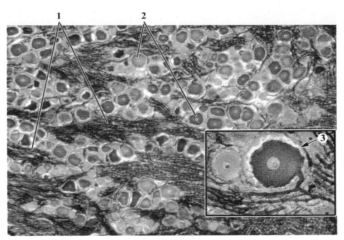

Figure 17.63

287 In reference to **Figure 17.63**, identify #1 - #3.
1 _____   3 _____
2 _____

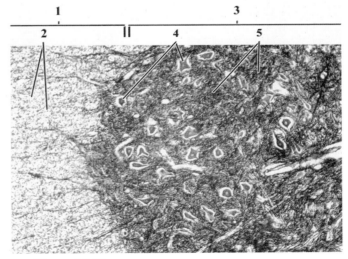

Figure 17.64

288 In reference to **Figure 17.64**, identify #1 - #5.
1 _____   4 _____
2 _____   5 _____
3 _____

268 Chapter 17 - Nervous System

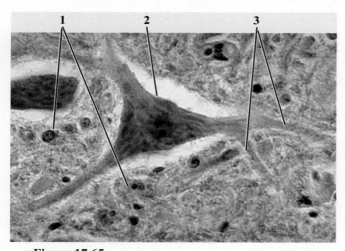

Figure 17.65

289 In reference to **Figure 17.65**, identify #1 - #3.
1 _____ 3 _____
2 _____

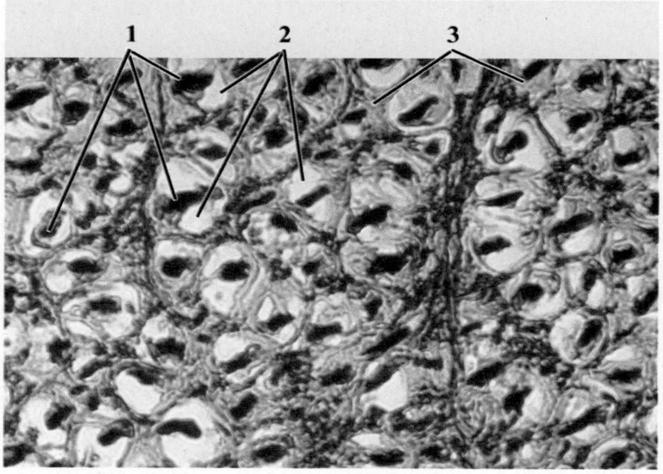

Figure 17.66

290 In reference to **Figure 17.66**, identify #1 - #3.
1 _____ 3 _____
2 _____

Name _____
Class _____

Chapter 18 - The Eye   269

The Eye - Worksheets

ACCESSORY STRUCTURES

1 What are the six accessory structures of the eye?
   1 _____ 4 _____
   2 _____ 5 _____
   3 _____ 6 _____

2 What are two functions of the eyebrows?
   1 _____
   2 _____

3 What are two functions of the eyelids (palpebrae)?
   1 _____
   2 _____

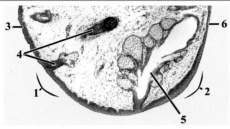

Figure 18.1

4 In reference to **Figure 18.1**, identify 1 - 6.
   1 _____ 4 _____
   2 _____ 5 _____
   3 _____ 6 _____

5 What is the name of the lining of the inner surface of the eyelids? _____

6 What are two functions of the eyelashes?
   1 _____
   2 _____

7 What two areas are covered by the conjunctiva?
   1 _____
   2 _____

8 What are two functions of the conjunctiva?
   1 _____
   2 _____

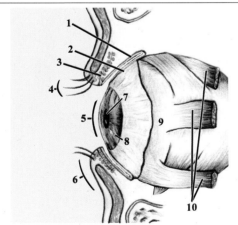

Figure 18.2

9 In reference to **Figure 18.2**, identify 1 - 10.
   1 _____ 6 _____
   2 _____ 7 _____
   3 _____ 8 _____
   4 _____ 9 _____
   5 _____ 10 _____

Lacrimal apparatus

10 Where are the lacrimal glands located? _____
   _____

11 Where do tears from the anterior surface of the eye drain?
   _____
   _____

12 Where to the lacrimal puncta drain? _____
   _____

13 Where do the lacrimal canals drain? _____
   _____

14 Where does the nasolacrimal duct drain? _____
   _____

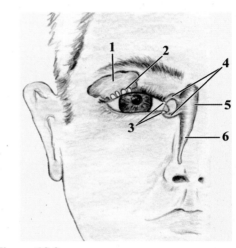

Figure 18.3

15 In reference to **Figure 18.3**, identify 1 - 6.
   1 _____ 4 _____
   2 _____ 5 _____
   3 _____ 6 _____

Extrinsic eye muscles

16 What are the extrinsic muscles of the eye? _____
   _____

17 What are the six extrinsic muscles of the eye?
   1 _____ 4 _____
   2 _____ 5 _____
   3 _____ 6 _____

18 The medial rectus muscle moves the eye _____
   _____.

19 The lateral rectus muscle moves the eye _____
   _____

20 The superior rectus muscle moves the eye _____
   _____

21 The inferior rectus muscle moves the eye _____
   _____

22 The superior oblique muscle moves the eye _____
   _____

23 The inferior oblique muscle moves the eye _____
   _____

270 Chapter 18 - The Eye

Name _____
Class _____

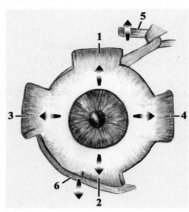

Figure 18.4

24 In reference to **Figure 18.4**, identify 1 - 6.
1 _____ 4 _____
2 _____ 5 _____
3 _____ 6 _____

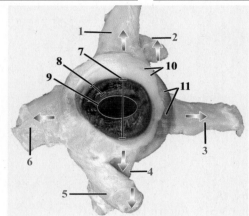

Figure 18.5

25 In reference to **Figure 18.5**, identify 1 - 11.
1 _____ 7 _____
2 _____ 8 _____
3 _____ 9 _____
4 _____ 10 _____
5 _____ 11 _____
6 _____

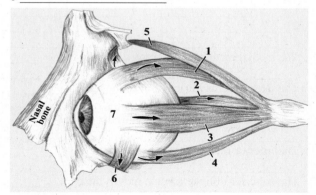

Figure 18.6

26 In reference to **Figure 18.6**, identify 1 - 6.
1 _____ 4 _____
2 _____ 5 _____
3 _____ 6 _____

STRUCTURE OF THE EYEBALL

27 What are the two cavities of the eye and what does each contain? _____
_____
_____

28 From outer to inner, what are the three layers (tunics) of the eye?
1 _____ 3 _____
2 _____

29 What are the two components of the fibrous tunic?
1 _____ 2 _____

30 What are the three components of the vascular tunic?
1 _____ 3 _____
2 _____

31 What comprises the neural tunic? _____

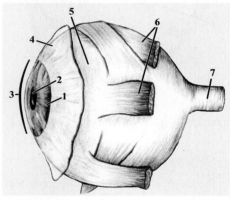

Figure 18.7

32 In reference to **Figure 18.7**, identify 1 - 7.
1 _____ 5 _____
2 _____ 6 _____
3 _____ 7 _____
4 _____

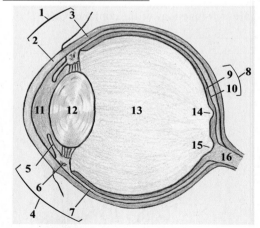

Figure 18.8

33 In reference to **Figure 18.8**, identify 1 - 15.
1 _____ 9 _____
2 _____ 10 _____
3 _____ 11 _____
4 _____ 12 _____
5 _____ 13 _____
6 _____ 14 _____
7 _____ 15 _____
8 _____ 16 _____

Name _____
Class _____

Chapter 18 - The Eye   271

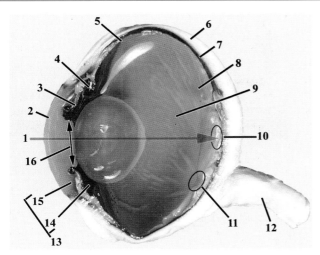

Figure 18.9

34 In reference to **Figure 18.9**, identify 1 - 15.

1 _____	9 _____
2 _____	10 _____
3 _____	11 _____
4 _____	12 _____
5 _____	13 _____
6 _____	14 _____
7 _____	15 _____
8 _____	16 _____

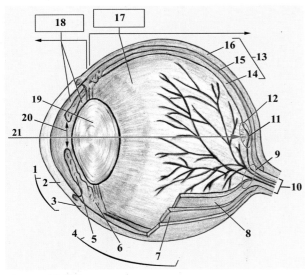

Figure 18.10

35 In reference to **Figure 18.10**, identify 1 - 21.

1 _____	12 _____
2 _____	13 _____
3 _____	14 _____
4 _____	15 _____
5 _____	16 _____
6 _____	17 _____
7 _____	18 _____
8 _____	19 _____
9 _____	20 _____
10 _____	21 _____
11 _____	

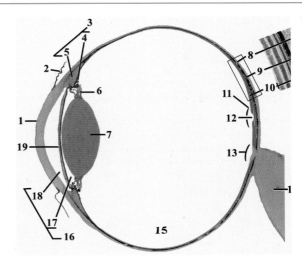

Figure 18.11

36 In reference to **Figure 18.11**, identify 1 - 19.

1 _____	11 _____
2 _____	12 _____
3 _____	13 _____
4 _____	14 _____
5 _____	15 _____
6 _____	16 _____
7 _____	17 _____
8 _____	18 _____
9 _____	19 _____
10 _____	

FIBROUS TUNIC

37 What are the two components of the fibrous tunic?
  1 _____ 2 _____

Sclera

38 What type of tissue forms the sclera? _____
  _____

39 Where does the sclera not cover the eyeball?
  1 _____
  2 _____

40 What are three functions of the sclera?
  1 _____
  2 _____
  3 _____

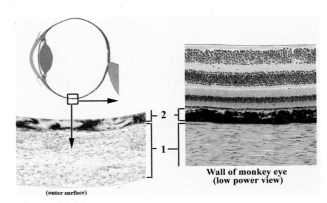

Figure 18.12

41 In reference to **Figure 18.12**, identify 1 - 2.
  1 _____ 2 _____

42 In reference to **Figure 18.12**, what is the function of #1?
_____
_____
_____

Cornea
43 What are three functions of the cornea?
1 _____
2 _____
3 _____
44 What type of tissue forms the cornea? _____
45 What is the limbus? _____
_____
46 What lines the surface of the cornea? _____
_____
47 What is the stroma? _____

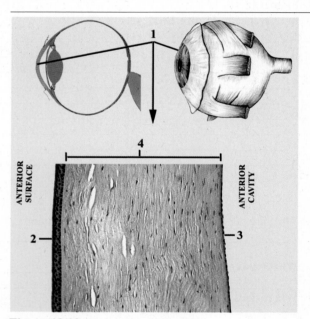

Figure 18.13
48 In reference to **Figure 18.13**, identify 1 - 4.
1 _____ 3 _____
2 _____ 4 _____
49 In reference to **Figure 18.13**, what is the function of #1?
_____
_____
_____

VASCULAR TUNIC
50 What are the three components of the vascular tunic?
1 _____ 3 _____
2 _____

Choroid
51 What is the position of the choroid in the wall of the eye?
_____
_____
52 What are two functions of the choroid?
1 _____
2 _____

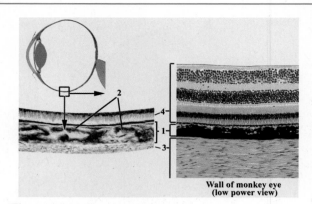

Figure 18.14
53 In reference to **Figure 18.14**, identify 1 - 4.
1 _____ 3 _____
2 _____ 4 _____
54 In reference to **Figure 18.14**, what is the function of #1?
_____
_____

Ciliary body
55 Where is the ciliary body located? _____
_____
56 What are two components of the ciliary body?
1 _____ 2 _____
57 What is the function of the ciliary muscle? _____
_____
58 What are two functions of the ciliary processes?
1 _____
_____
2 _____
_____

Suspensory ligament
59 Where is the suspensory ligament located? _____
_____
_____
60 What are two functions of the suspensory ligament?
1 _____
_____
2 _____
_____

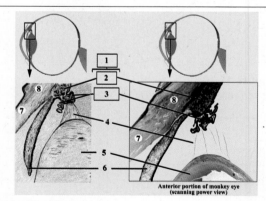

Figure 18.15
61 In reference to **Figure 18.15**, identify 1 - 8.
1 _____ 5 _____
2 _____ 6 _____
3 _____ 7 _____
4 _____ 8 _____

62. In reference to **Figure 18.15**, what is the function of #2? _____
63. In reference to **Figure 18.15**, what are the functions of #3? _____
64. In reference to **Figure 18.15**, what is the function of #4? _____

Iris

65. Where is the iris located? _____
66. What two compartments are formed by the iris?
    1 _____ 2 _____
67. What is the function of each muscle group of the iris? _____
68. What controls each muscle group of the iris? _____

Pupil

69. What is the pupil? _____

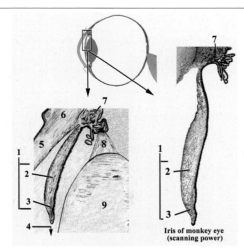

Figure 18.16

70. In reference to **Figure 18.16**, identify 1 - 9.
    1 _____ 6 _____
    2 _____ 7 _____
    3 _____ 8 _____
    4 _____ 9 _____
    5 _____
71. In reference to **Figure 18.16**, what is the function of #1? _____
72. In reference to **Figure 18.16**, what is the function of #2? _____
73. In reference to **Figure 18.16**, which division of the ANS controls #2? _____
74. In reference to **Figure 18.16**, what is the function of #3? _____
75. In reference to **Figure 18.16**, which division of the ANS controls #3? _____

SENSORY TUNIC (RETINA)

76. What is the sensory tunic? _____
77. What are the two divisions of the retina?
    1 _____ 2 _____
78. From outer to inner what are the three retinal cell layers?
    1 _____ 3 _____
    2 _____

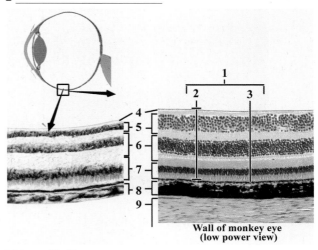

Figure 18.17

79. In reference to **Figure 18.17**, identify 1 - 9.
    1 _____ 6 _____
    2 _____ 7 _____
    3 _____ 8 _____
    4 _____ 9 _____
    5 _____

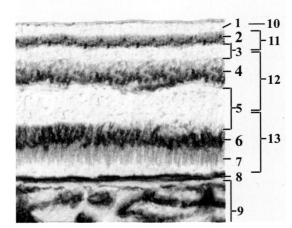

Figure 18.18

80. In reference to **Figure 18.18**, identify 1 - 13.
    1 _____ 8 _____
    2 _____ 9 _____
    3 _____ 10 _____
    4 _____ 11 _____
    5 _____ 12 _____
    6 _____ 13 _____
    7 _____

274 Chapter 18 - The Eye

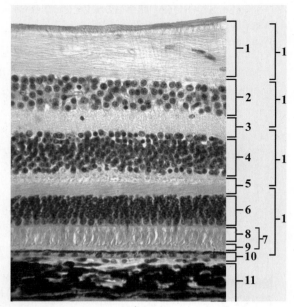

Figure 18.19

81 In reference to **Figure 18.19**, identify 1 - 15.
1 _____ 9 _____
2 _____ 10 _____
3 _____ 11 _____
4 _____ 12 _____
5 _____ 13 _____
6 _____ 14 _____
7 _____ 15 _____
8 _____

82 What do the plexiform layers of the retina contain? _____

83 What are the light sensitive elements of the retina called?
_____

84 Which layer of the retina contains the nuclei of the rods and cones? _____

85 Where do the photoreceptors terminate? _____
_____

86 What is the function of rods? _____
_____

87 What are the three varieties of cones?
1 _____ 3 _____
2 _____

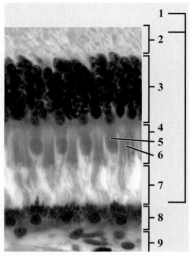

Figure 18.20

88 In reference to **Figure 18.20**, identify 1 - 9.
1 _____ 6 _____
2 _____ 7 _____
3 _____ 8 _____
4 _____ 9 _____
5 _____

89 In reference to **Figure 18.20**, what is the function of #5?
_____

90 In reference to **Figure 18.20**, what is the function of #6?
_____

91 Where are the receptive portions (dendrites) of the bipolar cells located? _____

92 Where do the bipolar cells terminate? _____

93 Where are the receptive portions of the ganglion cells? ___
_____

94 What do the axons of the ganglion cells form? _____
_____

Macula lutea

95 Which photoreceptors are located in the macula lutea?
_____

96 What structure is centrally located in the macula lutea?
_____

Chapter 18 - The Eye

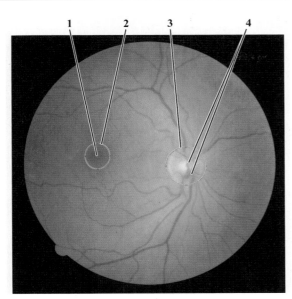

Figure 18.21

97 In reference to **Figure 18.21**, identify 1 - 4.
   1 _____ 3 _____
   2 _____ 4 _____

Fovea centralis

98 Where is the fovea centralis located and how is it described? _____

99 What is the function of the fovea centralis? _____

100 What is the visual axis and where does it fall? _____

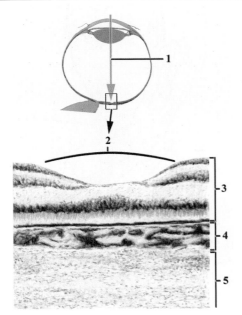

Figure 18.22

101 In reference to **Figure 18.22**, identify 1 - 5.
   1 _____ 4 _____
   2 _____ 5 _____
   3 _____

102 In reference to **Figure 18.22**, what is the function of #2? _____

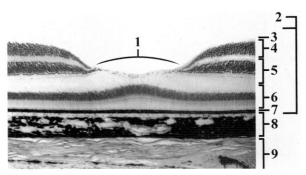

Figure 18.23

103 In reference to **Figure 18.23**, identify 1 - 9.
   1 _____ 6 _____
   2 _____ 7 _____
   3 _____ 8 _____
   4 _____ 9 _____
   5 _____

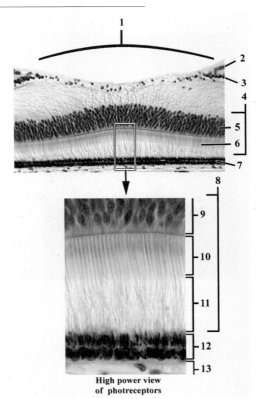

High power view of photoreceptors

Figure 18.24

104 In reference to **Figure 18.24**, identify 1 - 13.
   1 _____ 8 _____
   2 _____ 9 _____
   3 _____ 10 _____
   4 _____ 11 _____
   5 _____ 12 _____
   6 _____ 13 _____
   7 _____

Optic disc (Blind spot)

105 What forms the optic disc? _____
_____
_____

106 Why is the optic disc called the blind spot? _____
_____
_____

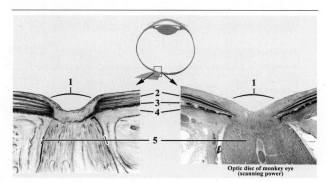

Figure 18.25

107 In reference to **Figure 18.25**, identify 1 - 5.
1 _____ 4 _____
2 _____ 5 _____
3 _____

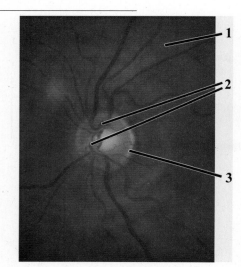

Figure 18.26

108 In reference to **Figure 18.26**, identify 1 - 3.
1 _____ 3 _____
2 _____

109 In reference to **Figure 18.26**, what is the function of #2?
_____
_____

CHAMPERS
Anterior cavity (segment)

110 Where is the anterior cavity located? _____
_____

111 What fluid does the anterior cavity contain? _____
_____

112 What are the two chambers of the anterior cavity?
1 _____ 2 _____

113 Where is the anterior chamber located? _____

114 Where is the posterior chamber located? _____
_____

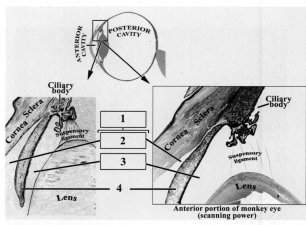

Figure 18.27

115 In reference to **Figure 18.27**, identify 1 - 4.
1 _____ 3 _____
2 _____ 4 _____

Aqueous humor - location, production, and reabsorption

116 Where is aqueous humor produced? _____
_____

117 What chamber does aqueous humor pass into from its site of production? _____

118 What happens to aqueous humor in the anterior chamber?
_____
_____

119 What is the function of the scleral venous sinus? _____
_____
_____

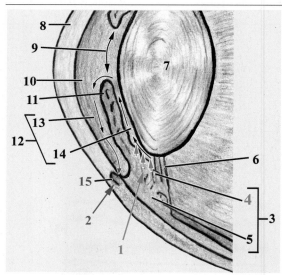

Figure 18.28

120 In reference to **Figure 18.28**, identify 1 - 15.
1 _____ 9 _____
2 _____ 10 _____
3 _____ 11 _____
4 _____ 12 _____
5 _____ 13 _____
6 _____ 14 _____
7 _____ 15 _____
8 _____

Name _____
Class _____

Chapter 18 - The Eye   277

121 In reference to **Figure 18.28**, what is the function of #4?
_____
_____

122 In reference to **Figure 18.28**, what is the function of #15?
_____
_____
_____

Posterior cavity (segment)

123 Where is the posterior cavity located? _____
_____

124 What does the posterior cavity contain? _____
_____

Lens

125 What is the shape of the lens? _____

126 What is the function of the lens? _____
_____

127 What structure attaches the lens to the ciliary muscle? ___
_____

128 What is the function of the ciliary muscle? _____
_____

129 Why is it necessary to change the curvature of the lens? ___
_____
_____

DISSECTION OF THE EYE - (SHEEP)
EXTERNAL ANATOMY

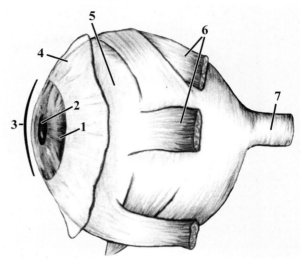

Figure 18.30

131 In reference to **Figure 18.30**, identify 1 - 7.
1 _____   5 _____
2 _____   6 _____
3 _____   7 _____
4 _____

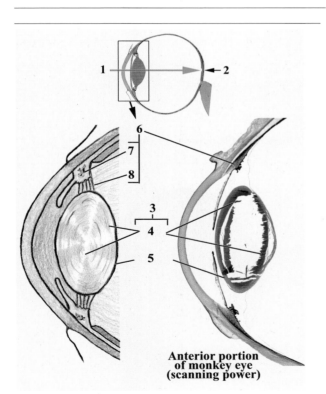

Figure 18.29

130 In reference to **Figure 18.29**, identify 1 - 8.
1 _____   5 _____
2 _____   6 _____
3 _____   7 _____
4 _____   8 _____

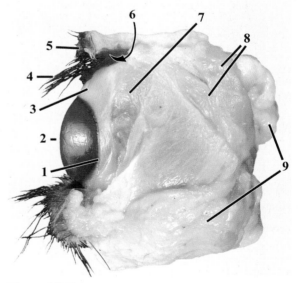

Figure 18.31

132 In reference to **Figure 18.31**, identify 1 - 9.
1 _____   6 _____
2 _____   7 _____
3 _____   8 _____
4 _____   9 _____
5 _____

278 Chapter 18 - The Eye

Name _____
Class _____

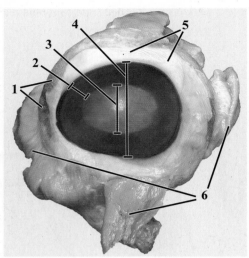

Figure 18.32

133 In reference to **Figure 18.32**, identify 1 - 6.
1 _____ 4 _____
2 _____ 5 _____
3 _____ 6 _____

Eyelashes (usually not present on dissection specimens)
134 What is the function of eyelashes? _____
_____
_____

Eyelids (usually not present on dissection specimens)
135 What covers the inner surface of the eyelids? _____
_____

136 What is the function of eyelids? _____
_____
_____

Cornea
137 What is the limbus? _____
_____

138 What are three functions of the cornea?
1 _____
2 _____
3 _____

Conjunctiva
139 Where is the conjunctiva located?
1 _____
_____
2 _____
_____

140 What are two functions of the conjunctiva?
1 _____
_____
2 _____
_____

Sclera
141 Where are the two locations that the sclera does not cover the surface of the eye?
1 _____
2 _____

142 What are three functions of the sclera?
1 _____
2 _____
3 _____

Extrinsic eye muscles
143 The muscles that move the eye are collectively called the
_____.

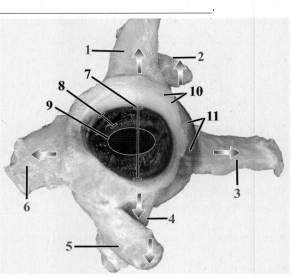

Figure 18.33
144 In reference to **Figure 18.33**, identify 1 - 11.
1 _____ 7 _____
2 _____ 8 _____
3 _____ 9 _____
4 _____ 10 _____
5 _____ 11 _____
6 _____

Optic nerve
145 Where does the optic nerve originate? _____
_____

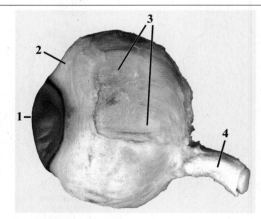

Figure 18.34
146 In reference to **Figure 18.34**, identify 1 - 4.
1 _____ 3 _____
2 _____ 4 _____

Chapter 18 - The Eye   279

ANTERIOR PORTION OF THE CUT EYE

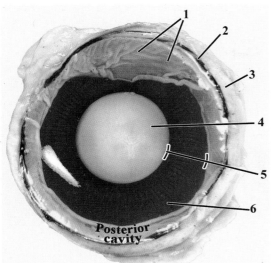

Figure 18.35

147 In reference to **Figure 18.35**, identify 1 - 6.
  1 _____ 4 _____
  2 _____ 5 _____
  3 _____ 6 _____

148 In reference to **Figure 18.35**, what are three functions of #6?
  1 _____
  2 _____
  3 _____

Lens

149 What is the function of the lens? _____
_____

150 What is the function of the ciliary muscle and the suspensory ligaments? _____
_____
_____
_____

151 The focal point of the lens is at the _____ of the retina.

Ciliary body

152 What are the two components of the ciliary body?
  1 _____ 2 _____

153 What is the function of the ciliary muscle? _____
_____

154 What are two functions of the ciliary processes?
  1 _____
  2 _____

Cut Wall of the Eye

155 From inner to outer, what are the three layers of the wall of the eye?
  1 _____ 3 _____
  2 _____

156 What is the function of the retina? _____
_____
_____

157 What are two functions of the choroid?
  1 _____
  2 _____

158 What is the name of the modified iridescent blue region of the choroid? _____

159 What is the function of the tapetum lucidum? _____
_____

160 What are two functions of the sclera?
  1 _____
  2 _____

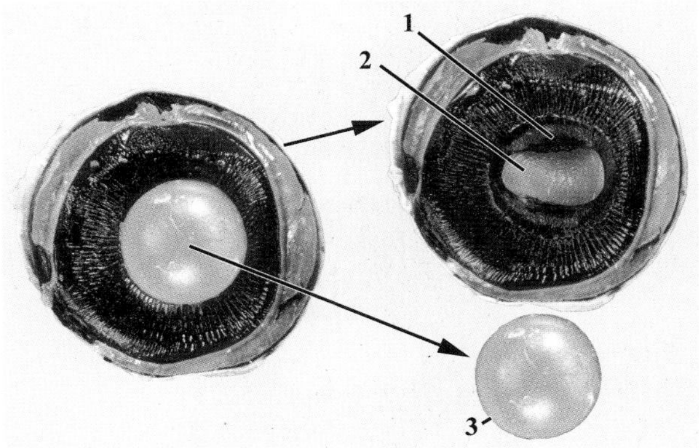

Figure 18.36

161 In reference to **Figure 18.36**, identify 1 - 3.
  1 _____ 3 _____
  2 _____

Iris

162 What is the function of the iris? _____
_____

Anterior cavity (segment)

163 What is the location of the anterior cavity? _____
_____

164 What is the name of the fluid found in the anterior cavity? _____

165 What are the two divisions of the anterior cavity?
  1 _____ 2 _____

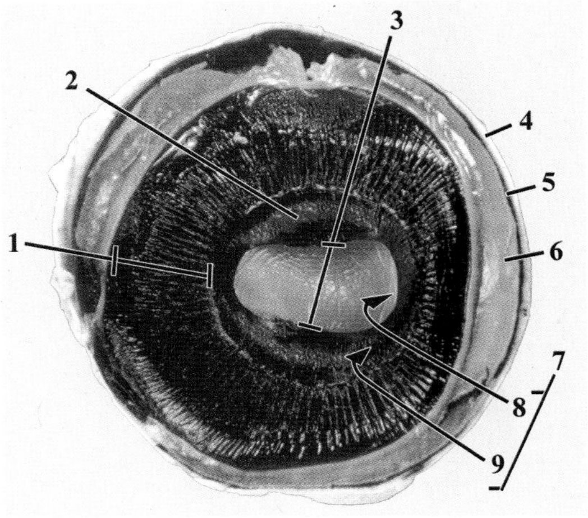

Figure 18.37

166 In reference to **Figure 18.37**, identify 1 - 9.
  1 _____ 6 _____
  2 _____ 7 _____
  3 _____ 8 _____
  4 _____ 9 _____
  5 _____

280   Chapter 18 - The Eye

167 In reference to **Figure 18.37**, what is the function of #1?
_____
_____

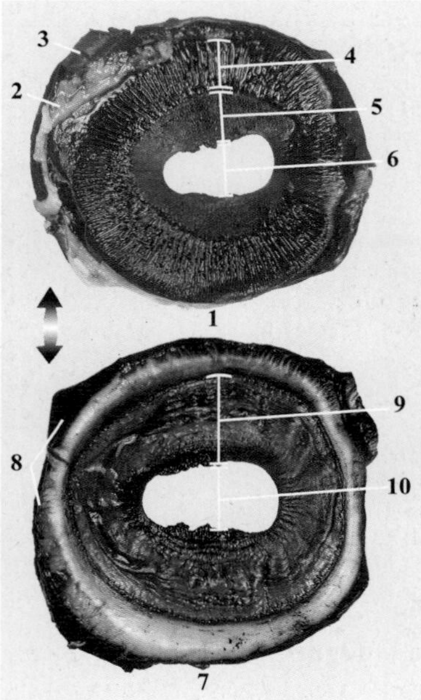

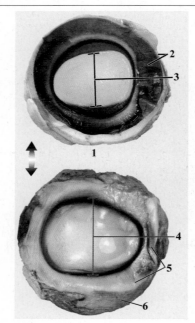

Figure 18.38

168 In reference to **Figure 18.38**, identify 1 - 9.
1 _____ 6 _____
2 _____ 7 _____
3 _____ 8 _____
4 _____ 9 _____
5 _____ 10 _____

169 In reference to **Figure 18.38**, what is the function of #9?
_____
_____

Figure 18.39

170 In reference to **Figure 18.39**, identify 1 - 6.
1 _____ 4 _____
2 _____ 5 _____
3 _____ 6 _____

POSTERIOR PORTION OF THE CUT EYE

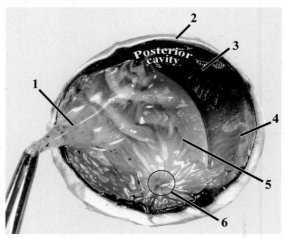

Figure 18.40

171 In reference to **Figure 18.40**, identify 1 - 6.
1 _____ 4 _____
2 _____ 5 _____
3 _____ 6 _____

Retina
172 What is the function of the retina? _____
_____

173 What is the optic disc? _____
_____

Choroid
174 What is the function of the choroid? _____
_____
_____

175 In relation to the retina, where is the choroid located? ____
_____

176 What is the function of the tapetum lucidum? _____
_____
_____

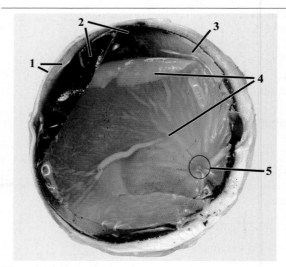

Figure 18.41

177 In reference to **Figure 18.41**, identify 1 - 5.
1 _____ 4 _____
2 _____ 5 _____
3 _____

Name _____
Class _____

Chapter 18 - The Eye   **281**

178 In reference to **Figure 18.41**, what is the function of #1?
_____

179 In reference to **Figure 18.41**, what is the function of #2?
_____

180 In reference to **Figure 18.41**, what is the function of #3?
_____

181 In reference to **Figure 18.41**, what is the function of #4?
_____

182 In reference to **Figure 18.41**, what is the function of #5?
_____

LIGHT
183 The energy of light is called _____.
184 What is the color spectrum? _____
_____
185 What is reflection? _____
_____
186 What gives an object its color? _____
_____
187 What are the names of the three color sensitive cones?
   1 _____   3 _____
   2 _____

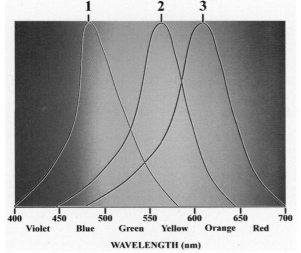

Figure 18.42
188 In reference to **Figure 18.42**, identify 1 - 3.
   1 _____   3 _____
   2 _____

Color Blindness and the Color Blindness Test
189 What is color blindness? _____
_____
190 What causes color blindness? _____
_____

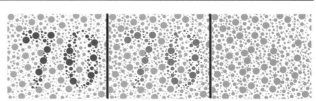

Figure 18.43
191 In reference to **Figure 18.43**, a subject unable to read the number in the middle test plate would have nonfunctional ____ sensitive cones.

LIGHT REFRACTION
192 Besides converting light energy into electrical energy, nerve impulses, what is another primary function of the eye? _____
_____

193 What is refraction? _____

194 When does refraction of light occur? _____
_____

195 The more the surface of a lens is curved, the (more or less) the light is refracted.

Figure 18.44
196 In reference to **Figure 18.44**, explain what is happening to make the straw change in its appearance? _____
_____
_____

Convex lens
197 Describe the shape of a convex lens. _____
_____

198 What effect do convex lenses have on light? ____
_____

199 How does increasing the convexity of the lens affect refraction? _____
_____

282 Chapter 18 - The Eye

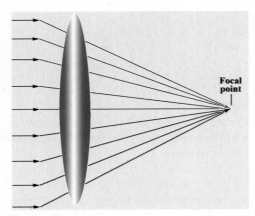

Figure 18.45

200 In reference to **Figure 18.45**, what type of lens is shown? _____

201 In reference to **Figure 18.45**, how would increasing the convexity of the lens affect refraction? _____

Concave lens
202 Describe the shape of a concave lens. _____

203 What effect does a concave lens have on light? _____

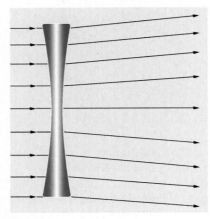

Figure 18.46

204 In reference to **Figure 18.46**, what type of lens is shown? _____

205 In reference to **Figure 18.46**, how would increasing the concavity of the lens affect refraction? _____

LIGHT REFRACTION ONTO THE RETINA
206 What structure is the primary refractory structure of the eye? _____

207 What structure of the eye has the ability to change convexity, thus, change its ability to refract light? _____

208 What is accommodation of the eye? _____

209 What two structures function in eye accommodation?
1 _____ 2 _____

210 What is the far point of vision? _____

211 What distance is the far point of vision for the human eye? _____

212 What is the near point of vision? _____

213 How does the lens change in shape to adjust for the near point of vision? _____

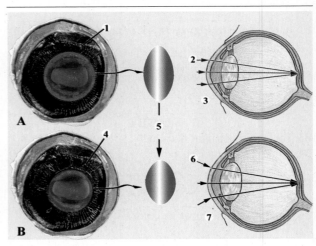

Figure 18.47

214 In reference to **Figure 18.47**, identify 1 - 7.
1 _____
2 _____
3 _____
4 _____
5 _____
6 _____
7 _____

215 In reference to **Figure 18.47**, more refraction of light is produced by the lens shown in 'A' or 'B.' _____ Explain your answer. _____

VISUAL ACUITY
216 What is emmetropic? _____

217 What causes abnormal visual acuity? _____

218 What are two common structural features that cause abnormal visual acuity? _____

219 What are two common visual abnormalities that involve an abnormally shaped eyeball?
1 _____ 2 _____

220 What is a common visual abnormality that involves an abnormally shaped cornea and/or lens? _____

Hyperopia
221 What is another name for hyperopia? _____

222 What causes hyperopia? _____

223 What type of lens is commonly used to correct hyperopia? _____

Myopia
224 What is another name for myopia? _____

225 What causes myopia? _____

226 What type of lens is commonly used to correct myopia? _____

Astigmatism
227 What causes astigmatism? _____

228 How is astigmatism commonly corrected? _____

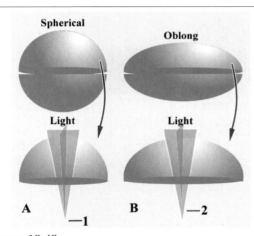

Figure 18.48

229 In reference to **Figure 18.48**, identify #1 - 2.
 1 _____ 2 _____
230 In reference to **Figure 18.48**, which figure shows astigmatism, 'A' or 'B'? _____

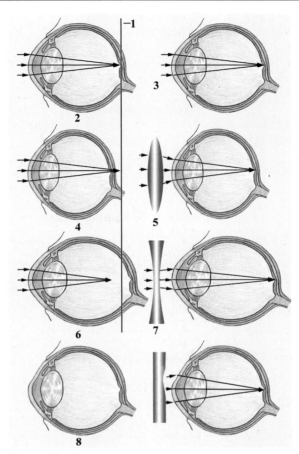

Figure 18.49

231 In reference to **Figure 18.49**, identify 1 - 8.
 1 _____ 5 _____
 2 _____ 6 _____
 3 _____ 7 _____
 4 _____ 8 _____

232 In reference to **Figure 18.49**, the converging lens is '<u>5</u>' or '<u>7</u>', the diverging lens is '<u>5</u>' or '<u>7</u>'.

VISUAL TESTS
233 What test distance is used when determining visual acuity? _____ Why is this distance used? _____

234 What does the top number on the visual acuity reading represent? _____

235 What does the bottom number on the visual acuity reading represent? _____

236 Explain the visual acuity of 20/100. _____

237 A person with a visual acuity reading of 20/70 is <u>myopic or hyperopic</u>? _____

Chapter 18 - The Eye

Test for Visual Acuity
238 Record your visual acuity for each eye:
Right eye _____ Left eye _____

239 If acuity is not 20/20 for each eye, what type of lens is needed to correct each eye?
Right eye _____ Left eye _____

Test for Astigmatism
240 Astigmatism test for each eye (present or absent):
Right eye _____ Left eye _____

Test for the Optic Disc (blind spot)
241 What causes the blind spot? _____
_____
_____

Test for the Near Point of Vision
242 What is the near point of vision? _____
_____

243 What happens to the lens that results in an increasing distance for the near point of vision? _____
_____

244 What is presbyopia? _____
_____

245 At what age range does the most dramatic change in the near point of vision usually occur? _____
_____

RETINA
246 What are the two layers of the retina? _____
_____

247 From inner to outer, what are the three cellular layers of the neural retina?
1 _____ 3 _____
2 _____

248 What is the function of the pigmented layer of the retina?
_____
_____

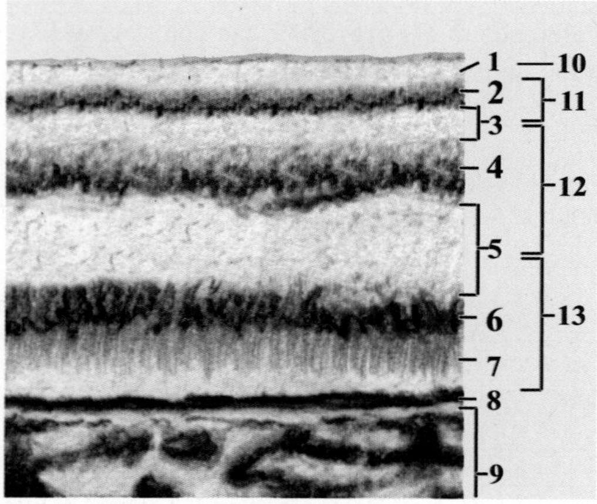

Figure 18.50
249 In reference to **Figure 18.50**, identify 1 - 13.
1 _____ 8 _____
2 _____ 9 _____
3 _____ 10 _____
4 _____ 11 _____
5 _____ 12 _____
6 _____ 13 _____
7 _____

PHOTORECEPTORS
250 From outer to inner what are the four retinal divisions of the photoreceptors?
1 _____ 3 _____
2 _____ 4 _____

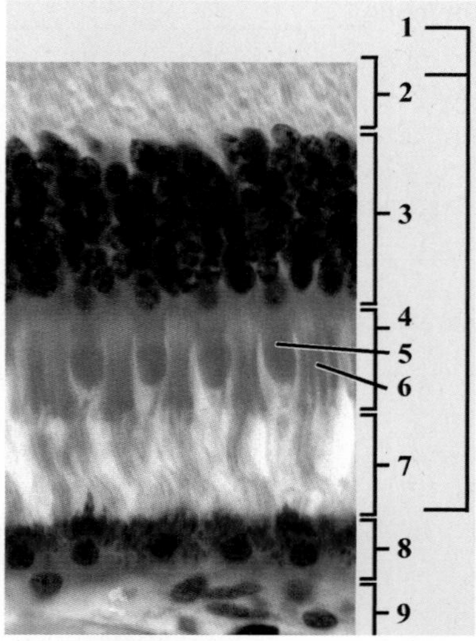

Figure 18.51
251 In reference to **Figure 18.51**, identify 1 - 9.
1 _____ 6 _____
2 _____ 7 _____
3 _____ 8 _____
4 _____ 9 _____
5 _____

VISUAL PIGMENTS
252 What is the light absorbing purple pigment of rods and cones? _____

253 What are the two components of rhodopsin?
1 _____ 2 _____

254 Which component of rhodopsin functions in the absorption of light? _____
_____

255 What is the function of opsin? _____
_____
_____

256 How many forms of opsin exist in cones? _____
_____

257 What are the two forms of retinal?
1 _____ 3 _____

258 What is the form of retinal when attached to opsin? ____
_____

259 What is the form of retinal when unattached to opsin? ___
_____

Chapter 18 - The Eye 285

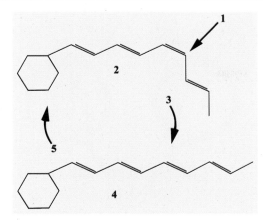

Figure 18.52

260 In reference to **Figure 18.52**, identify 1 - 5.
   1 _____ 4 _____
   2 _____ 5 _____
   3 _____

261 What causes the change of 11-cis retinal to all-trans retinal? _____

262 When retinal is changed from to 11-cis retinal to all-trans retinal, what happens to its association with opsin? _____

263 What is "bleaching of the pigment?" _____

264 How is all-trans retinal converted to 11-cis retinal? _____

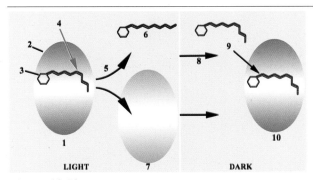

Figure 18.53

265 In reference to **Figure 18.53**, identify 1 - 10.
   1 _____ 6 _____
   2 _____ 7 _____
   3 _____ 8 _____
   4 _____ 9 _____
   5 _____ 10 _____

TRANSDUCTION OF LIGHT

266 What begins the process of converting the energy of light into a nervous signal, nerve impulses? _____

267 The conversion of 11-cis retinal to _____ retinal results in the separation of all-trans retinal from _____.

268 Free opsin functions as an enzyme that results in the ____ _____ of the photoreceptors.

269 Hyperpolarization of the photoreceptors _____ the release of neurotransmitter at the (<u>inhibitory</u> or <u>excitatory</u>) bipolar synapses.

270 With the reduction of inhibition, the bipolar cells are (<u>inhibited</u> or <u>excited</u>) and (<u>stimulate</u> or <u>inhibit</u>) the ganglion cells.

271 Once stimulated, the ganglion cells generate action potentials (nerve impulses) that travel through the _____.

OPTIC NERVE PATHWAY TO THE BRAIN

272 Where do the optic nerves originate? _____

273 What is the optic chiasma? _____

274 What fibers does each optic tract carry? _____

275 What are three destinations for fibers from the optic tracts?
   1 _____ 3 _____
   2 _____

276 What is the destination of fibers that enter the thalamus? _____

277 Fibers that enter the midbrain mostly function in _____

278 Fibers that enter the hypothalamus mostly function to _____

286 Chapter 18 - The Eye

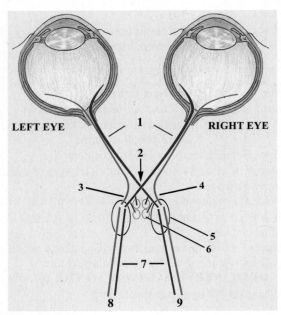

Figure 18.54
279 In reference to **Figure 18.54**, identify 1 - 9.
1 _____ 6 _____
2 _____ 7 _____
3 _____ 8 _____
4 _____ 9 _____
5 _____

VISUAL FIELDS

280 The right optic tract carries visual information from _____
_____
_____

281 The left optic tract carries visual information from _____
_____
_____
_____

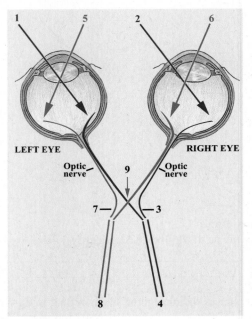

Figure 18.55
282 In reference to **Figure 18.55**, identify 1 - 9.
1 _____ 6 _____
2 _____ 7 _____
3 _____ 8 _____
4 _____ 9 _____
5 _____

Name _____
Class _____

Chapter 19 - The Ear 287

The Ear - Worksheets

1. What are two functions of the ear?
   1 _____ 2 _____
2. What are the three regions of the ear?
   1 _____ 3 _____
   2 _____
3. All portions of the ear are surrounded by the _____ bone, except the skin-covered flap, the _____.

STRUCTURE OF THE EAR

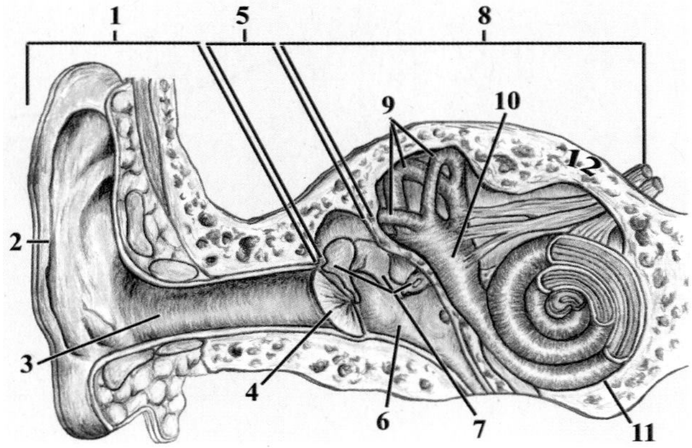

Figure 19.1

4. In reference to **Figure 19.1**, identify #1 - 11.
   1 _____ 7 _____
   2 _____ 8 _____
   3 _____ 9 _____
   4 _____ 10 _____
   5 _____ 11 _____
   6 _____

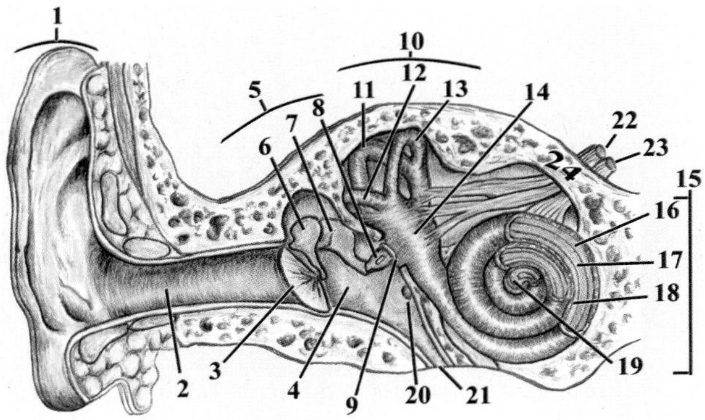

Figure 19.2

5. In reference to **Figure 19.2**, identify #1 - 24.
   1 _____ 13 _____
   2 _____ 14 _____
   3 _____ 15 _____
   4 _____ 16 _____
   5 _____ 17 _____
   6 _____ 18 _____
   7 _____ 19 _____
   8 _____ 20 _____
   9 _____ 21 _____
   10 _____ 22 _____
   11 _____ 23 _____
   12 _____ 24 _____

OUTER EAR

6. What are the two divisions of the outer ear?
   1 _____ 2 _____
7. What are the location and function of the auricle? _____
8. What are the location and function of the external auditory canal? _____
9. What are the location and function of the tympanic membrane? _____

MIDDLE EAR

10. What forms the middle ear? _____
11. What does the middle ear house? _____
12. What are the location and function of the pharyngotympanic tube (Eustachian, or auditory, tube)? _____

Ossicles

13. What are the ossicles? _____
14. From the tympanic membrane, name the three ossicles.
    1 _____ 3 _____
    2 _____
15. Where does the stapes terminate? _____
16. Where is the oval window located and what is its function? _____
17. What happens to the vibrations as they are conducted along the ossicles? _____

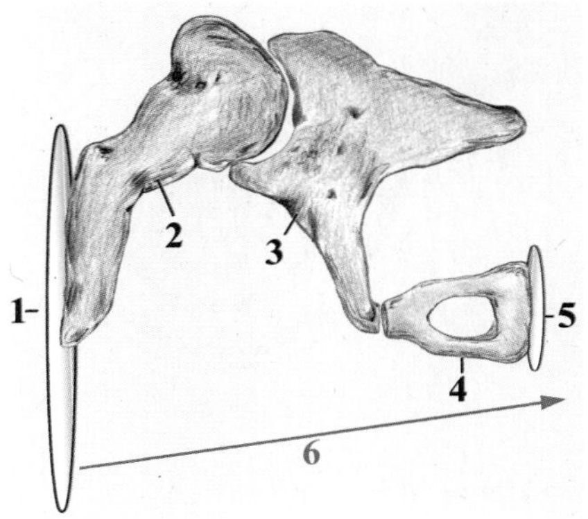

Figure 19.3

18. In reference to **Figure 19.3**, identify #1 - 6.
    1 _____ 4 _____
    2 _____ 5 _____
    3 _____ 6 _____

288 Chapter 19 - The Ear

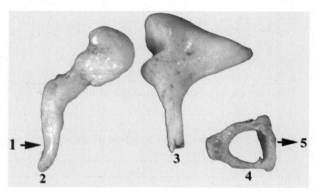

Figure 19.4

19 In reference to **Figure 19.4**, identify #1 - 5.
   1 _____ 4 _____
   2 _____ 5 _____
   3 _____

20 In reference to **Figure 19.4**, what structure transfers sound vibrations to #2? _____

21 In reference to **Figure 19.4**, #4 terminates at the _____.

INNER EAR

22 What are the three divisions of the inner ear?
   1 _____ 3 _____
   2 _____

23 What is the osseous labyrinth? _____
   _____

24 What fluid does the osseous labyrinth contain? _____

25 What is the membranous labyrinth? _____

26 What fluid does the membranous labyrinth contain? _____
   _____

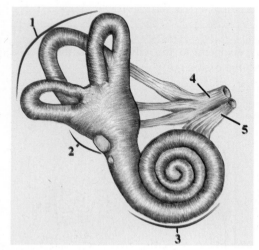

Figure 19.5

27 In reference to **Figure 19.5**, identify #1 - 5.
   1 _____ 4 _____
   2 _____ 5 _____
   3 _____

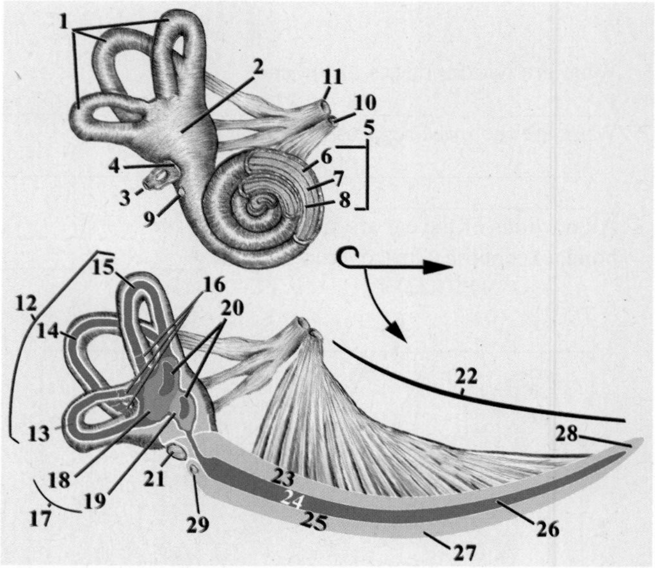

Figure 19.6

28 In reference to **Figure 19.6**, identify #1 - 29.
   1 _____ 16 _____
   2 _____ 17 _____
   3 _____ 18 _____
   4 _____ 19 _____
   5 _____ 20 _____
   6 _____ 21 _____
   7 _____ 22 _____
   8 _____ 23 _____
   9 _____ 24 _____
   10 _____ 25 _____
   11 _____ 26 _____
   12 _____ 27 _____
   13 _____ 28 _____
   14 _____ 29 _____
   15 _____

Cochlea

29 What are the three chambers of the cochlea?
   1 _____ 3 _____
   2 _____

30 Which chamber begins at the oval window? _____
   _____

31 At the apex of the cochlea the scala vestibuli joins with the scala _____ at the region called the _____.

32 Which chamber ends at the round window? _____
   _____

33 Both the scala vestibuli and scala tympani contain a fluid called _____.

34 Where is the cochlea duct located? _____
   _____

35 What is the name of the receptor found in the cochlear duct? _____

36 What is the name of the fluid found in the cochlear duct?
   _____

Chapter 19 - The Ear   289

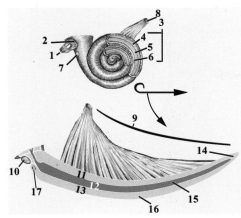

Figure 19.7

37 In reference to **Figure 19.7**, identify #1 - 17.
1 _____ 10 _____
2 _____ 11 _____
3 _____ 12 _____
4 _____ 13 _____
5 _____ 14 _____
6 _____ 15 _____
7 _____ 16 _____
8 _____ 17 _____
9 _____

Vestibule

38 Where is the vestibule located? _____

39 What are the two regions of the vestibule?
1 _____ 2 _____

40 What is the name of the receptors located in the vestibule?
_____
_____

41 The maculae respond to _____.

42 What is the function of the maculae? _____

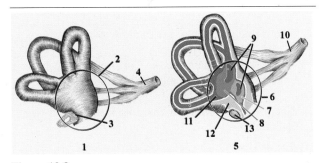

Figure 19.8

43 In reference to **Figure 19.8**, identify #1 - 13.
1 _____ 8 _____
2 _____ 9 _____
3 _____ 10 _____
4 _____ 11 _____
5 _____ 12 _____
6 _____ 13 _____
7 _____

44 In reference to **Figure 19.8**, what is the function of #9? _____
_____

Semicircular canals

45 What are the names of the three semicircular canals?
1 _____ 3 _____
2 _____

46 What is the ampulla? _____

47 What is the name of the receptor located in the ampulla?
_____
_____

48 What type of motion stimulates the crista ampullaris? _____
_____

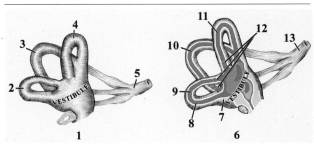

Figure 19.9

49 In reference to **Figure 19.9**, identify #1 - 13.
1 _____ 8 _____
2 _____ 9 _____
3 _____ 10 _____
4 _____ 11 _____
5 _____ 12 _____
6 _____ 13 _____
7 _____

50 Which semicircular canal responds to motion in the sagittal plane (front-back rotation)? _____

51 Which semicircular canal responds to motion in the frontal plane (side-side rotation)? _____

52 Which semicircular canal responds to motion in the horizontal plane (horizontal rotation)? _____

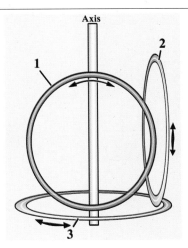

Figure 19.10

53 In reference to **Figure 19.10**, identify #1 - 3.
1 _____ 3 _____
2 _____

290 Chapter 19 - The Ear

Name _____
Class _____

COCHLEA

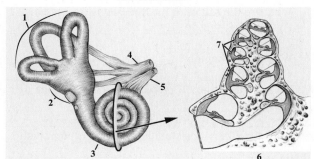

Figure 19.11

54 In reference to **Figure 19.11**, identify #1 - 7.
1 _____ 5 _____
2 _____ 6 _____
3 _____ 7 _____
4 _____

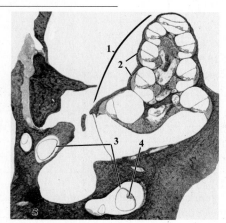

Figure 19.12

55 In reference to **Figure 19.12**, identify #1 - 5.
1 _____ 4 _____
2 _____ 5 _____
3 _____

56 What is the modiolus? _____

57 What is the spiral ganglion? _____

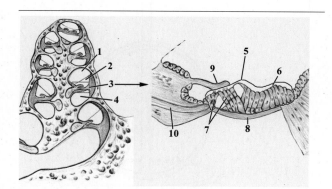

Figure 19.13

58 In reference to **Figure 19.13**, identify #1 - 10.
1 _____ 6 _____
2 _____ 7 _____
3 _____ 8 _____
4 _____ 9 _____
5 _____ 10 _____

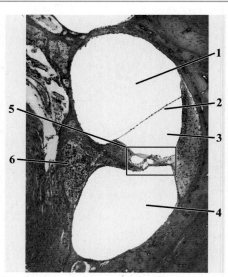

Figure 19.14

59 In reference to **Figure 19.14**, identify #1 - 6.
1 _____ 4 _____
2 _____ 5 _____
3 _____ 6 _____

60 Which chamber originates at the oval window? _____

61 What is the name of the membrane that forms the floor of the scala vestibuli? _____

62 What is the name of the membrane that forms the floor of the cochlear duct? _____

63 What is the name of the structure that sits on the basilar membrane? _____

64 How is the basilar membrane set into motion? _____

65 Which chamber terminates at the round window? _____

66 What is the function of the round window? _____

67 Which cells of the organ of Corti respond to movements against the tectorial membrane? _____

68 Where is the tectorial membrane located, and what is its function? _____

69 What do the fibers that leave the spiral ganglion form? ___

70 The cochlear nerve merges with the _____ nerve to form the _____ nerve.

Name _____
Class _____

Chapter 19 - The Ear   **291**

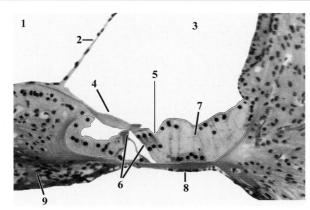

Figure 19.15
71 In reference to **Figure 19.15**, identify #1 - 9.
1 _____   6 _____
2 _____   7 _____
3 _____   8 _____
4 _____   9 _____
5 _____

MECHANISM OF HEARING
Sound
72 What is sound? _____
_____
_____

73 How is a sound wave produced? _____
_____

74 What is wavelength? _____
_____

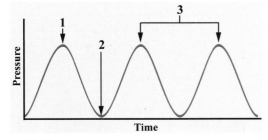

Figure 19.16
75 In reference to **Figure 19.16**, identify #1 - 3.
1 _____   3 _____
2 _____

Frequency
76 What is frequency? _____
_____

77 What is hertz? _____
_____

78 The greater the number of cycles (waves) per second, the _____ the wavelength and the _____ the frequency.

79 The higher the pitch the _____ the frequency.

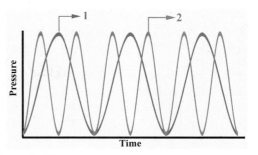

Figure 19.17
80 In reference to **Figure 19.17**, identify #1 - 2.
1 _____   2 _____

Amplitude
81 What is amplitude? _____
_____

82 The greater the amplitude, the _____ the sound.
83 What is a decibel? _____

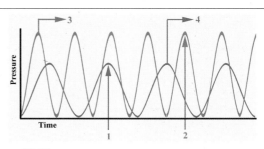

Figure 19.18
84 In reference to **Figure 19.18**, identify #1 - 4.
1 _____   3 _____
2 _____   4 _____

PATHWAYS OF SOUND CONDUCTION
Sound Conduction to the Cochlea
85 Sound waves from the auditory canal strike the _____ _____, which transfers sound vibrations to the first ossicle, the _____.

86 From the tympanic membrane, the three ossicles are the
1 _____   3 _____
2 _____

87 The ossicular chain conducts and _____ the sound vibrations.

88 The oval window transfers the sound vibrations as pressure waves into the fluid, the _____, in the cochlear chamber called the _____.

89 The pressure waves of the scala vestibuli pass into the chamber called the _____.

90 The pressure waves are absorbed by the _____ window, located at the _____ of the scala tympani.

91 Pressure waves that pass along the perilymph set into motion regions of the _____ membrane.

92 The basilar membrane that is closest to the oval window responds to _____ frequency.

93 Moving toward the apex of the cochlea (away from the oval window) the basilar membrane is set into motion by increasingly _____ frequencies.

94 The amplitude (intensity) of the pressure wave determines how much the basilar membrane is _____.

292 Chapter 19 - The Ear

95 Movement of the basilar membrane results in movement of the hair cells of the _____.
96 Movement of the hair cells results in their electrical change called a _____.
97 Depolarization of the hair cells leads to depolarization of the bipolar cells, and the generation of _____ potentials that are transmitted by the cochlear nerve.

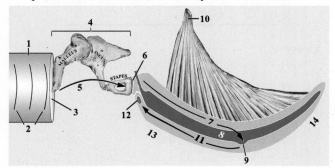

Figure 19.19
98 In reference to **Figure 19.19**, identify #1 - 14.
   1 _____   8 _____
   2 _____   9 _____
   3 _____ 10 _____
   4 _____ 11 _____
   5 _____ 12 _____
   6 _____ 13 _____
   7 _____ 14 _____

Neural Conduction to the Brain
99 What region of the brain receives the cochlear fibers? _____
_____
100 Most fibers from the cochlear nucleus crossover to the opposite side of the brain and enter the _____.
101 The midbrain uses auditory information for auditory _____.
102 Fibers from the midbrain enter the brain's sensory relay center, the _____, which distributes the auditory information to the auditory cortex of the _____ _____.

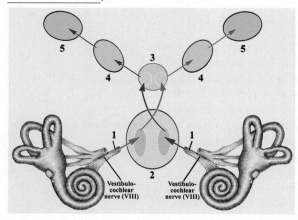

Figure 19.20
103 In reference to **Figure 19.20**, identify #1 - 5.
   1 _____   4 _____
   2 _____   5 _____
   3 _____

HEARING TESTS
104 What are two types of deafness?
   1 _____   2 _____
105 What is conduction deafness? _____
_____
_____
106 What are some causes of conduction deafness? _____
_____
_____
107 What is sensorineural deafness? _____
_____
_____

Weber's Test
108 When performing Weber's test for conductive hearing loss, the ear (with or without) conductive hearing loss shows lateralization.
109 When performing Weber's test for sensorineural hearing loss, lateralization occurs to the (better or worse) ear.

Rinne's Test
110 Hearing by conduction of sound by air is about _____ times better than hearing by bone conduction.
111 When performing Rinne's test for conductive hearing loss, in an ear with conductive hearing loss sound is heard better by _____ conduction than _____ conduction.
112 If an ear has sensorineural hearing loss, hearing is reduced with _____ conduction being better than _____ conduction.

EQUILIBRIUM
113 The regions of the ear that function in equilibrium are the _____ and the _____.

Vestibule
114 What are the two regions of the vestibule and what do they contain? _____
_____
_____
115 What types of motion do the maculae respond? _____
_____
_____

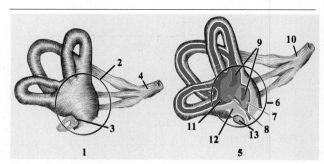

Figure 19.21
116 In reference to **Figure 19.21**, identify #1 - 13.
   1 _____   8 _____
   2 _____   9 _____
   3 _____ 10 _____
   4 _____ 11 _____
   5 _____ 12 _____
   6 _____ 13 _____
   7 _____

Name _____
Class _____

Chapter 19 - The Ear   293

Maculae

117 Describe the structure of the maculae. _____

118 What types of motion influence the movement of the otoliths and gelatinous material over the hair cells? _____

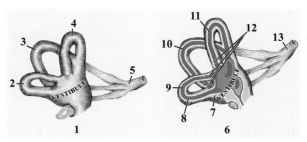

Figure 19.24

125 In reference to **Figure 19.24**, identify #1 - 13.
1 _____ 8 _____
2 _____ 9 _____
3 _____ 10 _____
4 _____ 11 _____
5 _____ 12 _____
6 _____ 13 _____
7 _____

126 Which semicircular canal responds to motion in the sagittal plane (front-back rotation)? _____

127 Which semicircular canal responds to motion in the frontal plane (side-side rotation)? _____

128 Which semicircular canal responds to motion in the horizontal plane (horizontal rotation)? _____

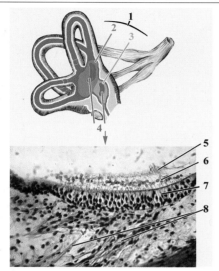

Figure 19.22

119 In reference to **Figure 19.22**, identify #1 - 8.
1 _____ 5 _____
2 _____ 6 _____
3 _____ 7 _____
4 _____ 8 _____

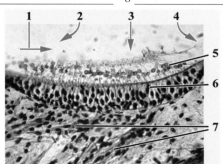

Figure 19.23

120 In reference to **Figure 19.23**, identify #1 - 7.
1 _____ 5 _____
2 _____ 6 _____
3 _____ 7 _____
4 _____

Semicircular canals

121 What are the names of the three semicircular canals?
1 _____ 3 _____
2 _____

122 What is the ampulla? _____

123 What is the name of the receptor located in the ampulla? _____

124 What type of motion stimulates the crista ampullaris? _____

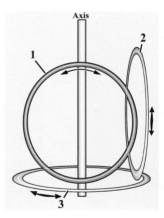

Figure 19.25

129 In reference to **Figure 19.25**, identify #1 - 3.
1 _____ 3 _____
2 _____

294 Chapter 19 - The Ear

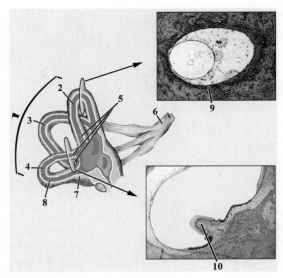

Figure 19.26
130 In reference to **Figure 19.26**, identify #1 - 10.
1 _____ 6 _____
2 _____ 7 _____
3 _____ 8 _____
4 _____ 9 _____
5 _____ 10 _____

131 Rotational movement of the head causes the fluid (_____) in a semicircular duct to move against its _____.
132 The cupula stimulates the _____ of the crista ampullaris.
133 Hair cells provide the electrical stimulus that results in the stimulation of the fibers of the _____ nerve.

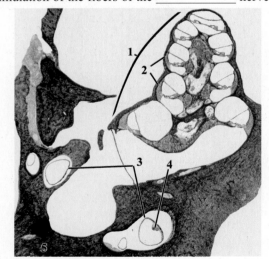

Figure 19.27
134 In reference to **Figure 19.27**, identify #1 - 5.
1 _____ 4 _____
2 _____ 5 _____
3 _____

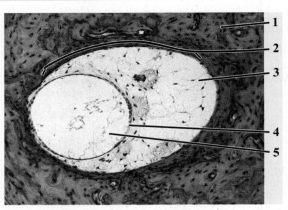

Figure 19.28
135 In reference to **Figure 19.28**, identify #1 - 4.
1 _____ 4 _____
2 _____ 5 _____
3 _____

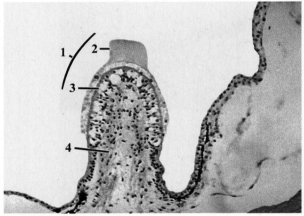

136 **Figure 19.29**
In reference to **Figure 19.29**, identify #1 - 4.
1 _____ 3 _____
2 _____ 4 _____

PATHWAY TO THE BRAIN
137 The vestibular nerve leaves the vestibule and joins with the _____ nerve to form the _____ _____ nerve.
138 The vestibulocochlear nerve enters the brain stem and enters the _____ nuclei.
139 What are two functions of the vestibular nuclei? _____
_____
_____

Name _____
Class _____

Chapter 20 - Endocrine System   295

Endocrine System - Worksheets

1 What forms the endocrine system? _____
   _____
2 What are hormones? _____
   _____
   _____
3 Where do endocrine glands release hormones? _____
   _____
4 What must cells have in order to respond to hormones? __
   _____
5 Once bound to a receptor, what does a hormone do? ____
   _____
6 What are ten endocrine glands?
   1 _____   6 _____
   2 _____   7 _____
   3 _____   8 _____
   4 _____   9 _____
   5 _____   10 _____

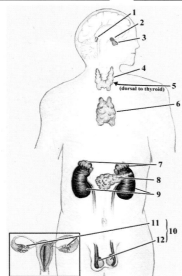

Figure 20.1
7 In reference to **Figure 20.1**, identify #1 - #12.
   1 _____   7 _____
   2 _____   8 _____
   3 _____   9 _____
   4 _____   10 _____
   5 _____   11 _____
   6 _____   12 _____

PITUITARY GLAND (HYPOPHYSIS)

8 What is another name for the pituitary gland? _____
   _____
9 Where is the pituitary gland located? _____
   _____
   _____
10 What attaches the pituitary gland and the hypothalamus? __
   _____
11 What are the two regions of the pituitary gland? _____
   _____

12 What is another name for the anterior pituitary? _____
13 What is another name for the posterior pituitary? _____
14 What are the three regions of the anterior pituitary?
   1 _____   3 _____
   2 _____

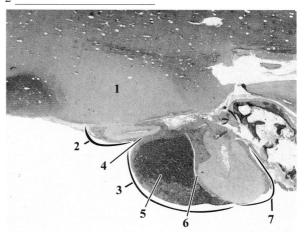

Figure 20.2
15 In reference to **Figure 20.2**, identify #1 - #7.
   1 _____   5 _____
   2 _____   6 _____
   3 _____   7 _____
   4 _____

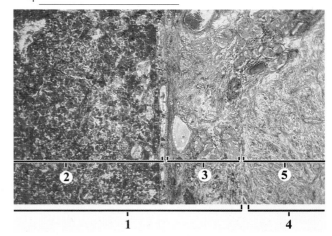

Figure 20.3
16 In reference to **Figure 20.3**, identify #1 - #5.
   1 _____   4 _____
   2 _____   5 _____
   3 _____

ANTERIOR PITUITARY (ADENOHYPOPHYSIS)
PARS DISTALIS

17 What are two populations of cells found in the pars distalis? 1 _____   2 _____
18 What are two types of chromophil cells? _____
   _____

Chapter 20 - Endocrine System

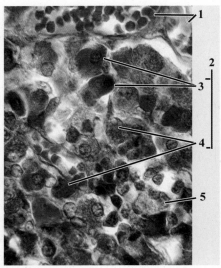

Figure 20.4

19 In reference to **Figure 20.4**, identify #1 - #5.
    1 _____ 4 _____
    2 _____ 5 _____
    3 _____

20 What are six hormones (and abbreviations) produced by the pars distalis?
    1 _____
    2 _____
    3 _____
    4 _____
    5 _____
    6 _____

21 What are tropic hormones? _____
    _____

22 What are the four tropic hormones?
    1 _____
    2 _____
    3 _____
    4 _____

Thyroid-stimulating hormone (TSH)

23 What is the target and effect of thyroid-stimulating hormone? _____
    _____
    _____

24 How is the secretion of thyroid-stimulating hormone regulated? _____
    _____

25 Where is TRH produced? _____
    _____

26 What happens to levels of TRH when levels of thyroid hormones increase? _____
    _____

27 What is the result of decreased levels of TRH? _____
    _____

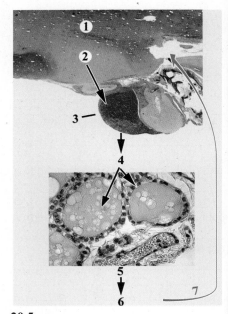

Figure 20.5

28 In reference to **Figure 20.5**, identify #1 - #7.
    1 _____ 5 _____
    2 _____ 6 _____
    3 _____ 7 _____
    4 _____

Adrenocorticotropic hormone (ACTH)

29 What is the target for ACTH? _____

30 What is the function of ACTH? _____
    _____

31 The release of ACTH is in response to _____
    _____

32 Where is CRH produced? _____
    _____

33 What mechanism controls the secretion of ACTH? _____
    _____

34 What happens to levels of CRH when levels of ACTH increase? _____
    _____

35 What is the result of decreased levels of CRH? _____
    _____

Name _____
Class _____

Chapter 20 - Endocrine System   297

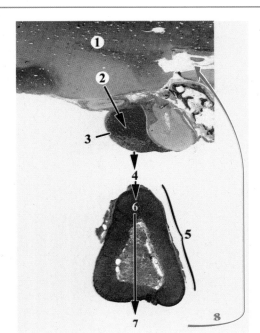

Figure 20.6

36 In reference to **Figure 20.6**, identify #1 - #8.
    1 _____   5 _____
    2 _____   6 _____
    3 _____   7 _____
    4 _____   8 _____

Follicle-stimulating hormone (FSH)

37 For females, what is the target and effect of FSH? _____
    _____
    _____

38 For males, what is the target and effect of FSH? _____
    _____
    _____

39 What hormone is produced in response to FSH? _____
    _____

40 What mechanism controls the secretion of FSH? _____
    _____

41 Where is GnRH produced? _____

42 What happens to levels of GnRH when levels of inhibin increase? _____

43 What is the result of decreased levels of GnRH? _____
    _____

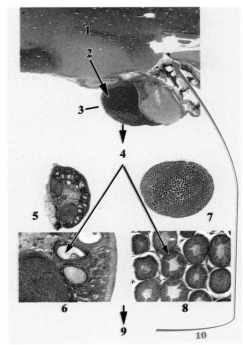

Figure 20.7

44 In reference to **Figure 20.7**, identify #1 - #10.
    1 _____   6 _____
    2 _____   7 _____
    3 _____   8 _____
    4 _____   9 _____
    5 _____   10 _____

Luteinizing hormone (LH)

45 For females, what is the target and effect of LH ? _____
    _____
    _____

46 For males, what is the target and effect of LH? _____
    _____

47 What hormones are produced in response to FSH?
    Females _____
    Males _____

48 What mechanism controls the secretion of LH? _____
    _____

49 Where is GnRH produced? _____

50 What happens to levels of GnRH when levels of estrogen or testosterone increase? _____
    _____

51 What is the result of decreased levels of GnRH? _____
    _____

298 Chapter 20 - Endocrine System

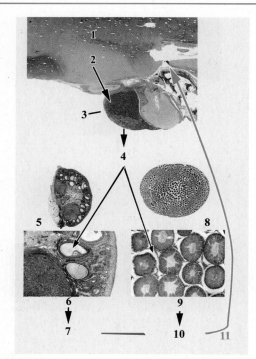

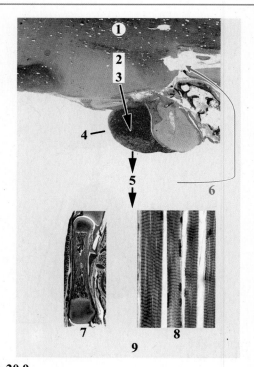

Figure 20.8
52 In reference to **Figure 20.8**, identify #1 - #11.
1 _____ 7 _____
2 _____ 8 _____
3 _____ 9 _____
4 _____ 10 _____
5 _____ 11 _____
6 _____

Growth hormone (GH)
53 What does GH target? _____
_____

54 The release of GH is in response to _____
_____

55 Where are GHIH and GHRH produced? _____
_____

56 What mechanism controls the secretion of GH? _____

57 Increased levels of GH result in the release of GHIH or GHRH? _____

58 What is the result of the release of GHIH? _____
_____

59 Decreased levels of GH result in the release of GHIH or GHRH? _____

60 What is the result of the release of GHRH? _____
_____

Figure 20.9
61 In reference to **Figure 20.9**, identify #1 - #9.
1 _____ 6 _____
2 _____ 7 _____
3 _____ 8 _____
4 _____ 9 _____
5 _____

Prolactin (PRL)
62 What is the target for PRL? _____

63 What is the effect of PRL? _____
_____

64 The release of PRL is in response to _____
_____

65 Where are PIH and PRH produced? _____
_____

66 Increased levels of estrogen result in the release of PIH or PRH? _____

67 Increased levels of PRH result in _____
_____

68 Decreased levels of estrogen result in the release of PIH or PRH? _____

69 Increased levels of PIH result in _____
_____

Chapter 20 - Endocrine System

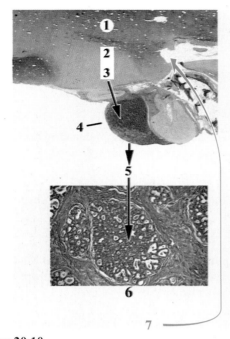

Figure 20.10

70 In reference to **Figure 20.10**, identify #1 - #7.
1 _____ 5 _____
2 _____ 6 _____
3 _____ 7 _____
4 _____

PARS INTERMEDIA

71 Where is the pars intermedia located? _____
_____

72 What is the target and effect of melanocyte-stimulating hormone (MSH)? _____

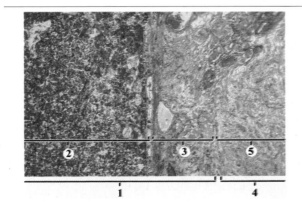

Figure 20.11

73 In reference to **Figure 20.11**, identify #1 - #5.
1 _____ 4 _____
2 _____ 5 _____
3 _____

POSTERIOR PITUITARY
(NEUROHYPOPHYSIS, PARS NERVOSA)

74 What forms the posterior pituitary? _____
_____

75 Where do the nerve fibers of the posterior pituitary originate? _____
_____

76 What hormones are released at the posterior pituitary? ___
_____

77 What is another name for ADH? _____
_____

Oxytocin

78 What are two targets and effects of oxytocin? _____
_____
_____
_____

79 The release of OT is a _____ feedback response to _____ stimuli.

80 During birth oxytocin is produced in response to _____ of the uterus and promotes uterine _____.

81 During nursing oxytocin is released in response to _____. Oxytocin promotes milk _____.

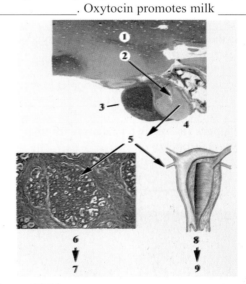

Figure 20.12

82 In reference to **Figure 20.12**, identify #1 - #9.
1 _____
2 _____
3 _____
4 _____
5 _____
6 _____
7 _____
8 _____
9 _____

Antidiuretic hormone (ADH)

83 What are two targets and effects of ADH? _____
_____
_____
_____

Chapter 20 - Endocrine System

HYPOTHALAMUS

84 What are three functions of the hypothalamus?
   1 _____
   _____
   2 _____
   _____
   3 _____
   _____

Hypothalamus - Pituitary Relationships

85 What are two functions of the hypothalamus in relation to the pituitary gland? _____

Regulatory Hormones and Vascularity

86 Where are the regulatory hormones for the anterior pituitary gland produced? _____

87 How are the regulatory hormones moved to the anterior pituitary? _____

88 What is a portal system? _____

89 Where is the capillary network called the hypophyseal primary plexus located? _____

90 Where is the hypophyseal secondary plexus located? _____

91 What happens to regulatory hormones that enter the hypophyseal secondary plexus? _____

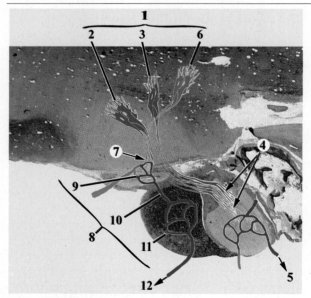

Figure 20.13

92 In reference to **Figure 20.13**, identify #1 - #12.
   1 _____
   2 _____
   3 _____
   4 _____
   5 _____
   6 _____
   7 _____
   8 _____
   9 _____
   10 _____
   11 _____
   12 _____

93 What are the two hormones produced by the hypothalamus and released at the posterior pituitary? _____

94 What regulates the secretion of the two hypothalmic hormones? _____

Hypothalamus - Control of Adrenal Medullae

95 What response is mediated by the sympathetic nuclei of the hypothalamus? _____

96 What portion of the adrenal gland is activated by the hypothalamus? _____

97 What hormones are released upon activation of the hypothalmic sympathetic centers? _____

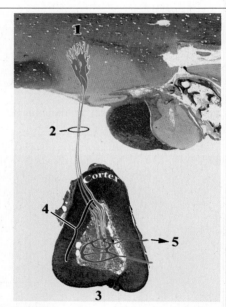

Figure 20.14

98 In reference to **Figure 20.14**, identify #1 - #5.
   1 _____ 4 _____
   2 _____ 5 _____
   3 _____

THYROID GLAND

99 Where is the thyroid gland located? _____

100 The thyroid gland is mostly composed of spherical structures called _____.

101 Thyroid follicles are lined with cuboidal-shaped _____.

Chapter 20 - Endocrine System   301

102 Located among the follicular cells are a small population of cells called _____.

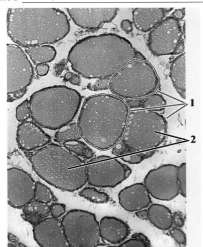

Figure 20.15

103 In reference to **Figure 20.15**, identify #1 - #2.
1 _____ 2 _____

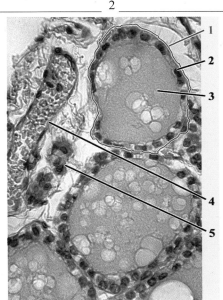

Figure 20.16

104 In reference to **Figure 20.16**, identify #1 - #5.
1 _____ 4 _____
2 _____ 5 _____
3 _____

Follicular Cells

105 What hormones are produced by thyroid follicular cells? _____

106 What mechanism regulates calcitonin secretion? _____

107 What are the targets of thyroid hormones? _____

108 What are two functions of thyroid hormones? _____

Parafollicular cells (C cells)

109 What hormone is produced by the parafollicular cells (C cells)? _____

110 What mechanism regulates parathyroid secretion? _____

111 What is the stimulus for increased secretion of calcitonin? _____

112 What inhibits the secretion of calcitonin? _____

113 What are three targets and effects of calcitonin?
1 _____
2 _____
3 _____

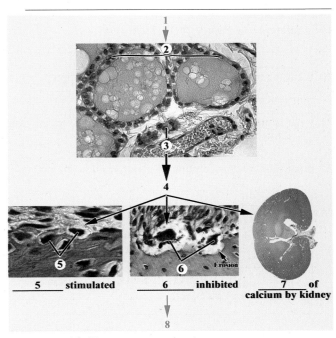

Figure 20.17

114 In reference to **Figure 20.17**, identify #1 - #8.
1 _____ 5 _____
2 _____ 6 _____
3 _____ 7 _____
4 _____ 8 _____

PARATHYROID GLANDS

115 Where are the parathyroids located? _____

116 What types of cells form the parathyroid glands? _____

117 What is the name of the endocrine cells of the parathyroid? _____

118 What hormone is produced by the principal (chief) cells? _____

302 Chapter 20 - Endocrine System

Name _____
Class _____

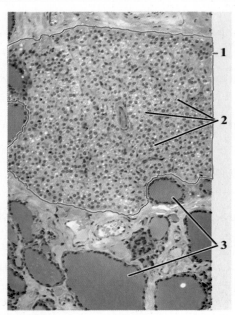

Figure 20.18

119 In reference to **Figure 20.18**, identify #1 - #3.
   1 _____ 3 _____
   2 _____

120 What mechanism regulates parathyroid secretion? _____
   _____

121 What is the stimulus for increased secretion of parathyroid hormone? _____

122 What inhibits the secretion of parathyroid hormone? _____
   _____

123 What are four targets and effects of parathyroid hormone?
   1 _____
   _____
   2 _____
   _____
   3 _____
   _____
   4 _____
   _____

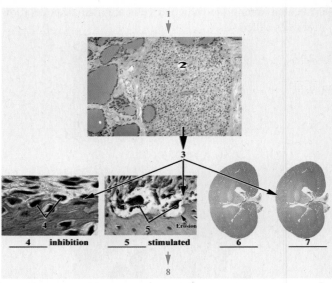

Figure 20.19

124 In reference to **Figure 20.19**, identify #1 - #8.
   1 _____ 5 _____
   2 _____ 6 _____
   3 _____ 7 _____
   4 _____ 8 _____

ADRENAL GLAND

125 Where is the adrenal gland located? _____
   _____

126 What are the two regions of the adrenal gland? _____
   _____

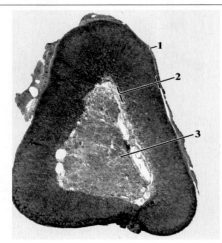

Figure 20.20

127 In reference to **Figure 20.20**, identify #1 - #3.
   1 _____ 3 _____
   2 _____

128 How many cellular regions are found in the adrenal cortex? _____

129 What forms the adrenal medulla? _____
   _____

130 What division of the autonomic nervous system innervates the chromaffin cells? _____
   _____

131 Where are the autonomic centers that control secretion of the chromaffin cells located? _____

132 What hormones are released by the adrenal medulla? _____

133 What are some of the adjustments produced by the "fight-or-flight response." _____

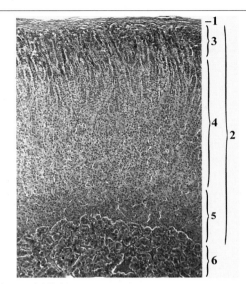

Figure 20.21

134 In reference to **Figure 20.21**, identify #1 - #6.
1 _____ 4 _____
2 _____ 5 _____
3 _____ 6 _____

135 What are the three regions of the adrenal cortex?
1 _____ 3 _____
2 _____

Zona glomerulosa

136 What is the principal hormone produced by the zona glomerulosa? _____

137 What mechanism regulates the release of aldosterone? ___

138 What promotes the release of aldosterone? _____

139 What does aldosterone target and effect? _____

Zona fasciculata

140 What is the primary glucocorticoid produced by the zona fasciculata? _____

141 What is the target and effect of cortisol? _____

Zona reticularis

142 What hormones are produced by the zona reticularis? ___

PANCREAS

143 Where is the pancreas located? _____

144 What forms the major portion of the pancreas? _____

145 What is the name of the endocrine units of the pancreas? _

146 What are the two pancreatic hormones that function in the regulation of blood sugar levels? _____

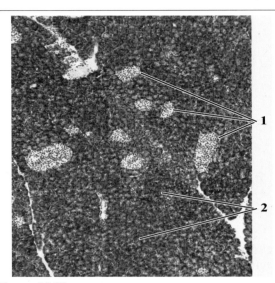

Figure 20.22

147 In reference to **Figure 20.22**, identify #1 - #2.
1 _____ 2 _____

Figure 20.23

148 In reference to **Figure 20.23**, identify #1 - #2.
1 _____ 2 _____

Glucagon

149 What hormone is produced by the alpha cells of the pancreatic islets? _____

304 Chapter 20 - Endocrine System

150 What mechanism regulates blood glucose levels? _____

151 What is the stimulus for increased secretion of glucagon? _____

152 What inhibits the secretion of glucagon? _____

153 What are three targets and associated effects of glucagon?
1 _____
2 _____
3 _____

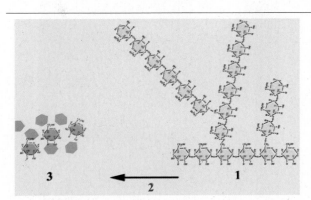

Figure 20.24

154 In reference to **Figure 20.24**, identify #1 - #3.
1 _____ 3 _____
2 _____

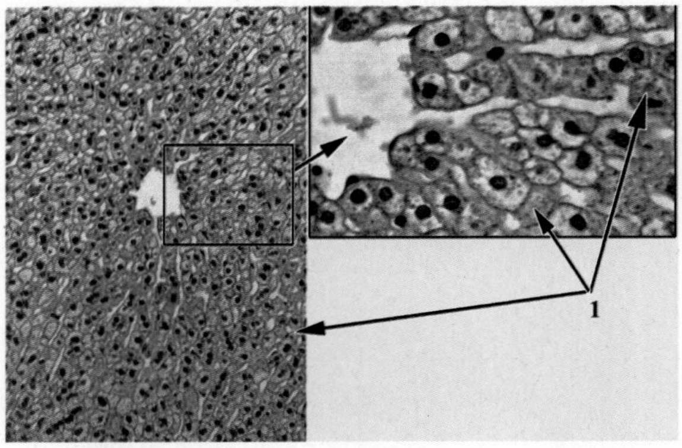

Figure 20.25

155 In reference to **Figure 20.25**, what is shown at #1? _____

Insulin

156 What hormone is produced by the beta cells of the pancreatic islets? _____

157 What mechanism regulates the release of insulin? _____

158 What is the stimulus for increased secretion of insulin? _____

159 What inhibits the secretion of insulin? _____

160 What are three targets and associated effects of glucagon?
1 _____
2 _____
3 _____

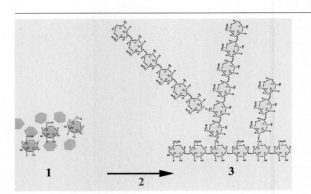

Figure 20.26

161 In reference to **Figure 20.26**, identify #1 - #3.
1 _____ 3 _____
2 _____

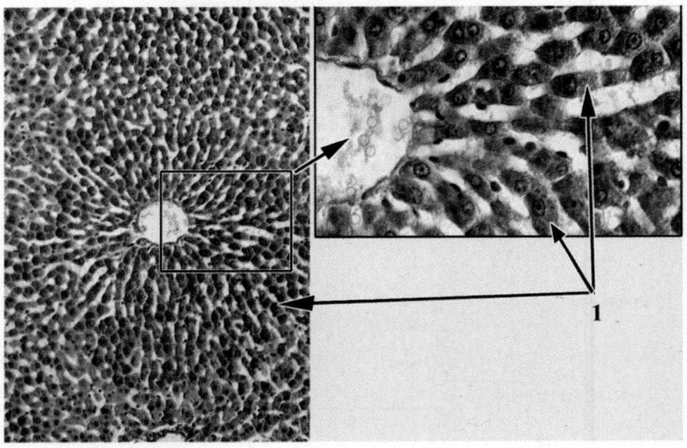

Figure 20.27

162 In reference to **Figure 20.27**, what is shown at #1? _____

ENDOCRINE TISSUES OF REPRODUCTIVE ORGANS

TESTES (SINGULAR, TESTIS)
Interstitial cells

163 Where are the interstitial cells of the testes located? _____

164 What is the most significant hormone produced by the interstitial cells of the testes? _____

165 What are three effects of androgens?
1 _____
2 _____
3 _____

166 Where are sustentacular cells located? _____

167 What hormone is produced by the sustentacular cells? _____

168 What are the effects of increased levels of inhibin? _____

169 What is the effect of reduced FSH? _____

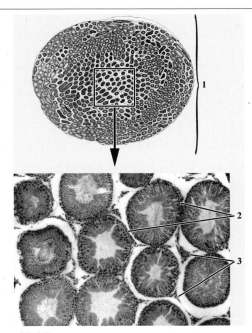

Figure 20.28
170 In reference to **Figure 20.28**, identify #1 - #3.
   1 _____   3 _____
   2 _____

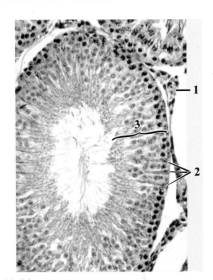

Figure 20.29
171 In reference to **Figure 20.29**, identify #1 - #3.
   1 _____   3 _____
   2 _____

Sustentacular cells (Sertoli cells)
172 What hormone targets the sustentacular cells (Sertoli cells)? _____

173 What is the effect of increased levels of inhibin? _____

174 What is the effect of decreased levels of inhibin? _____

Interstitial cells (Leydig cells)
175 What hormone targets the interstitial cells of the testes? _____

176 What is the effect of increased levels of testosterone? _____

177 What is the effect of decreased LH? _____

178 What is the effect of decreased levels of testosterone? _____

179 What is the effect of increased LH? _____

OVARIES
Follicles
180 What hormones are produced by the follicles of the ovary? _____

181 What hormone targets receptive follicles of the ovary? _____

182 What is the effect of FSH? _____

183 What effect does follicular development have on levels of estrogens and inhibin? _____

184 What are two functions of estrogens?
   1 _____
   2 _____

185 What does inhibin target? _____

186 What is the effect of increased levels of inhibin? _____

Corpus luteum
187 The corpus luteum develops from _____.

188 What hormones does the corpus luteum produce? _____

189 What are two effects of progesterone? _____

306 Chapter 20 - Endocrine System

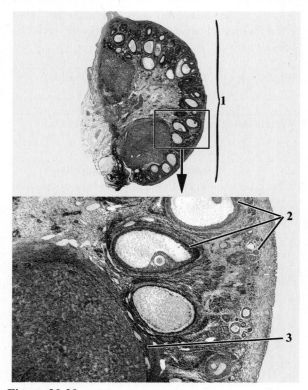

Figure 20.30
190 In reference to **Figure 20.30**, identify #1 - #3.
  1 _____   3 _____
  2 _____

THYMUS
191 Where is the thymus located? _____
_____
_____

192 What are the two regions of thymic lobules? _____
_____

193 What is the principal hormone produced by the thymus? __
_____

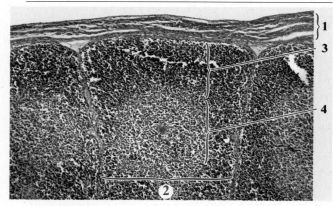

Figure 20.31
194 In reference to **Figure 20.31**, identify #1 - #3.
  1 _____   3 _____
  2 _____   4 _____

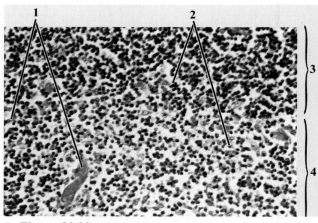

Figure 20.32
195 In reference to **Figure 20.32**, identify #1 - #3.
  1 _____   3 _____
  2 _____   4 _____

KIDNEYS
196 Where are the kidneys located? _____
_____
_____

197 What are the two hormones produced by the kidneys? ___
_____

198 What is the enzyme produced by the kidneys? _____
_____

199 What is the target and effect of erythropoietin? _____
_____

200 What is the stimulus for the release of erythropoietin? ___

201 What hormone targets the kidneys and promotes the secretion of calcitriol? _____

202 What is the target and effect of calcitriol? _____
_____

203 What is the stimulus for the secretion of the enzyme renin?
_____

204 What is the function of renin? _____
_____

205 What is the function of angiotensin II? _____
_____
_____

Blood - Worksheets

1. What are four functions of blood?
   1. _____
   2. _____
   3. _____
   4. _____

2. What are the two major constituents of blood?
   1. _____ 2. _____

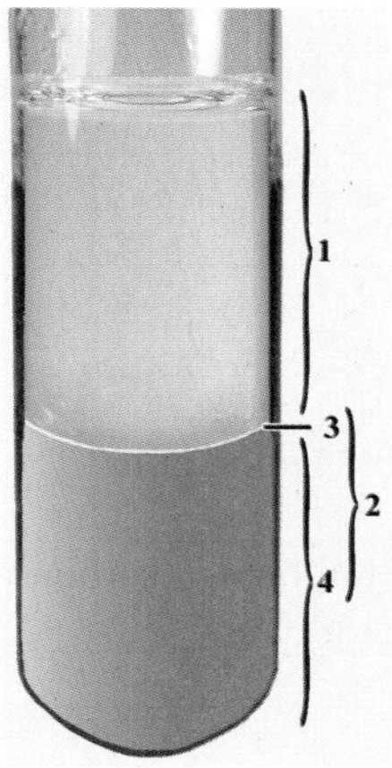

Figure 21.1

3. In reference to **Figure 21.1**, identify and give the percentages for #1 - #4.
   1. _____ 3. _____
   2. _____ 4. _____

4. What is the difference between plasma and serum? _____

5. What are the three formed elements of blood?
   1. _____ 3. _____
   2. _____

Percentage by Volume of Red Blood Cells

6. What is the hematocrit? _____

7. What are the hematocrit values for males and females? _____

8. What is anemia? _____

9. What are three types of anemia that produce a decrease in the number of RBCs? _____

10. What is polycythemia? _____

Locations and Functions of Blood

11. Where is blood located? _____

12. From the heart, what is the conduction pathway for the vascular system? _____

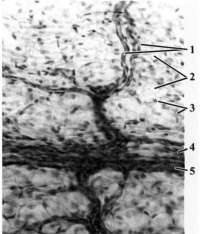

Figure 21.2

13. In reference to **Figure 21.2**, identify #1 - #5.
    1. _____ 4. _____
    2. _____ 5. _____
    3. _____

14. What is the pH range of blood? _____

BLOOD COMPOSITION AND FUNCTION
PLASMA

15. What percentage of plasma is water? _____ What percentage of the plasma is solute? _____

16. What are some of the functions of plasma proteins? _____

17. What are the most abundant plasma proteins? _____

18. Where are albumins produced? _____

19. What is a function of plasma albumins? _____

20. What are the second most abundant plasma proteins? _____

21. Where are globulins produced? _____

22. What is a function of globulins produced by the liver? ___

23. What is the function of globulins produced by the B lymphocytes? _____

24. Where is fibrinogen produced? _____

25. What is the function of fibrinogen? _____

26. What is interstitial fluid? _____

308 Chapter 21 - Blood

27 What is a major difference between interstitial fluid and plasma? _____

FORMED ELEMENTS

28 What are the formed elements? _____

29 Where are the formed elements produced? _____

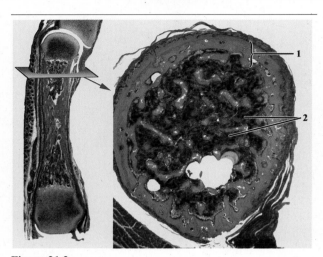

Figure 21.3

30 In reference to **Figure 21.3**, identify #1 - #2.
1 _____ 2 _____

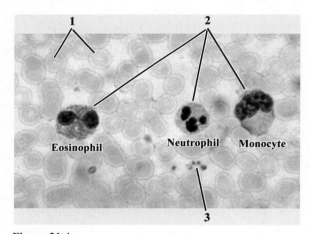

Figure 21.4

31 In reference to **Figure 21.4**, identify #1 - #3.
1 _____ 3 _____
2 _____

Erythrocytes (Red Blood Cells)

32 Where are erythrocytes produced? _____

33 When erythrocytes enter blood capillaries in the marrow, what happens to their nuclei? _____

34 What do erythrocytes in circulation mostly contain? _____

Hemoglobin

35 What is the function of hemoglobin? _____

36 What is the normal range of hemoglobin for women? _____

37 What is the normal range of hemoglobin for men? _____

38 How many polypeptide chains are organized into a molecule of hemoglobin? _____

39 What is the heme group? _____

40 What is oxyhemoglobin, HbO_2? _____

41 What percent of the body's oxygen is transported as oxyhemoglobin, HbO_2? _____

42 What is carbaminohemoglobin? _____

43 What percent of the body's carbon dioxide is transported as carbaminohemoglobin? _____

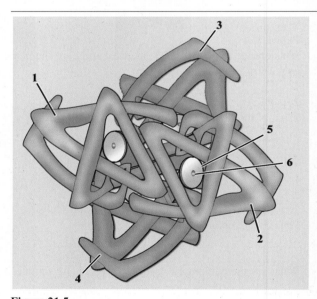

Figure 21.5

44 In reference to **Figure 21.5**, identify #1 - #6.
1 _____ 4 _____
2 _____ 5 _____
3 _____ 6 _____

45 In reference to **Figure 21.5**, what is the function of #6? _____

46 What are some of the causes of iron deficiency anemia? _____

47 What is sickle cell anemia? _____

48 The abnormal sickle shape of the RBCs results in _____

Chapter 21 - Blood **309**

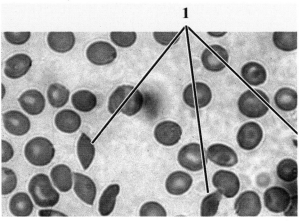

Figure 21.6

49 In reference to **Figure 21.6**, identify #1.
1 _____

Leukocytes (White Blood Cells)

50 What are granulocytes? _____
_____

51 What are the three granulocytes? _____
_____

52 What are agranulocytes? _____
_____

53 What are the two agranulocytes? _____
_____

Granulocytes

54 What is the percent range of neutrophils? _____
_____

55 Describe the general shape of the nuclei of neutrophils.
_____
_____

56 What are two functions of neutrophils? _____
_____

57 What is the percent range of eosinophils? _____
_____

58 Describe the general shape of the nuclei of eosinophils.
_____
_____

59 Describe the staining of the cytoplasmic granules of the eosinophil. _____
_____

60 What are two functions of eosinophils? _____
_____

61 What is the percent range of basophils? _____
_____

62 Describe the general shape of the nuclei of a basophils.
_____
_____

63 Describe the staining of the cytoplasmic granules of the basophil. _____
_____

64 What is a function of basophils? _____
_____

Agranulocytes

65 What is the percent range of lymphocytes? _____
_____

66 Describe the general shape of the nuclei of lymphocytes.
_____
_____

67 What is a function of lymphocytes? _____
_____

68 What is the percent range of monocytes? _____
_____

69 Describe the general shape of the nuclei of a monocytes.
_____
_____

70 What is a function of monocytes? _____
_____

Platelets

71 What produces platelets? _____

72 What are two functions of platelets? _____
_____

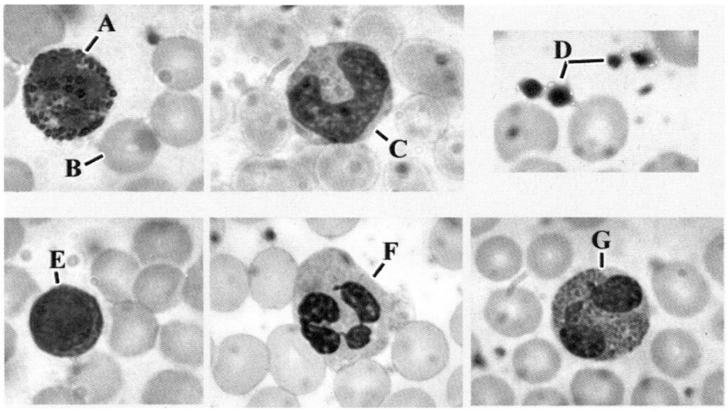

Figure 21.7

73 In reference to **Figure 21.7**, identify A - G.
A _____ E _____
B _____ F _____
C _____ G _____
D _____

DIFFERENTIAL COUNT

74 What is a differential count? _____
_____

75 What are the differential percentages for:
neutrophils - _____   eosinophils- _____
lymphocytes- _____   basophils- _____
monocytes- _____

76 Complete the following, **Figure 21.8**, with the results of your differential count.

Leukocyte	Number of Cells Counted	Percentage
Neutrophil		
Eosinophil		
Basophil		
Lymphocyte		
Monocyte		

Figure 21.8

310 Chapter 21 - Blood

77 Define the following terms:
neutropenia _____
neutrophilia _____
lymphopenia _____
lymphocytosis _____
monocytopenia _____
monocytosis _____
eosinopenia _____
eosinophilia _____
basopenia _____
basocytosis _____

78 Which, if any, of the above conditions were indicated on your differential count? _____
_____
_____

ABO BLOOD GROUPS

79 Immunologically, what are antigens? _____
_____
_____

80 Immunologically, what is the function of antibodies? _____
_____
_____

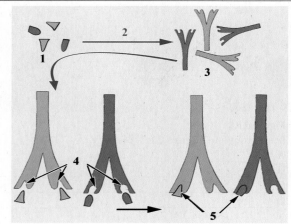

Figure 21.9

81 In reference to **Figure 21.9**, identify #1 - #5.
1 _____ 4 _____
2 _____ 5 _____
3 _____

82 What are the three most significant red blood cell surface antigens considered in blood transfusions? _____
_____
_____

83 Why is ABO blood typing performed? _____
_____
_____

84 Why is Rh blood typing performed? _____
_____

85 Type A blood has RBC surface antigen _____.
86 Type A blood has anti- _____ antibody.
87 Type B blood has RBC surface antigen _____.
88 Type B blood has anti- _____ antibody
89 Type AB blood has RBC surface antigens _____.
90 Type AB blood has _____ antibodies.
91 Type O blood has _____ antigens.
92 Type O blood has _____ antibodies.

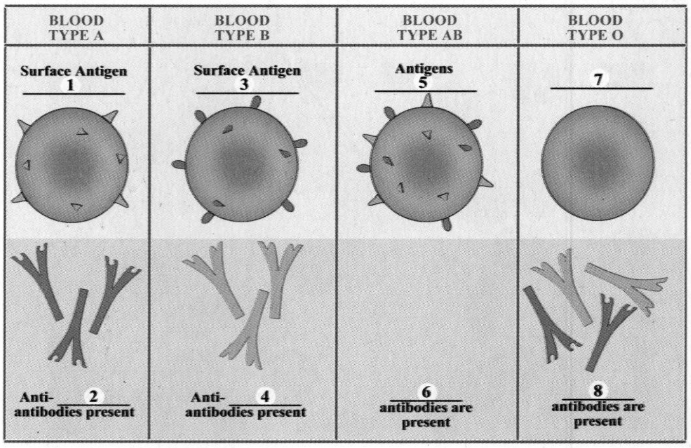

Figure 21.10

93 In reference to **Figure 21.10**, identify #1 - #8.
1 _____ 5 _____
2 _____ 6 _____
3 _____ 7 _____
4 _____ 8 _____

94 When do incompatible ABO transfusion reactions result in agglutination? _____
_____

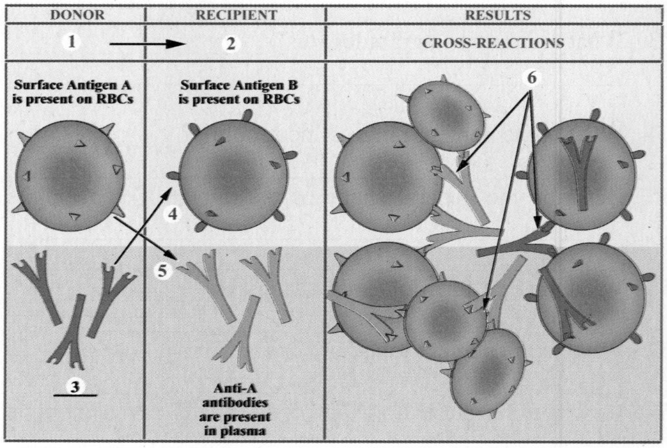

Figure 21.11

95 In reference to **Figure 21.11**, identify #1 - #6.
1 Blood type _____
2 Blood type _____
3 _____
4 Reaction between donated _____ and recipient's RBCs surface _____
5 Reaction between donated surface _____ and recipients _____
6 _____
In reference to **Figure 21.11**, which results in the most significant agglutination, #4 or #5? _____

96 Two ways to prevent agglutination are: (1) the donor's blood must have _____
_____
or (2) the donor's blood must lack _____
_____
_____

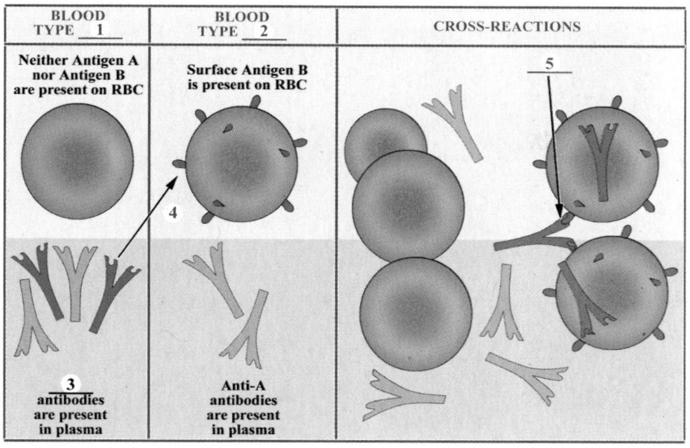

Figure 21.12

97 In reference to **Figure 21.12**, identify #1 - #5.
   1 _____
   2 _____
   3 _____
   4 Reaction between donated _____ and recipient's RBCs surface _____
   5 _____

98 In reference to **Figure 21.12**, is the agglutination considered significant or insignificant? _____

99 Complete the following table by indicating with a check the recipients that can receive from the donors.

	DONOR			
RECIPIENT	Type O	Type A	Type B	Type AB
Type AB				
Type B				
Type A				
Type O				

Figure 21.13

100 What ABO blood type is considered as the universal donor? _____

101 What ABO blood type is considered as the universal recipient? _____

Rh BLOOD TYPING

102 Why is Rh blood typing performed? _____

103 A person that has antigen D is described as _____

104 When are Rh (D) antibodies present in the plasma? _____

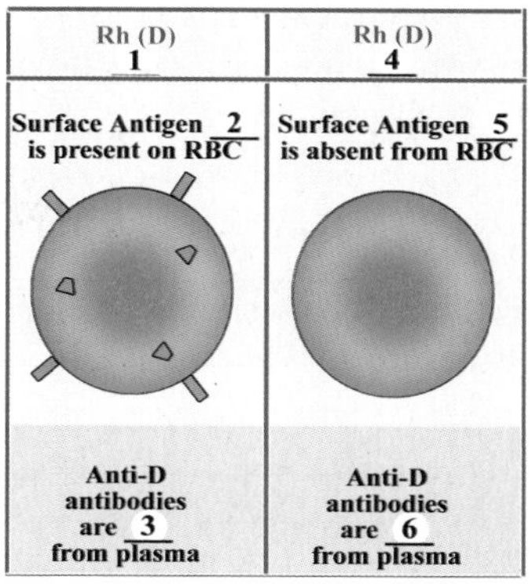

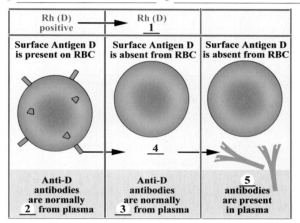

Figure 21.14

105 In reference to **Figure 21.14**, identify #1 - #6.
   1 _____ 4 _____
   2 _____ 5 _____
   3 _____ 6 _____

Figure 21.15

106 In reference to **Figure 21.15**, identify #1 - #5.
   1 _____ 4 _____
   2 _____ 5 _____
   3 _____

107 In reference to **Figure 21.15**, does this illustration show first time exposure to antigen D, or subsequent exposure?

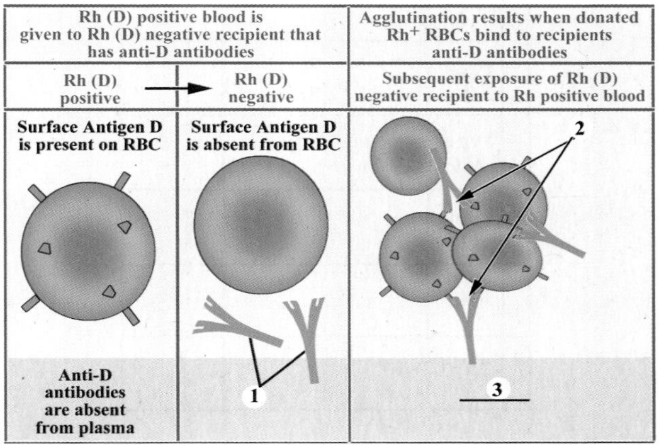

Figure 21.16

108 In reference to **Figure 21.16**, identify #1 - #3.
   1 _____ 2 _____
   3 _____

ERYTHROBLASTOSIS FETALIS
(HEMOLYTIC DISEASE OF THE NEWBORN)

109 What is erythroblastosis fetalis? _____

110 Why does erythroblastosis fetalis develop as a result of a subsequent exposure to Rh positive blood, not the first exposure? _____

111 Considering erythroblastosis fetalis the mother is Rh _____ and the fetus is Rh _____.

112 What is the reactive component of RhoGam? _____

113 What is the function of RhoGam? _____

BLOOD TYPING

114 What is the reacting substance found in the bottle of anti-A sera? _____

115 Agglutination resulting from anti-A sera means that _____ are present.

116 What is the reacting substance found in the bottle of anti-B sera? _____

117 Agglutination resulting from anti-B sera means that _____ are present.

118 If two samples of blood show no agglutination when mixed with anti-A sera and anti-B sera, respectively, then the blood type is _____.

119 Three samples of blood show the following: agglutination with anti-A sera, no agglutination with anti-B sera, and agglutination with anti-D sera. The blood type is _____.

120 Complete **Figure 21.17** by placing a check in the rows for each blood type's matching antigens.

		Results of Agglutination		
		Antigen A	Antigen B	Antigen D
Blood Type	Type A			
	Type B			
	Type AB			
	Type O			
	Rh⁺			
	Rh⁻			

Figure 21.17

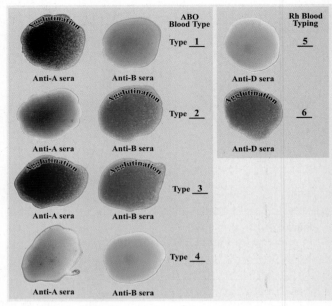

Figure 21.18

121 In reference to **Figure 21.18**, identify #1 - #6.
1 _____ 4 _____
2 _____ 5 _____
3 _____ 6 _____

122 The results from typing of blood are:
Anti-A sera: Is agglutination positive or negative?___
Anti-B sera: Is agglutination positive or negative?___
Anti-D sera: Is agglutination positive or negative?___
ABO blood type _____ Rh _____

Lymphatic System - Worksheets

1. What are two components of the lymphatic system? _____

2. What is a function of the lymphatic vessels? _____

3. What is a function of the lymphatic tissues and organs? _____

LYMPHATIC VESSELS

4. Where do the lymphatic vessels begin? _____

5. Where do the lymphatic vessels ultimately empty? _____

6. What fluid forms lymph? _____

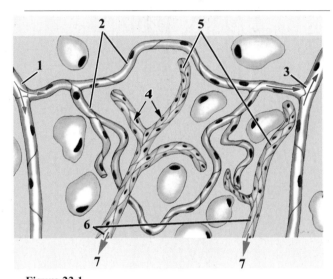

Figure 22.1

7. In reference to **Figure 22.1**, identify #1 - #7.
   1 _____ 5 _____
   2 _____ 6 _____
   3 _____ 7 _____
   4 _____

8. What is the name of the lymph capillaries that are located in the mucosa of the small intestine? _____

9. What is chyle? _____

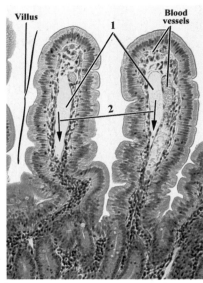

Figure 22.2

10. In reference to **Figure 22.2**, identify #1 - #2.
    1 _____ 2 _____

11. From inner to outer, what are the three layers of the wall of a collecting vessel? _____

12. What is the function of the skeletal muscle pumping mechanism? _____

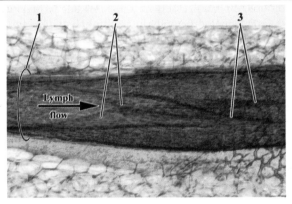

Figure 22.3

13. In reference to **Figure 22.3**, identify #1 - #3.
    1 _____ 3 _____
    2 _____

14. When does the skeletal muscle pumping mechanism operate? _____

15. When does the distal one-way valve open? _____

16. When does the proximal one-way valve open? _____

Chapter 22 - Lymphatic System

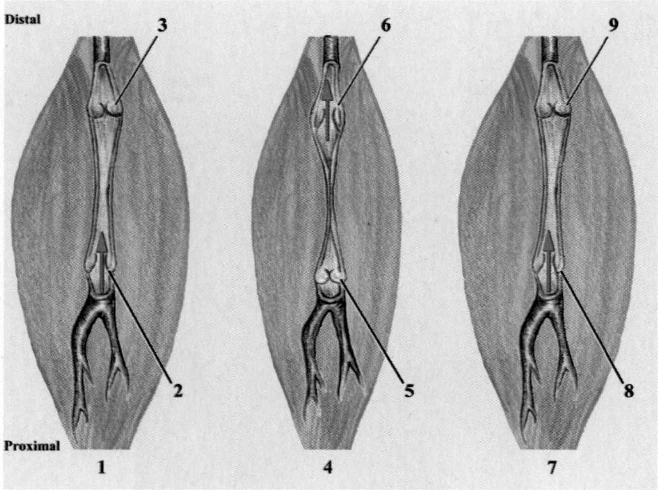

Figure 22.4

17 In reference to **Figure 22.4**, identify #1 - #9.
   1 _____   6 _____
   2 _____   7 _____
   3 _____   8 _____
   4 _____   9 _____
   5 _____

PATHWAY OF LYMPHATIC VESSELS

18 The _____ is the terminus of the lymphatic vessels from the right side of the body superior to the diaphragm and the right upper extremity.

19 The _____ is the terminus of the lymphatic vessels from the left side of the body and the right side of the body inferior to the diaphragm.

20 Both the right lymphatic duct and the thoracic duct empty into _____ circulation.

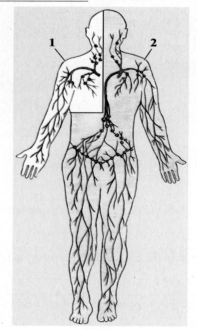

Figure 22.5

21 In reference to **Figure 22.5**, identify #1 - #2.
   1 _____
   2 _____

22 What are the three trunks that enter the right lymphatic duct? _____

23 What vessels enter the thoracic duct? _____

24 What is the cisterna chyli? _____

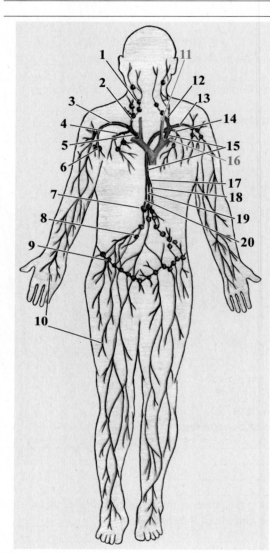

Figure 22.6

25 In reference to **Figure 22.6**, identify #1 - #20.
   1 _____   11 _____
   2 _____   12 _____
   3 _____   13 _____
   4 _____   14 _____
   5 _____   15 _____
   6 _____   16 _____
   7 _____   17 _____
   8 _____   18 _____
   9 _____   19 _____
   10 _____   20 _____

LYMPHATIC TISSUES AND ORGANS

26 Lymphatic mostly consists of a framework of _____ connective tissue that houses _____ and _____.

27 Lymphatic tissue is mostly organized into spherical structures called _____.

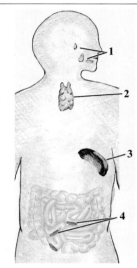

Figure 22.7

28 In reference to **Figure 22.7**, identify #1 - #4.
   1 _____ 3 _____
   2 _____ 4 _____

TONSILS

29 The tonsils include the _____.

30 The paired palatine tonsils are located _____.

31 The lingual tonsils are located _____.

32 The pharyngeal tonsil (_____) is located _____.

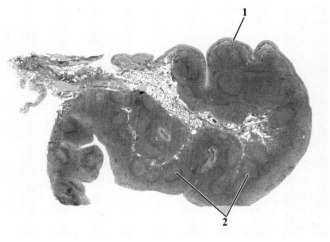

Figure 22.8

33 In reference to **Figure 22.8**, identify #1 - #2.
   1 _____ 2 _____

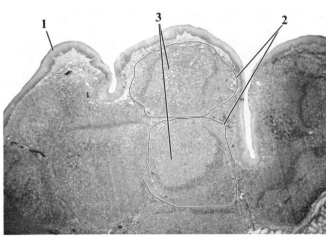

Figure 22.9

34 In reference to **Figure 22.9**, identify #1 - #3.
   1 _____ 3 _____
   2 _____ 4 _____

35 What are the dominate cells of the tonsil's germinal center? _____

36 What surrounds the germinal center? _____

316 Chapter 22 - Lymphatic System

LYMPH NODE

37. Lymph nodes are located along the _____ vessels. What are five regions that have abundant lymph nodes?
   1 _____   4 _____
   2 _____   5 _____
   3 _____

38. What is the function of lymph nodes? _____

39. Vessels that enter the lymph nodes are called _____ vessels, and vessels that exit the lymph nodes are called the _____ vessels.

40. Most of the lymphatic nodules of the lymph node are located in the _____.

41. Extending from the lymphatic nodules are _____ cords, which are surrounded by the _____ sinuses.

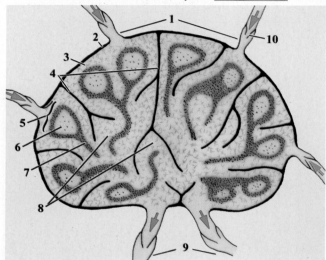

Figure 22.10

42. In reference to **Figure 22.10**, identify #1 - #10.
   1 _____   6 _____
   2 _____   7 _____
   3 _____   8 _____
   4 _____   9 _____
   5 _____   10 _____

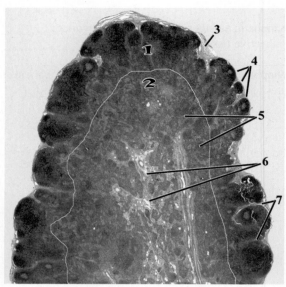

Figure 22.11

43. In reference to **Figure 22.11**, identify #1 - #7.
   1 _____   5 _____
   2 _____   6 _____
   3 _____   7 _____
   4 _____

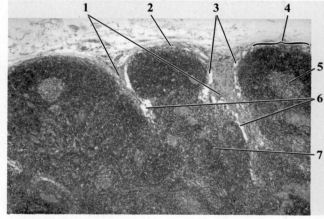

Figure 22.12

44. In reference to **Figure 22.12**, identify #1 - #7.
   1 _____   5 _____
   2 _____   6 _____
   3 _____   7 _____
   4 _____

Spleen

45 Where is the spleen located? _____
_____

46 What are three functions of the spleen? _____
_____
_____

47 Internally, the spleen is divided into _____ nodules surrounded by _____.

48 Red pulp consists of _____
_____

49 The white pulp of the lymphatic nodules consists of _____
_____

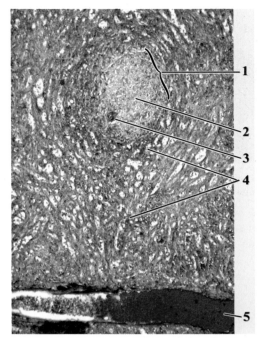

Figure 22.14

51 In reference to **Figure 22.14**, identify #1 - #5.
1 _____   4 _____
2 _____   5 _____
3 _____

Thymus

52 Where is the thymus located? _____
_____

53 The thymus is divided by connective tissue into small units called _____, each consisting of an outer _____ and an inner _____.

54 The cortex of the thymus consists mostly of dividing _____.

55 The major hormone produced by the thymus is _____, which functions to _____
_____.

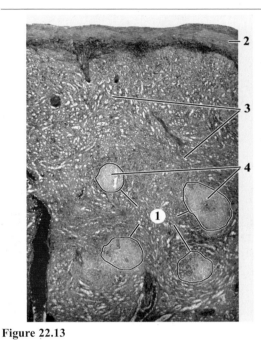

Figure 22.13

50 In reference to **Figure 22.13**, identify #1 - #4.
1 _____   3 _____
2 _____   4 _____

318 Chapter 22 - Lymphatic System

Name _____
Class _____

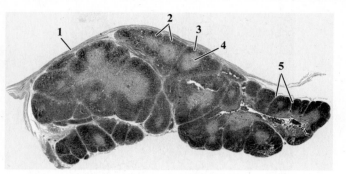

Figure 22.15

56 In reference to **Figure 22.15**, identify #1 - #5.
   1 _____ 4 _____
   2 _____ 5 _____
   3 _____

PEYER'S PATCHES

59 What are Peyer's patches? _____
_____

60 Where are Peyer's patches located? _____
_____

61 What is the function of Peyer's patches? _____
_____
_____

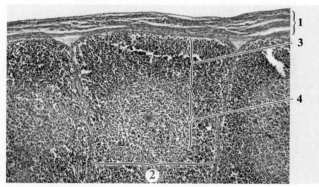

Figure 22.16

57 In reference to **Figure 22.16**, identify #1 - #4.
   1 _____ 3 _____
   2 _____ 4 _____

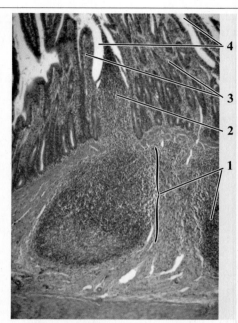

Figure 22.18

62 In reference to **Figure 22.8**, identify #1 - #4.
   1 _____ 3 _____
   2 _____ 4 _____

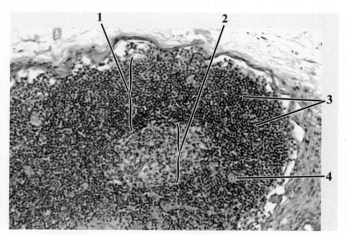

Figure 22.17

58 In reference to **Figure 22.17**, identify #1 - #4.
   1 _____ 3 _____
   2 _____ 4 _____

The Heart - Worksheets

1. What is the name of the central region of the thorax? _____

2. About how much of the heart is to the left of the mid-sternal line? _____

3. Where is the location of the apex of the heart? _____

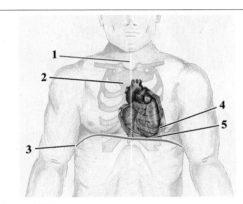

Figure 23.1

4. In reference to **Figure 23.1**, identify #1 - #5.
   1 _____ 4 _____
   2 _____ 5 _____
   3 _____

COVERINGS OF THE HEART

5. What is the name of the covering of the heart? _____

6. What is the function of the fibrous layer of the pericardium? _____

7. What is the name of the inner layer of the pericardium? _____

8. What is the name of the serous membrane on the surface of the heart? _____

9. What forms the pericardial cavity? _____

10. What is the function of the serous membrane? _____

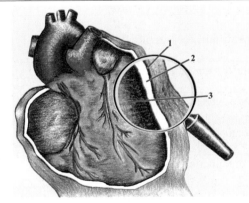

Figure 23.2

11. In reference to **Figure 23.2**, identify #1 - #3.
    1 _____ 3 _____
    2 _____

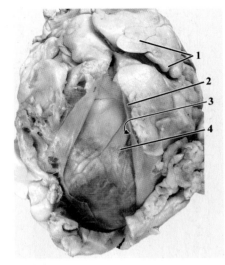

Figure 23.3

12. In reference to **Figure 23.3**, identify #1 - #4.
    1 _____ 3 _____
    2 _____ 4 _____

GROSS ANATOMY OF THE HEART

13. What is the name of the upper chambers of the heart? _____

14. What is the name of the lower chambers of the heart? _____

15. What divides the right and the left chambers of the heart? _____

16. What is the function of the right side of the heart? _____

17. What is the function of the left side of the heart? _____

18. What is the name of the muscle of the heart? _____

19. What is an auricle of the heart? _____

20. Which side of the heart is thicker? _____ Why? _____

Valves of the Heart

21. Where are the atrioventricular valves located? _____

22. What are the names of the right and the left atrioventricular valves? _____

23. What is the function of the atrioventricular valves? _____

24. What is the name of the exiting vessel from the right ventricle? _____

25. What is the name of the exiting vessel from the left ventricle? _____

26. What is the name of the valve located at the base of each exiting vessel? _____

320 Chapter 23 - The Heart

Name _____
Class _____

27 What is the function of the aortic and pulmonary valves?
_____
_____

1 _____ 8 _____
2 _____ 9 _____
3 _____ 10 _____
4 _____ 11 _____
5 _____ 12 _____
6 _____ 13 _____
7 _____ 14 _____

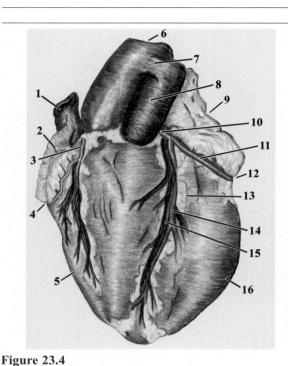

Figure 23.4

28 In reference to **Figure 23.4**, identify #1 - #16.

1 _____ 9 _____
2 _____ 10 _____
3 _____ 11 _____
4 _____ 12 _____
5 _____ 13 _____
6 _____ 14 _____
7 _____ 15 _____
8 _____ 16 _____

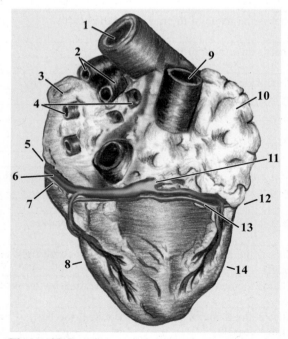

Figure 23.5

29 In reference to **Figure 23.5**, identify #1 - #14.

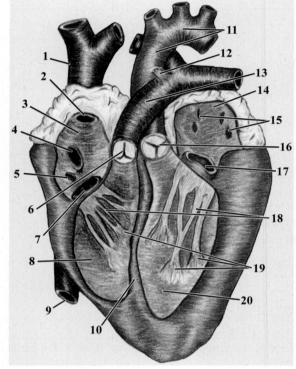

Figure 23.6

30 In reference to **Figure 23.6**, identify #1 - #20.

1 _____ 11 _____
2 _____ 12 _____
3 _____ 13 _____
4 _____ 14 _____
5 _____ 15 _____
6 _____ 16 _____
7 _____ 17 _____
8 _____ 18 _____
9 _____ 19 _____
10 _____ 20 _____

31 The right atrium receives blood from _____

32 The superior vena cava returns blood from the _____

33 The inferior vena cava returns blood from the _____

34 The coronary sinus returns blood from the _____

35 What is the function of the tricuspid valve? _____

36 What is the function of the right ventricle? _____
_____

37 What is the function of the pulmonary valve? _____
_____

38 What is the function of the pulmonary trunk? _____
_____
39 The left atrium receives blood from _____
_____
40 The pulmonary veins carry _____
_____
41 What is the function of the mitral (bicuspid) valve? _____
_____
42 What is the function of the left ventricle? _____
_____
_____
43 What is the function of the aortic (semilunar) valve? _____
_____
44 What vessel exits the heart to feed the systemic circuit?
_____
45 What is the ligamentum arteriosum? _____
_____
46 What are the chordae tendineae? _____
_____
_____
47 What are the papillary muscles? _____
_____

Route of Blood Flow through the Heart
Right Side of the Heart
48 Blood low in oxygen enters the right _____ from the
_____ and _____ cavae and the
_____.
49 Blood leaves the right atrium by way of the
_____ and enters the
_____.
50 Blood leaves the _____
by way of the _____ valve and enters
the _____ trunk.
51 The pulmonary truck directs blood into the _____
_____ of the lungs for gas exchange.

52 In reference to **Figure 23.7**, identify #1 - #10.
1 _____ 6 _____
2 _____ 7 _____
3 _____ 8 _____
4 _____ 9 _____
5 _____ 10 _____

Left Side of the Heart
53 Oxygen rich blood enters the left _____ from the
_____ veins.
54 Blood leaves the left atrium by way of the
_____ valve and enters the left
_____.
55 Blood leaves the left ventricle by way of the _____
_____ and enters the _____.
56 The aorta directs blood into the _____
of the body.

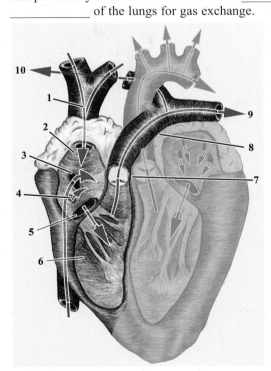

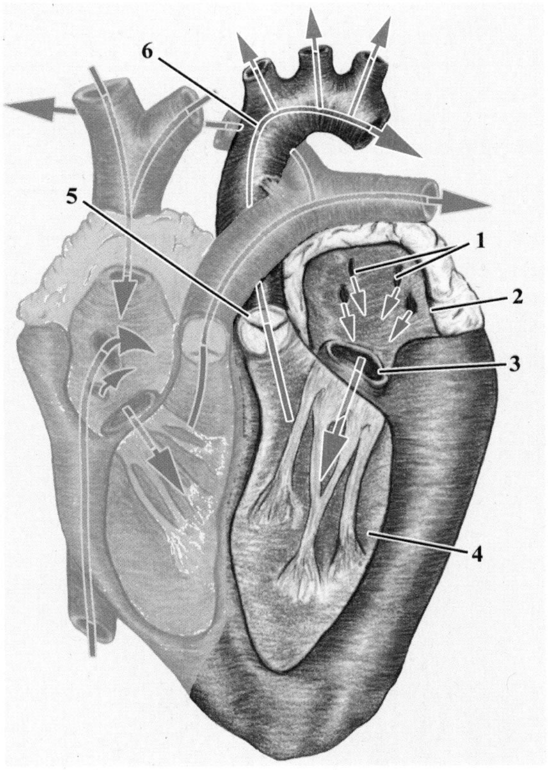

Figure 23.8
57 In reference to **Figure 23.8**, identify #1 - #6.
1 _____ 4 _____
2 _____ 5 _____
3 _____ 6 _____

Coronary Circulation
Coronary Arteries
58 Oxygen rich blood leaves the aorta and enters the two
coronary arteries, the _____ arteries.
59 What are the aortic sinuses? _____
_____
60 Which sinuses give rise to the coronary arteries? _____
_____

Figure 23.7

322 Chapter 23 - The Heart

Name _____
Class _____

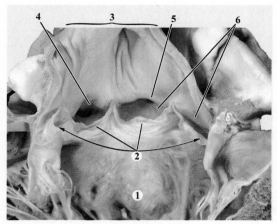

Figure 23.9

61 In reference to **Figure 23.9**, identify #1 - #6.
    1 _____ 4 _____
    2 _____ 5 _____
    3 _____ 6 _____

Right Coronary Artery

62 The right coronary artery follows the right _____ sulcus to the posterior _____ sulcus.

63 At the posterior interventricular sulcus, the right coronary artery continues as the _____.

64 At the margin of the right ventricle, a branch called the right _____ artery is formed from the right coronary artery.

Left coronary artery

65 The left coronary artery branches into the _____ artery and the _____.

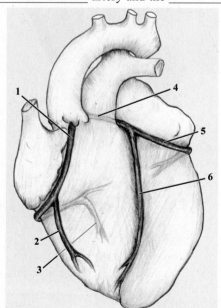

Figure 23.10

66 In reference to **Figure 23.10**, identify #1 - #6.
    1 _____ 4 _____
    2 _____ 5 _____
    3 _____ 6 _____

Coronary Veins

67 The three major veins that enter the coronary sinus are the: _____
    _____

68 What is the coronary sinus? _____
    _____

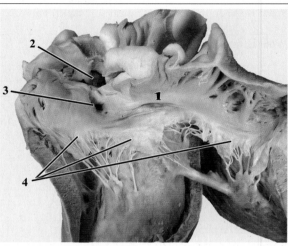

Figure 23.11

69 In reference to **Figure 23.11**, identify #1 - #4.
    1 _____ 3 _____
    2 _____ 4 _____

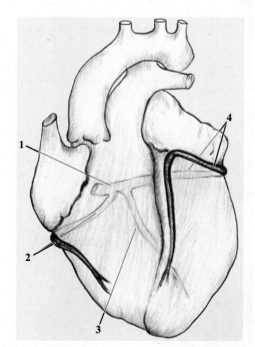

Figure 23.12

70 In reference to **Figure 23.12**, identify #1 - #4.
    1 _____ 3 _____
    2 _____ 4 _____

Name _____
Class _____

Chapter 23 - The Heart 323

SHEEP HEART DISSECTION

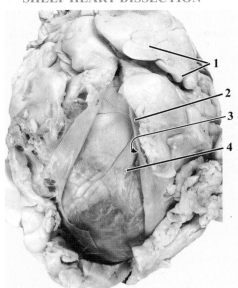

Figure 23.13
71 In reference to **Figure 23.13**, identify #1 - #4.
1 _____ 3 _____
2 _____ 4 _____

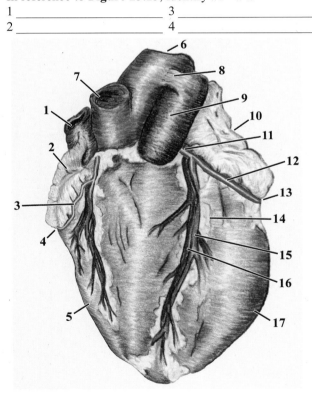

Figure 23.14
72 In reference to **Figure 23.14**, identify #1 - #17.
1 _____ 10 _____
2 _____ 11 _____
3 _____ 12 _____
4 _____ 13 _____
5 _____ 14 _____
6 _____ 15 _____
7 _____ 16 _____
8 _____ 17 _____
9 _____

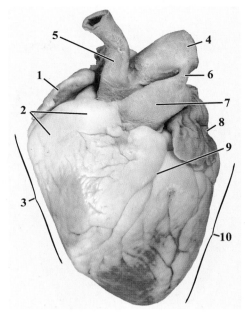

Figure 23.15
73 In reference to **Figure 23.15**, identify #1 - #10.
1 _____ 6 _____
2 _____ 7 _____
3 _____ 8 _____
4 _____ 9 _____
5 _____ 10 _____

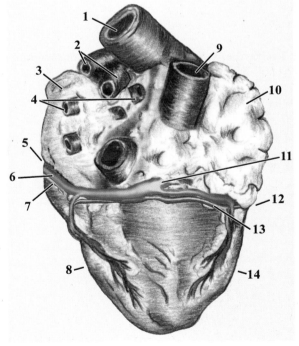

Figure 23.16
74 In reference to **Figure 23.16**, identify #1 - #14.
1 _____ 8 _____
2 _____ 9 _____
3 _____ 10 _____
4 _____ 11 _____
5 _____ 12 _____
6 _____ 13 _____
7 _____ 14 _____

324 Chapter 23 - The Heart

Name _____
Class _____

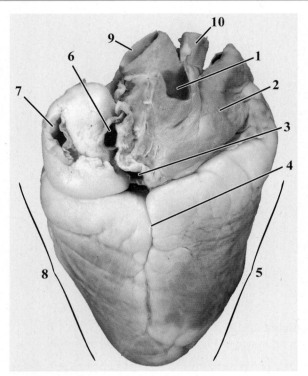

Figure 23.17

76 In reference to **Figure 23.17**, identify #1 - #10.
   1 _____   6 _____
   2 _____   7 _____
   3 _____   8 _____
   4 _____   9 _____
   5 _____ 10 _____

Pericardial sac, Epicardium, and Endocardium

76 What is the pericardium? _____

77 What is the inner layer of the pericardium? _____

78 What is the epicardium? _____

79 What is the myocardium? _____

80 What is the endocardium? _____

81 Internally, what does the interventricular sulcus follow? ___

82 Which side of the heart is more muscular? _____
    Why? _____

BLOOD VESSELS ASSOCIATED WITH THE HEART

83 Do arteries or veins have thicker walls? _____ Why? _____

84 What is the function of the superior and inferior vena cavae? _____

85 Where is the pulmonary trunk located? _____

86 What is the function of the pulmonary trunk? _____

87 Where are the pulmonary veins located? _____

88 What is the function of the pulmonary veins? _____

89 Where is the aorta located? _____

90 What is the function of the aorta? _____

FRONTAL SECTION

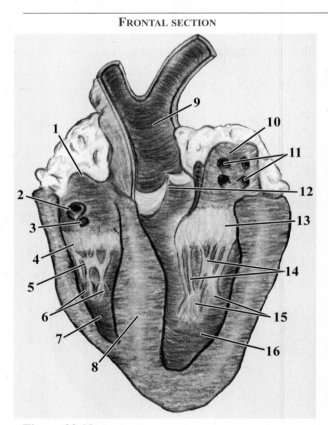

Figure 23.18

91 In reference to **Figure 23.18**, identify #1 - #16.
   1 _____   9 _____
   2 _____ 10 _____
   3 _____ 11 _____
   4 _____ 12 _____
   5 _____ 13 _____
   6 _____ 14 _____
   7 _____ 15 _____
   8 _____ 16 _____

Name _____
Class _____

Chapter 23 - The Heart   325

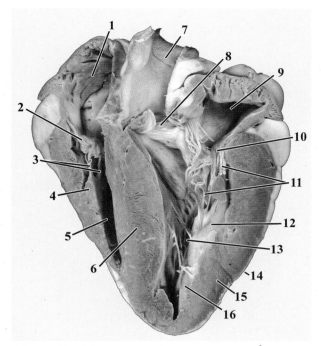

Figure 23.19
92 In reference to **Figure 23.19**, identify #1 - #16.
    1 _____ 9 _____
    2 _____ 10 _____
    3 _____ 11 _____
    4 _____ 12 _____
    5 _____ 13 _____
    6 _____ 14 _____
    7 _____ 15 _____
    8 _____ 16 _____
    DISSECTION OF RIGHT SIDE OF HEART

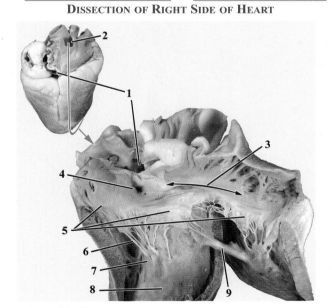

Figure 23.20
93 In reference to **Figure 23.20**, identify #1 - #9.
    1 _____ 6 _____
    2 _____ 7 _____
    3 _____ 8 _____
    4 _____ 9 _____
    5 _____

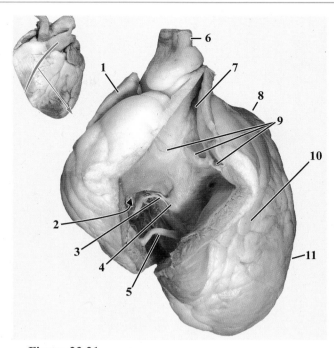

Figure 23.21
94 In reference to **Figure 23.21**, identify #1 - #11.
    1 _____ 7 _____
    2 _____ 8 _____
    3 _____ 9 _____
    4 _____ 10 _____
    5 _____ 11 _____
    6 _____

95 What is attached to the free-edges of the cusps of the tricuspid valve? _____

96 The chordae tendineae attached to the myocardium at modified sites called _____ muscles.

97 When does the tricuspid valve close? _____

98 What prevents the cusps from moving into the atrium? _____

99 What is the function of the moderator band? _____

100 Where is the pulmonary (semilunar) valve located? _____

101 When does the pulmonary (semilunar) valve open? _____

102 What is the function of the pulmonary valve? _____

103 When does the pulmonary valve close? _____

104 Where is blood in the pulmonary trunk directed? _____

326 Chapter 23 - The Heart

DISSECTION OF THE LEFT SIDE OF HEART

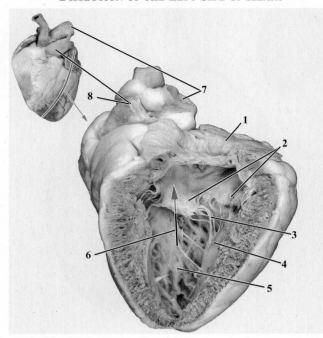

Figure 23.22

105 In reference to **Figure 23.22**, identify #1 - #8.
1 _____   5 _____
2 _____   6 _____
3 _____   7 _____
4 _____   8 _____

106 What is attached to the free-edges of the cusps of the bicuspid valve? _____

107 The chordae tendineae attached to the myocardium at modified sites called _____ muscles.

108 When does the mitral valve close? _____

109 What is the function of the chordae tendineae and papillary muscles? _____

110 Where is the aortic (semilunar) valve located? _____

111 When does the aortic (semilunar) valve open? _____

112 What is the function of the aortic valve? _____

113 When does the aortic valve close? _____

114 Where is blood in the aorta trunk directed? _____

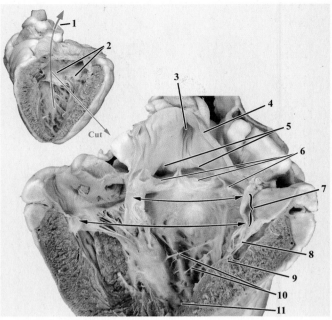

Figure 23.23

115 In reference to **Figure 23.23**, identify #1 - #11.
1 _____   7 _____
2 _____   8 _____
3 _____   9 _____
4 _____   10 _____
5 _____   11 _____
6 _____

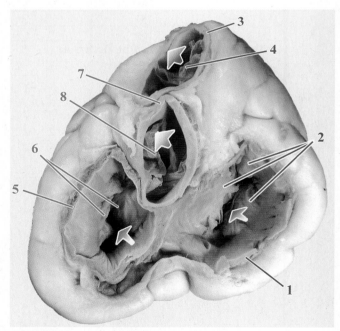

Figure 23.24

116 In reference to **Figure 23.24**, identify #1 - #8.
1 _____   5 _____
2 _____   6 _____
3 _____   7 _____
4 _____   8 _____

117 What causes the opening and closing of the valves of the heart? _____

118 The free-edges of the atrioventricular valves point into the _____.

119 When are the atrioventricular valves open? _____

120 When are the atrioventricular valves closed? _____

121 As one-way valves, the atrioventricular valves prevent the back flow of blood into the _____.

122 What is atrioventricular valve prolapse? _____

123 Which of the two atrioventricular valves is mostly involved in prolapse? _____

124 The free-edges of the pulmonary and aortic valves point into the _____.

125 When are the pulmonary and aortic valves open? _____

126 When are the pulmonary and aortic valves closed? _____

127 As one-way valves, the pulmonary and aortic valves prevent the back flow of blood into the _____.

VENTRICULAR WALLS

128 Which ventricle has the thickest wall? _____

129 Do the right and left ventricles normally pump the same volume of blood? _____

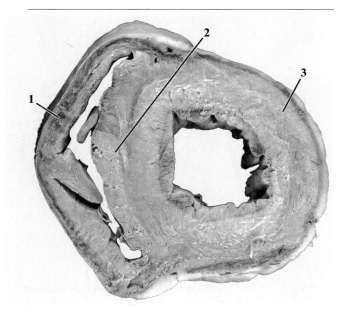

Figure 23.25

130 In reference to **Figure 23.25**, identify #1 - #3.
1 _____ 3 _____
2 _____

ELECTRICAL EVENTS OF THE HEART
Electrocardiogram

131 What is an electrocardiogram? _____

132 How many leads are normally used in a diagnostic ECG? _____

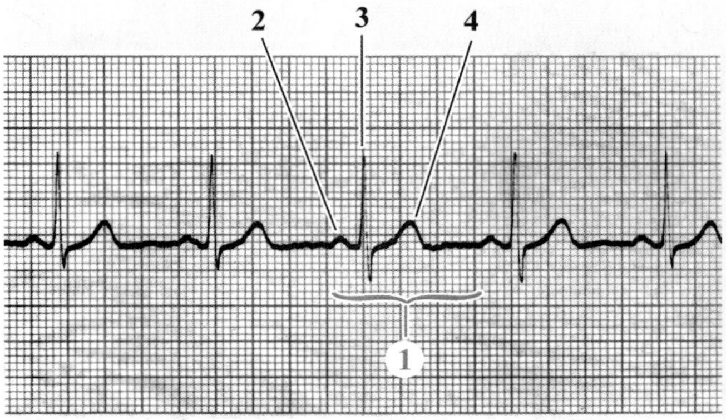

Figure 23.26

133 In reference to **Figure 23.26**, identify #1 - #4.
1 _____ 3 _____
2 _____ 4 _____

Conduction System of the Heart

134 Where is the sinoatrial node (SA node) located? _____

135 What is the function of the SA node? _____

136 What is a cardiac cycle? _____

137 What depolarizes after the SA node? _____

138 What results from atrial depolarization? _____

139 Where is the atrioventricular (AV) node located? _____

140 When does the AV node depolarize? _____

141 From the AV node list in sequence the structures that depolarize. _____

142 What is the result of the depolarization of the ventricular myocardium? _____

328 Chapter 23 - The Heart

Name _____
Class _____

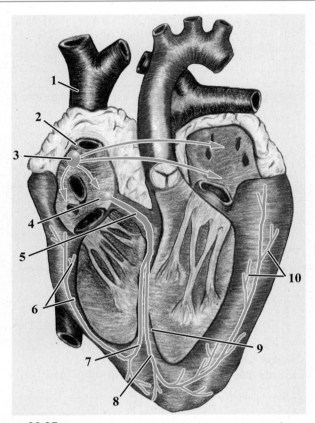

Figure 23.27

143 In reference to **Figure 23.27**, identify #1 - #10.
1 _____ 6 _____
2 _____ 7 _____
3 _____ 8 _____
4 _____ 9 _____
5 _____ 10 _____

ECG Waves

144 What are three distinctive waves of a normal ECG? _____

145 When does the P wave begin, and what does it represent?
_____
_____

146 What does the QRS complex represent? _____
_____

147 What does the T wave represent? _____

148 What is the P-Q interval? _____
_____

149 What is the Q-T interval? _____
_____

150 What is the S-T segment? _____
_____

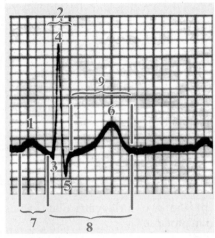

Figure 23.28

151 In reference to **Figure 23.28**, identify #1 - #9.
1 _____ 6 _____
2 _____ 7 _____
3 _____ 8 _____
4 _____ 9 _____
5 _____

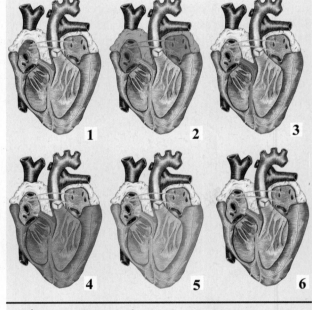

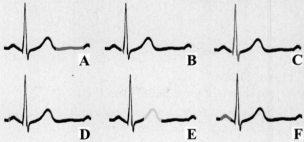

Figure 23.29

152 Match the numbers of the heart illustrations with their ECG tracing (A - F).
1 _____ 4 _____
2 _____ 5 _____
3 _____ 6 _____

CARDIAC CYCLE

153 What characterizes a normal sinus rhythm? _____
_____
_____

154 The average heart rate is _____ beats per minute, taking _____ second per beat.

155 During the end of the previous cardiac cycle (following T wave) the heart is in its _____ period.

156 During the resting period both the atria and the ventricles are _____ with blood.

157 During the resting period the atrioventricular valves are _____ and the semilunar valves are _____.

158 Atrial depolarization, shown on the ECG by the _____ wave, results in atrial _____.

159 How long is atria systole? _____

160 How long is atrial diastole? _____

161 The QRS complex represents the _____ of the ventricles and results in ventricular _____.

162 The atrioventricular valves close when pressure in the ventricles is _____ than pressure in the atria, and produces the _____.

163 The pulmonary and aortic valves open when pressure in the ventricles is _____ than pressure in the pulmonary trunk and aorta.

164 Blood ejection causes the elastic arteries to _____ to accommodate the increased blood volume.

165 The T wave represents ventricular _____ and results in ventricular _____.

166 The pulmonary and aortic valves close when ventricular pressure is _____ than pressure in the pulmonary trunk and aorta, and produces the _____.

167 When ventricular pressure is _____ than atrial pressure, the atrioventricular valves _____.

168 The elastic arteries _____ and provide the force to continue driving blood through the circulatory system.

169 The expansion and recoil of the elastic arteries are described as the _____.

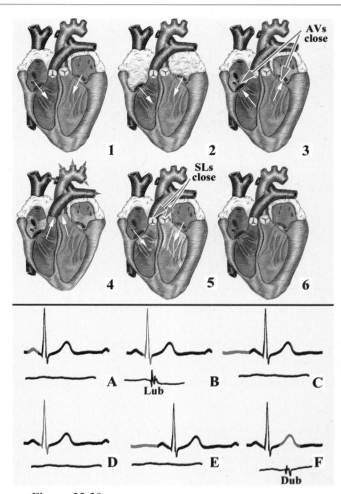

Figure 23.30

170 Match the numbers of the heart illustrations showing blood flow with their ECG and heart sound tracings (A - F).

1 _____   4 _____
2 _____   5 _____
3 _____   6 _____

AUSCULTATION OF HEART

171 What is auscultation? _____
_____

172 What site is the best site for auscultation of the aortic semilunar valve? _____

173 What site is the best site for auscultation of the pulmonary semilunar valve? _____

174 What site is the best site for auscultation of the mitral (bicuspid) valve? _____

175 What site is the best site for auscultation of the tricuspid valve? _____

330 Chapter 23 - The Heart

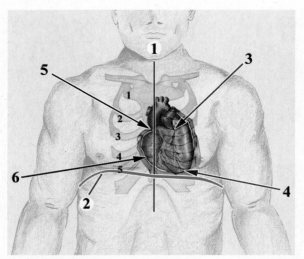

Figure 23.31
176 In reference to **Figure 23.31**, identify #1 - #6.
1 _____ 4 _____
2 _____ 5 _____
3 _____ 6 _____

CARDIAC MUSCLE

177 What are several characteristics of cardiac muscle fibers?
_____
_____

178 The T tubes of cardiac muscle are _____, and cardiac muscle does not have _____.

179 Contraction of cardiac muscle is dependent upon the presence of _____ ions.

180 Calcium ions are delivered from the _____ _____ and the _____ environment.

181 Contraction is terminated by the removal of _____ ions.

182 Intercalated discs contain ____ junctions and _____.

183 Gap junctions function as _____
_____.

184 Desmosomes function in providing _____
_____

185 The pacemaker, or sinoatrial node, is controlled by the ___
_____.

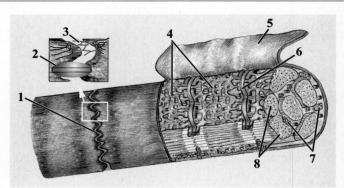

Figure 23.32
186 In reference to **Figure 23.32**, identify #1 - #8.
1 _____ 5 _____
2 _____ 6 _____
3 _____ 7 _____
4 _____ 8 _____

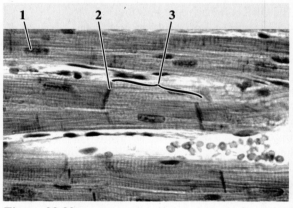

Figure 23.33
187 In reference to **Figure 23.33**, identify #1 - #3.
1 _____ 3 _____
2 _____

Cardiac Muscle Cells - Action Potential

188 The cardiac action potential is divided into three segments, _____
_____

189 The segment of rapid depolarization is due to the opening of _____ channels.

190 The segment of rapid depolarization results in the beginning of muscle _____.

191 The segment of slow depolarization is due to the opening of _____ channels.

192 During the segment of slow depolarization, cross-bridge activation _____ producing increased tension.

193 The segment of repolarization is due to the opening of _____.

194 During the segment of repolarization, the sarcolemma is retuned to is resting potential by the _____
_____
_____.

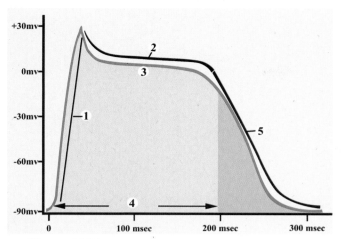

Figure 23.34
195 In reference to **Figure 23.34**, identify #1 - #5.
1 _____
2 _____
3 _____
4 _____
5 _____

CARDIAC CONDUCTION FIBERS
196 What are some of the conduction pathways? _____
_____

197 What is the function of conduction pathways? _____

198 What are Purkinje fibers? _____
_____
_____

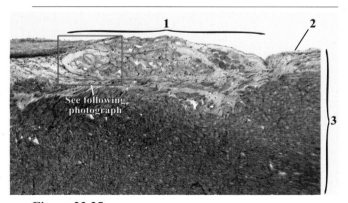

Figure 23.35
199 In reference to **Figure 23.35**, identify #1 - #3.
1 _____ 3 _____
2 _____

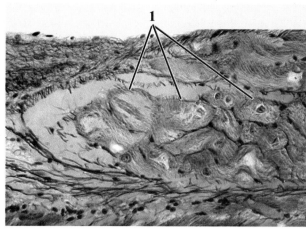

Figure 23.36
200 In reference to **Figure 23.36**, identify #1.
1 _____

CARDIAC OUTPUT
201 What is cardiac output? _____

202 How is cardiac output calculated? _____

STROKE VOLUME
203 What is stroke volume? _____

204 How is stroke volume calculated? _____

205 What is end diastolic volume? _____
_____

206 What is end systolic volume? _____
_____

Changes affecting the End Diastolic Volume
207 What are two factors that affect the end diastolic volume?
_____

208 How does fill time affect the end diastolic volume? _____
_____

209 How does venous return affect the end diastolic volume?
_____
_____

Chapter 23 - The Heart

Changes affecting the End Systolic Volume

210 What are three factors that influence the end systolic volume? _____

211 What is preload? _____

212 Stretching of the myocardium (within limits) produces a _____ alignment between the thin filaments and the thick filaments.

213 According to _____ law of the heart (<u>more</u> / <u>less</u>) blood is ejected when the myocardium is stretched.

214 What is contractility? _____

215 What are three factors that affect contractility? _____

216 What is afterload? _____

217 What determines afterload? _____

218 The greater the pressure in the exiting vessels of the ventricles, the _____ the afterload.

219 What is the primary factor that influences afterload? _____

HEART RATE

220 Heart rate is primarily controlled by the _____.

221 The parasympathetic division _____ heart rate.

222 The sympathetic division _____ heart rate.

223 In addition to the ANS other factors that influence heart rate are _____

224 The normal range for the heart rate is _____, with most people averaging _____.

225 Bradycardia is _____

226 Tachycardia is _____

Blood Vessels and Circulation - Worksheets

1. What are the two components of the cardiovascular system? _____

2. What are two functions of blood vessels? _____

3. Starting with blood flow away from the heart, list the sequence of vessels back to the heart. _____

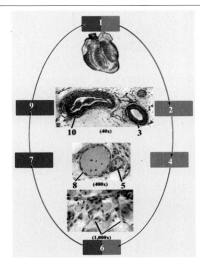

Figure 24.1

4. In reference to **Figure 24.1**, identify #1 - #10.
   1 _____   6 _____
   2 _____   7 _____
   3 _____   8 _____
   4 _____   9 _____
   5 _____   10 _____

ARTERIES

5. Typically, arteries are described as the vessels with the (thickest/thinnest) walls and having the (highest/lowest) blood pressure.

6. All arteries carry oxygenated blood except those of the _____.

7. Two types of arteries are the _____ and the _____ arteries.

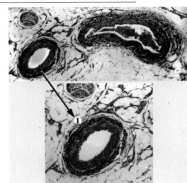

Figure 24.2

8. In reference to **Figure 24.2**, identify #1.
   1 _____

Elastic arteries

9. Elastic arteries are also called _____ arteries and have abundant _____ fibers in their middle and outer walls.

10. During the resting period of the heart, the elastic arteries function to _____

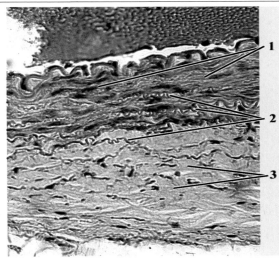

Figure 24.3

11. In reference to **Figure 24.3**, identify #1 - #3.
    1 _____   3 _____
    2 _____

Muscular arteries

12. Muscular arteries are also called _____ arteries and have abundant _____ in their middle wall.

13. The muscular arteries function by _____

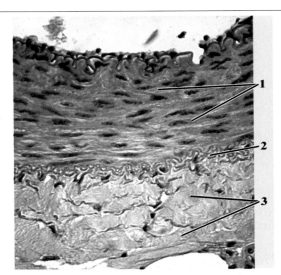

Figure 24.4

14. In reference to **Figure 24.4**, identify #1 - #3.
    1 _____   3 _____
    2 _____

334 Chapter 24 - Blood Vessels and Circulation

Name _____
Class _____

ARTERIOLES

15 What is the function of the smooth muscle of arterioles? ___
_____
_____

16 Vasoconstriction (increases / decreases) resistance and (increases / decreases) blood flow to distal tissues.

17 Vasoconstriction of arterioles supplying the body's peripheral tissues (increases / decreases) systemic blood pressure.

18 Vasodilation (increases / decreases) resistance and (increases / decreases) blood flow to distal tissues.

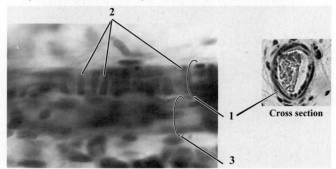

Figure 24.5

19 In reference to **Figure 24.5**, identify #1 - #3.
    1 _____ 3 _____
    2 _____

CAPILLARIES

20 Capillaries function as the _____
    _____

21 What are four mechanisms that provide for movement of substances across the capillary wall? _____
    _____
    _____
    _____

22 What are three types of capillaries? _____
    _____

23 What characterizes continuous capillaries? _____
    _____
    _____

24 Diffusion moves small substances that can move through the _____ and if permeable across the _____.

25 The driving force for filtration is _____.

26 What feature of the capillary determines which substances are allowed to pass by filtration? _____
    _____

27 What moves water back into the capillary? _____
    _____

28 What are fenestrations? _____
    _____

29 What is the function of fenestrations? _____
    _____

30 Which is the most permeable of the capillary types? _____
    _____

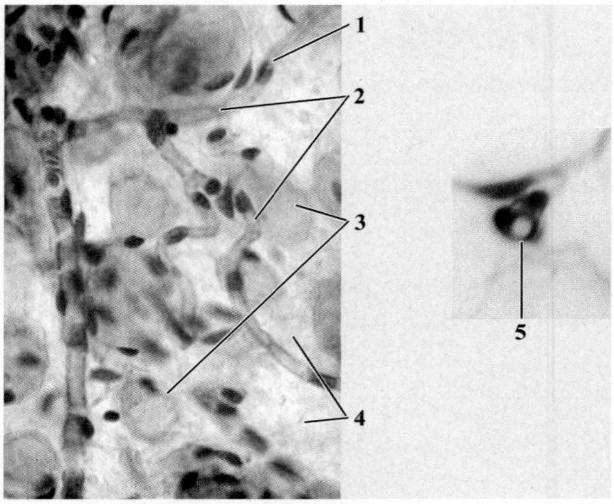

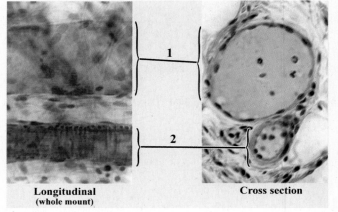

Figure 24.6

31 In reference to **Figure 24.6**, identify #1 - #5.
    1 _____ 4 _____
    2 _____ 5 _____
    3 _____

VENULES

32 Venules are located between _____ and _____.

Figure 24.7

33 In reference to **Figure 24.7**, identify #1 - #2.
    1 _____ 2 _____

Name _____
Class _____

Chapter 24 - Blood Vessels and Circulation   335

VEINS

34 Veins carry oxygen poor blood toward the _____, except for the _____.

35 The pulmonary venous circuit carries _____ rich blood from the _____ to the _____.

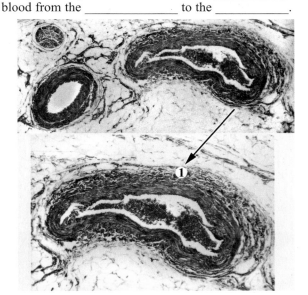

Figure 24.8

36 In reference to **Figure 24.8**, identify #1.
    1 _____

37 What is the function of vascular anastomoses? _____
    _____
    _____

ARTERY AND VEIN STRUCTURE

38 From outer to inner, what are the three layers of arteries and veins? _____
    _____

39 What forms the surface layer of the tunica interna? _____

40 Why do preparations of large arteries have a wavy, or scalloped, surface? _____
    _____

41 What are the components of the tunica media? _____
    _____

42 What type of vessel has the most smooth muscle in the tunica media? _____

43 Contraction of the smooth muscle of the tunica media produces _____, and relaxation of the smooth muscle produces _____.

44 Why is elasticity important in blood vessels? _____
    _____

45 What is the function of collagen fibers? _____
    _____

46 What are three functions of the tunica externa? _____
    _____
    _____

LAB ACTIVITY 1

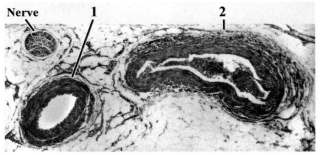

Figure 24.9

47 In reference to **Figure 24.9**, identify #1 - #2.
    1 _____   2 _____

STRUCTURE OF AN ARTERY

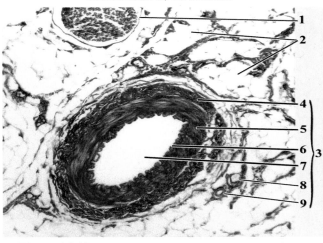

Figure 24.10

48 In reference to **Figure 24.10**, identify #1 - #9.
    1 _____   6 _____
    2 _____   7 _____
    3 _____   8 _____
    4 _____   9 _____
    5 _____

336 Chapter 24 - Blood Vessels and Circulation

Name _____
Class _____

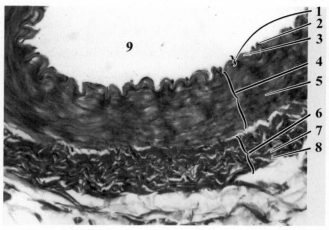

Figure 24.11
49 In reference to **Figure 24.11**, identify #1 - #9.
    1 _____ 6 _____
    2 _____ 7 _____
    3 _____ 8 _____
    4 _____ 9 _____
    5 _____

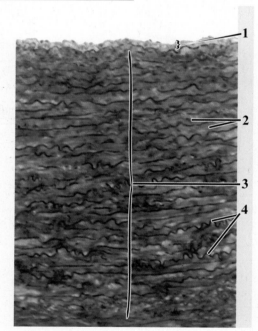

Figure 24.12
50 In reference to **Figure 24.12**, identify the type of vessel.
    _____

51 In reference to **Figure 24.12**, identify #1 - #4.
    1 _____ 3 _____
    2 _____ 4 _____

52 In reference to **Figure 24.12**, what is the function of #4?
    _____
    _____

STRUCTURE OF A VEIN

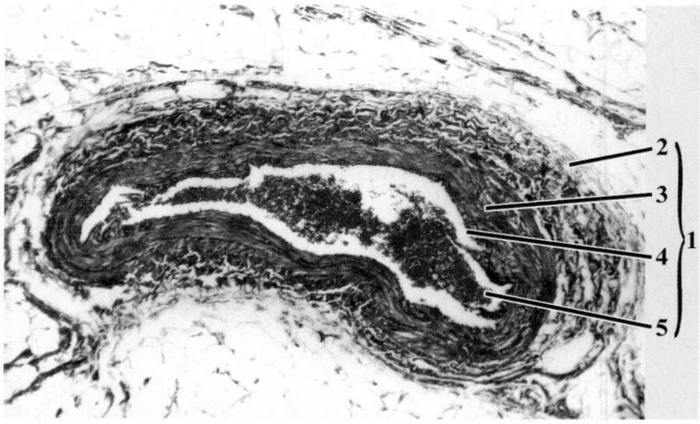

Figure 24.13
53 In reference to **Figure 24.13**, identify #1 - #5.
    1 _____ 4 _____
    2 _____ 5 _____
    3 _____

Figure 24.14
54 In reference to **Figure 24.14**, identify #1 - #6.
    1 _____ 4 _____
    2 _____ 5 _____
    3 _____ 6 _____

VALVES

55 Vascular valves are found in medium sized _____, and are formed from the tunica _____.
56 Vascular valves allow blood to flow in only one direction, toward the _____.
57 Venous return is increased by the _____ _____.

LAB ACTIVITY 2

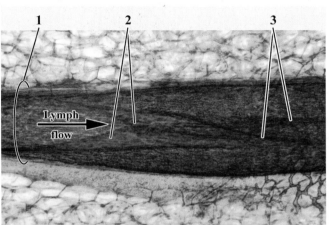

Figure 24.15

58 In reference to **Figure 24.15**, identify #1 - #3.
    1 _____   3 _____
    2 _____
59 The skeletal muscle pumping mechanism operates when skeletal muscles _____.
60 During increased pressure, the proximal valve (toward the capillaries) _____ and the distal valve _____.
61 With reduced pressure, the distal valve _____, preventing back flow of blood, and the proximal valve _____.

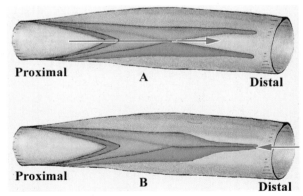

Figure 24.16

62 **Figure 24.16** shows the function of _____.
63 **Figure 24.16 -A** shows valve opening due to _____
64 **Figure 24.16 -B** shows valve closure due to _____

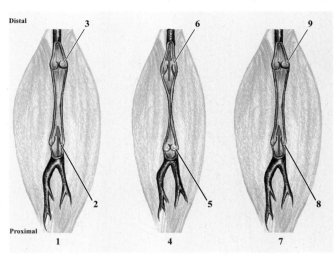

Figure 24.17

65 In reference to **Figure 24.17**, identify #1 - #9.
    1 _____   6 _____
    2 _____   7 _____
    3 _____   8 _____
    4 _____   9 _____
    5 _____

CIRCULATION PATHWAYS

66 Name two of the major circulation pathways. _____
67 What does the systemic circulation pathway supply? _____
68 Where does the systemic pathway begin? _____
69 What are three regional pathways of systemic circulation?
70 What does the pulmonary circulation pathway supply? _____
71 What route supplies oxygen poor blood to the lungs? _____
72 What route supplies oxygen rich blood to the left atrium?

338 Chapter 24 - Blood Vessels and Circulation

Name _____
Class _____

ARTERIES

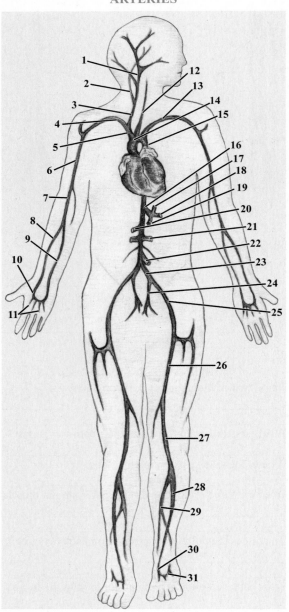

Figure 24.18

73 In reference to **Figure 24.18**, identify #1 - #31.

1 _____	17 _____
2 _____	18 _____
3 _____	19 _____
4 _____	20 _____
5 _____	21 _____
6 _____	22 _____
7 _____	23 _____
8 _____	24 _____
9 _____	25 _____
10 _____	26 _____
11 _____	27 _____
12 _____	28 _____
13 _____	29 _____
14 _____	30 _____
15 _____	31 _____
16 _____	

VEINS

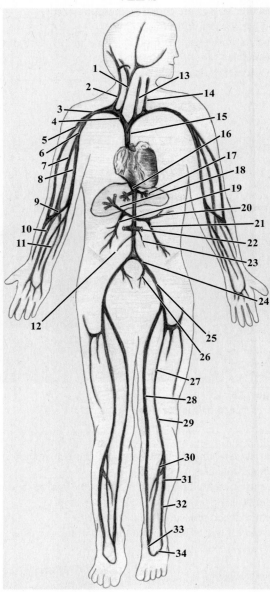

Figure 24.19

74 In reference to **Figure 24.19**, identify #1 - #31.

1 _____	18 _____
2 _____	19 _____
3 _____	20 _____
4 _____	21 _____
5 _____	22 _____
6 _____	23 _____
7 _____	24 _____
8 _____	25 _____
9 _____	26 _____
10 _____	27 _____
11 _____	28 _____
12 _____	29 _____
13 _____	30 _____
14 _____	31 _____
15 _____	32 _____
16 _____	33 _____
17 _____	34 _____

Name _____
Class _____

Chapter 24 - Blood Vessels and Circulation   **339**

SYSTEMIC BLOOD PRESSURE

75 What produces vascular blood pressure? _____
_____

76 What is systolic pressure? _____
_____

77 What is diastolic pressure? _____
_____

78 What is a sphygmomanometer? _____
_____

79 Blood pressure is measured in _____ of mercury (Hg).

80 When recording blood pressure, the number written first is the _____ pressure, and the number written second is the _____ pressure.

MEASURING BLOOD PRESSURE

81 When is systolic pressure determined? _____
_____

82 What produces Karotkov's sounds? _____
_____

83 How long will Karotkov's sounds continue? ____
_____

84 When is diastolic pressure determined? _____
_____

85 What reading is considered to the "normal" blood pressure?
86 What readings are indications of high blood pressure? ___
_____

87 What is a stethoscope? _____

88 What are two common types of sphygmomanometers? ___
_____

LAB ACTIVITY 3

89 Where is the sphygmomanometer's cuff arrow properly positioned on the arm? _____
_____

90 Where is the diaphragm of the stethoscope placed? ___
_____

91 The cuff should initially be inflated to _____.
92 The cuff should be deflated at a rate of _____.
93 If Karotkov's sounds are heard at the initial inflation pressure of 140 mm Hg., what must be done? _____
_____

94 When is systolic pressure recorded? _____
_____

95 When is diastolic pressure recorded? _____
_____

BLOOD PRESSURE	Systolic pressure	Diastolic pressure

Figure 24.20

96 Record your blood pressure in **Figure 24.20**.

REGULATION OF BLOOD PRESSURE AND FLOW

97 What are three major factors that influence blood pressure and flow? _____
_____

98 What is cardiac output? _____
_____

99 What effect does increased cardiac output have on blood pressure and flow? _____
_____

100 What effect does blood volume have on cardiac output? ___
_____

101 What are two factors that determine cardiac output? ___

102 How is stroke volume determined? _____
_____

103 What effect does increased venous return have on stroke volume? _____
_____

104 What effect does increased heart rate have on blood pressure and flow? _____
_____

105 What primarily controls heart rate? _____
_____

PERIPHERAL RESISTANCE

106 What is peripheral resistance of blood vessels? ___
_____

107 What are three factors that influence peripheral resistance?
_____
_____
_____

340 Chapter 24 - Blood Vessels and Circulation

Blood viscosity
108 What is viscosity? _____
109 What is the effect of increased viscosity? _____
_____
110 What is a blood disease that results in increased blood viscosity? _____
_____

Total Length of Blood Vessels
111 As blood moves through a blood vessel, what produces vascular resistance? _____
_____
112 The further blood travels through the vascular system, the (more / less) resistance is encountered.

Blood Vessel Diameter
113 Immediate changes to vascular resistance are by regulation of the _____ of blood vessels.
114 The blood vessels that produce the greatest control of resistance are called the _____.
115 Vasoconstriction _____ resistance, while vasodilation _____ resistance.

Regulation of Peripheral Resistance
116 Regulation of peripheral resistance by controlling blood vessel diameter is by the _____ located in _____ of the brain stem.
117 The vasomotor center is part of the _____ center.
118 The activity of the vasomotor center is primarily regulated by _____
_____
119 What are baroreceptors? _____
_____
120 Where are baroreceptors located? _____
_____
_____

Vasodilation
121 Increased stimulation of the baroreceptors by (increased / decreased) blood pressure (stimulates / inhibits) the vasomotor center and results in _____.
122 Vasodilation results in (increased / decreased) blood flow into the capillaries and into the _____ system, which functions as a _____.
123 Increased stimulation of the baroreceptors also targets the cardiac centers resulting in a _____ of heart rate and force of contraction.

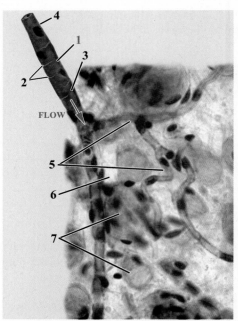

Figure 24.21
124 In reference to **Figure 24.21**, identify #1 - #7.
1 _____ 5 _____
2 _____ 6 _____
3 _____ 7 _____
4 _____
125 In reference to **Figure 24.21**, what controls #2? _____
_____
_____

Vasoconstriction
126 Decreased stimulation of the baroreceptors by (increased / decreased) blood pressure results in (stimulation / inhibition) of the vasomotor center, which produces _____ _____.
127 Vasoconstriction results in (increased / decreased) blood flow into the capillaries, which increases resistance and results in (increased / decreased) blood pressure.
128 Decreased stimulation of the baroreceptors also targets the cardioacceleratory center resulting in _____ heart rate and force of contraction.

Chapter 24 - Blood Vessels and Circulation   341

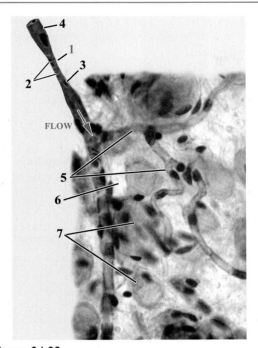

Figure 24.22

129 In reference to **Figure 24.22**, identify #1 - #7.
1 _____   5 _____
2 _____   6 _____
3 _____   7 _____
4 _____

Chemoreceptors

130 What are vascular chemoreceptors? _____
_____

131 Where are vascular chemoreceptors located? _____
_____

132 What do the vascular chemoreceptors target? _____
_____

133 What stimuli influence the vascular chemoreceptors? ____
_____
_____

134 Stimulation of the vascular chemoreceptors results in (decreased / increased) blood pressure and _____ of the stimuli.

ADDITIONAL FACTORS INFLUENCING BLOOD PRESSURE

135 In addition to baroreceptors and chemoreceptors, what are four other factors that influence blood pressure? _____
_____
_____

Stress (epinephrine and norepinephrine)

136 Stress and the "fight-or-flight" response are mediated through the hypothalamic control of the _____.
137 Epinephrine and norepinephrine _____ blood pressure.
138 Norepinephrine is a powerful _____.

Atrial Natriuretic Peptide (ANP)

139 Atrial natriuretic peptide is a hormone produced in the _____ of the heart and functions to _____ blood pressure by reduction of blood _____ and by general _____.

Antidiuretic Hormone (ADH)

140 Antidiuretic hormone is released when the volume of water in the blood _____.
141 Antidiuretic hormone is a powerful _____, hence its other common name of _____.

Angiotensin II

142 Angiotensin II is a powerful hormone that is produced as a result of the kidneys releasing the enzyme renin when blood pressure is _____.
143 As sodium moves into the blood, water osmotically follows and _____.
144 Angiotensin II is a powerful _____.

FLUID MOVEMENT AT THE CAPILLARY

145 Capillaries consist of a layer of _____ and basement membrane.
146 What are three ways substances move across the capillary wall? _____
_____

Diffusion

147 What is diffusion? _____
_____

148 A substance will diffuse across the capillary wall if the substance is permeable to the _____, has a facilitated _____, or is small enough to pass through the _____.

Chapter 24 - Blood Vessels and Circulation

Filtration

149 What is filtration? _____
_____

150 What features of capillaries allow them to function in filtration? _____

151 What is the driving force for filtration? _____

152 Is blood (hydrostatic) pressure higher at the arteriole end or the venous end of the capillary? _____
_____

153 The hydrostatic pressure of interstitial fluid is (greater, less) than capillary blood pressure.

Reabsorption

154 What is capillary reabsorption? _____

155 What process drives the reabsorption of water by the capillary? _____

156 What is osmosis? _____
_____

157 What is a hypertonic solution? _____

158 What is a hypotonic solution? _____
_____

159 Is the concentration of solutes greater in blood or in interstitial fluid? _____

160 Which solution has the greatest osmotic pressure, a solution of 50% solute and 50% water, or a solution of 10% solute and 90% water? _____

161 Two solutions, solution A of 50% solute and solution B of 10% solute, are separated by a semipermeable membrane (where only water can diffuse). Water will move from solution _____ into solution _____.

162 What is the primary solute of blood that contributes to the osmotic movement of water? _____
_____

Forces at the Arterial End of the Capillary

163 How is the net filtration force at the arterial end of the capillary calculated? _____
_____
_____

164 Which direction does water move at the arterial end of the capillary? _____
Why? _____
_____

Forces at the Venous End of the Capillary

165 How is the net filtration force at the venous end of the capillary calculated? _____
_____
_____

166 Which direction does water move at the venous end of the capillary? _____
Why? _____
_____

167 What happens to excess interstitial fluid? _____
_____
_____

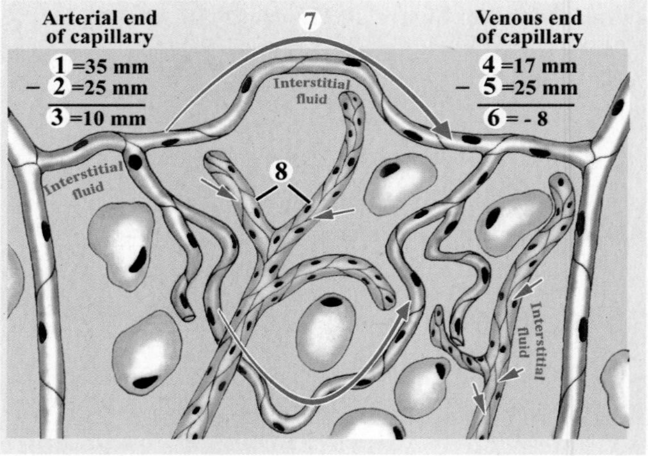

Figure 24.23

168 In reference to **Figure 24.23**, identify #1 - #8.

1 _____ 5 _____
2 _____ 6 _____
3 _____ 7 _____
4 _____ 8 _____

Name _____
Class _____

Chapter 25 - Respiratory System **343**

The Respiratory System - Worksheets

1. What are four functions of the respiratory system? _____
   _____
   _____

2. What are the components of the respiratory system? _____
   _____
   _____
   _____

3. What forms the upper respiratory tract? _____

4. What forms the lower respiratory tract? _____

LAB ACTIVITY 1

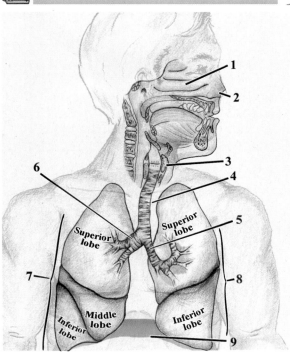

Figure 25.1

5. In reference to **Figure 25.1**, identify #1 - #9.
   1 _____ 6 _____
   2 _____ 7 _____
   3 _____ 8 _____
   4 _____ 9 _____
   5 _____

   NOSE

6. What are the functions of the nose? _____
   _____

7. What are the two portions of the nose? _____
   _____

8. What are the external nares? _____
   _____

9. What is the nasal cavity? _____
   _____

10. What is the nasal vestibule? _____

11. Where is the nasal septum located? _____
    _____

12. What are the nasal conchae? _____
    _____

13. What are the nasal meatuses? _____
    _____

14. What are the internal nares? _____
    _____
    _____

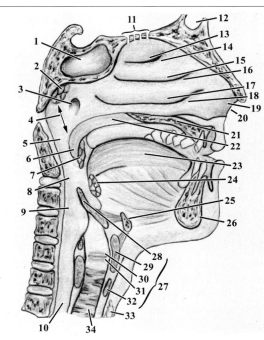

Figure 25.2

15. In reference to **Figure 25.2**, identify #1 - #34.
    1 _____ 18 _____
    2 _____ 19 _____
    3 _____ 20 _____
    4 _____ 21 _____
    5 _____ 22 _____
    6 _____ 23 _____
    7 _____ 24 _____
    8 _____ 25 _____
    9 _____ 26 _____
    10 _____ 27 _____
    11 _____ 28 _____
    12 _____ 29 _____
    13 _____ 30 _____
    14 _____ 31 _____
    15 _____ 32 _____
    16 _____ 33 _____
    17 _____ 34 _____

344 Chapter 25 - Respiratory System

Name _____
Class _____

PHARYNX

16 What is the location of the pharynx? _____
_____

17 What are the three divisions of the pharynx? _____
_____

18 What is the location of the nasopharynx? _____
_____

19 Where is the pharyngotympanic tube located? _____
_____

20 Where is the pharyngeal tonsil located? _____

21 What is the location of the oropharynx? _____
_____

22 Where is the uvula located? _____
_____

23 What is the location of the laryngopharynx? _____
_____

LARYNX

24 What is the location of the larynx? _____

25 Which cartilage forms the upper portion of the larynx? _____
_____
_____

26 What is the function of the epiglottis? _____
_____

27 What is the function of the upper vestibular folds? _____
_____

28 What is the function of the lower vocal folds (true vocal cords)? _____
_____
_____

29 What is the glottis? _____

TRACHEA

30 What is the trachea? _____

31 The trachea branches into the _____
_____.

32 From inner to outer, what are the three regions of the trachea's wall? _____
_____
_____

33 What type of epithelium lines the mucosa? _____
_____

34 Where are the seromucous glands located, and what is their function? _____
_____

35 What supports the adventitia? _____

36 Why are the hyaline cartilage rings described as "C" rings? _____
_____

37 Where is the trachealis muscle located, and what is its function? _____
_____
_____

LAB ACTIVITY 2

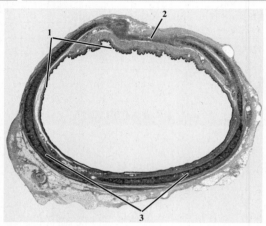

Figure 25.3

38 In reference to **Figure 25.3**, identify #1 - #3.
1 _____ 3 _____
2 _____

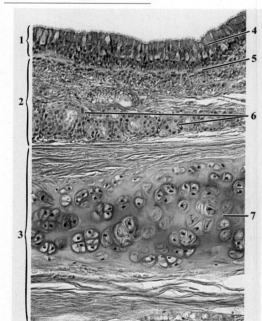

Figure 25.4

39 In reference to **Figure 25.4**, identify #1 - #7.
1 _____ 5 _____
2 _____ 6 _____
3 _____ 7 _____
4 _____

Figure 25.5

40 In reference to **Figure 25.5**, identify #1 - #5.
1 _____ 4 _____
2 _____ 5 _____
3 _____

41 What is the function of cilia? _____

42 What is the function of the goblet cells? _____

LUNGS

43 Where are the lungs located? _____

44 What are the names of the three lobes of the right lung? _____

45 What are the names of the two lobes of the left lung? _____

46 What are the names of the inner and the outer portions of the lung's serous membrane? _____

47 What occupies the pleural cavity? _____

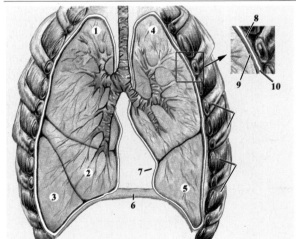

48 **Figure 25.6**
In reference to **Figure 25.6**, identify #1 - #10.
1 _____ 6 _____
2 _____ 7 _____
3 _____ 8 _____
4 _____ 9 _____
5 _____ 10 _____

TRACHEA AND BRONCHIAL TREE

49 The trachea branches into the _____.

50 The right primary bronchus branches into _____

51 The left primary bronchus branches into _____

52 Secondary (lobar) bronchi branch into _____

53 The tertiary (segmental) bronchi branch into _____.

54 Bronchioles branch into _____ lead into the _____ ducts.

55 Alveolar ducts terminate with the _____.

Structure of Trachea and Bronchial Tree

56 From the trachea and primary bronchi the amount of ____ _____ continues to decrease until it is completely lacking in the _____.

57 The respiratory tubes with the greatest amount of smooth muscle are the _____.

58 Pseudostratified ciliated columnar epithelium changes to ciliated _____ epithelium in the bronchioles.

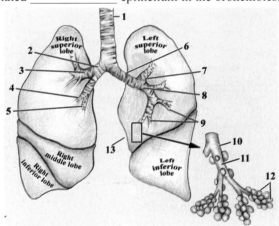

59 **Figure 25.7**
In reference to **Figure 25.7**, identify #1 - #12.
1 _____ 7 _____
2 _____ 8 _____
3 _____ 9 _____
4 _____ 10 _____
5 _____ 11 _____
6 _____ 12 _____

346 Chapter 25 - Respiratory System

Name _____
Class _____

LAB ACTIVITY 3

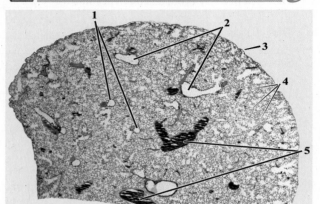

Figure 25.8
60 In reference to **Figure 25.8**, identify #1 - #5.
1 _____ 4 _____
2 _____ 5 _____
3 _____

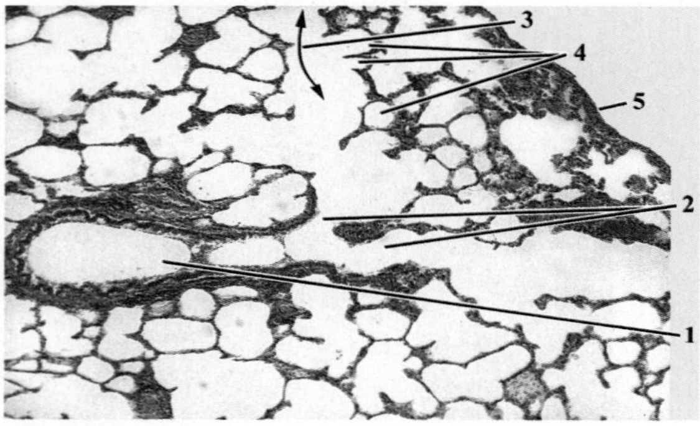

Figure 25.9
61 In reference to **Figure 25.9**, identify #1 - #5.
1 _____ 4 _____
2 _____ 5 _____
3 _____

62 What are bronchi? _____

63 The amount of smooth muscle _____ as the bronchi become smaller, and the amount of hyaline cartilage _____.

64 Bronchioles are called the airways of _____.
65 Alveoli are the sites of _____.
66 Alveoli are surrounded by a network of _____ and _____.
67 What is the function of the elastic fibers of the alveoli? _____

68 What are two functions of the simple squamous cells (type I) of the alveoli? _____

69 What is the function of the cuboidal cells (type II) of the alveolus? _____

70 What is the function of lung surfactant? _____

71 What is lung compliance? _____

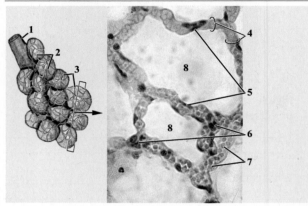

Figure 25.10
72 In reference to **Figure 25.10**, identify #1 - #7.
1 _____ 5 _____
2 _____ 6 _____
3 _____ 7 _____
4 _____ 8 _____

LAB ACTIVITY 4

73 What is the function of alveolar macrophages? _____

74 What characterizes emphysema? _____

75 What do the alveolar macrophages of smokers' lungs usually contain? _____

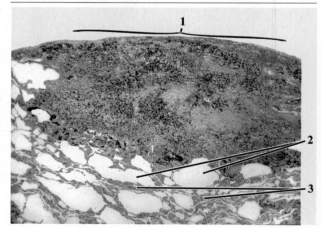

Figure 25.11
76 In reference to **Figure 25.11**, identify #1 - #3.
1 _____
2 _____
3 _____

Chapter 25 - Respiratory System   347

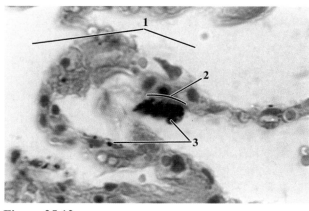

Figure 25.12
77 In reference to **Figure 25.12**, identify #1 - #3.
   1 _____ 3 _____
   2 _____

Respiratory membrane
78 What structures form the respiratory membrane? _____
   _____
   _____

79 What is the function of the respiratory membrane? _____
   _____
   _____

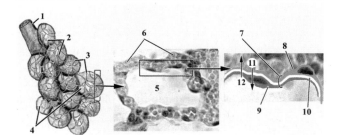

Figure 25.13
80 In reference to **Figure 25.13**, identify #1 - #12.
   1 _____ 7 _____
   2 _____ 8 _____
   3 _____ 9 _____
   4 _____ 10 _____
   5 _____ 11 _____
   6 _____ 12 _____

PULMONARY VENTILATION
81 What is pulmonary ventilation? _____
   _____

82 What is the driving force for pulmonary ventilation? ____

83 What is atmospheric pressure in millimeters of mercury (mm Hg.) at sea level? _____
   _____

84 What does Boyle's law state? _____
   _____
   _____

85 A change in pressure between intrapulmonary pressure and atmospheric pressure results in _____.

NORMAL QUIET BREATHING
86 What two types of breathing occur in normal quiet breathing? _____
   _____

87 Contraction of the diaphragm results in intrapulmonary pressure (decreasing / increasing) and results in (inhalation / exhalation).

88 Contraction of the external intercostal muscles results in intrapulmonary pressure (decreasing / increasing) and results in (inhalation / exhalation).

89 Expiration in normal quiet breathing is _____
   _____

90 Increased thoracic volume results in (inhalation / exhalation).

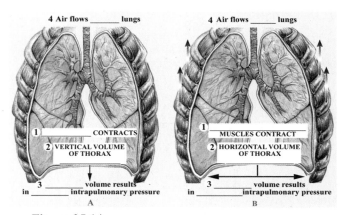

Figure 25.14
91 In reference to **Figure 25.14**, identify A & B, #1 - #4.
   A _____
   1 _____ 3 _____
   2 _____ 4 _____
   B _____
   1 _____ 3 _____
   2 _____ 4 _____

FORCED BREATHING
92 What muscles are involved in forced inspiration? _____
   _____

93 What muscles are involved in forced expiration? _____
   _____

RESPIRATORY VOLUMES and CAPACITIES
94 What are two variables in the exchange of air? _____
   _____

95 What is a spirometer? _____
   _____

96 What is a respiratory volume? _____
   _____

97 What is a respiratory capacity? _____
   _____

98 Define tidal volume (TV). _____
   _____

99 Define inspiratory reserve volume (IRV) _____
   _____

348 Chapter 25 - Respiratory System

100 Define expiratory reserve volume (ERV). _____
_____

101 Define residual volume (RV). _____
_____

102 Define inspiratory capacity (IC). _____
_____

103 Define vital capacity (VC). _____
_____

104 Define total lung capacity (TLC). _____
_____

105 Define functional residual capacity (FRC). _____
_____

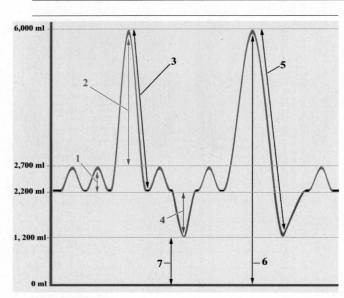

Figure 25.15

106 In reference to **Figure 25.15**, identify #1 - #7.
1 _____ 5 _____
2 _____ 6 _____
3 _____ 7 _____
4 _____

107 Match the following with their equations:
Respiratory Capacities
    IC
    VC
    FRC
    TLC
Equations for Respiratory Capacities
_____ = TV + IRV
_____ = ERV + TV + IRV
_____ = ERV + IC
_____ = RV + ERV + TV + IRV
_____ = RV + ERV
_____ = RV + VC

LAB ACTIVITY 5
MEASURING RESPIRATORY VOLUMES

108 **RESULTS FROM HAND-HELD SPIROMETER**
   Tidal Volume (TV)
      First exhalation volume _____
      Second exhalation volume _____
      Third exhalation volume _____
      Average TV _____
   Inspiratory Capacity (IC)
      First exhalation volume _____
      Second exhalation volume _____
      Third exhalation volume _____
      Average IC _____
   Inspiratory Reserve Volume (IRV)
      IRV = _____
   Expiratory Reserve Volume (ERV)
      First exhalation volume _____
      Second exhalation volume _____
      Third exhalation volume _____
      Average ERV _____
   Vital Capacity (VC)
      First exhalation volume _____
      Second exhalation volume _____
      Third exhalation volume _____
      Average VC _____
   Minute Respiratory Volume (MRV)
      MRV = _____

FORCED EXPIRATORY VOLUME TIMED (FEV$_T$)

109 What does the Forced Expiratory Volume Test measure? _____
_____

110 What respiratory disorders are indicated by decreased forced expiratory volume? _____
_____

111 What percentage of the forced vital capacity should be exhaled within the first second interval? _____
_____

LAB ACTIVITY 6

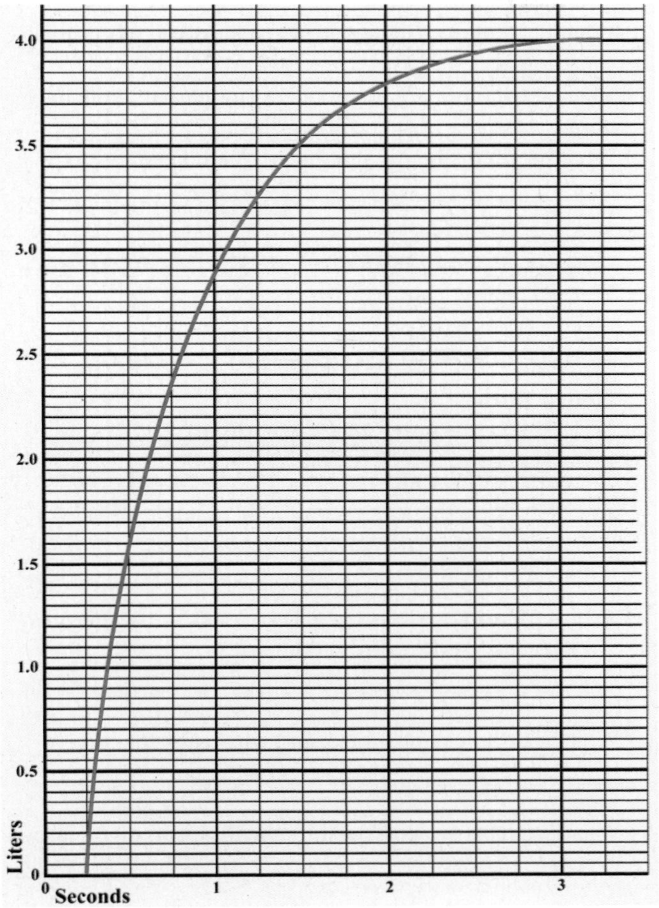

Figure 25.16

112 In reference to **Figure 25.16**, what is the vital capacity? ___

113 In reference to **Figure 25.16**, what is the one second volume? ___

114 In reference to **Figure 25.16**, the one second volume is what percentage of the vital capacity? ___

115 In reference to **Figure 25.16**, the person's predicted vital capacity is 4, 500 ml. The vital capacity is what percentage of the predicted vital capacity? ___

GAS MOVEMENT - RESPIRATORY MEMBRANE

116 How does the concentration of gases of atmospheric air compare to the concentration of air in the alveoli? ___

117 What are the percentages of the gases of atmospheric air at sea level? ___

118 What is partial pressure? ___

119 What is Dalton's Law? ___

120 What is atmospheric pressure (in mm Hg.) at sea level? ___

121 What is the partial pressure of each atmospheric gas at sea level? ___

122 What is the driving force for movement of respiratory gases? ___

Partial pressures at the Alveoli

123 The P_{O_2} in the alveoli is (greater / less) than the P_{O_2} of the blood.

124 The P_{CO_2} of the alveoli is (greater / less) than the P_{CO_2} of blood.

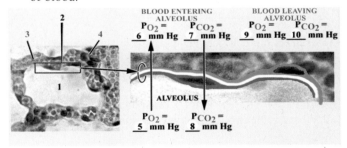

Figure 25.17

125 In reference to **Figure 25.17**, identify #1 - #10.

1 _____ 6 _____
2 _____ 7 _____
3 _____ 8 _____
4 _____ 9 _____
5 _____ 10 _____

Partial pressures at the Tissues

126. The P_{O_2} of tissue blood capillaries is (greater / less) than the P_{O_2} of the tissue cells.

127. The P_{CO_2} of tissue blood capillaries is (greater / less) than the P_{CO_2} of the tissue cells.

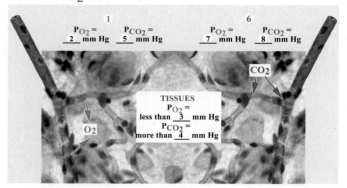

Figure 25.18

128. In reference to **Figure 25.18**, identify #1 - #8.

1 _____ 5 _____
2 _____ 6 _____
3 _____ 7 _____
4 _____ 8 _____

Transport of Respiratory Gases

129. What is Henry's Law? _____

130. Rank the respiratory gases in order of decreasing solubility in water. _____

131. What is the solution for the inadequate transport of dissolved respiratory gases? _____

Oxygen Transport

132. What are the two ways oxygen is transported in the blood? _____

133. What percent of oxygen is transported dissolved in the plasma? _____

134. What percent of oxygen is transported as HbO_2? _____

135. In the lungs deoxyhemoglobin releases _____ ions and combines with oxygen to form _____.

136. In the tissues oxyhemoglobin releases _____ and binds _____ ions.

Carbon Dioxide Transport

137. What are the three ways carbon dioxide is transported in the blood? _____

138. What percent of carbon dioxide is dissolved in plasma? ___

139. What percent of carbon dioxide forms carbaminohemoglobin? _____

140. What percent is bound into bicarbonate ions? _____

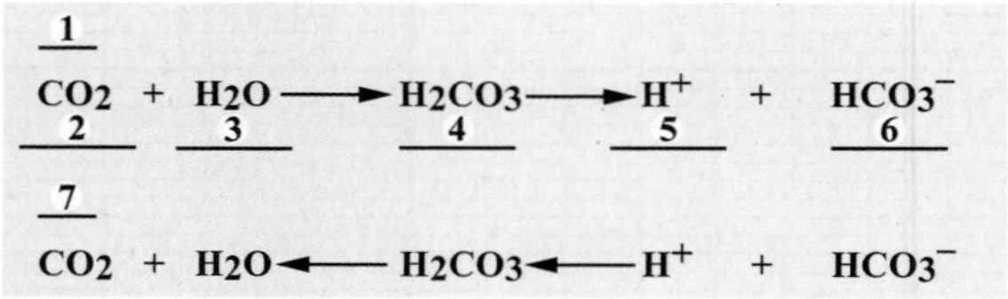

$$\underset{2}{\overset{1}{CO_2}} + \underset{3}{H_2O} \longrightarrow \underset{4}{H_2CO_3} \longrightarrow \underset{5}{H^+} + \underset{6}{HCO_3^-}$$

$$\underset{}{\overset{7}{CO_2}} + H_2O \longleftarrow H_2CO_3 \longleftarrow H^+ + HCO_3^-$$

Figure 25.19

141. In reference to **Figure 25.19**, identify #1 - #7.

1 _____ 5 _____
2 _____ 6 _____
3 _____ 7 _____
4 _____

Carbon dioxide and Oxygen Transport at the Tissues

142. Why do cell produce carbon dioxide? _____

143. What is carbonic anhydrase, and what is its function? _____

144. What products are produced by the dissociation of carbonic acid? _____

145. What happens to the bicarbonate ions produced in the RBCs? _____

146. What happens to the hydrogen ions produced in the RBCs? _____

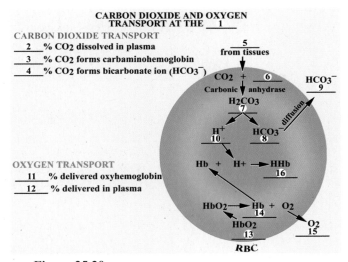

Figure 25.20

147 In reference to **Figure 25.20**, identify #1 - #16.

1 _____ 9 _____
2 _____ 10 _____
3 _____ 11 _____
4 _____ 12 _____
5 _____ 13 _____
6 _____ 14 _____
7 _____ 15 _____
8 _____ 16 _____

Carbon dioxide and Oxygen Transport at the Lungs

148 In what form does most of the carbon dioxide arrive at the lungs? _____
_____

149 What happens to the bicarbonate ions that enter the RBCs?
_____
_____

150 What happens to the carbon dioxide that is produced by the dissociation of carbonic acid? _____
_____

151 Oxygen diffuses into the RBCs and forms _____.

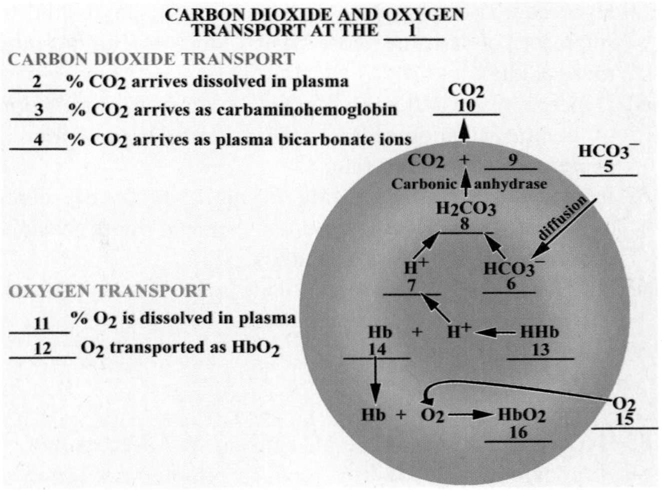

Figure 25.21

152 In reference to **Figure 25.21**, identify #1 - #16.

1 _____ 9 _____
2 _____ 10 _____
3 _____ 11 _____
4 _____ 12 _____
5 _____ 13 _____
6 _____ 14 _____
7 _____ 15 _____
8 _____ 16 _____

MECHANISMS CONTROLLING RESPIRATION

153 The respiratory centers are located in the _____ _____ and _____ of the brain stem.

154 The respiratory center that controls rate is the (dorsal / ventral) respiratory group.

155 The respiratory center is mostly controlled by input from _
_____
_____

156 Peripheral chemoreceptors are mostly located in the _____
_____
_____ and are sensitive to changes in blood _____.

157 Central chemoreceptors are mostly located in the _____
_____
_____ and are sensitive to changes in blood _____.

158 Increased input from proprioceptors results in (increasing / decreasing) respiratory rate.

159 Hypercapnia is defined as _____
_____

160 Hypercapnia causes the pH of the cerebrospinal fluid to (increase, or become more basic / decrease, or become more acidic).
161 Hypercapnia results in (increased / decreased) stimulation of central chemoreceptors and results in (increased / decreased) respiratory rate.
162 Increased respiration results in the pH of the cerebrospinal fluid to (increase, or become more basic / decrease, or become more acidic).
163 Rapid removal of carbon dioxide results in (increased/ decreased) levels of carbonic acid.
164 What is hypocapnia? _____
_____
_____
165 Hypocapnia results in the pH (increasing / decreasing).
166 Hypocapnia can be reversed by (hyperventilation / hypoventilation).
167 Rebreathing into a bag to (increase / decrease) carbon dioxide levels (increases / decreases) pH.

Name _____
Class _____

Chapter 26 - Digestive System 353

The Digestive System - Worksheets

1. What is digestion? _____
_____

2. What are the two major divisions of the digestive system?
_____
_____

3. What is the digestive tract (gastrointestinal, or alimentary canal)? _____
_____

4. What are six components of the digestive tract?
   1 _____ 4 _____
   2 _____ 5 _____
   3 _____ 6 _____

5. What are six accessory organs?
   1 _____ 4 _____
   2 _____ 5 _____
   3 _____ 6 _____

LAB ACTIVITY 1

DIGESTIVE SYSTEM OVERVIEW

26

7. What are two structures contained within the oral cavity?
_____

8. Where do the salivary glands empty? _____
_____

9. What are the three major salivary glands? _____
_____

10. What structure is located behind the mouth? _____
_____

11. What two parts of the pharynx function in the digestive system? _____
_____

12. Food moves from the pharynx into the _____.
13. Food moves from the esophagus into the _____.
14. What are the four major regions of the stomach?
    1 _____ 3 _____
    2 _____ 4 _____

15. The part of the stomach that is continuous with the esophagus is the _____.

16. The pyloric region of the stomach is continuous with the _____.

17. The _____ sphincter is located between the stomach and the first region of the small intestine, the _____.

18. The two curvatures of the stomach are the lateral _____ and medial _____ curvatures.

19. The three regions of the small intestine are the _____.

20. The duodenum receives a liquid mixture of food from the stomach called _____.

21. The duodenum receives secretions from the _____ and the _____.

22. Bile contains _____, which function in the emulsification of _____.

23. The pancreas produces _____, which contains two main components, _____ and _____.

24. The common bile duct and the pancreatic duct join at the _____ ampulla, which enters the _____.

25. The ileum terminates with its junction with the portion of the large intestine called the _____.

26. The valve between the ileum and the cecum is the _____ valve.

27. The large intestine is divided into the _____
_____

28. The appendix is continuous with the _____.

29. The four divisions of the colon are the _____
_____

30. The short anal canal terminates at the _____, the terminal opening of the alimentary canal to the outside.

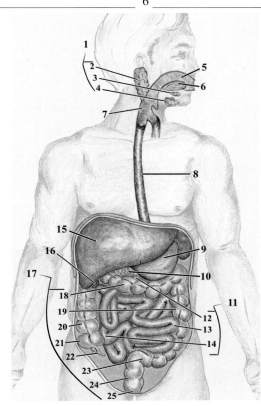

6. In reference to **Figure 26.1**, identify #1 - #25.
   Figure 26.1
   1 _____ 14 _____
   2 _____ 15 _____
   3 _____ 16 _____
   4 _____ 17 _____
   5 _____ 18 _____
   6 _____ 19 _____
   7 _____ 20 _____
   8 _____ 21 _____
   9 _____ 22 _____
   10 _____ 23 _____
   11 _____ 24 _____
   12 _____ 25 _____
   13 _____

354 Chapter 26 - Digestive System

Name _____
Class _____

31 From inner to outer, the four layers of the alimentary canal are the _____ _____.

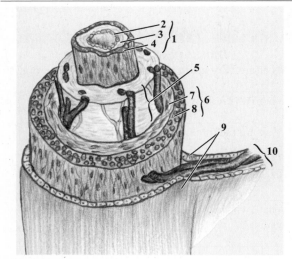

Figure 26.2

32 In reference to **Figure 26.2**, identify #1 - #10.
1 _____   6 _____
2 _____   7 _____
3 _____   8 _____
4 _____   9 _____
5 _____ 10 _____

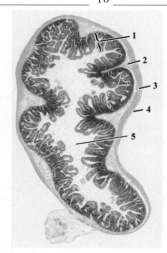

Figure 26.3

33 In reference to **Figure 26.3**, identify #1 - #5.
1 _____   4 _____
2 _____   5 _____
3 _____

MUCOSA

34 What is the mucosa of the digestive tract? _____ _____ _____

35 What are the three components of the mucosa? _____ _____

36 What are three functions of the mucosa of the digestive tract? _____ _____ _____

Epithelia of the Mucosa

37 What are the two types of epithelia found in the mucosa of the digestive tract? _____ _____

38 What regions of the digestive tract are lined with stratified squamous epithelium? _____ _____

39 What is the function of stratified squamous epithelium? ___ _____

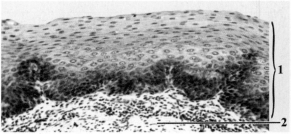

Figure 26.4

40 In reference to **Figure 26.4**, identify #1 - #2.
1 _____   2 _____

41 What regions of the digestive tract are lined with simple columnar epithelium? _____ _____

42 What is the function of simple columnar epithelium? ___ _____

43 What are microvilli? _____ _____

44 What is the function of microvilli? _____ _____

45 What is the function of goblet cells? _____ _____

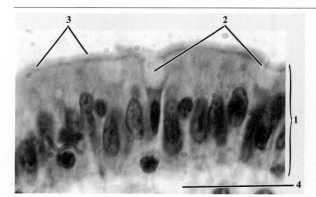

Figure 26.5

46 In reference to **Figure 26.5**, identify #1 - #4.
1 _____   3 _____
2 _____   4 _____

MODIFICATIONS OF THE MUCOSA
Mucosa of the Stomach

47 What is a modification of the mucosa of the stomach? ___ _____ _____

48 Gastric glands produce _____ juice and _____.

Chapter 26 - Digestive System 355

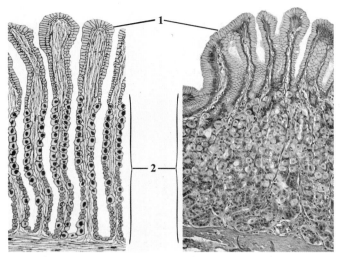

Figure 26.6

49 In reference to **Figure 26.6**, identify #1 - #2.
   1 _____ 2 _____

Mucosa of the Small Intestine

50 What are two modifications of the mucosa of the small intestine? _____

51 What is the function of villi? _____

52 What is the function of intestinal glands (crypts)? _____

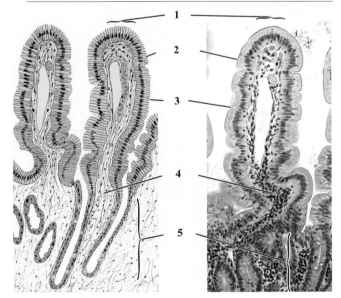

Figure 26.7

53 In reference to **Figure 26.7**, identify #1 - #5.
   1 _____ 4 _____
   2 _____ 5 _____
   3 _____

Mucosa of the Large Intestine

54 What is a modification of the mucosa of the large intestine? _____

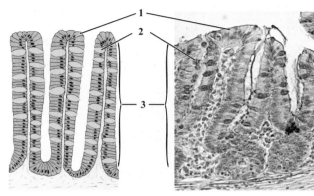

Figure 26.8

55 In reference to **Figure 26.8**, identify #1 - #3.
   1 _____ 3 _____
   2 _____

SUBMUCOSA

56 What type of tissue forms the submucosa? _____

MUSCULARIS EXTERNA

57 Where is the muscularis externa located? _____

58 What is the function of the muscularis externa? _____

59 Various percentages of _____ muscle tissue are found in the mouth, pharynx, and upper esophagus.

60 From the lower esophagus to the anus, the muscularis externa is formed from _____ muscle tissue.

61 Except for the stomach, the inner layer of the muscularis externa is the _____ layer and the outer layer is the _____ layer.

62 The stomach has an additional muscular layer called the _____ layer, which functions in the _____ movements of the stomach.

SMALL INTESTINE

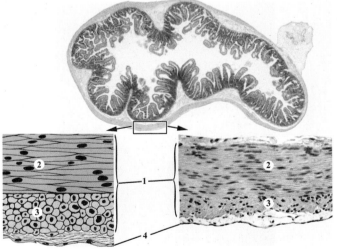

Figure 26.9

63 In reference to **Figure 26.9**, identify #1 - #3.
   1 _____ 3 _____
   2 _____ 4 _____

356 Chapter 26 - Digestive System

Peristalsis

64 What is peristalsis? _____

65 The contraction of which muscle layer results in the constriction the lumen? _____

66 The contraction of which muscle layer results in the dilation of the lumen? _____

Segmentation

67 What is segmentation? _____

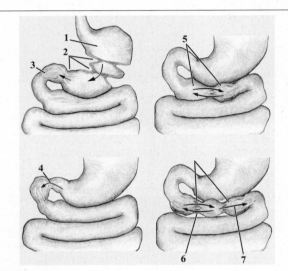

Figure 26.10

68 In reference to **Figure 26.10**, identify #1 - #7.
1 _____ 5 _____
2 _____ 6 _____
3 _____ 7 _____
4 _____

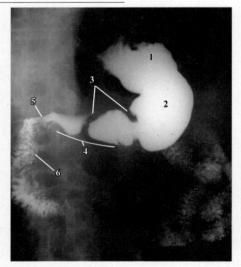

Figure 26.11

69 In reference to **Figure 26.11**, identify #1 - #6.
1 _____ 4 _____
2 _____ 5 _____
3 _____ 6 _____

SEROSA

70 Except for the esophagus, the outermost layer of the alimentary canal is called the _____.

71 The outer layer of the serosa, the visceral peritoneum, functions in the production of the lubricating _____ fluid.

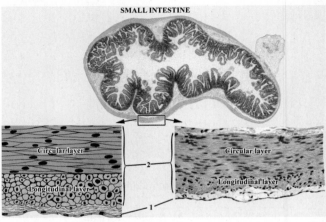

Figure 26.12

72 In reference to **Figure 26.12**, identify #1 - #2.
1 _____ 2 _____

THE PERITONEUM AND ITS CAVITY

73 What is the peritoneum? _____

74 What are the two layers of the peritoneum and where is each located? _____

75 What is the name of the fluid located between the two layers of the peritoneum? _____

76 How is a mesentery formed? _____

77 What does the term retroperitoneal mean? _____

78 What are two locations of the mesenteries? _____

79 What is the omentum? _____

80 What are the two divisions of the omentum? _____

81 What does the lesser omentum connect? _____

82 What does the greater omentum connect? _____

Name _____
Class _____

Chapter 26 - Digestive System **357**

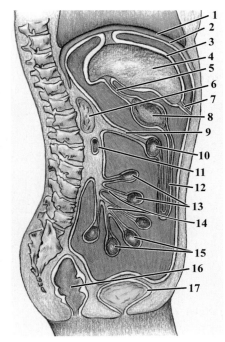

Figure 26.13
83 In reference to **Figure 26.13**, identify #1 - #17.
  1 _____ 10 _____
  2 _____ 11 _____
  3 _____ 12 _____
  4 _____ 13 _____
  5 _____ 14 _____
  6 _____ 15 _____
  7 _____ 16 _____
  8 _____ 17 _____
  9 _____

ORGANS OF THE DIGESTIVE SYSTEM

MOUTH

84 The anterior region of the mouth, the area between the lips and cheeks, is called the _____.
85 The cavity of the mouth, the oral cavity proper, is located _____
86 The pharynx extends from the mouth and superior nasal cavity to the _____.
87 The pharynx is divided into three regions, the _____ _____
88 The roof of the mouth is called the _____, which is divided into two regions the _____ palate and the _____ palate.
89 The _____ is a fleshy appendage hanging from the posterior boundary of the _____ palate.
90 During swallowing, the soft palate and uvula function to _____

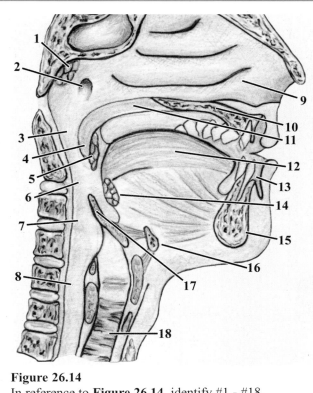

Figure 26.14
91 In reference to **Figure 26.14**, identify #1 - #18.
  1 _____ 10 _____
  2 _____ 11 _____
  3 _____ 12 _____
  4 _____ 13 _____
  5 _____ 14 _____
  6 _____ 15 _____
  7 _____ 16 _____
  8 _____ 17 _____
  9 _____ 18 _____

TONGUE

92 The functions of the tongue include _____
93 The epithelial lining of the tongue is _____.
94 The small protuberances on the surface of the tongue are called _____.
95 The three types of papillae on the surface of the tongue are the _____.
96 The tongue is attached to the floor of the oral cavity by the _____.
97 The filiform papillae make the surface of the tongue rough, which aids in _____.
98 The two types of papillae that house taste buds are the ____ _____.

358 Chapter 26 - Digestive System

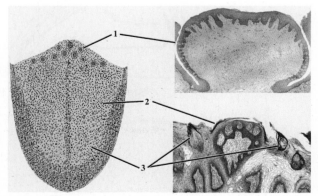

Figure 26.15

99 In reference to **Figure 26.15**, identify #1 - #3.
  1 _____ 3 _____
  2 _____

LAB ACTIVITY 2

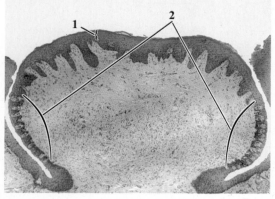

Figure 26.16

100 In reference to **Figure 26.16**, identify #1 - #2.
  1 _____ 2 _____

LAB ACTIVITY 3

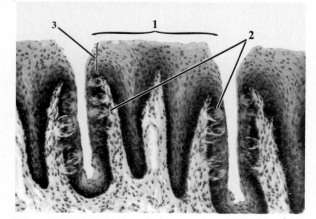

Figure 26.17

101 In reference to **Figure 26.17**, identify #1 - #3.
  1 _____ 3 _____
  2 _____

Taste buds

102 What are the two types of cells found within tastebuds? ___
_____

103 What is the function of "taste hairs?" _____
_____

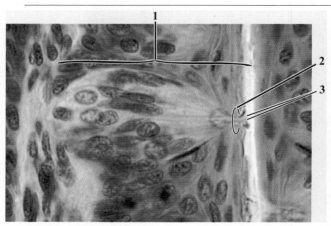

Figure 26.18

104 In reference to **Figure 26.18**, identify #1 - #3.
  1 _____ 3 _____
  2 _____

SALIVARY GLANDS

105 What are the three major salivary glands? _____
_____
_____

LAB ACTIVITY 4

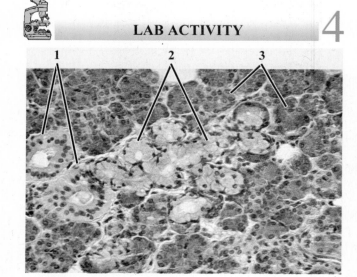

Figure 26.19

106 In reference to **Figure 26.19**, identify #1 - #3.
  1 _____ 3 _____
  2 _____

107 What is the function of mucous cells? _____
_____

108 What is the function of serous cells? _____
_____

109 What is the function of salivary amylase? _____
_____
_____

PHARYNX

110 What is the pharynx? _____

111 What are the three regions of the pharynx? _____

112 Which region of the pharynx is not a division of the digestive system? _____

113 The region of the pharynx that extends from the soft palate to the hyoid bone is the _____, and from the hyoid bone to the cricoid cartilage of the larynx is the _____.

114 What is the uvula? _____

ESOPHAGUS

115 The esophagus extends from the _____ to the _____ region of the stomach.

116 What is the esophageal hiatus? _____

117 The sphincter that surrounds the cardiac orifice is called the _____.

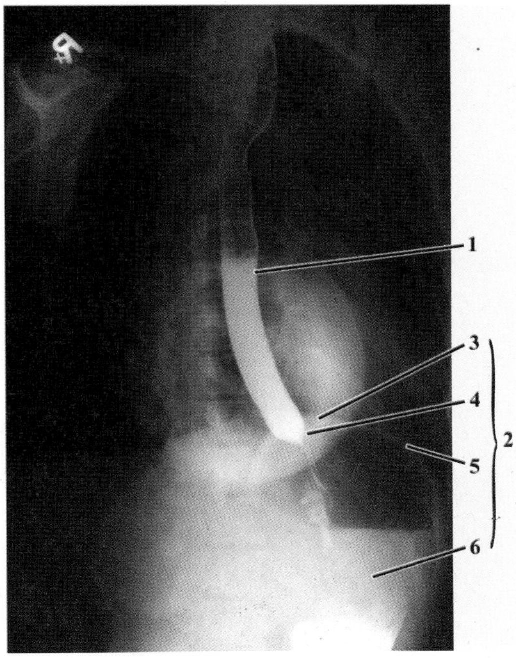

Figure 26.20

118 In reference to **Figure 26.20**, identify #1 - #6.
1 _____ 4 _____
2 _____ 5 _____
3 _____ 6 _____

119 From inner to outer, what are the four layers of the esophagus? _____

120 How does the muscularis externa of the esophagus differ from muscularis externa of other regions of the digestive tract? _____

121 What is the function of the epithelium of the esophagus? _____

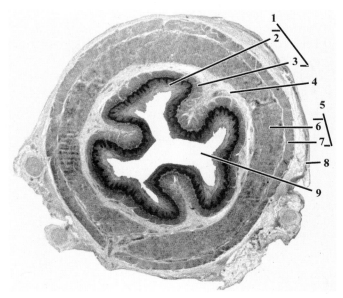

Figure 26.21

122 In reference to **Figure 26.21**, identify #1 - #9.
1 _____ 6 _____
2 _____ 7 _____
3 _____ 8 _____
4 _____ 9 _____
5 _____

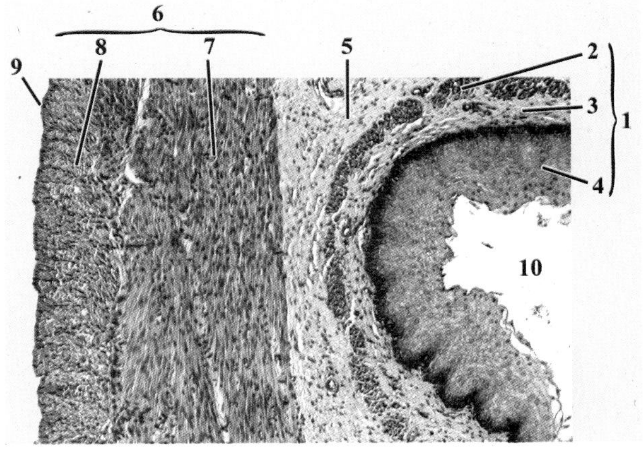

Figure 26.22

123 In reference to **Figure 26.22**, identify #1 - #9.
1 _____ 6 _____
2 _____ 7 _____
3 _____ 8 _____
4 _____ 9 _____
5 _____

STOMACH

124 The stomach is located between the _____

125 Three functions of the stomach are _____

126 The four regions of the stomach are the _____

127 The superior medial portion of the stomach that houses the cardiac orifice is the _____.

360   Chapter 26 - Digestive System

128 The upper portion of the stomach is called the _____.
129 The lower portion of the stomach is called the _____.
130 The region of the stomach between the fundus and the pylorus is the _____.
131 The stomach terminates with the _____ sphincter.
132 What is the function of the pyloric sphincter? _____

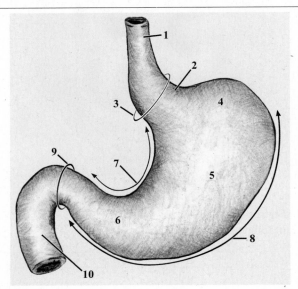

Figure 26.23
133 In reference to **Figure 26.23**, identify #1 - #10.
1 _____   6 _____
2 _____   7 _____
3 _____   8 _____
4 _____   9 _____
5 _____   10 _____

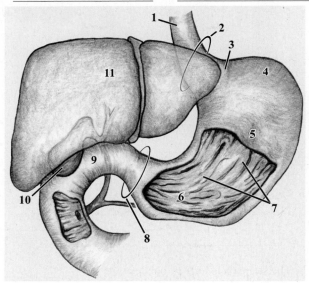

Figure 26.24
134 In reference to **Figure 26.24**, identify #1 - #11.
1 _____   7 _____
2 _____   8 _____
3 _____   9 _____
4 _____   10 _____
5 _____   11 _____
6 _____

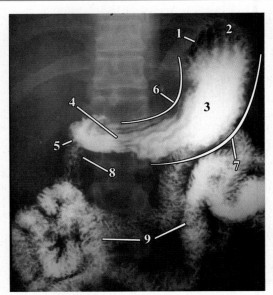

Figure 26.25
135 In reference to **Figure 26.25**, identify #1 - #9.
1 _____   6 _____
2 _____   7 _____
3 _____   8 _____
4 _____   9 _____
5 _____
136 What are the four layers of the wall of the stomach? _____

 LAB ACTIVITY 5

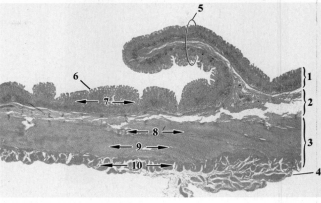

Figure 26.26
137 In reference to **Figure 26.26**, identify #1 - #10.
1 _____   6 _____
2 _____   7 _____
3 _____   8 _____
4 _____   9 _____
5 _____   10 _____

MUCOSA

138 What are two modifications of the stomach's mucosa? _____

Chapter 26 - Digestive System   361

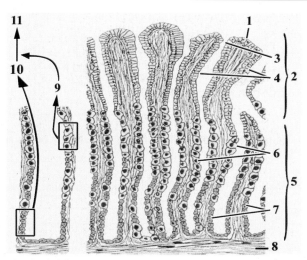

Figure 26.27

139 In reference to **Figure 26.27**, identify #1 - #11.
1 _____ 7 _____
2 _____ 8 _____
3 _____ 9 _____
4 _____ 10 _____
5 _____ 11 _____
6 _____

140 What is the function of the stomach's rugae? _____

141 What are gastric pits? _____
_____

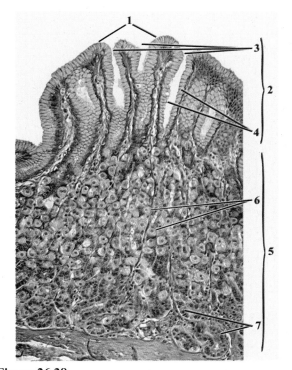

Figure 26.28

142 In reference to **Figure 26.28**, identify #1 - #7.
1 _____ 5 _____
2 _____ 6 _____
3 _____ 7 _____
4 _____

143 What are three substances produced by the gastric glands?
_____
_____

144 What two types of cells contribute to the production of gastric juice? _____
_____

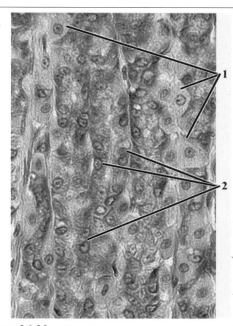

Figure 26.29

145 In reference to **Figure 26.29**, identify #1 - #2.
1 _____ 2 _____

146 What is the function of parietal cells? _____
_____

147 What is the function of hydrochloric acid (HCl)? _____
_____

148 What is the function of intrinsic factor? _____

149 What is the function of chief cells? _____

150 What is the name of the active protein digesting enzyme of gastric juice? _____

151 What is the function of enteroendocrine cells? _____
_____
_____

152 What triggers the release of gastrin? _____
_____

153 What is the function of gastrin? _____
_____

CONTROL OF GASTRIC ACTIVITY

154 What are three phases of gastric secretion? _____
_____

155 What triggers the cephalic phase? _____

Chapter 26 - Digestive System

156 What is the function of the cephalic phase? _____

157 When does the gastric phase begin? _____

158 What are three stimuli for increased gastric secretion? _____

159 What is the primary hormone produced in the gastric phase? _____

160 What is the affect of gastrin on the stomach's gastric glands? _____

161 What is the primary mechanism that decreases the secretion of gastric juice? _____

162 What is the affect of gastrin on the small intestine? _____

163 What is the affect of gastrin on the ileocecal valve? _____

164 What is the affect of gastrin on the large intestine? _____

165 How is the intestinal phase regulated? _____

166 What is the result of distension of the duodenum by incoming chyme? _____

167 What is the duodenum's response to the presence of lipid and protein? _____

168 What is the affect of cholecystokinin (CCK) on the gastric glands? _____

169 What is the affect of CCK on the pancreas? _____

170 What is the affect of CCK on the gallbladder? _____

171 What is the affect of CCK on the hepatopancreatic sphincter? _____

172 What is the function of intestinal gastrin? _____

173 What hormone is produced with the arriving chyme has a pH below 4.5? _____

174 What is the affect of secretin on the stomach? _____

175 What is the affect of secretin on the pancreas? _____

176 What is the affect of secretin on the liver? _____

177 What is the function of the bicarbonate ions in pancreatic juice? _____

Submucosa
178 What forms the submucosa? _____

Muscularis
179 From inner to outer, what are the three layers of the stomach's muscularis externa? _____

180 What is the function of the stomach's muscularis externa? _____

Serosa
181 What is another name for the stomach's serosa? _____

Small Intestine
182 From the stomach, what are the three regions of the small intestine? _____

183 What are three substances received by the duodenum? _____

184 Where is the hepatopancreatic ampulla? _____

185 What is the hepatopancreatic (duodenal) papilla? _____

186 What ducts merge to form the hepatopancreatic ampulla? _____

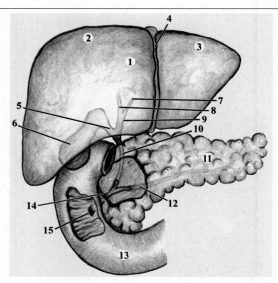

Figure 26.30

187 In reference to **Figure 26.30**, identify #1 - #15.

1 _____ 8 _____
2 _____ 9 _____
3 _____ 10 _____
4 _____ 11 _____
5 _____ 12 _____
6 _____ 13 _____
7 _____ 14 _____
            15 _____

Name _____
Class _____

Chapter 26 - Digestive System   363

188 What hormone controls the hepatopancreatic sphincter? ___
_____

189 What is the function of the jejunum? _____

190 What is the name of the valve at the terminus of the ileum?
_____

191 What is the function of the ileocecal valve? _____
_____

192 What hormone targets the ileocecal valve? _____
_____

193 From inner to outer what are the four layers of the wall of the small intestine? _____
_____

LAB ACTIVITY 6

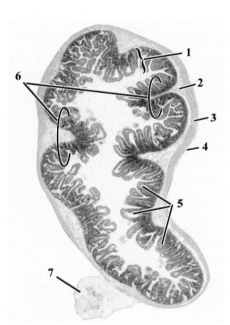

Figure 26.31

194 In reference to **Figure 26.31**, identify #1 - #7.
   1 _____   5 _____
   2 _____   6 _____
   3 _____   7 _____
   4 _____

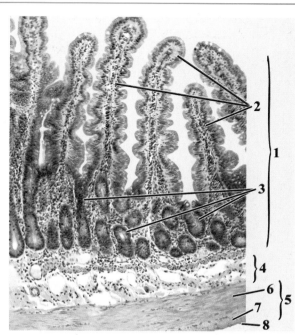

Figure 26.32

195 In reference to **Figure 26.32**, identify #1 - #8.
   1 _____   5 _____
   2 _____   6 _____
   3 _____   7 _____
   4 _____   8 _____

MUCOSA

196 The mucosa of the small intestine is modified for _____ surface area, and _____ of intestinal juice.

197 Two ways the mucosa is modified to increase surface are is by _____ and _____.

198 Intestinal glands produce _____.

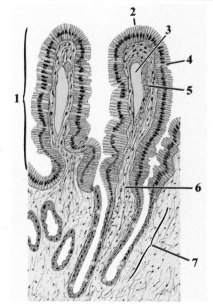

Figure 26.33

199 In reference to **Figure 26.33**, identify #1 - #7.
   1 _____   5 _____
   2 _____   6 _____
   3 _____   7 _____
   4 _____

Chapter 26 - Digestive System

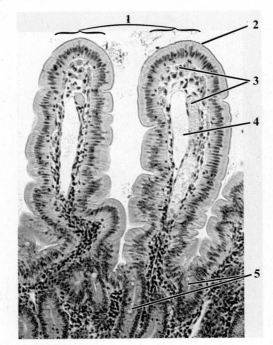

Figure 26.34

200 In reference to **Figure 26.34**, identify #1 - #5.
1 _____ 4 _____
2 _____ 5 _____
3 _____

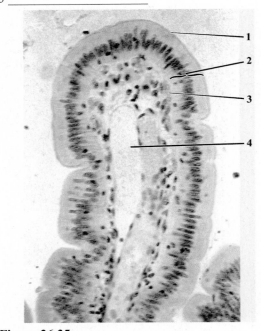

Figure 26.35

201 In reference to **Figure 26.35**, identify #1 - #4.
1 _____ 3 _____
2 _____ 4 _____

202 What is the function of villi? _____
_____

203 What are microvilli? _____
_____

204 What are two functions of microvilli? _____
_____
_____

205 What is a lacteal? _____
_____

206 What enters the capillaries of the villi? _____
_____

207 What enters the lacteals? _____
_____

208 What epithelium lines the small intestine? _____
_____

209 How is the surface (apical) plasma membranes of the columnar cells modified? _____
_____

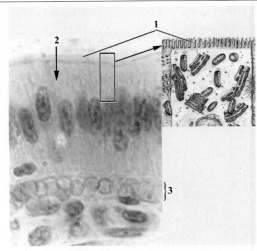

Figure 26.36

210 In reference to **Figure 26.36**, identify #1 - #3.
1 _____ 3 _____
2 _____

211 Where are the intestinal glands (crypts) located? _____
_____

212 What is the function of the intestinal glands (crypts)? _____
_____

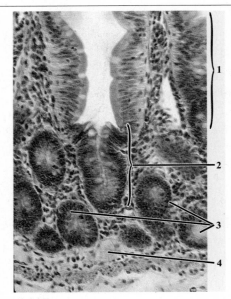

Figure 26.37

213 In reference to **Figure 26.37**, identify #1 - #4.
1 _____ 3 _____
2 _____ 4 _____

SUBMUCOSA

214 What forms the submucosa of the small intestine? _____
_____

MUSCULARIS EXTERNA

215 From inner to outer, what are the two layers of the muscularis externa of the small intestine? _____
_____

216 Which layer contracts to produce a dilation of the small intestine's internal cavity? _____
_____

217 Which layer contracts to produce a constriction of the small intestine's internal cavity? _____
_____

218 What is segmentation? _____
_____
_____

219 What is peristalsis? _____
_____
_____

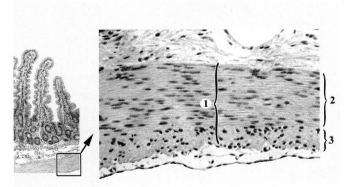

Figure 26.38

220 In reference to **Figure 26.38**, identify #1 - #3.
1 _____ 3 _____
2 _____

SEROSA

221 What is another name for the serosa? _____
_____

REGIONS OF THE SMALL INTESTINE
DUODENUM - DUODENAL (BRUNNER'S) GLANDS

222 What three substances are mixed in the duodenum? _____
_____
_____

223 What layer of the duodenum contains the duodenal (Brunner's) glands? _____
_____

224 What is the function of duodenal (Brunner's) glands? _____
_____

LAB ACTIVITY 7

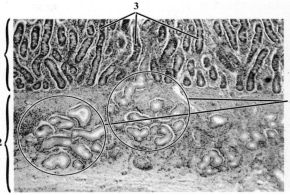

Figure 26.39

225 In reference to **Figure 26.39**, identify #1 - #4.
1 _____ 3 _____
2 _____ 4 _____

ILEUM - PEYER'S PATCHES

226 What are Peyer's patches? _____
_____

227 What is the function of Peyer's patches? _____
_____

LAB ACTIVITY 8

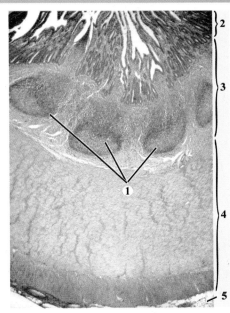

Figure 26.40

228 In reference to **Figure 26.40**, identify #1 - #5.
1 _____ 4 _____
2 _____ 5 _____
3 _____

Chapter 26 - Digestive System

PANCREAS

229 The pancreas functions both as an _____ and an _____ gland.

230 What is the function of the pancreatic islets (islets of Langerhans)? _____

231 What are pancreatic acini? _____

232 What is the function of pancreatic acini? _____

233 What is the hepatopancreatic ampulla? _____

234 What hormone controls the relaxation of the hepatopancreatic sphincter? _____

235 Where is cholecystokinin produced? _____

236 Which division of the autonomic nervous system stimulates pancreatic secretion? _____

237 What two hormones influence the secretion of pancreatic juice? _____

Secretin

238 What is the stimulus for the release of secretin? _____

239 How does secretin affect the secretion of pancreatic juice? _____

240 What is the function of bicarbonate ions in pancreatic juice? _____

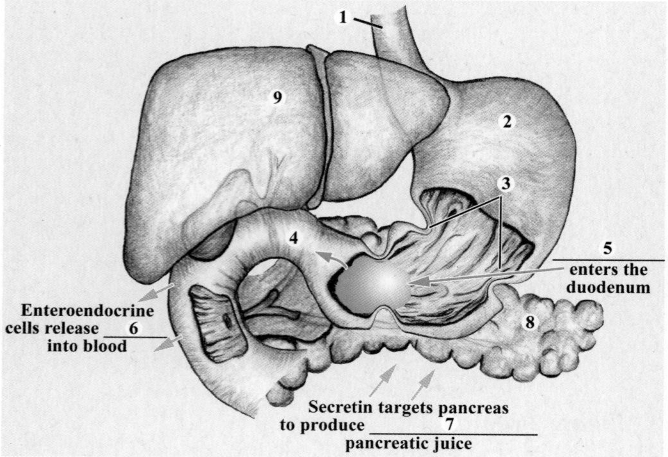

Figure 26.41

241 In reference to **Figure 26.41**, identify #1 - #9.
1 _____ 6 _____
2 _____ 7 _____
3 _____ 8 _____
4 _____ 9 _____
5 _____

Cholecystokinin (CCK)

242 What is the stimulus for the release of cholecystokinin? _____

243 How does cholecystokinin affect the secretion of pancreatic juice? _____

244 What is the function of pancreatic enzymes? _____

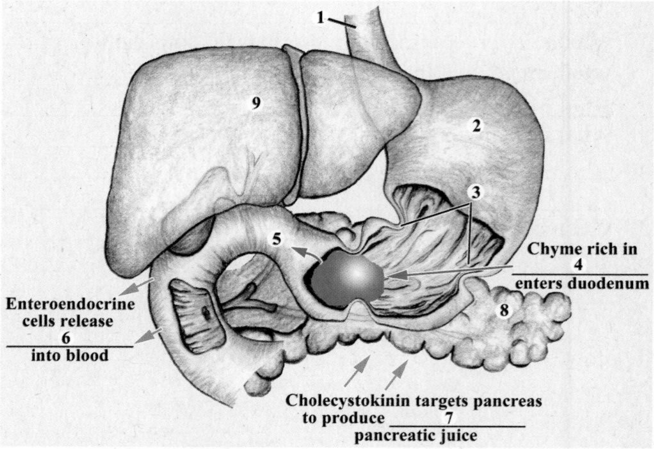

Figure 26.42

245 In reference to **Figure 26.42**, identify #1 - #9.
1 _____ 6 _____
2 _____ 7 _____
3 _____ 8 _____
4 _____ 9 _____
5 _____

LAB ACTIVITY 9

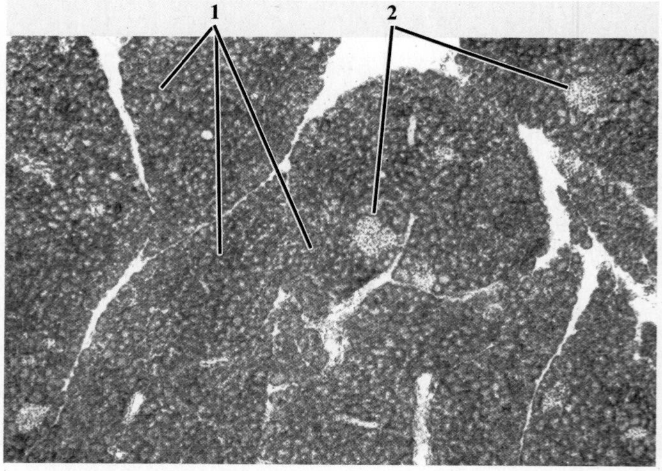

Figure 26.43

246 In reference to **Figure 26.43**, identify #1 - #2.
1 _____ 2 _____

Chapter 26 - Digestive System   367

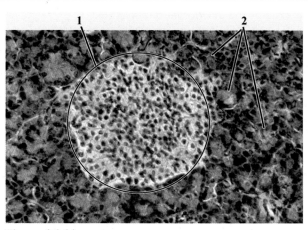

Figure 26.44

247 In reference to **Figure 26.44**, identify #1 - #2.
   1 _____ 2 _____

LIVER

248 Where is the liver located? _____
_____

249 How many lobes form the liver? _____

250 Which lobe is the largest? _____

251 What is the falciform ligament? _____

252 What is the ligamentum teres? _____

253 Where is the gallbladder located? _____
_____

254 What are the two major vessels that enter the liver? ___

255 What is the function of the hepatic artery? _____
_____

256 What is the function of the hepatic portal vein? _____
_____

257 What are two factors that function to increase the production of bile? _____
_____
_____

258 What is the function of bile salts? _____
_____

259 What are two liver related functions of cholecystokinin?
_____
_____
_____

260 What is the stimulus for the secretion of cholecystokinin?
_____

261 What is the affect of increased blood levels of bile salts? _
_____
_____

262 How does secretin affect bile production? _____
_____

263 What is the stimulus for the secretion of secretin? _____
_____
_____

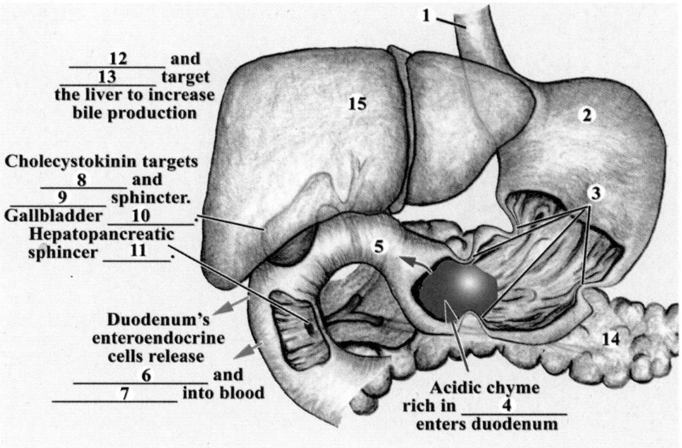

Figure 26.45

264 In reference to **Figure 26.45**, identify #1 - #15.
   1 _____ 9 _____
   2 _____ 10 _____
   3 _____ 11 _____
   4 _____ 12 _____
   5 _____ 13 _____
   6 _____ 14 _____
   7 _____ 15 _____
   8 _____

265 What is the name of the functional units of the liver? ___
_____

266 Lobules consists of _____
_____

267 What are sinusoids? _____
_____

268 What are the two sources of blood that enter the sinusoids?
_____
_____

269 What is the function of Kupffer cells? _____
_____

270 What is the destination of blood leaving the central veins?
_____
_____

Chapter 26 - Digestive System

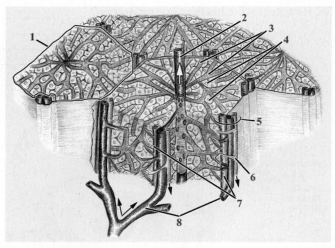

Figure 26.46

271 In reference to **Figure 26.46**, identify #1 - #8.
1 _____   5 _____
2 _____   6 _____
3 _____   7 _____
4 _____   8 _____

 LAB ACTIVITY 10

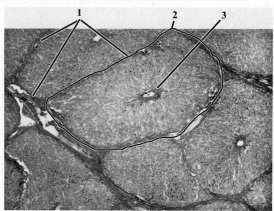

Figure 26.47

272 In reference to **Figure 26.47**, identify #1 - #3.
1 _____   3 _____
2 _____

Figure 26.48

273 In reference to **Figure 26.48**, identify #1 - #4.
1 _____   3 _____
2 _____   4 _____

274 What is the function of bile salts? _____

275 What are two waste products found in bile? _____

276 What is the function of the hepatic portal system? _____

277 What vein carries nutrient rich blood to the liver? _____

278 What pancreatic hormones target the liver? _____

279 What affect does insulin have on the liver? _____

LAB ACTIVITY 11

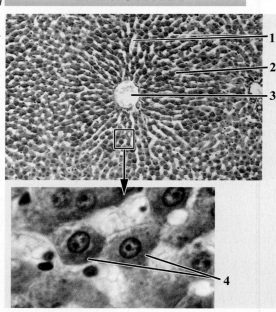

Figure 26.49

280 In reference to **Figure 26.49**, identify #1 - #4.
1 _____   3 _____
2 _____   4 _____

281 What affect does glucagon have on the liver? _____

Name _____
Class _____

Chapter 26 - Digestive System   **369**

LAB ACTIVITY 12

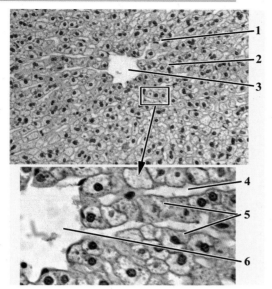

Figure 26.50

282 In reference to **Figure 26.50**, identify #1 - #6.
1 _____ 4 _____
2 _____ 5 _____
3 _____ 6 _____

KUPFFER CELLS
283 What is the function of Kupffer cells? _____
_____

LAB ACTIVITY 13

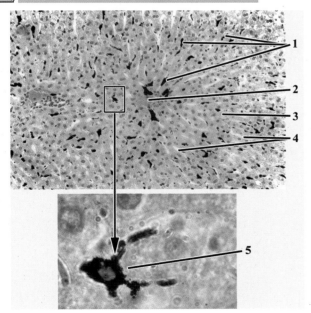

Figure 26.51
284 In reference to **Figure 26.51**, identify #1 - #5.
1 _____ 4 _____
2 _____ 5 _____
3 _____

LARGE INTESTINE
285 What are three functions of the large intestine? _____
_____
_____

286 What are three ways the gross anatomy of the large intestine differs from that of the small intestine? _____
_____
_____

287 What are haustra? _____
_____

288 What are epiploic appendages? _____
_____

289 What are the five regions of the large intestine? _____
_____

290 What is the name of the valve at the junction of the ileum with the cecum? _____
_____

291 What hormone targets the ileocecal valve and results in its relaxation? _____
_____

292 Where is gastrin produced? _____
_____

293 Where is the appendix located? _____

294 What is housed in the walls of the appendix? _____
_____

295 What are the four divisions of the colon? _____
_____

296 The last portion of the colon, the sigmoid colon joins with the _____, which terminates at the _____ canal.

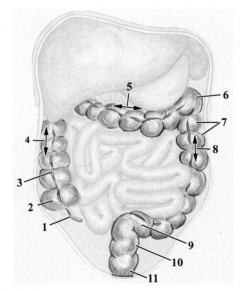

Figure 26.52
297 In reference to **Figure 26.52**, identify #1 - #11.
1 _____ 7 _____
2 _____ 8 _____
3 _____ 9 _____

370 Chapter 26 - Digestive System

Name _____
Class _____

4 _____ 10 _____
5 _____ 11 _____
6 _____

1 _____ 4 _____
2 _____ 5 _____
3 _____

298 From inner to outer, what are the four layers of the wall of the large intestine? _____

LAB ACTIVITY 14

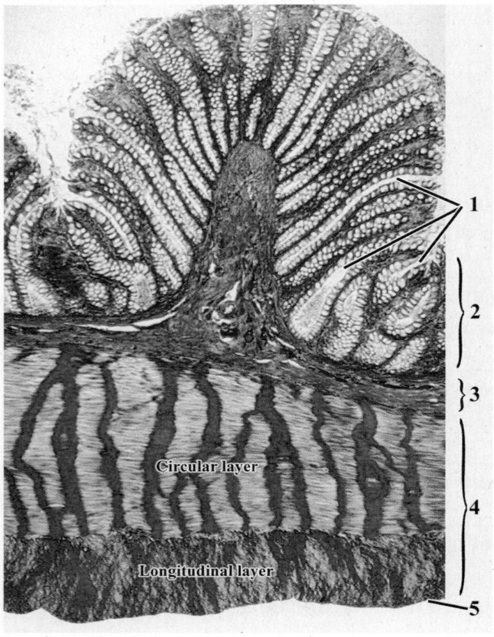

Figure 26.53

299 In reference to **Figure 26.53**, identify #1 - #5.
1 _____ 4 _____
2 _____ 5 _____
3 _____

MUCOSA

300 What are two modifications of the mucosa of the large intestine? _____

301 What is the function of the goblet cells of the intestinal glands? _____

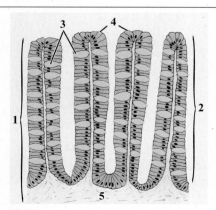

Figure 26.54

302 In reference to **Figure 26.54**, identify #1 - #5.

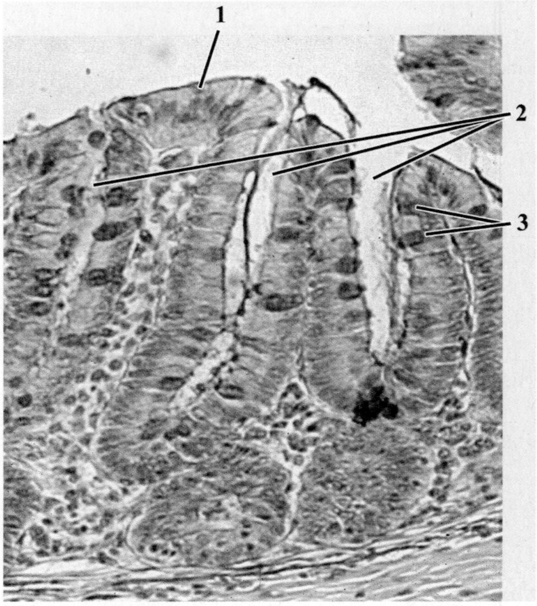

Figure 26.55

303 In reference to **Figure 26.55**, identify #1 - #3.
1 _____ 3 _____
2 _____

SUBMUCOSA

304 What forms the submucosa of the large intestine? _____

MUSCULARIS EXTERNA

305 What forms the muscularis externa of the large intestine? _____

306 What are the teniae coli? _____

307 What are haustra? _____

308 What is the function of the muscularis externa? _____

SEROSA

309 What is another name for the serosa? _____

Urinary System - Worksheets

1. What are four components of the urinary system?
   1. _____ 3. _____
   2. _____ 4. _____
2. What are four functions of the urinary system?
   1. _____ 3. _____
   2. _____ 4. _____
3. What are three processes involved in formation of urine?
   1. _____ 3. _____
   2. _____
4. What is the function of filtration? _____
5. What is the function of reabsorption? _____
6. What is the function of secretion? _____
7. What is the function of the ureters? _____
8. What is the function of the urinary bladder? _____
9. What is the function of the urethra? _____
10. What is micturition? _____

ANATOMY OF THE KIDNEY

LAB ACTIVITY 1

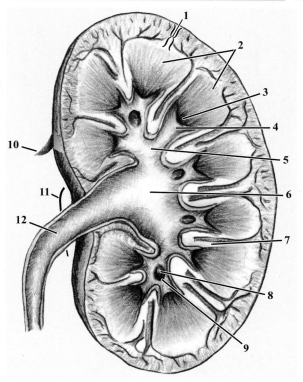

Figure 27.2

12. In reference to **Figure 27.2**, identify #1 - #12.
    1. _____ 7. _____
    2. _____ 8. _____
    3. _____ 9. _____
    4. _____ 10. _____
    5. _____ 11. _____
    6. _____ 12. _____

LAB ACTIVITY 2

KIDNEY DISSECTION
EXTERNAL ANATOMY

13. What is the hilus? _____
14. What passes through the hilus? _____
15. What is the renal capsule? _____

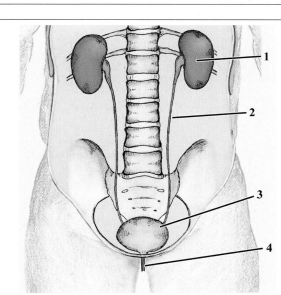

Figure 27.1

11. In reference to **Figure 27.1**, identify #1 - #4.
    1. _____ 3. _____
    2. _____ 4. _____

372   Chapter 27 - Urinary System

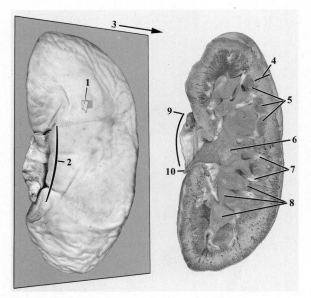

Figure 27.3

16 In reference to **Figure 27.3**, identify #1 - #10.

1 _____   6 _____
2 _____   7 _____
3 _____   8 _____
4 _____   9 _____
5 _____   10 _____

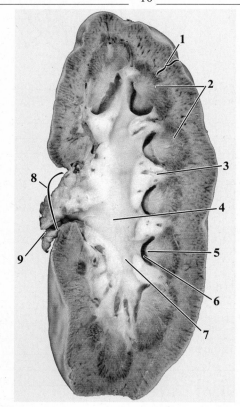

Figure 27.4

17 In reference to **Figure 27.4**, identify #1 - #9.

1 _____   6 _____
2 _____   7 _____
3 _____   8 _____
4 _____   9 _____
5 _____

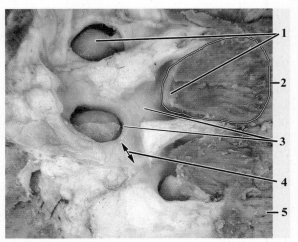

Figure 27.5

18 In reference to **Figure 27.5**, identify #1 - #5.

1 _____   4 _____
2 _____   5 _____
3 _____

19 What is the renal sinus? _____

20 Where is the renal cortex located? _____

21 What gives the renal cortex a granular appearance? _____

22 Where is the renal medulla located? _____

23 What forms the renal medulla? _____

24 What produces the striations of the renal pyramids? _____

25 What are the apices of the renal pyramids called? _____

26 What do the apices of the renal pyramids enter? _____

27 What are renal columns? _____

28 What is the function of renal columns? _____

29 What are minor calyces? _____

30 What are major calyces? _____

31 Where do the major calyces converge? _____

31 The renal pelvis is formed at the convergence of the _____ calyces and is continuous with the _____.

33 What is the function of the ureter? _____

NEPHRONS

34 What are the functional units of the kidney? _____

35 In what region of the kidney do nephrons originate? _____

Name _____
Class _____

Chapter 27 - Urinary System   373

36 What are the two major regions of a nephron? _____
    _____
37 What are the four divisions of the renal tubule? _____
    _____
38 What portion of the nephrons unites with the collecting ducts? _____
39 Where do the collecting ducts lead? _____
    _____
40 Where do the papillary ducts lead? _____
    _____
41 What are the two major classifications of nephrons? ___
    _____
42 The loop of Henle of cortical nephrons is mostly located in the _____ and is surround with capillaries called _____.
43 The loop of Henle of juxtamedullary nephrons is mostly located in the _____ and is surrounded with capillaries called the _____.
44 In juxtaglomerular nephrons, both the _____ capillaries and the _____ originate from the efferent arteriole.

LAB ACTIVITY 3

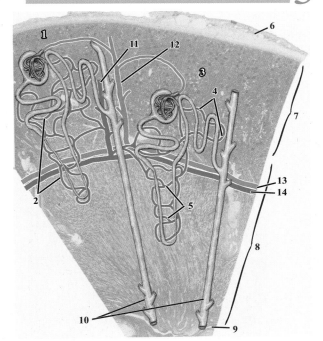

Figure 27.6
45 In reference to **Figure 27.6**, identify #1 - #14.
    1 _____   8 _____
    2 _____   9 _____
    3 _____   10 _____
    4 _____   11 _____
    5 _____   12 _____
    6 _____   13 _____
    7 _____   14 _____

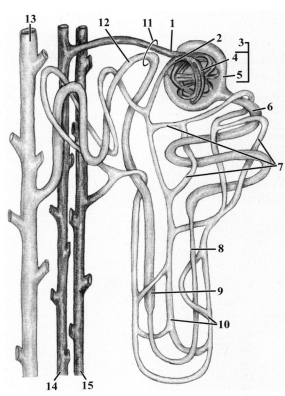

Figure 27.7
46 In reference to **Figure 27.7**, identify #1 - #15.
    1 _____   9 _____
    2 _____   10 _____
    3 _____   11 _____
    4 _____   12 _____
    5 _____   13 _____
    6 _____   14 _____
    7 _____   15 _____
    8 _____
47 What is the renal corpuscle? _____
    _____
48 What forms the glomerulus? _____
    _____
49 What are fenestrated capillaries? _____
    _____
50 What cells cover the glomerular capillaries? _____
    _____
51 What is the glomerular (Bowman's) capsule? _____
    _____
52 What is the name of the outer layer of the glomerular (Bowman's) capsule? _____
53 What is the name of the inner layer of the glomerular (Bowman's) capsule? _____
54 What cells form the visceral layer of the glomerular (Bowman's) capsule? _____

374 Chapter 27 - Urinary System

55 What are filtration slits? _____

56 What forms the filtration membrane? _____

57 Where is the capsular space located? _____

58 What does the capsular space receive through the filtration membrane? _____

59 Where does the capsular space lead? _____

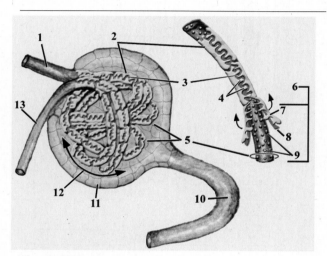

Figure 27.8

60 In reference to **Figure 27.8**, identify #1 - #13.
1 _____   8 _____
2 _____   9 _____
3 _____   10 _____
4 _____   11 _____
5 _____   12 _____
6 _____   13 _____
7 _____

61 Which tubule receives filtrate from the glomerular capsule? _____

62 What portion of the nephron follows the proximal convoluted tubule? _____

63 What are the two regions of the loop of Henle? _____

64 Which region of the loop of Henle leads to the distal convoluted tubule? _____

65 Which region of the nephron leads to the collecting duct? _____

66 The collecting ducts merge to form the _____.

67 The papillary ducts empty into the _____.

68 The arteriole that arises from the interlobular arteries and leads to the glomerulus is the _____.

69 The afferent arteriole is (smaller / larger) than the efferent arteriole.

70 The efferent arteriole of cortical nephrons leads to the _____.

71 The efferent arteriole of juxtamedullary nephrons leads to the _____.

72 The smaller diameter of the efferent arteriole (increases / decreases) blood pressure in the glomerular capillaries.

73 Peritubular capillaries surround the _____ of the renal cortex.

74 Peritubular capillaries function in _____.

75 The vasa recta are capillaries that parallel the _____.

76 The vasa recta functions in _____.

77 The juxtaglomerular apparatus is formed by the association of the _____.

78 The modified cells of the afferent arteriole are called the _____.

79 The modified cells of the distal convoluted tubule are called the _____.

80 The juxtaglomerular apparatus produces the hormone _____, and the enzyme _____.

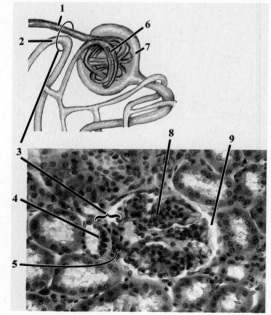

Figure 27.9

81 In reference to **Figure 27.9**, identify #1 - #9.
1 _____   6 _____
2 _____   7 _____
3 _____   8 _____
4 _____   9 _____
5 _____

Name _____
Class _____

Chapter 27 - Urinary System 375

LAB ACTIVITY 4

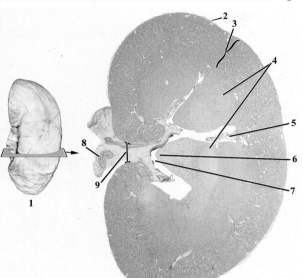

Figure 27.10
82 In reference to **Figure 27.10**, identify #1 - #9.
1 _____ 6 _____
2 _____ 7 _____
3 _____ 8 _____
4 _____ 9 _____
5 _____

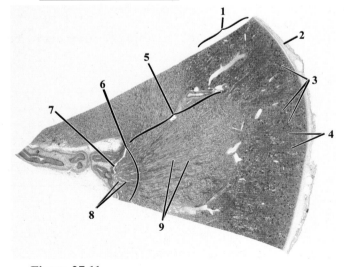

Figure 27.11
83 In reference to **Figure 27.11**, identify #1 - #9.
1 _____ 6 _____
2 _____ 7 _____
3 _____ 8 _____
4 _____ 9 _____,
5 _____ _____

84 Histologically, what are two features of the cortex which are used in its identification? _____
_____
_____

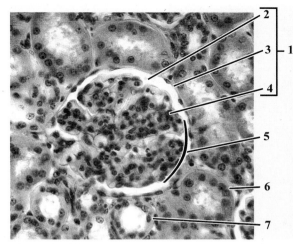

Figure 27.12
85 In reference to **Figure 27.12**, identify #1 - #7.
1 _____ 5 _____
2 _____ 6 _____
3 _____ 7 _____
4 _____

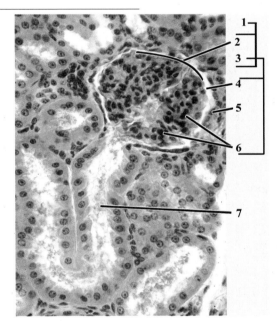

Figure 27.13
86 In reference to **Figure 27.13**, identify #1 - #7.
1 _____ 5 _____
2 _____ 6 _____
3 _____ 7 _____
4 _____

87 What forms the renal medulla? _____
_____

88 Histologically, how are the renal pyramids identified? ___
_____
_____

376 Chapter 27 - Urinary System

Name _____
Class _____

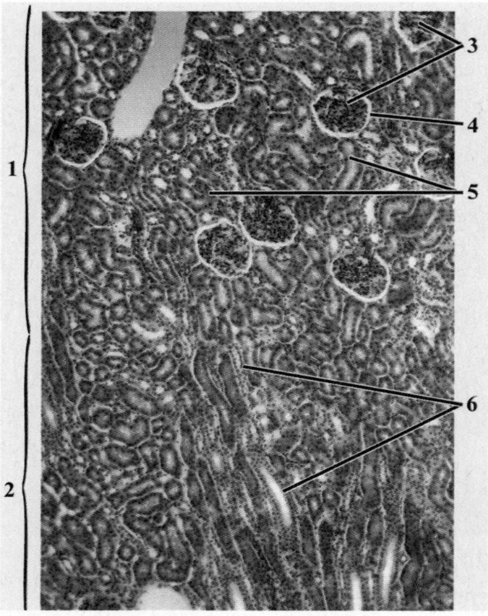

Figure 27.14

89 In reference to **Figure 27.14**, identify #1 - #11.
1 _____ 7 _____
2 _____ 8 _____
3 _____ 9 _____

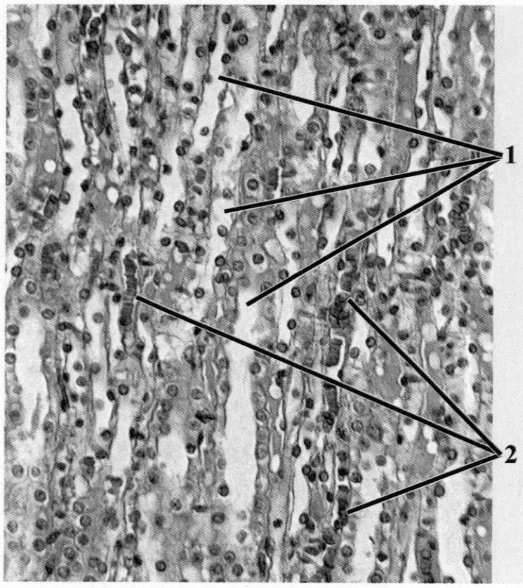

Figure 27.15

90 In reference to **Figure 27.15**, identify #1 - #2.
1 _____
2 _____

PROCESSES OF URINE FORMATION

91 What are the three processes involved in the production of urine? _____

92 What is filtration? _____

93 What functions as the kidney's filters? _____

94 What is reabsorption? _____

95 If substances in the tubular fluid are not reabsorbed they become a component of _____.

96 What is secretion? _____

97 What two structures of the kidney function in secretion? _____

98 Secreted substances become part of the _____.

99 What is excretion? _____

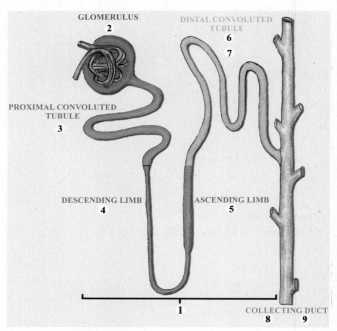

Figure 27.16

100 In reference to **Figure 27.16**, identify functions of #1 - #9.
1 _____
2 _____
3 _____
4 _____
5 _____
6 Reabsorption of _____
7 Secretion of _____
8 Reabsorption of _____
9 Secretion of _____

FILTRATION AT THE GLOMERULUS
FILTRATION MEMBRANE

101 What forms the kidney's filtration membranes? _____

102 What does the filtration membrane allow to pass into the capsular space as filtrate? _____

Chapter 27 - Urinary System

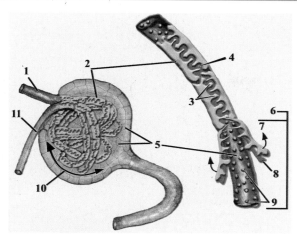

Figure 27.17

103 In reference to **Figure 27.17**, identify #1 - #11.
1 _____   7 _____
2 _____   8 _____
3 _____   9 _____
4 _____ 10 _____
5 _____ 11 _____
6 _____

FILTRATION PRESSURE

104 What is the driving force for filtration? _____

105 What two forces oppose filtration? _____

106 How is net filtration pressure (NFP) determined? _____

107 Why is glomerular hydrostatic pressure relatively high? _____

108 What produces glomerular osmotic pressure? _____

109 What produces capsular hydrostatic pressure? _____

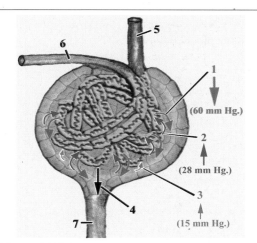

Figure 27.18

110 In reference to **Figure 27.18**, identify #1 - #7.
1 _____   5 _____
2 _____   6 _____
3 _____   7 _____
4 _____

GLOMERULAR FILTRATION

111 What is glomerular filtration rate? _____

112 Of the factors that influence the function of the filtration membrane in producing the glomerular filtration rate, which one has the tendency to show the highest degree of variation? _____

MAINTENANCE OF GLOMERULAR FILTRATION

113 What are two ways normal glomerular filtration rate is maintained? _____

Myogenic mechanism

114 Where does the myogenic mechanism originate? _____

115 What is the basis for the function of the myogenic mechanism? _____

116 What is the response of the vascular smooth muscle of the afferent arteriole to increased blood pressure? _____

117 What is the result of contraction of the vascular smooth muscle of the afferent arteriole? _____

Tubuloglomerular mechanism

118 Where does the tubuloglomerular mechanism originate? _____

119 What is the name of the group of cells of the distal convoluted tubule that operate the tubuloglomerular mechanism? _____

120 What do the macula densa cells respond to? _____

121 What is the result of the chemical mediator released by the macula densa cells? _____

122 How does vasoconstriction of the afferent arteriole reduce the osmolarity (solute concentration) of the tubular fluid? _____

378 Chapter 27 - Urinary System

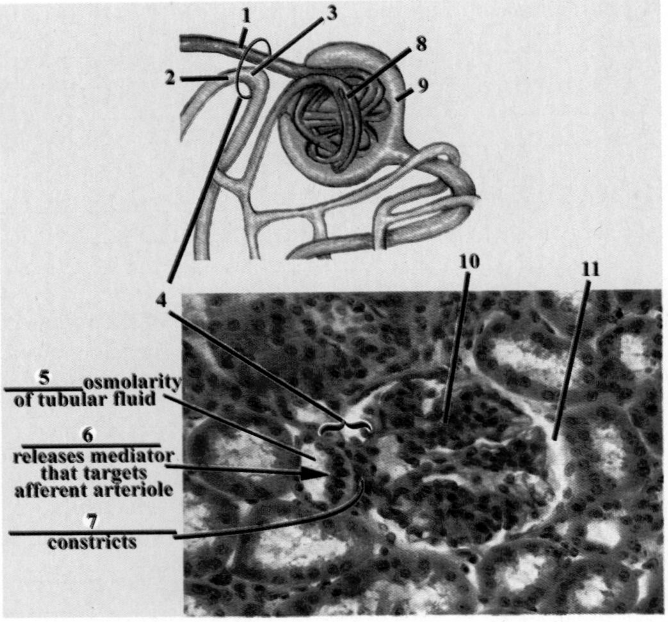

Figure 27.19

123 In reference to **Figure 27.19**, identify #1 - #11.

1 _____ 7 _____
2 _____ 8 _____
3 _____ 9 _____
4 _____ 10 _____
5 _____ 11 _____
6 _____

Renin-Angiotensin Mechanism

124 What is renin? _____

125 When is renin increasingly released into the blood? ____

126 What is the function of renin? _____

127 How is angiotensin I converted to angiotensin II? ____

128 What is the effect of angiotensin II on systemic blood pressure? _____

129 What is the effect of angiotensin II on the adrenal cortex? _____

130 What is the effect of aldosterone on the distal convoluted tubules? _____

131 What is the osmotic effect of the reabsorption of sodium ions? _____

132 What effect does increased blood volume have upon systemic blood pressure? _____

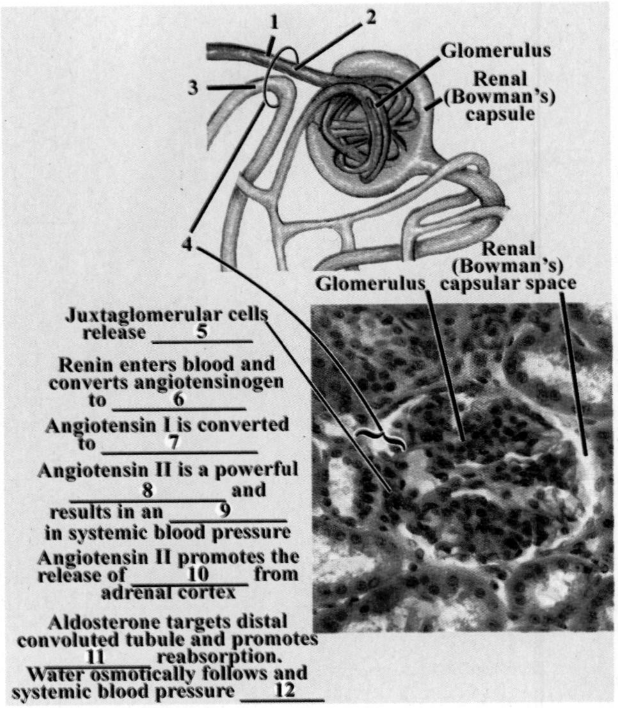

Figure 27.20

133 In reference to **Figure 27.20**, identify #1 - #12.

1 _____ 7 _____
2 _____ 8 _____
3 _____ 9 _____
4 _____ 10 _____
5 _____ 11 _____
6 _____ 12 _____

Sympathetic Nervous System

134 What effect does decreased systemic blood pressure have on arterioles controlled by the sympathetic nervous system? _____

135 What effect does constriction of arterioles have on systemic blood pressure? _____

136 What effect does the sympathetic nervous system have upon the juxtaglomerular cells? _____

137 Increased blood levels of the enzyme renin ultimately results in the production of _____.

138 What effect does angiotensin II have upon the adrenal cortex? _____

139 What is the osmotic effect of the reabsorption of sodium ions? _____

REABSORPTION

140 What is reabsorption? _____

141 Substances that are not reabsorbed become a component of _____.

140 Substances that are reabsorbed enter the _____.

Name _____
Class _____

PROXIMAL CONVOLUTED TUBULE

143 Which portion of the nephron functions as the site where most reabsorption occurs? _____

144 What is reabsorbed at the proximal convoluted tubule? _____

145 What processes function in reabsorption? _____

146 About how much of the sodium ions are reabsorbed in the proximal convoluted tubule? _____

147 What is obligatory water reabsorption? _____

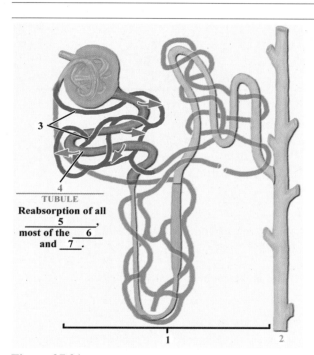

Figure 27.21

148 In reference to **Figure 27.21**, identify #1 - #7.
1 _____ 5 _____
2 _____ 6 _____
3 _____ 7 _____
4 _____

LOOP OF HENLE

149 The descending limb of the loop of Henle is permeable to _____.

150 In the cortical nephron water is reabsorbed into the _____ and in the juxtamedullary nephron water is reabsorbed into the _____.

151 The ascending limb of the loop of Henle is permeable to _____.

152 In the cortical nephron sodium and chloride ions are reabsorbed into the _____ and in the juxtamedullary nephron sodium and chloride ions are reabsorbed into the _____.

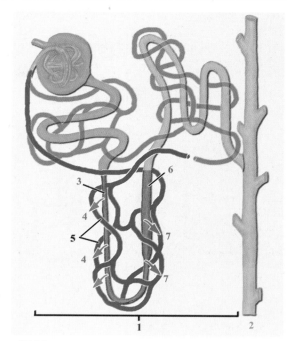

Figure 27.22

153 In reference to **Figure 27.22**, identify #1 - #7.
1 _____ 5 _____
2 _____ 6 _____
3 _____ 7 _____
4 _____

DISTAL CONVOLUTED TUBULE AND COLLECTING DUCT

154 Reabsorption in the distal convoluted tubule is mostly under _____ control.

155 Three major substances that are reabsorbed at the distal convoluted tubule and the collecting duct are _____

Water reabsorption and Antidiuretic Hormone (ADH)

156 The hormone that mostly regulates water reabsorption is _____.

157 Antidiuretic hormone is released from the posterior pituitary in response to _____

158 Antidiuretic hormone targets the _____ and (increases / decreases) water reabsorption.

159 Increased blood volume _____ blood pressure.

Sodium Reabsorption - Renin-Angiotensin Mechanism

160 Sodium reabsorption is mostly regulated by the hormone _____.

161 In response to low blood pressure, sodium ion reabsorption is enhanced by the _____ mechanism.

162 Juxtaglomerular cells of the afferent arteriole release _____ in response to reduced stretch (low blood pressure).

163 Renin converts angiotensinogen to _____.

164 Angiotensin II promotes the release of _____ from the adrenal cortex.

165 Aldosterone promotes the reabsorption of _____, which are followed by the osmotic reabsorption of _____.

380 Chapter 27 - Urinary System

166 The reabsorption of calcium ions is increased by the hormone _____.

167 Parathyroid hormone is released by the parathyroid glands in response to _____.

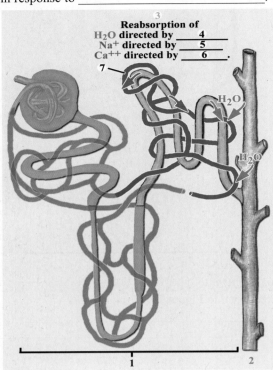

Figure 27.23

168 In reference to **Figure 27.23**, identify #1 - #7.

1 _____ 5 _____
2 _____ 6 _____
3 _____ 7 _____
4 _____

SECRETION

169 What is secretion? _____

170 Tubular secretion results in the movement of substances from the blood into the _____.

171 What several commonly secreted substances? _____

172 What portion of the nephron plays a major role in secretion? _____

173 What are two primary functions of secretion? _____

174 What substance is secreted with increasing acidosis? _____

175 What substance is secreted with increasing alkalosis? _____

176 How does the hormone aldosterone effect potassium ion secretion? _____

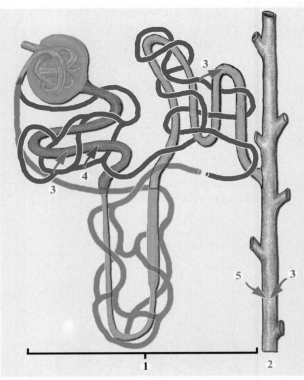

Figure 27.24

177 In reference to **Figure 27.24**, identify #1 - #7.

1 _____ 4 _____
2 _____ 5 _____
3 _____

PRODUCTION OF THE MEDULLARY OSMOTIC GRADIENT

178 Why is an osmotic gradient established in the kidney's medulla? _____

179 What is the countercurrent mechanism? _____

180 What are three factors that establish the medullary osmotic gradient? _____

181 The ascending limb of the loop of Henle pumps _____ into the interstitial fluid resulting in (increased / decreased) solute (salt) concentration of the interstitial fluid and (increased / decreased) solute (salt) concentration of the tubular fluid.

182 Due to the (increased / decreased) solute (salt) concentration of the interstitial fluid water moves out of the _____ of the loop of Henle resulting in (increased / decreased) solute (salt) in the descending limb.

183 Tubular fluid high in solute (salt) moves into the ascending limb where the tubular fluid serves additionally as the site of solute (salt) movement (out / in) of the tubule and (out of / into) the interstitial fluid.

184 Because the solute concentration of the tubular fluid in the descending limb becomes increasingly concentrated, the countercurrent mechanism is further described as the countercurrent _____ mechanism.

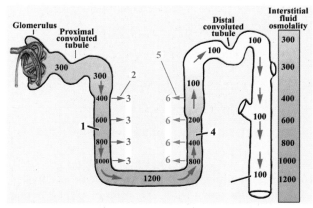

Figure 27.25

185 In reference to **Figure 27.25**, identify #1 - #6.
1 _____   4 _____
2 _____   5 _____
3 _____   6 _____

Countercurrent Exchange Mechanism

186 The blood supply to the medullary region must (remove / maintain) the osmotic gradient produced by the countercurrent mechanism.

187 The vasa recta of the juxtamedullary nephron is permeable to both _____ and _____ and is called the countercurrent _____.

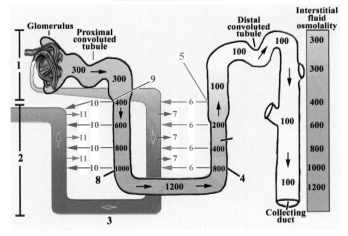

Figure 27.26

188 In reference to **Figure 27.26**, identify #1 - #11.
1 _____   7 _____
2 _____   8 _____
3 _____   9 _____
4 _____   10 _____
5 _____   11 _____
6 _____

FINAL REGULATION OF URINE WATER VOLUME

189 What two essential events have occurred by the time the tubular fluid has reached the distal convoluted tubule? ___
_____
_____
_____
_____
_____

WATER ELIMINATION BY FORMING DILUTE URINE

190 Dilute urine is produced when the body does not need to _____ water.

191 Dilute urine is mostly produced by allowing the already dilute tubular fluid to flow into the _____.

192 Very little _____ of water occurs from the distal convoluted tubule and the collecting duct.

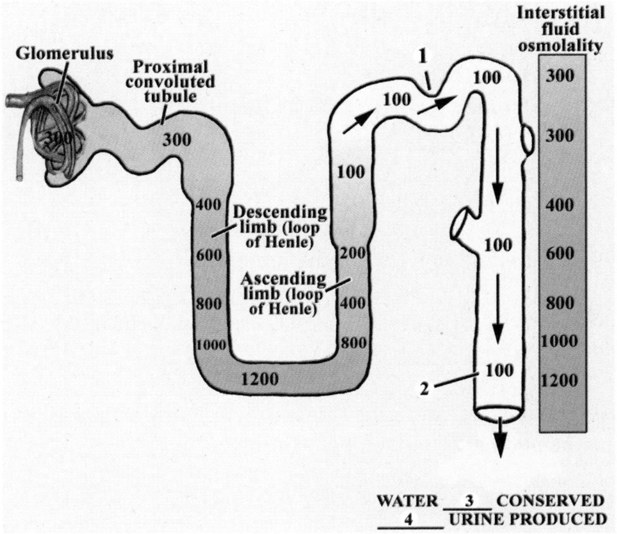

Figure 27.27

193 In reference to **Figure 27.27**, identify #1 - #4.
1 _____   3 _____
2 _____   4 _____

WATER CONSERVATION BY FORMING CONCENTRATED URINE

194 The hormone released from the posterior pituitary in response to increased osmolarity of the blood (low water volume) is _____.

195 ADH targets both the _____ convoluted tubule and the _____, resulting in (increased / decreased) water permeability.

196 The osmotic movement of water from the distal convoluted tubule and the collecting duct produces a tubular fluid (urine) that is _____.

197 The reabsorption of water is called _____ reabsorption because the amount of water that is reabsorbed is contingent upon the amount of _____.

382 Chapter 27 - Urinary System

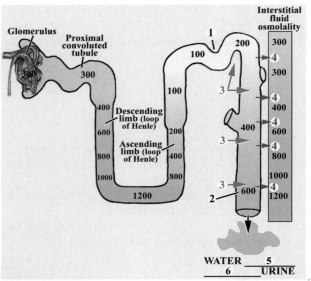

Figure 27.28

198 In reference to **Figure 27.28**, identify #1 - #6.
1 _____ 4 _____
2 _____ 5 _____
3 _____ 6 _____

URETERS

199 What is the function of the ureters? _____

200 Histologically, the wall of a ureter is composed of three layers, the _____

201 The primary stimulus for contraction of the muscularis and the peristaltic movement of urine is _____

URINARY BLADDER

202 What is the function of the urinary bladder? _____

203 The wall of the urinary bladder is composed of three layers, the _____

204 The muscularis of the urinary bladder is called the _____.

205 The area called the trigone is identified as the triangular region marked by the boundaries of the _____

206 The sphincter that provides for the involuntary control of passage of urine into the urethra is the _____.

URETHRA

207 What is the function of the urethra? _____ In the male, what is an additional function of the urethra? _____

208 What are the two sphincters that regulate the flow of urine into the urethra? _____

209 Which sphincter is under involuntary control? _____

210 Which sphincter is under voluntary control? _____

211 What are the three regions of the male urethra? _____

212 What is the name of the terminal opening of the urethra? _____

MICTURITION

213 What is micturition? _____

214 What are two other commonly used terms for the discharge of urine? _____

215 What are the two phases of the micturition reflex? _____

Filling Phase

216 During the filling phase a somatic circuit is activated that results in the (contraction / relaxation) of the external urethral sphincter.

217 A sympathetic circuit results in the (contraction / relaxation) of the internal urethral sphincter and the (contraction / relaxation) of the detrusor muscle.

218 Somatic fibers to the external urethral sphincter are (stimulated / inhibited) resulting in (contraction / relaxation) of the external urethral sphincter.

219 The conscious awareness of the need for micturition occurs when the bladder fills to about _____.

Emptying Phase

220 During the emptying phase, the micturition center in the _____ is activated.

221 Micturition is under _____ control.

222 Parasympathetic outflow results in (contraction / relaxation) of the internal urethral sphincter and (contraction / relaxation) of the detrusor muscle.

223 Somatic motor fibers to the external urethral sphincter are (stimulated / inhibited) resulting in (contraction / relaxation) of the external urethral sphincter.

224 Voluntarily withholding the micturition reflex results in (contraction / relaxation) of the detrusor muscle, until maximal filling is reached.

The Reproductive System - Worksheets

MALE REPRODUCTIVE SYSTEM

1. What are the primary sex organs of the male? _____
_____

2. What is produced by the testes? _____

3. What are two hormones produced by the testes? _____
_____

4. Starting with the testis, list in sequence the system of ducts of the male reproductive system. _____
_____

5. List the glands that contribute to the production of semen.
_____
_____

6. What are the external genitalia of the male? _____

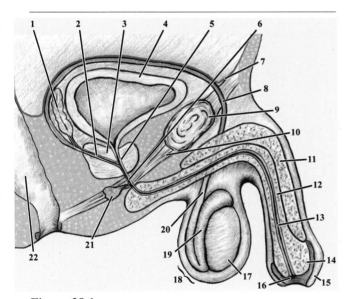

Figure 28.1

7. In reference to **Figure 28.1**, identify #1 - #22.
   1 _____ 12 _____
   2 _____ 13 _____
   3 _____ 14 _____
   4 _____ 15 _____
   5 _____ 16 _____
   6 _____ 17 _____
   7 _____ 18 _____
   8 _____ 19 _____
   9 _____ 20 _____
   10 _____ 21 _____
   11 _____ 22 _____

8. What is the scrotum? _____
_____

9. What do the testes produce? _____
_____

10. What is the name of the tubules of the testes? _____

11. What do the seminiferous tubules produce? _____
_____

12. Where are the interstitial (Leydig) cells located? _____

13. What is the function of the interstitial (Leydig) cells? ____
_____

14. Where are the epididymides located? _____
_____

15. What is the function of the epididymis? _____
_____

16. The ductus deferens transports sperm from the _____ to the _____ duct.

17. The ejaculatory duct is formed from the union of the _____
_____

18. What is the function of the ejaculatory duct? _____
_____

19. Where are the paired seminal vesicles located? _____
_____

20. Fluid from the seminal vesicles is mixed with _____ from the ductus deferens in the _____.

21. The urethra functions in both the _____ and the _____ systems.

22. What are the three regions of the urethra? _____
_____

23. Where is the prostatic urethra located? _____

24. Where is the membranous urethra located? _____

25. Where is the spongy urethra located? _____

26. Where is the prostate gland located? _____
_____

27. Prostatic fluid enters the _____
_____.

28. Where are the bulbourethral glands located? _____
_____

29. Fluid from the bulbourethral glands enters the _____.

30. The proximal portion of the penis is called the _____.

31. The middle region of the penis is called the _____, and the distal portion is called the _____.

32. The glans penis is surrounded by the _____, unless removed by the procedure called _____.

33. The three regions of erectile tissue of the body of the penis are the _____
_____.

34. The region of erectile tissue that surrounds the spongy urethra is the _____.

35. The terminus of the urethra is the _____
_____.

Chapter 28 - Reproductive System

384 Chapter 28 - Reproductive System

TESTES

36 The testes are housed in the _____ and are surrounded by two layers of connective tissue, the _____.

37 The connective tissue that forms a capsule around the testis is the _____.

38 The connective tissue that lines the surface of the testis is the _____.

39 Internally, the testis is organized into units called _____, with each unit containing _____ _____.

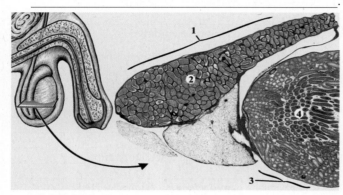

Figure 28.2

40 In reference to **Figure 28.2**, identify #1 - #4.
    1 _____   3 _____
    2 _____   4 _____

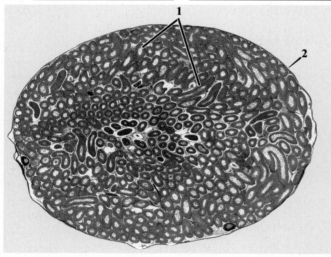

Figure 28.3

41 In reference to **Figure 28.3**, identify #1 - #2.
    1 _____   2 _____

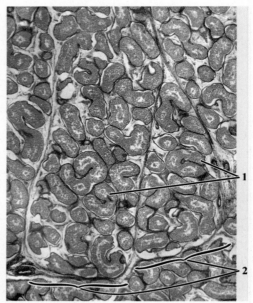

Figure 28.4

42 In reference to **Figure 28.4**, identify #1 - #2.
    1 _____   2 _____

43 The layer of connective tissue that covers the surface of the testis is the _____.

44 Where are the interstitial (Leydig) cells located? _____

45 What is the function of the interstitial (Leydig) cells? ____

46 What are two function of the seminiferous tubules? _____
_____
_____

47 What is spermatogenesis? _____
_____

48 What is spermiogenesis? _____
_____

MEIOSIS AND GAMETE PRODUCTION

49 What is meiosis? _____
_____

50 In meiosis, which chromosomes are distributed to the daughter cells? _____

51 What is the function of the haploid (n) cells formed by meiosis? _____
_____

52 What does the term 'diploid' mean? _____
_____

53 What is the diploid number of chromosomes for humans? _____

54 What is a homologous chromosome? _____

55 How many pairs of homologous chromosomes are found in humans? _____

56 How is the haploid number of chromosomes described? __
_____

Chapter 28 - Reproductive System

57 What is the haploid chromosome number in humans? ____
58 Before meiosis begins what happens to the chromosomes? _____
59 What are the two stages of meiosis? _____
60 In meiosis I, the replicated chromosomes are first paired, in an event called _____.
61 During synapsis _____ of portions of the chromosomes occurs.
62 After crossing-over, one replicated chromosome (a homologous chromosome) is distributed to _____.
63 The daughter cells that enter meiosis II have one of ____ _____ of chromosome in the _____ form.
64 In meiosis II the replicated chromosomes are _____ and one of each sister chromosome is distributed to a _____, which function as _____.
65 Fusion of gametes, a sperm and an egg, produces a ____ number of chromosomes in a cell called a _____.

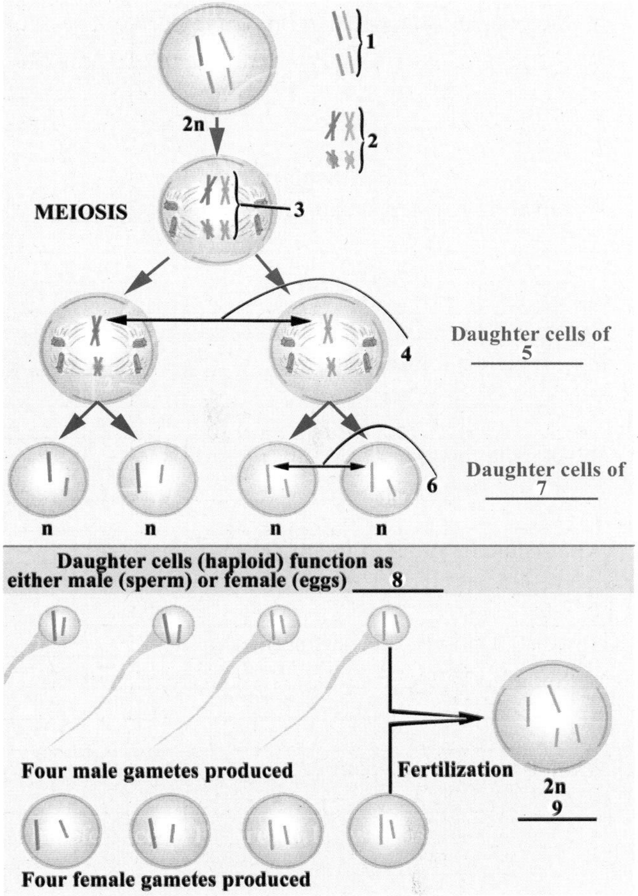

MEIOSIS - SPERMATOGENESIS

67 Precursor (stem) cells are located at the periphery of the _____ and undergo _____ divisions.
68 Some of the daughter spermatogonia differentiate into primary _____.
69 Primary spermatocytes enter the stage of division called _____ and produce two (haploid / diploid) daughter cells called _____.
70 The two secondary spermatocytes enter the stage of division called _____ and each produces two (haploid / diploid) cells called _____.
71 During _____ spermatids are transformed into _____.

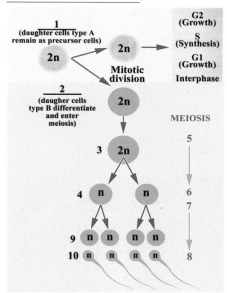

Figure 28.6

72 In reference to **Figure 28.6**, identify #1 - #10.
   1 _____ 6 _____
   2 _____ 7 _____
   3 _____ 8 _____
   4 _____ 9 _____
   5 _____ 10 _____

Figure 28.5

66 In reference to **Figure 28.5**, identify #1 - #9.
   1 _____ 6 _____
   2 _____ 7 _____
   3 _____ 8 _____
   4 _____ 9 _____
   5 _____

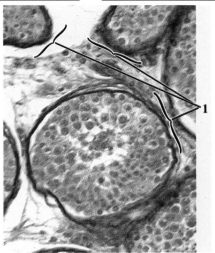

Figure 28.7

73 In reference to **Figure 28.7**, identify #1.
   1 _____

Chapter 28 - Reproductive System

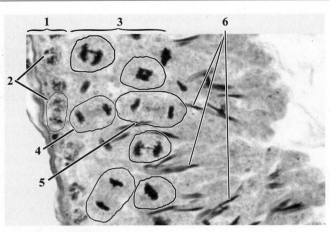

Figure 28.8

74 In reference to **Figure 28.8**, identify #1 - #6.
  1 _____ 4 _____
  2 _____ 5 _____
  3 _____ 6 _____

75 From the spermatogonium, list in sequence the stages of spermatogenesis. _____
_____
_____

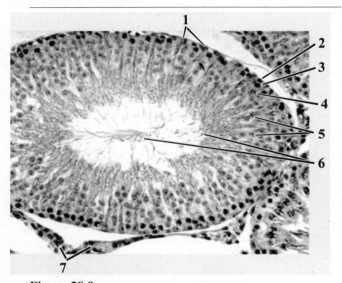

Figure 28.9

76 In reference to **Figure 28.9**, identify #1 - #7.
  1 _____ 5 _____
  2 _____ 6 _____
  3 _____ 7 _____
  4 _____

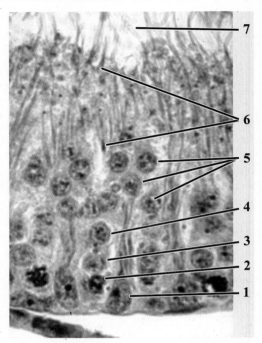

Figure 28.10

77 In reference to **Figure 28.10**, identify #1 - #7.
  1 _____ 5 _____
  2 _____ 6 _____
  3 _____ 7 _____
  4 _____

Sustentacular cells

78 Where are the sustentacular cells located? _____
_____
_____

79 In relation to the sustentacular cells, where are the spermatogenic cells located? _____
_____
_____

80 What is the function of sustentacular cells? _____
_____

81 What is the function of the blood-testis barrier? _____
_____

82 How is the blood-testis barrier formed? _____
_____
_____

83 What are the two compartments of the seminiferous tubule? _____
_____
_____

Chapter 28 - Reproductive System

MALE HORMONAL REGULATION
Follicle-stimulating hormone (FSH)

86 What hormone targets sustentacular cells? _____

87 Follicle stimulating hormone (FSH) stimulates sustentacular cells to release _____.

88 What does androgen binding protein (ABP) target? _____

89 What effect does an increased level of androgen binding protein have on spermatogenic cells? _____

90 What effect does an increased level of testosterone have on spermatogenic cells? _____

91 What cells release inhibin? _____

92 What is the function of inhibin? _____

93 What effect does an increased level of inhibin have on the release of follicle stimulating hormone (FSH)? _____

94 What effect does a decreased level of follicle stimulating hormone have on the release of inhibin and androgen binding protein? _____

95 What effect does a decreased level of inhibin have on the release of follicle stimulating hormone (FSH)? _____

96 What effect does an increased level of follicle stimulating hormone have on the release of inhibin and androgen binding protein? _____

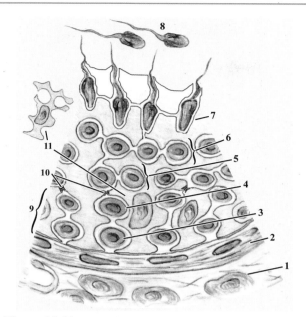

Figure 28.11

84 In reference to **Figure 28.11**, identify #1 - #11.
1 _____   7 _____
2 _____   8 _____
3 _____   9 _____
4 _____   10 _____
5 _____   11 _____
6 _____

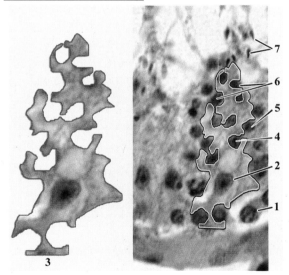

Figure 28.12

85 In reference to **Figure 28.12**, identify #1 - #7.
1 _____   5 _____
2 _____   6 _____
3 _____   7 _____
4 _____

388 Chapter 28 - Reproductive System

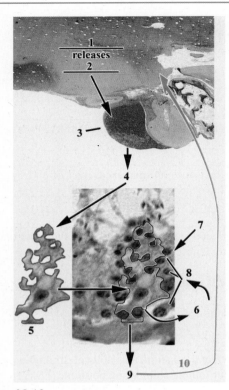

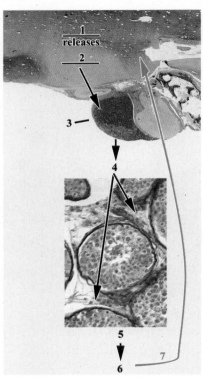

Figure 28.13

97 In reference to **Figure 28.13**, identify #1 - #10.
    1 _____   6 _____
    2 _____   7 _____
    3 _____   8 _____
    4 _____   9 _____
    5 _____   10 _____

Luteinizing hormone (LH)

98 What does luteinizing hormone target? _____

99 What is the effect of luteinizing hormone (LH) on the interstitial cells? _____

100 What is the effect of increased levels of testosterone on the hypothalamus? _____

101 What is the effect of decreased amounts of gonadothropin-releasing hormone (GnRH)? _____

102 What is the effect of decreased amounts of testosterone? _____

103 What is the effect of increased levels of testosterone on the hypothalamus? _____

104 What is the effect of increased amounts of gonadothropin-releasing hormone (GnRH)? _____

Figure 28.14

105 In reference to **Figure 28.14**, identify #1 - #7.
    1 _____   5 _____
    2 _____   6 _____
    3 _____   7 _____
    4 _____

SPERMATOZOA (SPERM)

106 What are the three regions of a spermatozoon? _____

107 Which and how many chromosomes are found in the nucleus of the spermatozoon? _____

108 What is the function of the acrosomal cap? _____

109 What is the function of the microtubules? _____

110 What is used as the primary fuel for the mitochondria? _____

Name _____
Class _____

Chapter 28 - Reproductive System 389

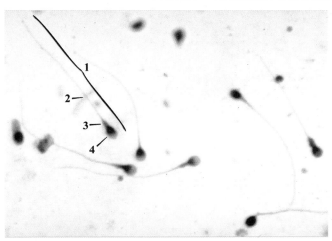

Figure 28.15

111 In reference to **Figure 28.15**, identify #1 - #4.
   1 _____ 3 _____
   2 _____ 4 _____

EPIDIDYMIS

112 What is the ductus epididymis? _____
_____

113 What is the function of the ductus epididymis? _____
_____

114 What structure does the ductus epididymis form? _____
_____

115 The ductus deferens is the tube located between the _____
_____ and the _____.

116 What is the function of the stereocilia (microvilli)? _____
_____

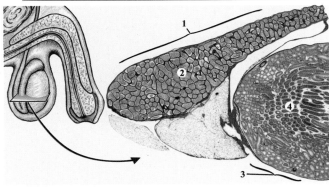

Figure 28.16

117 In reference to **Figure 28.16**, identify #1 - #4.
   1 _____ 3 _____
   2 _____ 4 _____

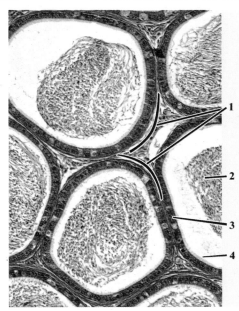

Figure 28.17

118 In reference to **Figure 28.17**, identify #1 - #4.
   1 _____ 3 _____
   2 _____ 4 _____

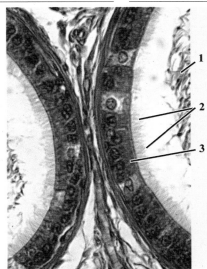

Figure 28.18

119 In reference to **Figure 28.18**, identify #1 - #3.
   1 _____ 3 _____
   2 _____

DUCTUS DEFERENS

120 Where does the ductus deferens originate? _____
_____

121 Where does the ductus deferens terminate? _____
_____

122 What is the function of the ductus deferens? _____
_____

123 What is the function of the muscularis of the ductus deferens? _____
_____

390 Chapter 28 - Reproductive System

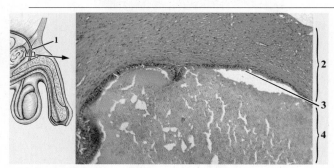

Figure 28.19

124 In reference to **Figure 28.19**, identify #1 - #4.
1 _____ 3 _____
2 _____ 4 _____

SEMINAL VESICLES

125 Where are the seminal vesicles located? _____
_____

126 The ducts from the seminal vesicles join with the _____
_____, and merge to form the _____.

127 What forms the seminal vesicle? _____

128 What is the function of the seminal vesicles? _____
_____
_____

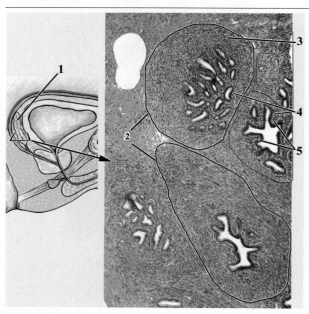

Figure 28.20

129 In reference to **Figure 28.20**, identify #1 - #5.
1 _____ 4 _____
2 _____ 5 _____
3 _____

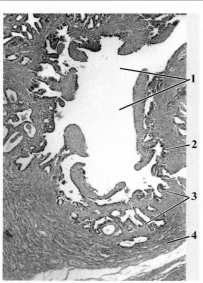

Figure 28.21

130 In reference to **Figure 28.21**, identify #1 - #4.
1 _____ 3 _____
2 _____ 4 _____

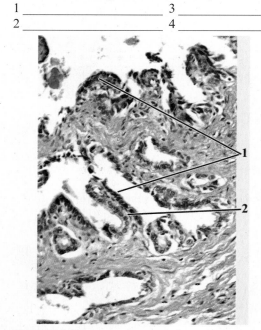

Figure 28.22

131 In reference to **Figure 28.22**, identify #1 - #2.
1 _____ 2 _____

PROSTATE GLAND

132 Where is the prostate gland located? _____
_____

133 What is the name of the duct that transverses the prostate gland? _____
_____

134 What is the name of the portion of the urethra that passes through the prostate? _____
_____

135 What forms the prostate gland? _____
_____
_____

136 Where is the prostatic secretion emptied? _____

137 What is a function of prostatic fluid? _____

138 How is prostatic fluid moved out of the prostate gland? _____

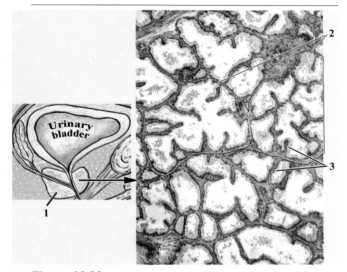

Figure 28.23
139 In reference to **Figure 28.23**, identify #1 - #3.
1 _____ 3 _____
2 _____

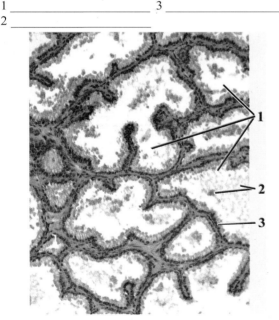

Figure 28.24
140 In reference to **Figure 28.24**, identify #1 - #3.
1 _____ 3 _____
2 _____

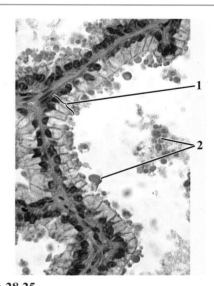

Figure 28.25
141 In reference to **Figure 28.25**, identify #1 - #2.
1 _____ 2 _____

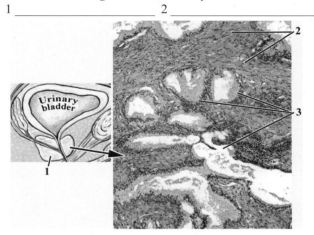

Figure 28.26
142 In reference to **Figure 28.26**, identify #1 - #3.
1 _____ 3 _____
2 _____

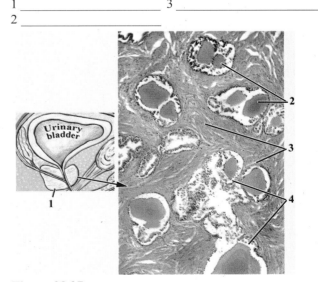

Figure 28.27
143 In reference to **Figure 28.27**, identify #1 - #4.
1 _____ 3 _____
2 _____ 4 _____

392 Chapter 28 - Reproductive System

PENIS

144 What three bodies of erectile, or cavernous, tissue form the penis? _____

145 Which erectile tissue houses the urethra? _____

146 What forms the glans penis? _____

147 What is the prepuce, or foreskin? _____

148 What is the name of the connective tissue that surrounds the cavernous tissue? _____

149 What are the cavernous spaces? _____

150 What are trabeculae? _____

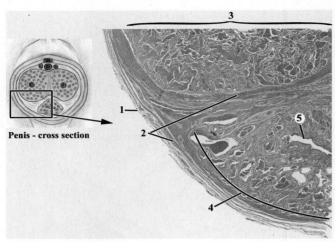

Figure 28.29

152 In reference to **Figure 28.29**, identify #1 - #5.
1 _____ 4 _____
2 _____ 5 _____
3 _____

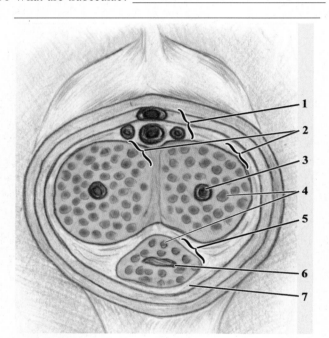

Figure 28.28

151 In reference to **Figure 28.28**, identify #1 - #7.
1 _____ 5 _____
2 _____ 6 _____
3 _____ 7 _____
4 _____

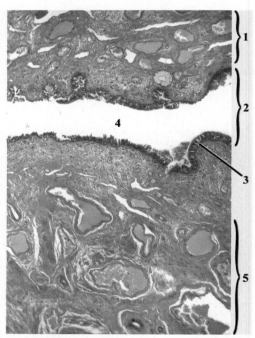

153 **Figure 28.30**
In reference to **Figure 28.30**, identify #1 - #5.
1 _____ 4 _____
2 _____ 5 _____
3 _____

Name _____
Class _____

Chapter 28 - Reproductive System 393

FEMALE REPRODUCTIVE SYSTEM

156 What are the primary sex organs of the female? _____

157 What do the ovaries produce? _____

158 What are two sex hormones produced by the ovaries? ___

159 Starting with the ovary, list in sequence the accessory sex organs of the female. _____

160 What are the external genitalia of the female? _____

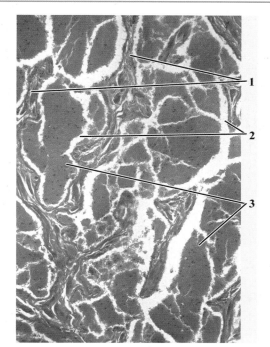

154 **Figure 28.31**
In reference to **Figure 28.31**, identify #1 - #3.
1 _____ 3 _____
2 _____

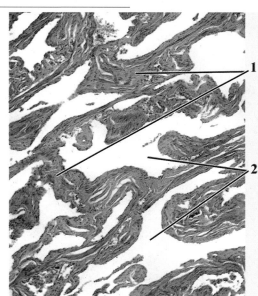

Figure 28.32
155 In reference to **Figure 28.32**, identify #1 - #2.
1 _____ 2 _____

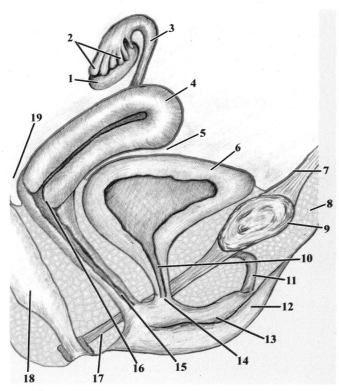

Figure 28.33

161 In reference to **Figure 28.33**, identify #1 - #19.
1 _____ 11 _____
2 _____ 12 _____
3 _____ 13 _____
4 _____ 14 _____
5 _____ 15 _____
6 _____ 16 _____
7 _____ 17 _____
8 _____ 18 _____
9 _____ 19 _____
10 _____

Chapter 28 - Reproductive System

EXTERNAL GENITALIA

162 Describe the structure and location of the labia majora. _____

163 Describe the structure and location of the labia majora. _____

164 What two structures open into the vestibule? _____

165 What is the clitoris? _____

166 What is the mons pubis? _____

167 Where are the greater vestibular glands located? _____

168 What is the function of the greater vestibular glands? _____

169 Where do the paraurethral glands open? _____

INTERNAL ORGANS

170 What forms the broad ligament? _____

171 What does the broad ligament anchor? _____

172 Where is the rectrouterine pouch located? _____

173 Where is the vesicouterine pouch located? _____

174 Where are the ovaries located? _____

175 What are the three structures that support and anchor each ovary? _____

176 What are the two internal regions of the ovary? _____

177 Which region of the ovary contains the follicles? _____

178 What is the function of the follicles? _____

179 What is the mesovarium? _____

180 What is the function of the ovarian ligament? _____

181 What is the function of the suspensory ligament? _____

182 Where are the uterine tubes located? _____

183 Where is the isthmus located? _____

184 What is the ampulla of the uterine tube? _____

185 What is the infundibulum of the uterine tube? _____

186 What are fimbriae? _____

187 Where is the uterus located? _____

188 What is the function of the uterus? _____

189 What are the three major regions of the uterus? _____

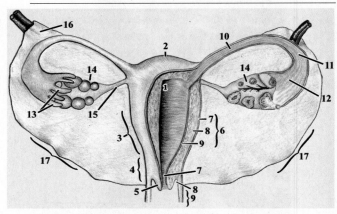

Figure 28.34

190 In reference to **Figure 28.34**, identify #1 - #17.

1 _____ 10 _____
2 _____ 11 _____
3 _____ 12 _____
4 _____ 13 _____
5 _____ 14 _____
6 _____ 15 _____
7 _____ 16 _____
8 _____ 17 _____
9 _____

191 For the outside inward, the three layers of the uterine wall are _____

192 What is the perimeterium? _____

193 What is the myometrium? _____

194 What is the endometrium? _____

195 What are the two layers of the endometrium? _____

196 Where is the vagina located? _____

Name _____
Class _____

Chapter 28 - Reproductive System 395

OVARY

197 The outer covering of the ovary is modified visceral _____ _____ and is called the _____.

198 The outer region of the ovary is called the _____ and contains _____ _____.

199 The inner region of the ovary is called the _____.

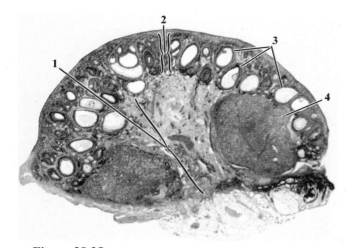

Figure 28.35

200 In reference to **Figure 28.35**, identify #1 - #4.
   1 _____   3 _____
   2 _____   4 _____

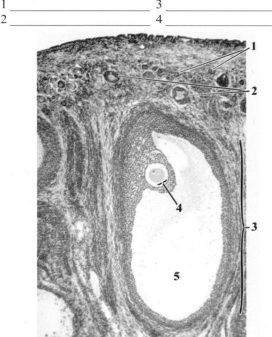

Figure 28.36

201 In reference to **Figure 28.36**, identify #1 - #5.
   1 _____   4 _____
   2 _____   5 _____
   3 _____

202 Depending upon the degree of development follicles are called _____

203 A primordial follicle consists of _____ _____

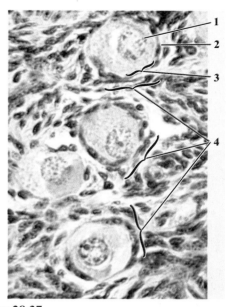

Figure 28.37

204 In reference to **Figure 28.37**, identify #1 - #4.
   1 _____   3 _____
   2 _____   4 _____

205 Primary follicles consist of _____ _____

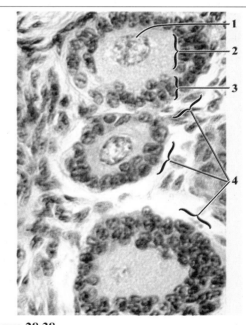

Figure 28.38

206 In reference to **Figure 28.38**, identify #1 - #4.
   1 _____   3 _____
   2 _____   4 _____

207 Secondary follicles are also called _____ _____.

208 Secondary follicles consist of _____ _____

209 What is the corona radiata? _____ _____

210 What is the zona pellucida? _____ _____

396 Chapter 28 - Reproductive System

Name _____
Class _____

211 What is the function of the microvilli of the zona pellucida? _____

212 Ovulation releases the oocyte surrounded by _____

213 A mature secondary follicle is commonly called a _____.

214 After ovulation, the follicle is transformed into the _____.

CORPUS LUTEUM

217 The corpus luteum develops from the _____

218 What is the function of the corpus luteum? _____

219 What develops if the corpus luteum degenerates? _____

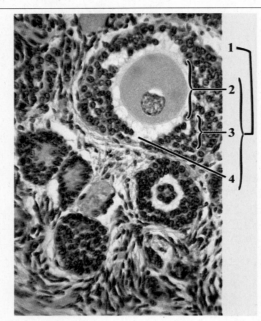

Figure 28.39
215 In reference to **Figure 28.39**, identify #1 - #4.
1 _____ 3 _____
2 _____ 4 _____

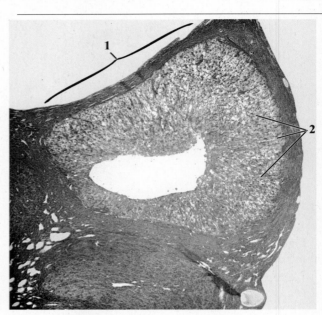

Figure 28.41
220 In reference to **Figure 28.41**, identify #1 - #2.
1 _____ 2 _____

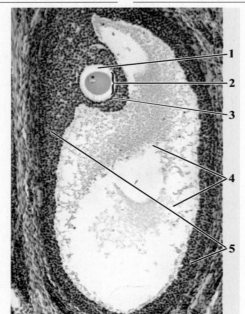

Figure 28.40
216 In reference to **Figure 28.40**, identify #1 - #5.
1 _____ 4 _____
2 _____ 5 _____
3 _____

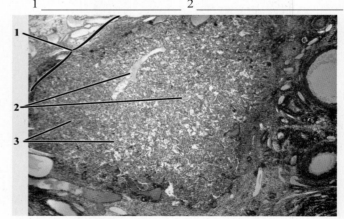

Figure 28.42
221 In reference to **Figure 28.42**, identify #1 - #3.
1 _____ 3 _____
2 _____

Name _____
Class _____

Chapter 28 - Reproductive System **397**

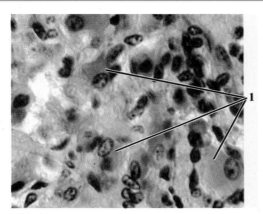

Figure 28.43
222 In reference to **Figure 28.43**, identify #1.
    1 _____

CORPUS ALBICANS
223 What is the corpus albicans? _____
_____
_____

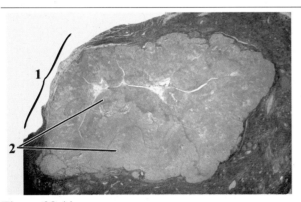

Figure 28.44
224 In reference to **Figure 28.44**, identify #1 - #2.
    1 _____ 2 _____

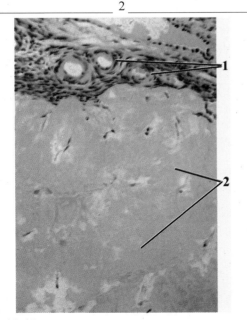

Figure 28.45
225 In reference to **Figure 28.45**, identify #1 - #2.
    1 _____ 2 _____

MEIOSIS - OOGENESIS
226 What is oogenesis? _____
_____
227 When does oogenesis begin? _____
_____
_____
228 At birth and until puberty, the ovary contains several million _____.
229 What hormones target receptive primordial follicles and stimulate their development? _____
_____
230 Primordial follicles develop into _____ follicles, characterized by several layers of _____ cells surrounding a _____ oocyte.
231 A primary oocyte completes meiosis I to form two cells, the first polar body and the _____ oocyte.
232 The secondary oocyte arrested in meiosis II is ovulated, and completes meiosis II if _____ occurs.
233 Nuclear union between the ovum and the sperm produces a _____.

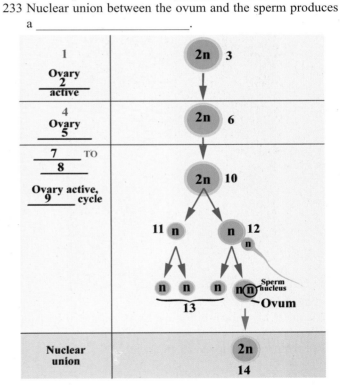

Figure 28.46
234 In reference to **Figure 28.46**, identify #1 - #14.
    1 _____ 8 _____
    2 _____ 9 _____
    3 _____ 10 _____
    4 _____ 11 _____
    5 _____ 12 _____
    6 _____ 13 _____
    7 _____ 14 _____

398 Chapter 28 - Reproductive System

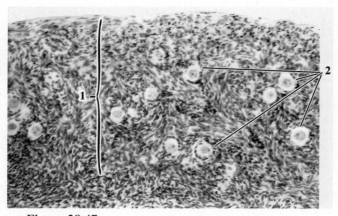

Figure 28.47

235 Is the ovary shown in **Figure 28.47** mature or immature?
_____

236 In reference to **Figure 28.47**, identify #1 - #2.
1 _____ 2 _____

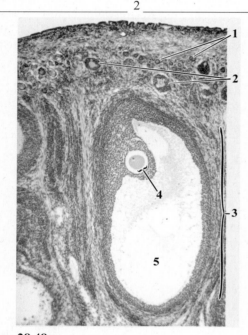

Figure 28.48

237 Is the ovary shown in **Figure 28.48** mature or immature?
_____

238 In reference to **Figure 28.48**, identify #1 - #5.
1 _____ 4 _____
2 _____ 5 _____
3 _____

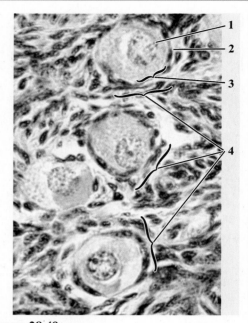

Figure 28.49

239 In reference to **Figure 28.49**, identify #1 - #4.
1 _____ 3 _____
2 _____ 4 _____

OVARIAN CYCLE

240 The ovarian cycle is approximately _____ days.

241 What regulates the ovarian cycle? _____
_____
_____
_____
_____

242 Where is follicle stimulating hormone (FSH) produced and released? _____
_____

243 What are the target and the effect of follicle stimulating hormone? _____
_____
_____

244 Follicles stimulated by FSH release the regulatory hormone called _____ and mostly the sex hormone called _____.

245 Increased secretion of inhibin results in hypothalamic _____ of gonadotropin-releasing hormone (GnRH).

246 Reduction of GnRH results in _____ secretion of FSH.

Chapter 28 - Reproductive System

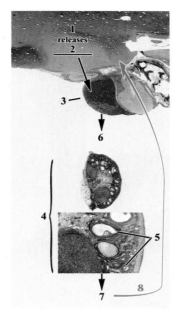

Figure 28.50

247 In reference to **Figure 28.50**, identify #1 - #8.

1 _____ 5 _____
2 _____ 6 _____
3 _____ 7 _____
4 _____ 8 _____

248 Where is luteinizing (LH) produced and released? _____
_____

249 What are the target and the effect of luteinizing hormone?
_____
_____
_____

250 What is the effect of increased levels of progesterone and estrogen on the release of luteinizing hormone? _____
_____
_____

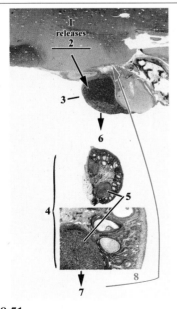

Figure 28.51

251 In reference to **Figure 28.51**, identify #1 - #8.

1 _____ 5 _____
2 _____ 6 _____
3 _____ 7 _____
4 _____ 8 _____

252 What are the two phases of the ovarian cycle? _____
_____

253 Describe the follicular phase of the ovarian cycle. _____
_____

254 Describe the luteal phase of the ovarian cycle. _____
_____

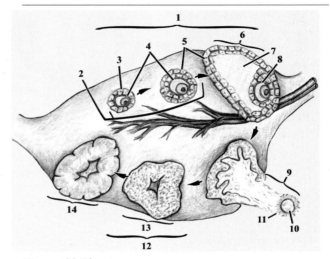

Figure 28.52

255 In reference to **Figure 28.52**, identify #1 - #14.

1 _____ 7 _____
2 _____ 8 _____
3 _____ 9 _____
4 _____ 10 _____
5 _____ 11 _____
6 _____ 12 _____

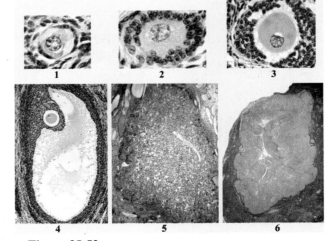

Figure 28.53

256 In reference to **Figure 28.53**, identify #1 - #6.

1 _____ 4 _____
2 _____ 5 _____
3 _____ 6 _____

400 Chapter 28 - Reproductive System

257 Describe the structure of a primordial follicle. _____

258 What hormone released from the anterior pituitary gland targets receptive primordial follicles? _____

259 What is the effect of FSH on primordial follicles? _____

260 Primordial follicles develop into _____ follicles.

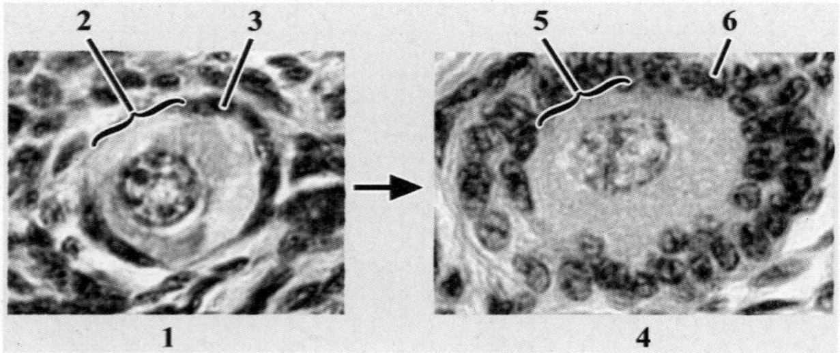

Figure 28.54

261 In reference to **Figure 28.54**, identify #1 - #6.
1 _____ 4 _____
2 _____ 5 _____
3 _____ 6 _____

262 Describe the structure of a primary follicle. _____

263 What two hormones are secreted by the primary follicles? _____

264 What hormone targets the uterus and what is its effect? _____

265 What does the hormone inhibin target and what is its effect? _____

266 Primary follicles develop into _____ follicles.

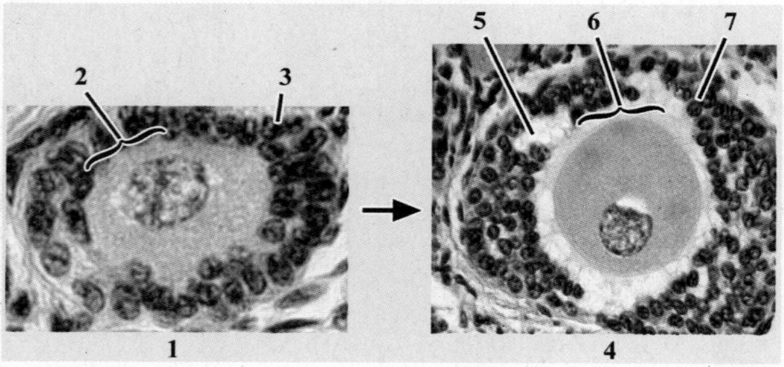

Figure 28.55

267 In reference to **Figure 28.55**, identify #1 - #7.
1 _____ 5 _____
2 _____ 6 _____
3 _____ 7 _____
4 _____

268 Describe the structure of a secondary follicle. _____

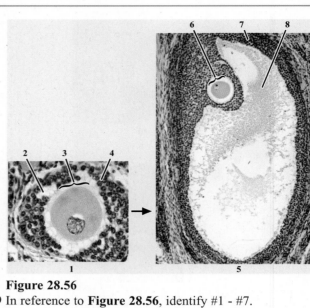

Figure 28.56

269 In reference to **Figure 28.56**, identify #1 - #7.
1 _____ 5 _____
2 _____ 6 _____
3 _____ 7 _____
4 _____

270 What is a mature secondary follicle called? _____

271 What is ovulation? _____

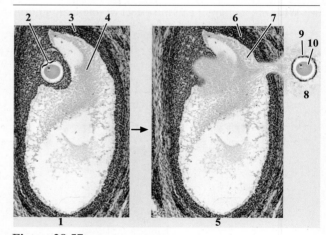

Figure 28.57

272 In reference to **Figure 28.57**, identify #1 - #10.
1 _____ 6 _____
2 _____ 7 _____
3 _____ 8 _____
4 _____ 9 _____
5 _____ 10 _____

273 What is the function of the corpus luteum? _____

274 If the corpus luteum is not maintained it degenerates into the _____.

Name _____
Class _____

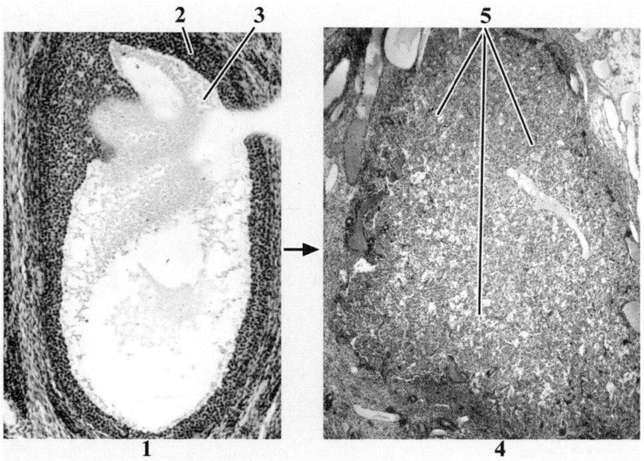

Chapter 28 - Reproductive System 401

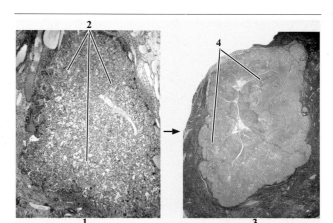

Figure 28.58
275 In reference to **Figure 28.58**, identify #1 - #5.
1 _____ 4 _____
2 _____ 5 _____
3 _____

276 Describe the structure of the corpus albicans. _____
_____

Figure 28.59
277 In reference to **Figure 28.59**, identify #1 - #4.
1 _____ 3 _____
2 _____ 4 _____

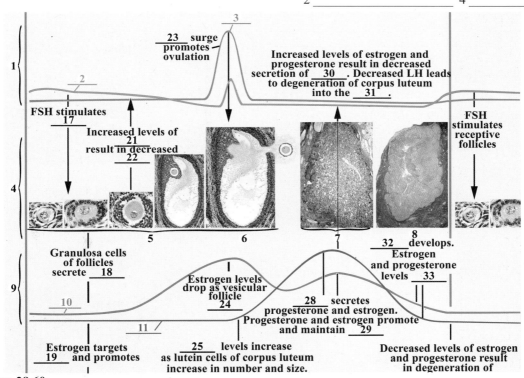

Figure 28.60
278 In reference to **Figure 28.60**, identify #1 - #35.

1 _____	13 _____	25 _____
2 _____	14 _____	26 _____
3 _____	15 _____	27 _____
4 _____	16 _____	28 _____
5 _____	17 _____	29 _____
6 _____	18 _____	30 _____
7 _____	19 _____	31 _____
8 _____	20 _____	32 _____
9 _____	21 _____	33 _____
10 _____	22 _____	34 _____
11 _____	23 _____	35 _____
12 _____	24 _____	36 _____

402 Chapter 28 - Reproductive System

UTERINE TUBE

279 What are two other names for the uterine tube? _____
_____

280 What are the three regions of the uterine tube? _____
_____
_____

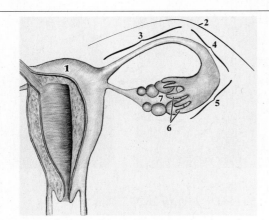

Figure 28.61

281 In reference to **Figure 28.61**, identify #1 - #6.
1 _____ 4 _____
2 _____ 5 _____
3 _____ 6 _____

282 What is the isthmus of the uterine tube? _____
_____

283 The ampulla of the uterine tube is located between the ____
_____ and the _____.

284 The region of the uterine tube that terminates at the ovary is the _____.

285 What are the fimbriae of the infundibulum? _____
_____
_____

286 What is the function of the fimbriae? _____
_____
_____

287 What type of epithelium lines the mucosa of the uterine tube? _____
_____

288 What is the function of the secretory cells of the mucosa?
_____
_____

289 What is the function of the muscularis? _____
_____
_____

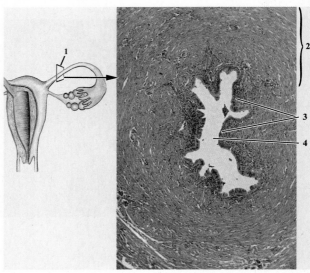

Figure 28.62

290 In reference to **Figure 28.62**, identify #1 - #4.
1 _____ 3 _____
2 _____ 4 _____

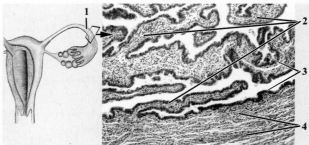

Figure 28.63

291 In reference to **Figure 28.63**, identify #1 - #4.
1 _____ 3 _____
2 _____ 4 _____

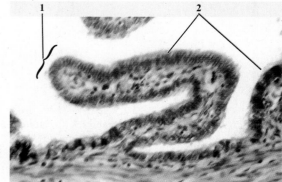

Figure 28.64

292 In reference to **Figure 28.64**, identify #1 - #2.
1 _____ 2 _____

UTERUS

293 Describe the location of the uterus. _____
_____
_____

294 What is the function of the uterus? _____
_____
_____

Name _____
Class _____

Chapter 28 - Reproductive System   403

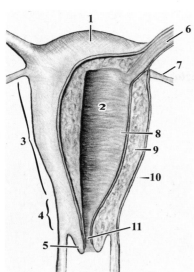

Figure 28.65
295 In reference to **Figure 28.65**, identify #1 - #11.
1 _____ 7 _____
2 _____ 8 _____
3 _____ 9 _____
4 _____ 10 _____
5 _____ 11 _____
6 _____

296 What is the fundus of the uterus? _____
_____

297 What is the body of the uterus? _____

298 Where is the isthmus of the uterus located? _____

299 What is the cervix of the uterus? _____
_____

300 What is the name of the external opening into the cervix?
_____

301 The external orifice opens into the _____.

UTERINE WALL

302 From outside inward what are the three regions of the uterine wall? _____
_____

303 What regions of the uterus are covered by the perimetrium? _____
_____

304 What type of muscle forms the myometrium? _____

305 The inner mucosa of the uterus is called the _____.

306 The two layers of the endometrium are the _____
_____.

307 The endometrium undergoes cyclic changes in response to _____ hormones.

308 The endometrium consists of numerous _____
_____

309 The uterine glands are targets of the hormones _____
_____.

310 What is the function of uterine glands? _____
_____

311 The blood vessels that deliver blood to the endometrium are called the _____ and their growth is promoted by the hormones _____.

312 The _____ is the connective tissue of the endometrium.

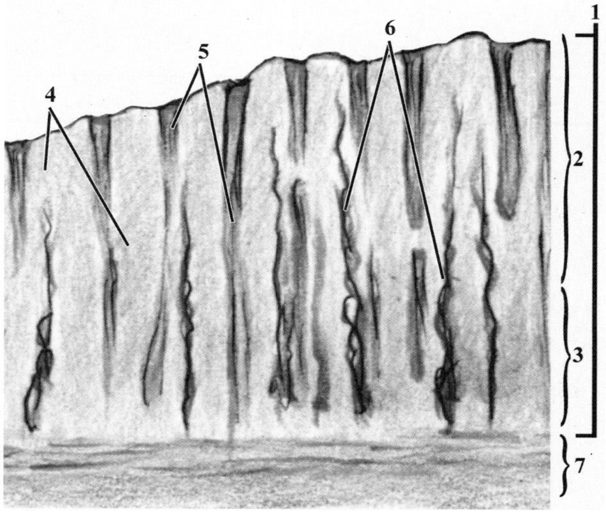

Figure 28.66
313 In reference to **Figure 28.66**, identify #1 - #5.
1 _____ 5 _____
2 _____ 6 _____
3 _____ 7 _____
4 _____

314 The layer of the endometrium closest to the muscularis is the _____.

315 The permanent layer of the endometrium is the _____.

316 The layer of the endometrium closest to the uterine cavity is the _____.

317 The layer that undergoes modifications in preparation for implantation of the fertilized egg is the _____.

UTERINE CYCLE

318 What are the three phases of the uterine (menstrual) cycle?
_____
_____

319 The uterine cycle is approximately _____ long, extending from one _____ phase to the next.

320 The menstrual phase occurs when levels of _____ are at their _____ levels.

321 The menstrual phase begins at the end of the _____ phase.

322 The proliferative phase is characterized by a _____ of the stratum _____.

323 The secretory phase is characterized by the continued rebuilding of the stratum functionalis and the _____.

324 The primary hormones that enhance the secretory phase are _____ released from the corpus _____.

404 Chapter 28 - Reproductive System

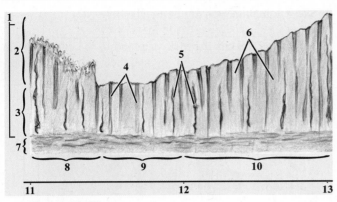

Figure 28.67

325 In reference to **Figure 28.67**, identify #1 - #13.
1 _____ 8 _____
2 _____ 9 _____
3 _____ 10 _____
4 _____ 11 _____
5 _____ 12 _____
6 _____ 13 _____
7 _____

Figure 28.68

326 In reference to **Figure 28.68**, match the numbers with the following descriptions:
___ Early Secretory Phase, Days 14 - 18
___ Late Secretory Phase, Days 25 - 28
___ Menstrual Phase, Days 1 - 4
___ Pregnant uterus
___ Proliferative Phase, Days 4 - 14
___ Secretory Phase, Days 18 - 23
___ Secretory Phase, Days 23 - 25

327 What characterizes the menstrual phase? _____
_____

328 During the menstrual phase which hormones are at their lowest? _____
_____

329 During the menstrual phase what happens to the spiral arteries of the stratum functionalis? _____
_____

330 What happens to the stratum functionalis when subjected to reduced blood flow? _____
_____
_____

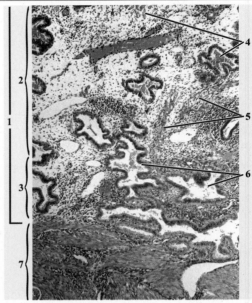

Figure 28.69

331 In reference to **Figure 28.69**, what phase is shown? _____

332 In reference to **Figure 28.69**, identify #1 - #7.
1 _____ 5 _____
2 _____ 6 _____
3 _____ 7 _____
4 _____

333 When does the proliferative phase begin? _____

334 What characterizes the proliferative phase? _____

335 What is the primary hormone that promotes the proliferative phase? _____
_____

Name _____
Class _____

Chapter 28 - Reproductive System   405

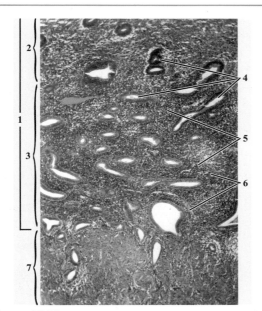

Figure 28.70

336 In reference to **Figure 28.70**, what phase is shown? _____
_____

337 In reference to **Figure 28.70**, identify #1 - #7.
   1 _____ 5 _____
   2 _____ 6 _____
   3 _____ 7 _____
   4 _____

338 Describe the appearance of the uterine glands during the secretory phase. _____
_____
_____

339 What hormones promote the secretory phase? _____
_____

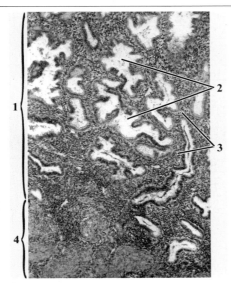

Figure 28.71

340 In reference to **Figure 28.71**, what phase is shown? _____
_____

341 In reference to **Figure 28.71**, identify #1 - #4.
   1 _____ 3 _____
   2 _____ 4 _____

342 **Figure 28.72**
In reference to **Figure 28.72** identify #1 - #40.
(**Figure is on the following page**)

VAGINA

343 Where is the vagina located? _____
_____
_____

344 What is the function of the vagina? _____
_____
_____

345 What tissue lines the mucosa of the vagina? _____
_____

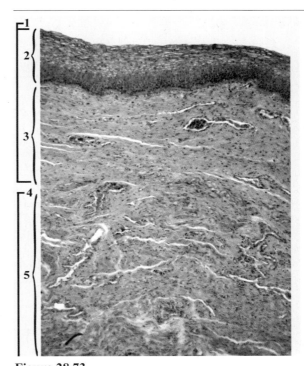

Figure 28.73

346 In reference to **Figure 28.73**, identify #1 - #5.
   1 _____ 4 _____
   2 _____ 5 _____
   3 _____

Chapter 28 - Reproductive System

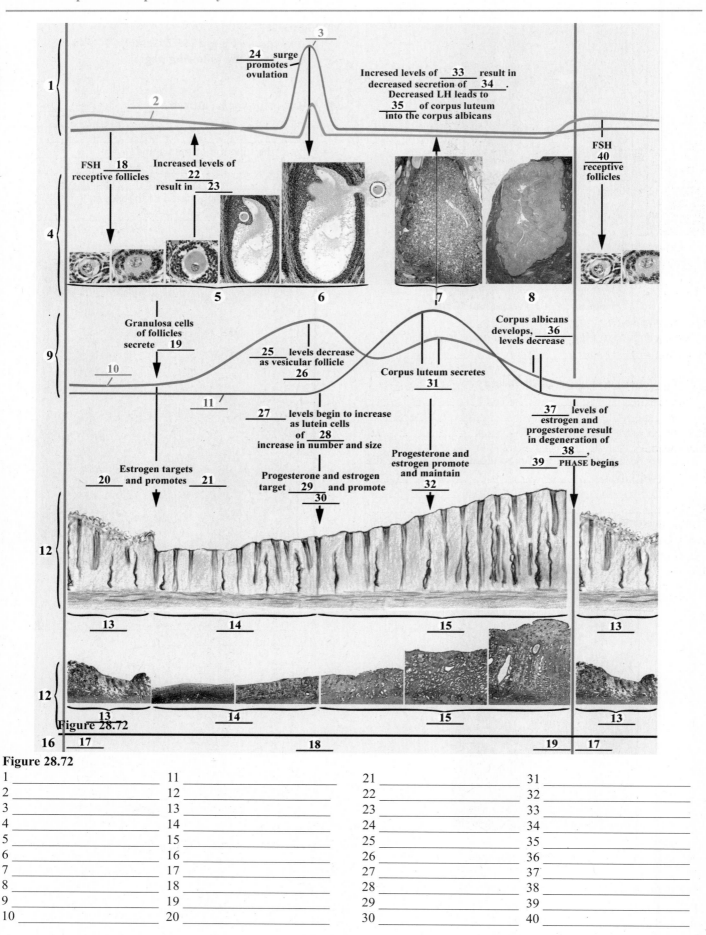

Figure 28.72

1 _____	11 _____	21 _____	31 _____
2 _____	12 _____	22 _____	32 _____
3 _____	13 _____	23 _____	33 _____
4 _____	14 _____	24 _____	34 _____
5 _____	15 _____	25 _____	35 _____
6 _____	16 _____	26 _____	36 _____
7 _____	17 _____	27 _____	37 _____
8 _____	18 _____	28 _____	38 _____
9 _____	19 _____	29 _____	39 _____
10 _____	20 _____	30 _____	40 _____

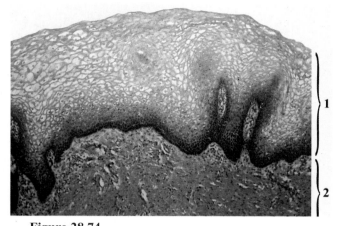

Figure 28.74

347 In reference to **Figure 28.74**, identify #1 - #2.
   1 _____ 2 _____

MAMMARY GLANDS

348 The mammary glands are modified _____ glands that function in the production of _____.
349 What is the areola? _____
_____
350 The glandular tissue of mammary glands is organized into units called _____, which are further divided into _____.
351 The secretory units of the mammary glands are _____.
352 The ducts that exit the lobes are called _____.
353 Lacteriferous ducts converge at the base of the nipple to form _____.
354 During pregnancy the primary hormones for mammary gland development are _____.
355 The hormone _____ is the primary hormone for the production of milk.
356 The hormones _____ have an inhibitory effect on prolactin.

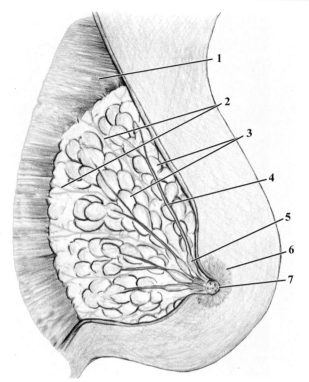

Figure 28.75

357 In reference to **Figure 28.75**, identify #1 - #7.
   1 _____ 5 _____
   2 _____ 6 _____
   3 _____ 7 _____
   4 _____

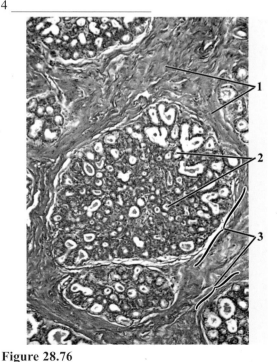

Figure 28.76

358 In reference to **Figure 28.76**, is the mammary gland active or inactive? _____
359 In reference to **Figure 28.76**, identify #1 - #3.
   1 _____ 3 _____
   2 _____

Chapter 28 - Reproductive System

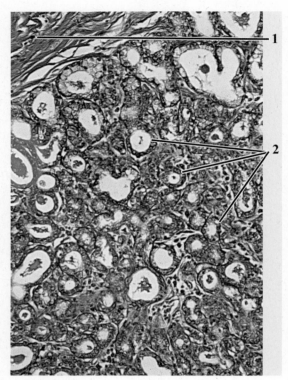

Figure 28.77

360 In reference to **Figure 28.77**, identify #1 - #2.
1 _____
2 _____

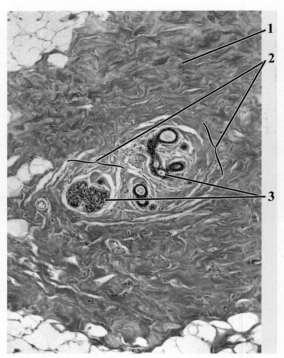

Figure 28.78

361 In reference to **Figure 28.78**, is the mammary gland active or inactive? _____

362 In reference to **Figure 28.78**, identify #1 - #3.
1 _____ 3 _____
2 _____

Name _____
Class _____

Chapter 29 - Dissection of the Fetal Pig **409**

Dissection of the Fetal Pig - Worksheets

DIRECTIONAL TERMINOLOGY

1. **Define the following terms:**
   Dorsal: _____

   Ventral: _____

   Cranial (anterior): _____

   Caudal (posterior): _____

   Medial: _____

   Lateral: _____

   Intermediate: _____

   Proximal: _____

   Distal: _____

   Superficial: _____

   Deep: _____

   Right and Left: _____

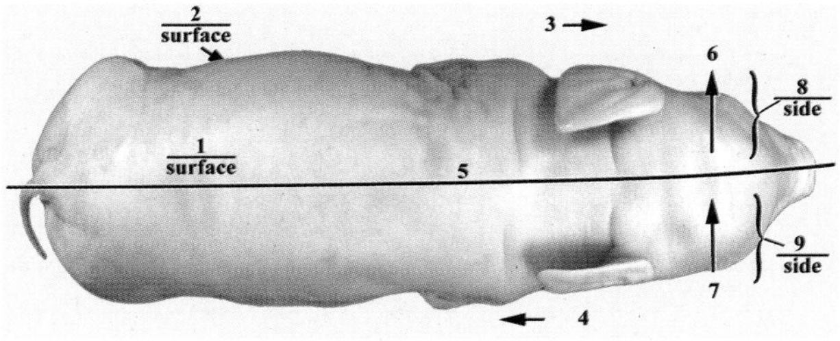

Figure 29.1

2. In reference to **Figure 29.1**, identify #1 - #9.
   1 _____ 6 _____
   2 _____ 7 _____
   3 _____ 8 _____
   4 _____ 9 _____
   5 _____

3. Describe the following sections:
   Sagittal section: _____

   Midsagittal (median) section: _____

   Frontal (coronal) section: _____

   Transverse (horizontal) plane: _____

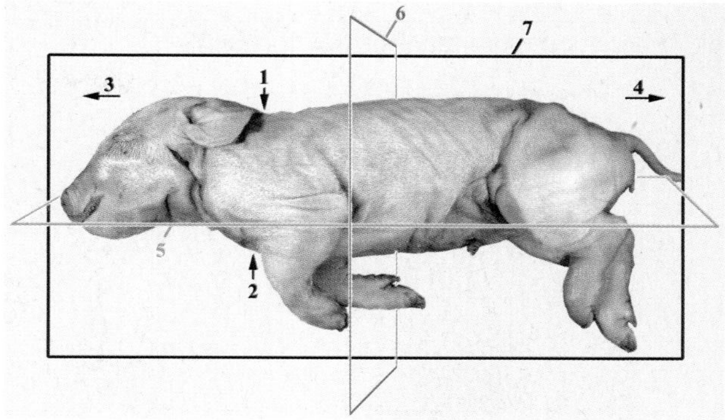

Figure 29.2

4. In reference to **Figure 29.2**, identify #1 - #7.
   1 _____ 5 _____
   2 _____ 6 _____
   3 _____ 7 _____
   4 _____

EXTERNAL ANATOMY

5. What are the three blood vessels located in the umbilical cord? _____

6. Where does the umbilical vein originate? _____

7. Blood carried by the umbilical vein is described as oxygen (rich / poor) and nutrient (rich / poor).

8. The umbilical vein terminates in the _____, where it merges with a shunt called the _____.

9. Blood from the ductus venosus empties into the _____.

10. Blood carried by the umbilical arteries is described as oxygen (rich / poor) and nutrient (rich / poor).

11. The umbilical arteries terminate at the _____.

13. The placenta functions to _____,
    (with / without) mixing of maternal and fetal blood.

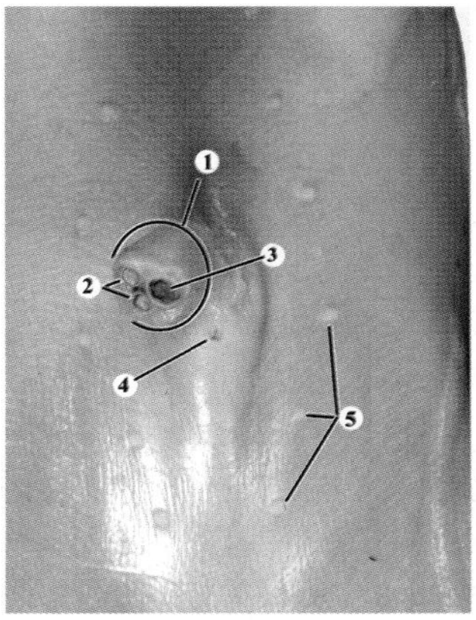

Figure 29.3

13. In reference to **Figure 29.3**, identify #1 - #5.
    1 _____ 4 _____
    2 _____ 5 _____
    3 _____

Chapter 29 - Dissection of the Fetal Pig

MALE FETAL PIG

14 What is the preputial orifice? _____

15 Where is the preputial orifice located? _____

16 What does the scrotum house? _____

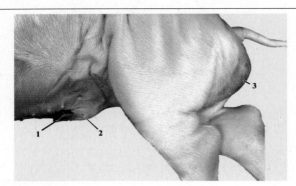

Figure 29.4

17 In reference to **Figure 29.4**, identify #1 - #3.
1 _____ 3 _____
2 _____

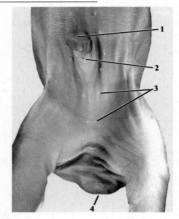

Figure 29.5

18 In reference to **Figure 29.5**, identify #1 - #3.
1 _____ 3 _____
2 _____

FEMALE FETAL PIG

19 What forms the genital papilla? _____

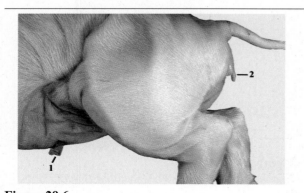

Figure 29.6

20 In reference to **Figure 29.6**, identify #1 - #2.
1 _____ 2 _____

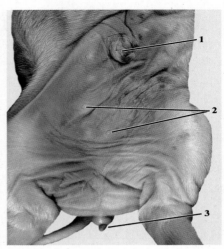

Figure 29.7

21 In reference to **Figure 29.7**, identify #1 - #3.
1 _____ 3 _____
2 _____

DISSECTION OF THE THORAX AND NECK

22 What is the larynx? _____

23 Internally what does the larynx house? _____

24 What is the function of the epiglottis? _____

25 Where does the trachea originate? _____

26 What is the function of the trachea? _____

27 What are "C rings" of the trachea? _____

28 Where is the thymus gland located? _____

29 What is the function of the thymus gland? _____

30 Where is the thyroid gland located? _____

31 What is the function of the thyroid gland? _____

32 Each lung is located within its _____ cavity.
Where is the heart located? _____

33 What is an auricle of the heart? _____

34 What separates the thoracic cavity from the abdominal cavity? _____

Name _____
Class _____

Chapter 29 - Dissection of the Fetal Pig   411

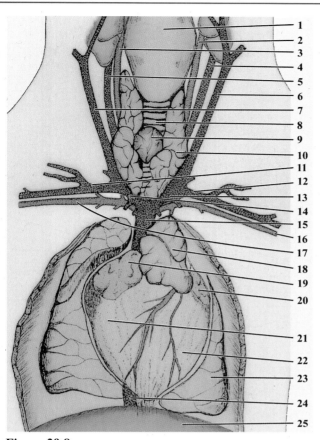

Figure 29.8

35 In reference to **Figure 29.8**, identify #1 - #25.

1 _____
2 _____
3 _____
4 _____
5 _____
6 _____
7 _____
8 _____
9 _____
10 _____
11 _____
12 _____
13 _____
14 _____
15 _____
16 _____
17 _____
18 _____ 22 _____
19 _____ 23 _____
20 _____ 24 _____
21 _____ 25 _____

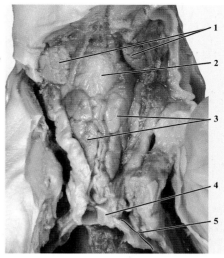

Figure 29.9

36 In reference to **Figure 29.9**, identify #1 - #5.

1 _____ 4 _____
2 _____ 5 _____
3 _____

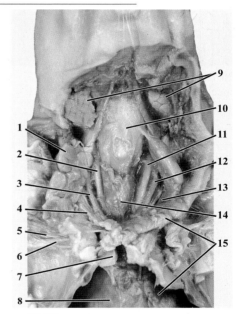

Figure 29.10

37 In reference to **Figure 29.10**, identify #1 - #15.

1 _____
2 _____
3 _____
4 _____
5 _____
6 _____
7 _____
8 _____
9 _____
10 _____
11 _____
12 _____
13 _____
14 _____
15 _____

412 Chapter 29 - Dissection of the Fetal Pig

Name _____
Class _____

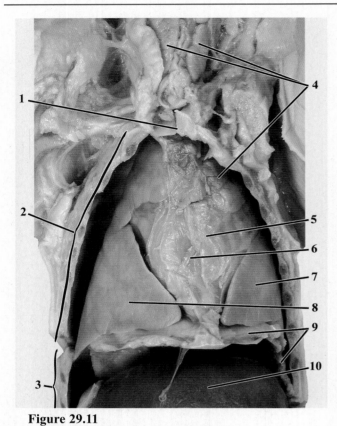

Figure 29.11

38 In reference to **Figure 29.11**, identify #1 - #10.
1 _____
2 _____
3 _____
4 _____
5 _____
6 _____
7 _____
8 _____
9 _____
10 _____

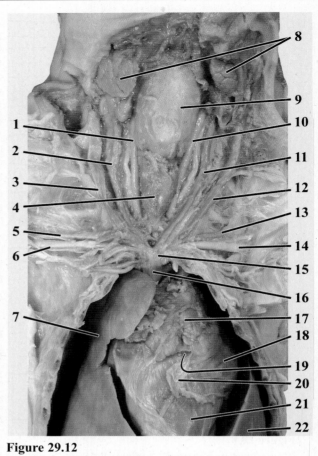

Figure 29.12

39 In reference to **Figure 29.12**, identify #1 - #22.
1 _____
2 _____
3 _____
4 _____
5 _____
6 _____
7 _____
8 _____
9 _____
10 _____
11 _____
12 _____
13 _____
14 _____
15 _____
16 _____
17 _____
18 _____
19 _____
20 _____
21 _____
22 _____

Name _____
Class _____

Chapter 29 - Dissection of the Fetal Pig   413

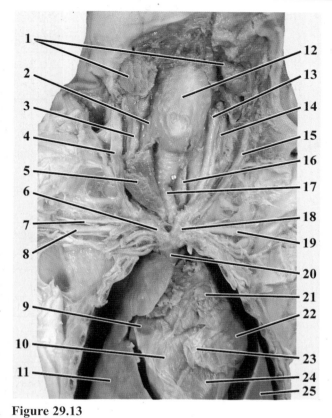

Figure 29.13
40 In reference to **Figure 29.13**, identify #1 - #25.
1 _____
2 _____
3 _____
4 _____
5 _____
6 _____
7 _____
8 _____
9 _____
10 _____
11 _____
12 _____
13 _____
14 _____
15 _____
16 _____
17 _____
18 _____
19 _____
20 _____
21 _____
22 _____
23 _____
24 _____
25 _____

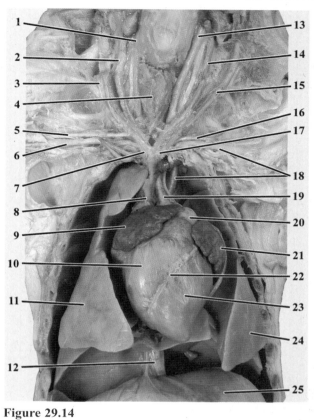

Figure 29.14
41 In reference to **Figure 29.14**, identify #1 - #25.
1 _____
2 _____
3 _____
4 _____
5 _____
6 _____
7 _____
8 _____
9 _____
10 _____
11 _____
12 _____
13 _____
14 _____
15 _____
16 _____
17 _____
18 _____
19 _____
20 _____
21 _____
22 _____
23 _____
24 _____
25 _____

414 Chapter 29 - Dissection of the Fetal Pig

Name _____
Class _____

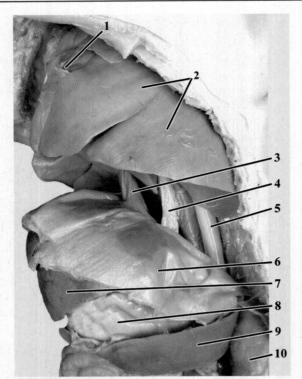

Figure 29.15

42 In reference to **Figure 29.15**, identify #1 - #10.
    1 _____   6 _____
    2 _____   7 _____
    3 _____   8 _____
    4 _____   9 _____
    5 _____ 10 _____

DISSECTION OF THE HEAD, NECK, AND ABDOMEN
for the study of the
ORGANS OF THE DIGESTIVE SYSTEM
DISSECTION OF THE LATERAL REGION OF THE HEAD AND NECK.

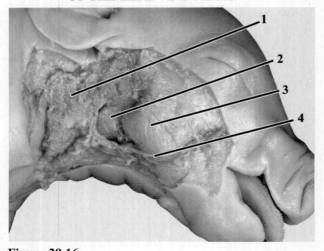

Figure 29.16

43 In reference to **Figure 29.16**, identify #1 - #4.
    1 _____   3 _____
    2 _____   4 _____

44 Describe the location of the parotid salivary gland. _____

45 Where is the oral vestibule located? _____

46 Describe the location of the mandibular gland. _____

47 What are the two divisions of the cavity of the mouth? __

48 Describe the location of the oral cavity. _____

49 What are the three divisions of the pharynx? _____

50 Where is the hard palate located? _____

51 Where is the soft palate located? _____

52 What is the nasopharynx? _____

53 What is the oropharynx? _____

54 What is the laryngopharynx? _____

55 What is the function of the epiglottis? _____

56 What is the function of the esophagus? _____

Name _____
Class _____

Chapter 29 - Dissection of the Fetal Pig 415

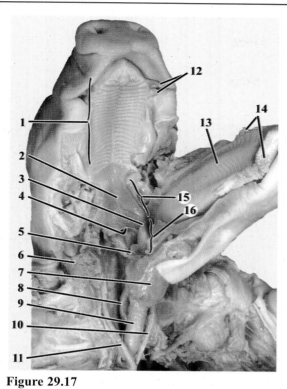

Figure 29.17

57 In reference to **Figure 29.17**, identify #1 - #16.
   1 _____   9 _____
   2 _____ 10 _____
   3 _____ 11 _____
   4 _____ 12 _____
   5 _____ 13 _____
   6 _____ 14 _____
   7 _____ 15 _____
   8 _____ 16 _____

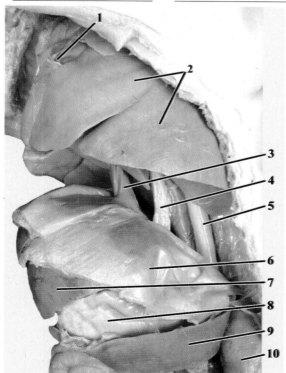

Figure 29.18

58 In reference to **Figure 29.18**, identify #1 - #10.
   1 _____   6 _____
   2 _____   7 _____
   3 _____   8 _____
   4 _____   9 _____
   5 _____ 10 _____

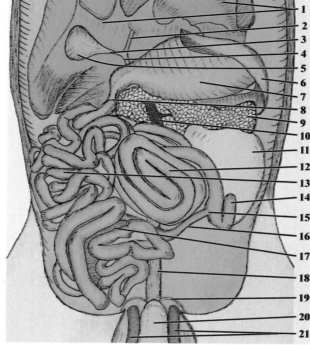

Figure 29.19

59 In reference to **Figure 29.19**, identify #1 - #21.
   1 _____ 12 _____
   2 _____ 13 _____
   3 _____ 14 _____
   4 _____ 15 _____
   5 _____ 16 _____
   6 _____ 17 _____
   7 _____ 18 _____
   8 _____ 19 _____
   9 _____ 20 _____
  10 _____ 21 _____
  11 _____

416 Chapter 29 - Dissection of the Fetal Pig

Name _____
Class _____

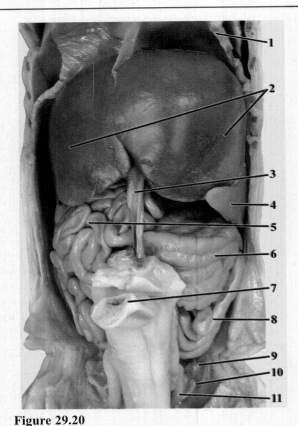

Figure 29.20

60 In reference to **Figure 29.20**, identify #1 - #11.
1 _____ 7 _____
2 _____ 8 _____
3 _____ 9 _____
4 _____ 10 _____
5 _____ 11 _____
6 _____

61 In reference to **Figure 29.21**, identify #1 - #11.
1 _____ 7 _____
2 _____ 8 _____
3 _____ 9 _____
4 _____ 10 _____
5 _____ 11 _____
6 _____

62 **Figure 29.22**
In reference to **Figure 29.22**, identify #1 - #20.
1 _____ 11 _____
2 _____ 12 _____
3 _____ 13 _____
4 _____ 14 _____
5 _____ 15 _____
6 _____ 16 _____
7 _____ 17 _____
8 _____ 18 _____
9 _____ 19 _____
10 _____ 20 _____

63 Describe the location of the stomach. _____

64 Where is the location of the cardia? _____

65 Where is the location of the pyloris? _____

66 What is the function of the pyloric sphincter? _____

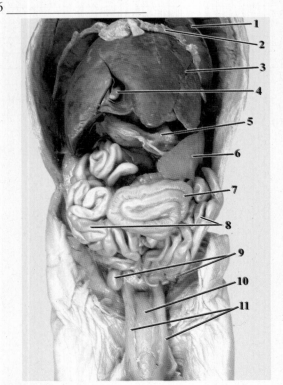

Figure 29.21

Name _____
Class _____

Chapter 29 - Dissection of the Fetal Pig   **417**

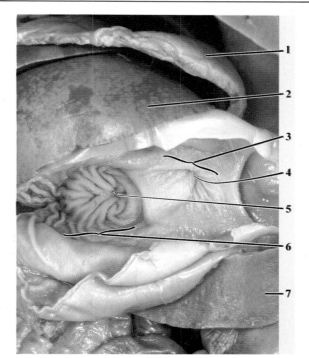

Figure 29.23

67 In reference to **Figure 29.23**, identify #1 - #7.
    1 _____ 5 _____
    2 _____ 6 _____
    3 _____ 7 _____
    4 _____

68 What is the proximal region of the small intestine? _____

69 What two organs deliver secretions into the duodenum? __
    _____

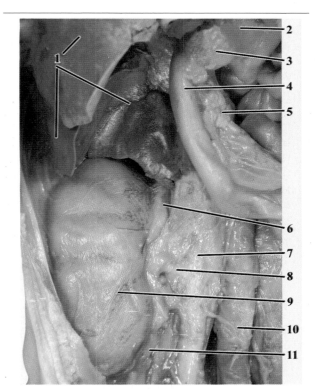

Figure 29.24

70 In reference to **Figure 29.24**, identify #1 - #11.
    1 _____ 7 _____
    2 _____ 8 _____
    3 _____ 9 _____
    4 _____ 10 _____
    5 _____ 11 _____
    6 _____

71 What are two functions of the liver? _____
    _____

72 What is the function of the gall bladder? _____
    _____

73 What duct connects to the gall bladder? _____

74 What are the three divisions of the pancreas? _____

75 What is the endocrine function of the pancreas? _____
    _____

76 What is the function of the pancreas as an accessory digestive organ? _____
    _____

77 What are the three regions of the small intestine? _____
    _____

78 What is the function of the small intestine? _____
    _____

79 What is the name of the middle region of the small intestine? _____

80 Where does the ileum terminate? _____

81 What is the function of the ileocecal valve? _____
    _____

82 What is the name of the first region of the colon? ____

83 What portion of the colon is formed after the spiral colon?
    _____

84 What is the function of the colon? _____
    _____

85 What is the function of the spleen? _____
    _____
    _____

418 Chapter 29 - Dissection of the Fetal Pig

Name _____
Class _____

DISSECTION OF THE NECK, ABDOMEN, AND PELVIS
for the study of the
ORGANS OF THE CIRCULATORY SYSTEM
HEART

86 What is the name of the serous membrane that surrounds the heart? _____
_____

87 Where is the visceral pericardium located? _____
_____

88 Where is the parietal pericardium located? _____
_____
_____

89 What forms the pericardial sac? _____

90 What is the function of pericardial fluid? _____
_____
_____

Right side of the heart

91 What forms the muscular right side of the heart? _____

92 What is the difference between an auricle and an atrium? ___
_____

93 What is the function of the right side of the heart? _____
_____

94 After birth, what is the function of pulmonary circulation?
_____
_____

95 Because the lungs are not functional as respiratory organs, some of the blood destined for pulmonary circulation is shunted into _____ circulation.

96 What vessels deliver blood to the right atrium? _____

97 In fetal circulation, blood from the right atrium is directed to both the _____
_____.

98 What vessel receives blood from the right ventricle? _____

99 In fetal circulation, the pulmonary trunk delivers blood to the _____
_____.

100 The ductus arteriosus is a _____

101 What is the function of the foramen ovale? _____
_____
_____

Left side of the heart

102 What is the function of the left side of the heart? _____

103 What forms the muscular left side of the heart? _____

104 Blood from the left atrium enters the _____

105 The left ventricle ejects blood into the _____.

106 What is the interventricular sulcus? _____
_____
_____

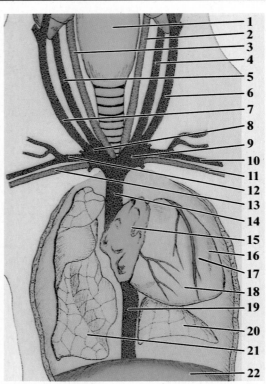

Figure 29.25

107 In reference to **Figure 29.25**, identify #1 - #22.

1 _____
2 _____
3 _____
4 _____
5 _____
6 _____
7 _____
8 _____
9 _____
10 _____
11 _____
12 _____
13 _____
14 _____
15 _____
16 _____
17 _____
18 _____
19 _____
20 _____
21 _____
22 _____

Name _____
Class _____

Chapter 29 - Dissection of the Fetal Pig **419**

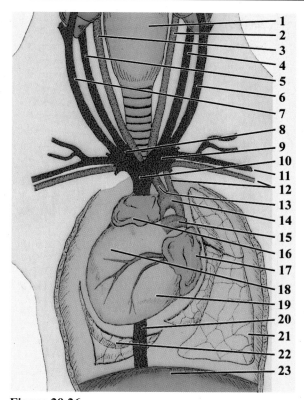

Figure 29.26
108 In reference to **Figure 29.26**, identify #1 - #23.
1 _____
2 _____
3 _____
4 _____
5 _____
6 _____
7 _____
8 _____
9 _____
10 _____
11 _____
12 _____
13 _____
14 _____
15 _____
16 _____
17 _____
18 _____
19 _____
20 _____
21 _____
22 _____
23 _____

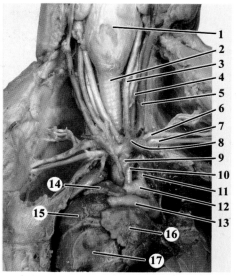

Figure 29.27
109 In reference to **Figure 29.27**, identify #1 - #17.
1 _____
2 _____
3 _____
4 _____
5 _____
6 _____
7 _____
8 _____
9 _____
10 _____
11 _____
12 _____
13 _____
14 _____
15 _____
16 _____
17 _____

420 Chapter 29 - Dissection of the Fetal Pig

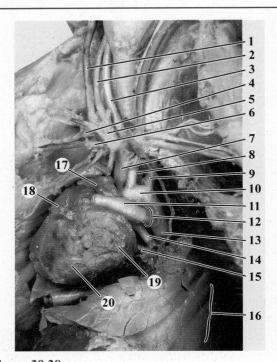

Figure 29.28
110 In reference to **Figure 29.28**, identify #1 - #20.

1 _____
2 _____
3 _____
4 _____
5 _____
6 _____
7 _____
8 _____
9 _____
10 _____
11 _____
12 _____
13 _____
14 _____
15 _____
16 _____
17 _____
18 _____
19 _____
20 _____

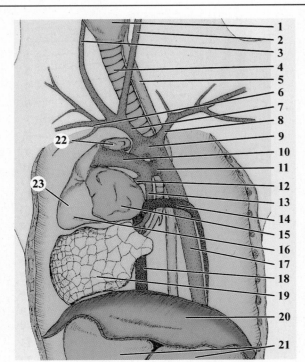

Figure 29.29
111 In reference to **Figure 29.29**, identify #1 - #23.

1 _____
2 _____
3 _____
4 _____
5 _____
6 _____
7 _____
8 _____
9 _____
10 _____
11 _____
12 _____
13 _____
14 _____
15 _____
16 _____
17 _____
18 _____
19 _____
20 _____
21 _____
22 _____
23 _____

Name _____
Class _____

Chapter 29 - Dissection of the Fetal Pig   **421**

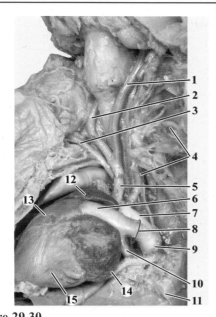

Figure 29.30

112 In reference to **Figure 29.30**, identify #1 - #15.
1 _____
2 _____
3 _____
4 _____
5 _____
6 _____
7 _____
8 _____
9 _____
10 _____
11 _____
12 _____
13 _____
14 _____
15 _____

113 In reference to **Figure 29.31**, identify #1 - #19.
1 _____
2 _____
3 _____
4 _____
5 _____
6 _____
7 _____
8 _____
9 _____
10 _____
11 _____
12 _____
13 _____
14 _____
15 _____
16 _____
17 _____
18 _____
19 _____

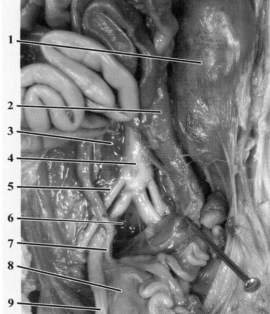

Figure 29.32

114 In reference to **Figure 29.32**, identify #1 - #9.
1 _____
2 _____
3 _____
4 _____
5 _____
6 _____
7 _____
8 _____
9 _____

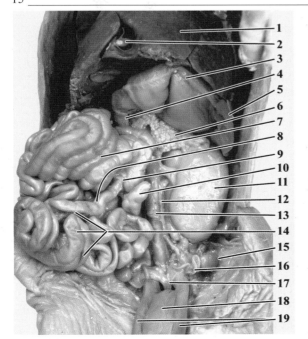

Figure 29.31

422 Chapter 29 - Dissection of the Fetal Pig

PATHWAY OF BLOOD FLOW IN FETAL CIRCULATION

115 What is the function of the umbilical vein? _____
_____

116 What vessel does the smaller branch of the umbilical vein merge with? _____
_____

117 What is the function of the hepatic portal vein? _____
_____

118 What vessel does the larger branch of the umbilical vein merge with? _____
_____

119 The ductus venosus functions as a _____ into the caudal vena cava.

120 Caudal to its merger with the ductus venosus, the caudal vena cava carries oxygen- _____ and nutrient- _____ blood.

121 The caudal vena cava returns blood to the right _____, blood that is oxygen- _____ and nutrient- _____.

122 Blood from the caudal vena cava is mostly directed into the shunt to the left atrium, the _____.

123 Blood from the left atrium enters the left _____.

124 Blood from the left ventricle enters the _____.

125 Blood from the aorta is delivered to the _____
_____

126 The cranial vena cava returns blood to the right _____, blood that is oxygen- _____ and nutrient- _____.

127 Blood from the cranial vena cava is mostly directed into the _____.

128 The right ventricle delivers blood to the _____ trunk.

129 Blood from the pulmonary trunk is mostly delivered to the _____, with only a small amount of blood directed to the _____.

130 The ductus arteriosus delivers oxygen- _____ and nutrient- _____ blood to the _____.

131 The caudal aorta delivers blood to the _____
_____

132 Blood directed to the placenta by the _____, is _____
_____

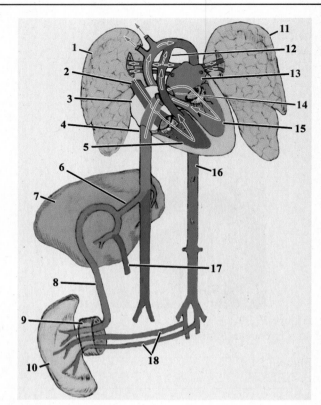

Figure 29.33

133 In reference to **Figure 29.33**, identify #1 - #18.

1 _____ 10 _____
2 _____ 11 _____
3 _____ 12 _____
4 _____ 13 _____
5 _____ 14 _____
6 _____ 15 _____
7 _____ 16 _____
8 _____ 17 _____
9 _____ 18 _____

GREAT VESSELS OF THE HEART

134 What is the function of the cranial vena cava? _____
_____

135 In fetal circulation, what is the function of the caudal vena? _____

136 In fetal circulation, how does oxygen-rich blood enter the caudal vena cava? _____
_____
_____

137 After birth, the caudal vena cava returns _____ blood to the right atrium.

138 The pulmonary trunk exits the _____.

139 In fetal circulation, the pulmonary trunk delivers blood to the _____ and the right and left _____.

140 The aorta exits the _____, and delivers blood into _____ circulation.

141 The two large branches from the aortic arch are the _____
_____

Chapter 29 - Dissection of the Fetal Pig   423

142 In fetal circulation, the pulmonary veins deliver _____ blood to the _____.

143 After birth the pulmonary veins deliver _____ blood to the _____.

147 When does the foramen ovale close? _____

148 What is the fossa ovalis? _____

149 What serves as the final bypass (shunt) from blood delivery to the pulmonary circuit (lungs)? _____

150 Describe the location of the ductus arteriosus. _____

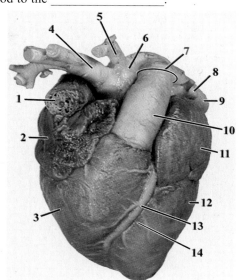

Figure 29.34
144 In reference to **Figure 29.34**, identify #1 - #14.
1 _____   8 _____
2 _____   9 _____
3 _____   10 _____
4 _____   11 _____
5 _____   12 _____
6 _____   13 _____
7 _____   14 _____

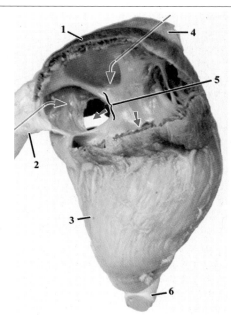

Figure 29.36
151 In reference to **Figure 29.36**, identify #1 - #6.
1 _____   4 _____
2 _____   5 _____
3 _____   6 _____

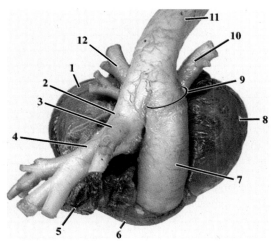

Figure 29.35
145 In reference to **Figure 29.35**, identify #1 - #12.
1 _____   7 _____
2 _____   8 _____
3 _____   9 _____
4 _____   10 _____
5 _____   11 _____
6 _____   12 _____

146 What is the function of the foramen ovale? _____

424 Chapter 29 - Dissection of the Fetal Pig

ARTERIES CAUDAL TO DIAPHRAGM

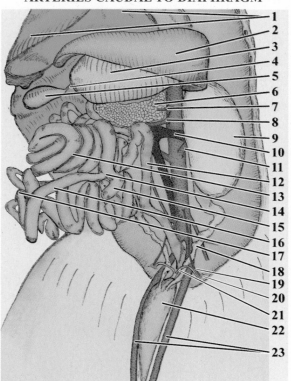

Figure 29.37

152 In reference to **Figure 29.37**, identify #1 - #23.

1 _____
2 _____
3 _____
4 _____
5 _____
6 _____
7 _____
8 _____
9 _____
10 _____
11 _____
12 _____
13 _____
14 _____
15 _____
16 _____
17 _____
18 _____
19 _____
20 _____
21 _____
22 _____
23 _____

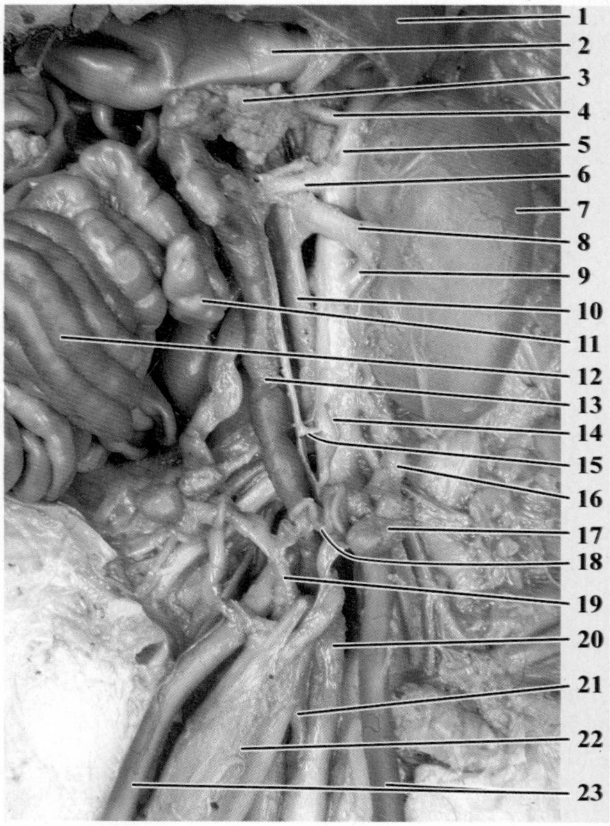

Figure 29.38

153 In reference to **Figure 29.38**, identify #1 - #23.

1 _____
2 _____
3 _____
4 _____
5 _____
6 _____
7 _____
8 _____
9 _____
10 _____
11 _____
12 _____
13 _____
14 _____
15 _____
16 _____
17 _____
18 _____
19 _____
20 _____
21 _____
22 _____
23 _____